Handbook of Heavy Oil Properties
and Analysis

Handbook of Heavy Oil Properties and Analysis

James G. Speight
CD & W Inc.
Laramie, WY, USA

WILEY

Published by John Wiley & Sons, Inc., Hoboken, New Jersey.
Published simultaneously in Canada.

For general information on our other products and services or for technical support, please contact our Customer Care Department within the United States at (800) 762-2974, outside the United States at (317) 572-3993 or fax (317) 572-4002.

Wiley also publishes its books in a variety of electronic formats. Some content that appears in print may not be available in electronic formats. For more information about Wiley products, visit our web site at www.wiley.com.

Library of Congress Cataloging-in-Publication Data:

Names: Speight, James G., author.
Title: Handbook of heavy oil properties and analysis / James G. Speight, CD & W Inc., Laramie, WY, USA.
Description: Hoboken, NJ, USA : John Wiley & Sons, Inc, 2023. | Includes bibliographical references and index.
Identifiers: LCCN 2023001790 (print) | LCCN 2023001791 (ebook) | ISBN 9781119577157 (hardback) | ISBN 9781119577126 (adobe pdf) | ISBN 9781119577102 (epub)
Subjects: LCSH: Heavy oil–Testing. | Analytical chemistry.
Classification: LCC TP691 .S6869 2023 (print) | LCC TP691 (ebook) | DDC 553.2/82–dc23/eng/20230213
LC record available at https://lccn.loc.gov/2023001790
LC ebook record available at https://lccn.loc.gov/2023001791

Cover Design: Wiley
Cover Images: © Adrienne Bresnahan/Getty Images

Set in 9.5/12.5pt STIXTwoText by Straive, Pondicherry, India

Contents

About the Author *xiii*
Preface *xv*

1 History and Terminology *1*
1.1 Introduction *1*
1.2 Historical Perspectives *8*
1.2.1 Pre-Christian Era Use of Heavy Oil and Bitumen *8*
1.2.2 Post-Christian Era Use of Heavy Oil and Bitumen *14*
1.3 Definitions and Terminology *15*
1.3.1 Nonviscous Feedstocks *17*
1.3.1.1 Crude Oil *17*
1.3.1.2 Opportunity Crude Oil *21*
1.3.1.3 High-Acid Crude Oil *25*
1.3.1.4 Foamy Oil *26*
1.3.2 Viscous Feedstocks *27*
1.3.2.1 Gas Oil *29*
1.3.2.2 Heavy Crude Oil *30*
1.3.2.3 Extra Heavy Crude Oil *33*
1.3.2.4 Tar Sand Bitumen *36*
1.3.2.5 Residuum *41*
1.3.2.6 Asphalt *45*
1.3.2.7 Tar and Pitch *49*
1.3.2.8 Sludge *50*
1.4 Classification *51*
1.5 Feedstock Evaluation *53*
1.6 Modern Analytical Perspectives *56*
References *58*

2 Sampling and Measurement *63*
2.1 Introduction *63*
2.2 Sampling *64*
2.2.1 Sampling Protocol *70*
2.2.1.1 Sampling Semi-volatile and Nonvolatile Compounds *71*
2.2.1.2 Solids *75*
2.2.1.3 Extract Concentration *77*
2.2.1.4 Sample Cleanup *80*
2.2.2 Representative Sample *80*
2.2.3 Sampling Error *82*
2.3 Measurement *82*
2.4 Method Validation *85*
2.4.1 Requirements *87*
2.4.2 Method Detection Limit *87*
2.4.3 Accuracy *88*
2.4.4 Precision *89*
2.5 Quality Control and Quality Assurance *90*
2.5.1 Quality Control *90*
2.5.2 Quality Assurance *92*
2.6 Assay and Specifications *93*
2.6.1 Assay *95*
2.6.2 Specifications *99*
2.6.3 Metallic Constituents *100*
2.6.4 Water Content *101*
2.7 Environmental Issues *102*
References *104*

3 Chemical Composition *109*
3.1 Introduction *109*
3.2 Elemental Composition *114*
3.3 Chemical Composition *120*
3.3.1 Hydrocarbon Constituents *122*
3.3.1.1 Paraffin Hydrocarbon Derivatives *123*
3.3.1.2 Cycloparaffin Hydrocarbon Derivatives *124*
3.3.1.3 Aromatic Hydrocarbon Derivatives *124*
3.3.1.4 Unsaturated Hydrocarbon Derivatives *124*
3.3.2 Non-hydrocarbon Constituents *125*
3.3.2.1 Sulfur Compounds *125*
3.3.2.2 Nitrogen Compounds *126*
3.3.2.3 Oxygen Compounds *126*
3.3.3 Metallic Constituents *127*
3.3.4 Porphyrins *128*

3.4 Chemical Composition by Distillation *131*
3.4.1 Vacuum Gas Oil *135*
3.4.2 Vacuum Residua *136*
3.5 Chemical Composition by Spectroscopy *137*
3.5.1 Infrared Spectroscopy *138*
3.5.2 Nuclear Magnetic Resonance Spectroscopy *138*
3.5.3 Mass Spectrometry *139*
3.5.4 Other Techniques *141*
References *142*

4 Fractional Composition *149*
4.1 Introduction *149*
4.2 Distillation *151*
4.3 Solvent Treatment *152*
4.3.1 Asphaltene Separation *156*
4.3.1.1 Influence of Solvent Type *158*
4.3.1.2 Influence of the Degree of Dilution *160*
4.3.1.3 Influence of Temperature *160*
4.3.1.4 Influence of Contact Time *161*
4.3.2 Fractionation *161*
4.3.3 Carbenes and Carboids *163*
4.4 Adsorption *164*
4.4.1 Chemical Factors *165*
4.4.2 Fractionation Methods *166*
4.4.2.1 General Methods *166*
4.4.2.2 ASTM Methods *172*
4.5 Chemical Methods *173*
4.5.1 Acid Treatment *173*
4.5.2 Molecular Complex Formation *174*
4.5.2.1 Urea Adduction *174*
4.5.2.2 Thiourea Adduction *176*
4.5.2.3 Adduct Composition *176*
4.5.2.4 Adduct Structure *176*
4.5.2.5 Adduct Properties *178*
4.6 The Asphaltene Fraction *179*
4.7 Carbenes and Carboids *180*
4.8 Use of the Data *182*
References *185*

5 Chemical Properties *191*
5.1 Introduction *191*
5.2 Acid Number *193*

5.3 Elemental Analysis and Metals *197*
5.4 Emulsion Formation *201*
5.5 Evaporation *202*
5.6 Flash Point and Fire Point *203*
5.7 Functional Group Analysis *204*
5.8 Halogenation *208*
5.9 Hydrogenation *210*
5.10 Oxidation *216*
5.11 Thermal Methods *219*
5.12 Miscellaneous Methods *222*
References *223*

6 Physical Properties, Electrical Properties, and Optical Properties *229*
6.1 Introduction *229*
6.2 Physical Properties *233*
6.2.1 Adhesion *234*
6.2.2 Density, Specific Gravity, and API Gravity *235*
6.2.3 Surface and Interfacial Tension *238*
6.2.4 Viscosity *239*
6.3 Electrical Properties *243*
6.3.1 Conductivity *243*
6.3.2 Dielectric Constant *244*
6.3.3 Dielectric Strength *244*
6.3.4 Dielectric Loss and Power Factor *245*
6.3.5 Static Electrification *246*
6.4 Optical Properties *246*
6.4.1 Optical Activity *248*
6.4.2 Refractive Index *249*
References *250*

7 Thermal Properties *255*
7.1 Introduction *255*
7.2 Ash Production *256*
7.3 Carbon Residue *258*
7.4 Critical Properties *260*
7.5 Enthalpy *262*
7.6 Heat of Combustion *263*
7.7 Latent Heat *264*
7.8 Liquefaction and Solidification *265*
7.9 Pour Point *267*
7.10 Pressure–Volume–Temperature Relationships *267*
7.11 Softening Point *269*

7.12 Specific Heat *269*
7.13 Thermal Conductivity *271*
7.14 Volatility *272*
References *278*

8 Chromatographic Properties *283*
8.1 Introduction *283*
8.2 Adsorption Chromatography *286*
8.3 Gas Chromatography *291*
8.4 Gel Permeation Chromatography *298*
8.5 High-Performance Liquid Chromatography *300*
8.6 Ion Exchange Chromatography *303*
8.7 Simulated Distillation *305*
8.8 Supercritical Fluid Chromatography *307*
8.9 Thin Layer Chromatography *309*
References *311*

9 Structural Group Analysis *317*
9.1 Introduction *317*
9.2 Physical Property Methods *320*
9.2.1 Density Method *321*
9.2.2 Density–Temperature Coefficient Method *321*
9.2.3 Direct Method *322*
9.2.4 Dispersion–Refraction Method *323*
9.2.5 Molecular Weight-Refractive Index Method *324*
9.2.6 *n-d-M* Method *325*
9.2.7 Waterman Ring Analysis *325*
9.2.8 Miscellaneous Methods *327*
9.3 Spectroscopic Methods *328*
9.3.1 Electron Spin Resonance *329*
9.3.2 Infrared Spectroscopy *329*
9.3.3 Mass Spectrometry *333*
9.3.4 Nuclear Magnetic Resonance Spectroscopy *339*
9.3.5 Ultraviolet Spectroscopy *345*
9.3.6 X-ray Diffraction *345*
9.4 Heteroatom Systems *347*
9.4.1 Nitrogen *347*
9.4.2 Oxygen *348*
9.4.3 Sulfur *348*
9.4.4 Metals *349*
9.5 Miscellaneous Methods *349*
References *350*

10 Molecular Weight Determination *357*
10.1 Introduction *357*
10.2 Methods for Molecular Weight Measurement *360*
10.2.1 Vapor Pressure Osmometry *361*
10.2.2 Freezing Point Depression *365*
10.2.3 Boiling Point Elevation *366*
10.2.4 Size Exclusion Chromatography *367*
10.2.5 Mass Spectrometry *369*
10.2.6 Nuclear Magnetic Resonance Spectroscopy *370*
10.3 Molecular Weights of Volatile Fractions *370*
10.4 Molecular Weights of Nonvolatile Fractions *371*
10.4.1 Resins *372*
10.4.2 Asphaltenes *372*
10.4.3 Carbenes and Carboids *378*
References *379*

11 Instability and Incompatibility *383*
11.1 Introduction *383*
11.2 Occurrence of Instability and Incompatibility *389*
11.3 Factors Influencing Instability and Incompatibility *394*
11.3.1 Acidity *395*
11.3.2 Asphaltene Content *395*
11.3.3 Density/Specific Gravity *398*
11.3.4 Elemental Analysis *398*
11.3.5 Metals Content *399*
11.3.6 Pour Point *400*
11.3.7 Viscosity *400*
11.3.8 Volatility *400*
11.3.9 Water Content, Salt Content, Bottom Sediment and Water (BS&W) *402*
11.4 Determination of Instability and Incompatibility *403*
References *406*

12 Use of the Data *413*
12.1 Introduction *413*
12.2 Use of the Data *414*
12.3 Process Analysis and Feedstock Mapping *416*
12.3.1 Property Predictions *418*
12.3.2 Predicting Separations *418*
12.3.3 Process Predictability *419*
12.4 Environmental Aspects of Processing *419*
12.4.1 Gaseous Emissions *422*

12.4.2 Liquid Effluents *428*
12.4.3 Solid Effluents *430*
12.5 Analytical Methods for Environmental Regulations *431*
12.5.1 Definitions *432*
12.5.2 Environmental Regulations *434*
12.5.3 Environmental Analysis *435*
References *437*

Glossary *441*
Conversion Factors *467*
Index *469*

About the Author

Dr. James G. Speight has a BSc and PhD in Chemistry; he also holds a DSc in Geological Sciences and a PhD in Petroleum Engineering. He has more than 50 years of experience in areas associated with (1) the properties, recovery, and refining of reservoir fluids, conventional petroleum, heavy oil, and tar sand bitumen; (2) the properties and refining of natural gas, gaseous fuels; (3) the properties and refining of biomass, biofuels, biogas; and (4) the generation of bioenergy. His work has also focused on environmental effects, environmental remediation, and safety issues associated with the production and use of fuels and biofuels. He is the author (and coauthor) of more than 100 books in petroleum science and engineering, biomass, biofuels, and environmental sciences.

Although he has always worked in private industry which focused on contract-based work, Dr. Speight has served as Adjunct Professor in the Department of Chemical and Fuels Engineering at the University of Utah and in the Departments of Chemistry and Chemical and Petroleum Engineering at the University of Wyoming. In addition, he was a Visiting Professor in the College of Science, University of Mosul (Iraq) and has also been a Visiting Professor in Chemical Engineering at the following universities: University of Akron (Ohio), University of Missouri-Columbia, Technical University of Denmark, and University of Trinidad and Tobago. He has served as a thesis examiner for more than 30 theses and has been an advisor/mentor to MSc and PhD students.

Dr. Speight has been honored as the recipient of the following awards:

- Diploma of Honor, United States National Petroleum Engineering Society. *For Outstanding Contributions to the Petroleum Industry,* 1995.
- Gold Medal of the Russian Academy of Sciences. *For Outstanding Work in Petroleum Science.* 1996.
- Einstein Medal of the Russian Academy of Sciences. *In recognition of Outstanding Contributions and Service in the field of Geologic Sciences.* 2001.

- Gold Medal—Scientists without Frontiers, Russian Academy of Sciences. *In recognition of His Continuous Encouragement of Scientists to Work Together across International Borders*. 2005.
- Methanex Distinguished Professor, University of Trinidad and Tobago. *In Recognition of Excellence in Research*. 2006.
- Gold Medal—Giants of Science and Engineering, Russian Academy of Sciences. *In recognition of Continued Excellence in Science and Engineering*. 2006.

Preface

The need for, and the acceptance of, the use of the viscous feedstocks as refinery feedstocks oil has increased substantially during the second half of the 20th century. Thus, attempts to understand and manipulate refinery processes have persisted over centuries, with an explosion of technological innovation and research occurring in the last 100 years. The majority of these efforts have focused on using the knowledge gained to produce a useful product and achieve a perceived improvement in the performance of that product.

The laws of science will ultimately dictate what can or cannot be done with feedstocks to provide the needed products. The science of analytical chemistry is at the core of understanding of both the problems of processing various feedstocks. This book will examine through a presentation discussion of the way that the analytical science has been applied to defining the properties and behavior of the different feedstocks that are used in the heavy crude oil, extra heavy crude oil, and tar sand bitumen (known collectively in this text as viscous feedstocks) refining industry.

In the 20th century and at the beginnings of the 21th century, scientists and engineers have become increasingly well-versed in utilizing chemical knowledge to better understand the nature of the feedstocks that arbitrarily fall under the term "heavy oil" but are more correctly known as viscous feedstocks (which include heavy crude oil, extra heavy crude oil, and tar sand bitumen) and the influence of each feedstock on refining scenarios and on product slate. Definitions of processing *do's* and *don'ts* abound in the scientific and engineering literature but the essence of these *rules* depends on analytical chemical measurements.

Heavy crude oil, extra heavy crude oil, and tar sand bitumen exhibit a wide range of physical properties and a wide range of tests have been (and continue to be) developed to provide an indication of the means by which a particular feedstock should be processed. Initial inspection of the nature of the viscous feedstocks will provide deductions about the most logical means of refining or correlation of various properties to structural types present and hence attempted classification of the viscous feedstocks. Proper interpretation of the data resulting from the inspection of crude oil requires an understanding of their significance.

Evaluation of heavy crude oil, extra heavy crude oil, and tar sand bitumen for use as refinery feedstocks usually involves an examination of one or more of the physical properties of the material. By this means, a set of basic characteristics can be obtained that can be correlated with utility. Consequently, various standards organizations, such as the American Society for Testing and Materials in North America have devoted considerable time and effort to the correlation and standardization of methods for the inspection and evaluation of the products from the viscous feedstocks.

The acceptance of the viscous feedstocks by refineries has meant that the analytical techniques used for the lighter feedstocks have had to evolve to produce meaningful data that can be employed to assist in defining refinery scenarios for processing the feedstocks. In addition, selection of the most appropriate analytical procedures will aid in the predictability of feedstock behavior during refining. This same rationale can also be applied to feedstocks behavior during recovery operations.

Because of the wide range of chemical and physical properties, a wide range of tests have been (and continue to be) developed to provide an indication of the means by which a particular feedstock should be processed. Initial inspection of the nature of the heavy crude oil, extra heavy crude oil, and tar sand bitumen will provide deductions about the most logical means of refining or correlation of various properties to structural types present and hence attempted classification of the viscous feedstocks. Proper interpretation of the data resulting from the inspection of crude oil requires careful consideration and an understanding of the significance of the data.

It is for these reasons that understanding the composition of the viscous feedstocks, as well as the chemical and physical properties of these feedstocks, is extremely important. Thus, an efficient evaluation of a feedstock requires the application of tests than specifications. These are then used to provide adequate control of product quality without being over restrictive with the minimum of testing effort.

Product quality is judged by the performance during service. The performance of any product in particular service applications is therefore the ultimate criterion of quality. It is therefore necessary to find properties that allow assessment of the service performance, especially those tests that correlate closely with the service conditions. Sometimes the inspection tests attempt to measure these properties, for example, the research octane number test that was devised to measure the antiknock performance of motor fuel or, in many cases, the significant property is obtained indirectly from the inspection test results.

However, where the specified property is not measured directly, it is important to ensure that a suitable combination of inspection tests is selected to give a high degree of correlation with the specified property.

Although the focus on this book is on the relevant ASTM test methods with the numbers given, where possible the corresponding IP test method number is also presented. As an aside, the ASTM or the IP may have withdrawn some of the tests noted herein. Nevertheless, the method is still included because of its continued use, for whatever reason, by analysts and also for historical (not hysterical) reference purposes!

Thus, this book will deal with the various aspects of the analysis (and properties) of the viscous feedstocks and will provide a detailed explanation of the necessary standard tests and procedures that are applicable to feedstocks in order to help define predictability of behavior. In addition, the application of test methods for determining instability and incompatibility as well as test methods related to environmental regulations will be described.

More important, the book will provide details of the meaning of the various test results and how they might be applied to predict feedstock behavior. In fact, analytical techniques for chemical analysis of crude oil and the viscous feedstocks as well as the transformation products formed by photochemical processes and biological processes is a necessary next step after a spill. In addition to the standard test method present elsewhere in this text, the spilled material can be investigated by advanced gas chromatography (GC) techniques such as comprehensive two-dimensional GC (GC-GC), pyrolysis GC with mass spectrometry (MS), and GC with tandem mass spectrometry (GC-MS/MS) which will provide a greater understanding at the molecular level of composition and complexity of the spilled material and any chemical changes that occur to the spilled material.

Analytical chemists represent a rich resource for the various aspects of feedstock refining and recognize that while the recent decades of experience have enhanced analytical knowledge, they have also revealed large gaps in the kind of data needed to thoroughly understand the nature of heavy crude oil, extra heavy crude oil, and tar sand bitumen as feedstocks for refineries.

The significance of a particular test is not always apparent by reading the procedure, and sometimes can only be gained through a working (i.e., a hands-on) familiarity with the test. The following tests are commonly used to characterize asphalts but these are not the only tests used for determining the property and behavior of an asphaltic binder. As in the petroleum industry, a variety of tests are employed having evolved through local, or company, use.

The development of laboratory instrumentation over the last 50 years has been one of the forces shaping analytical standards and improved instrumentation is already changing approaches to the analysis of viscous feedstocks. Like the twin strands of the famous DNA double helix, technology and analytical science and regulation will continue their closely linked relationship.

The prime focus of the book is the emphasis of the analytical methods applied to heavy crude oil, extra heavy crude oil, and tar sand bitumen with lesser emphasis on product analysis. Although product analysis is an extremely important aspect of the science and technology of the viscous feedstocks, it is felt that the move to these feedstocks for refinery operations requires that more emphasis be placed on the analysis and properties of refinery feedstocks.

Furthermore, while the focus of this chapter is on the sampling and preparation of the viscous feedstocks for refining, it is worthwhile (because of the similarity of the methods) to consider the application of the sampling and measurement methods to situations where the feedstock has been spilled into the environment. In fact, the spills of crude oil and viscous feedstocks can occur anywhere that the material is being stored, transported, transferred from one tank to another, or refined. Thus, methods that are applicable to both refining and environmental issues are cited in this chapter. This will save unnecessary repetition of the methods at other parts of the book.

This book will assist the reader to gain an understanding of the criteria for determining the quality and processability of heavy crude oil, extra heavy crude oil, and tar sand bitumen and the production of products from these viscous feedstocks as well as the appropriate test methods when one of these feedstocks is spilled into the environment.

Dr. James G. Speight,
Laramie, Wyoming
December 2022

1

History and Terminology

1.1 Introduction

Crude oil (also commonly referred to as "petroleum") is typically a mixture of hydrocarbon derivatives that occurs widely in the sedimentary rocks in the form of gases, liquids, semisolids, or solids. Crude oil provides not only raw materials for the ubiquitous plastics and other products but also fuel for energy, industry, heating, and transportation. Chemically, crude oil is an extremely complex mixture of organic compounds many of which (in conventional crude oil) are hydrocarbon derivatives along with minor amounts of nitrogen-containing compounds, oxygen-containing compounds, and sulfur-containing compounds as well as trace amounts of metal-containing compounds (Chapter 2) (Parkash, 2003; Gary et al., 2007; Speight, 2014a; Hsu and Robinson, 2017; Speight, 2017).

However, over the past several decades, the feedstocks available to refineries have generally decreased in API gravity and a major focus in refineries is to develop processing options for the viscous feedstocks (Khan and Patmore, 1997; Gary et al., 2007; Rana et al., 2007; Rispoli et al., 2009; Stratiev and Petkov, 2009; Stratiev et al., 2009; Motaghi et al., 2010a, 2010b; Speight, 2011a; Hsu and Robinson, 2017; Speight, 2017). Simultaneously, the changing crude oil properties are reflected in changes such as an increase in asphaltene constituents, and an increase in sulfur, metal, and nitrogen contents. Pretreatment processes for removing such constituents or at least negating their effect in thermal process would also play an important role.

The limitations of processing these heavy feedstocks depend to a large extent on the amount of higher molecular weight constituents (i.e., asphaltene constituents and resin constituents) that contain the majority of the heteroatom-containing compounds, which are responsible for high yields of thermal and catalytic coke. Be that as it may, the essential step required of a modern refinery is the upgrading

Handbook of Heavy Oil Properties and Analysis, First Edition. James G. Speight.

of heavy feedstocks, particularly the complex atmospheric residua and vacuum residua.

Upgrading feedstocks such as heavy oils and residua began with the introduction of hydrodesulfurization processes (Speight, 2000). In the early days, the goal was desulfurization but, in later years, the processes were adapted to a 10–30% partial conversion operation, as intended to achieve desulfurization and obtain low-boiling fractions simultaneously, by increasing severity in operating conditions. However, as refineries have evolved and feedstocks have changed, refining viscous feedstocks has become a major part of modern refinery practice and there has also been the evolution and the necessity of the development of process configurations to accommodate the viscous feedstocks (Khan and Patmore, 1997; Parkash, 2003; Gary et al., 2007; Speight, 2011a; Hsu and Robinson, 2017; Speight, 2017).

For example, hydrodesulfurization of the light (low-boiling) distillate (naphtha or kerosene) is one of the more common catalytic hydrodesulfurization processes since it is usually used as a pretreatment of such feedstocks prior to deep hydrodesulfurization or prior to catalytic reforming. A similar concept of pretreating residua prior to hydrocracking to improve the quality of the products is also practiced (Speight, 2011a). Hydrodesulfurization of such feedstocks is required because sulfur compounds poisoning the precious-metal catalysts used in the hydrocracking process can be achieved under relatively mild conditions. If the feedstock arises from a cracking operation (such as cracked residua), hydropretreatment will be accompanied by some degree of saturation resulting in increased hydrogen consumption.

In fact, since the so-called oil shocks of the 1970s, the refining industry been the subject of the forces which have hastened the development of refineries such as (1) the demand for products such as gasoline, diesel, fuel oil, and jet fuel; (2) feedstock supply, specifically the changing quality of crude oil and geopolitics between different countries; (3) the emergence of alternate feed supplies such as heavy crude oil, extra heavy crude oil, and tar sand—collectively known as viscous feedstocks; (4) technology development such as new catalysts and processes, especially processes involving the use of hydrogen; and (5) environmental regulations that include more stringent regulations in relation to sulfur in gasoline and diesel (Parkash, 2003; Gary et al., 2007; Speight, 2011a; Hsu and Robinson, 2017; Speight, 2017).

The viscous feedstocks by virtue of the generation and maturation process that contribute to the formation of these feedstocks have lesser amounts of hydrocarbon derivatives and have properties characteristics that affect their flowability in the processes of recovery, pipeline transportation, and in the refinery. Moreover, the key to feedstock refining is, as in any industrial process, knowledge of the character of the feedstock. Chemical composition is the starting point for the oil characterization and it has a major impact on other properties, including key properties for their dynamics, such as density and viscosity (Chapter 2).

The particular characteristics of the viscous oils are mainly attributed to a biodegradation process in which microorganisms on a geological time scale degrade light and medium hydrocarbons, making the reserves rich in polynuclear aromatic derivatives, resin constituents, and asphaltene constituents (Parkash, 2003; Gary et al., 2007; Speight, 2014a; Hsu and Robinson, 2017; Speight, 2017). Microbial degradation reaches optimal temperatures below 80 °C (715 °F), promoting oil oxidation, reduction of the gas/oil ratio (GOR), and increasing density, acidity, and viscosity as well as the relative proportion of sulfur and heavy metals. In addition to bio-processes, the formation of heavy crude oil (as well as extra heavy crude oil and tar sand bitumen) occurs through mechanisms such as water washing and phase fractionation, which are based on the loss of a significant fraction of the original source mass as well as removal of lower boiling constituents (and fractions) by physical rather than bio-processes.

The viscous feedstocks display a high content of high molecular weight constituents that contain elevated levels of heteroatom constituents including sulfur, nitrogen, oxygen, and metals that require modification of the refinery processing options (Speight, 2011a, 2017). Typically, the molecules present in the viscous oils have more than 15 carbon atoms in the chain which can lead to the generation of products (in the refinery) such as a low yield of content of high-octane naphtha and kerosene. Although the amount of compounds containing heteroatom constituents is relatively small, the effect of these compounds on the refining processes and the properties of the products are usually significant.

Chemical species containing sulfur atoms are often regarded as harmful for their effects on the refining process. The most common types are thiol derivatives (RSH), sulfide derivatives (R^1SR^2, where R^1 and R^2 are the same or different organic moieties), and thiophene and derivatives.

By way of definition, thiophene (C_4H_4S) is a heterocyclic compound (the sulfur is in a ring structure) that is a colorless liquid with a benzene-like odor that is only one of several sulfur-containing derivatives that occur in crude oil and the viscous oils (Figure 1.1).

These derivatives consist of thiol mercaptan derivatives (RSH, where R is an alkyl moiety or an aromatic moiety), sulfide derivatives (R^1SR^2, where R^1 and R^2 are the same or different moieties), disulfide derivatives (R^1SSR^2, where R^1 and R^2 are the same or different moieties), and cyclic sulfide derivatives where the sulfur atom occurs in a six-membered saturated ring system consisting of five carbon atoms and one sulfur atom. There are also examples where the sulfur atom occurs in a

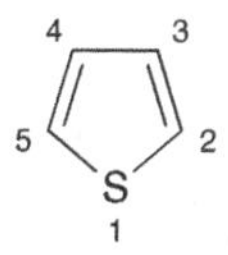

Figure 1.1 Thiophene (the numbers, 2–5, indicate the positions of the substitutable hydrogen atoms).

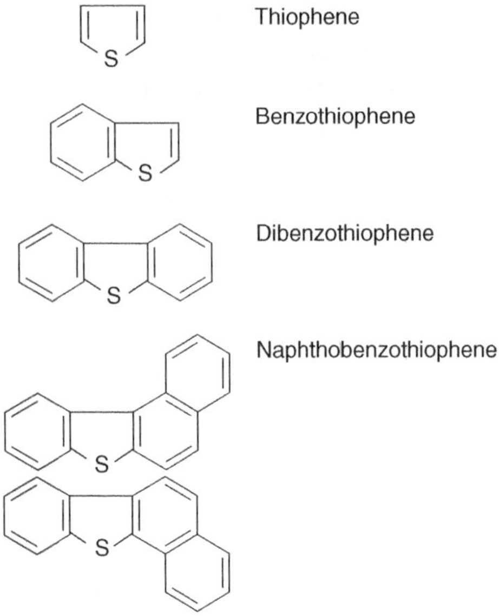

Figure 1.2 Examples of organic sulfur derivative that occur in crude oil and the viscous oils.

five-membered saturated ring system consisting of four carbon atoms and one sulfur atom. The remaining derivatives may be (arbitrarily or simple) considered to be derivatives of thiophene (Figure 1.2).

Thiophene is a planar five-membered ring and reacts very much in the manner of benzene derivatives.

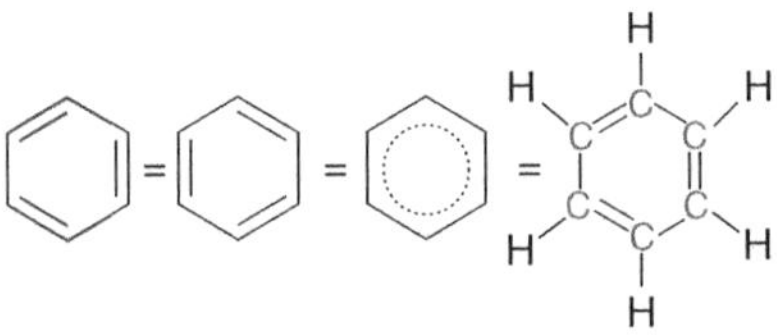

Figure 1.3 Benzene (showing the various alternate structures; in terms of reactivity, the structures are equivalent).

Benzene is an organic hydrocarbon (C_6H_6), that is, the benzene molecule is composed of six carbon atoms joined in a planar ring with one hydrogen atom attached to each carbon (Figure 1.3).

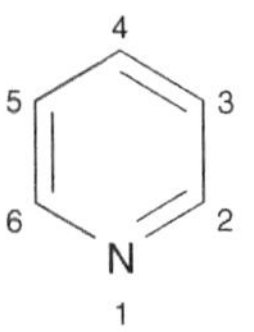

Figure 1.4 Pyridine (the numbers, 2–6, indicate the positions of the substitutable hydrogen atoms).

The nitrogen compounds are generally basic, formed by pyridine and its homologs (Figure 1.4). Pyridine is a basic heterocyclic organic compound (C_5H_5N) which is structurally related to benzene, with one methine (—C—H) group replaced by a nitrogen atom. It is a weakly alkaline, water-miscible liquid.

However, nitrogen compounds can also occur in nonbasic forms, formed by species including pyrrole derivatives, indole derivatives, and carbazole derivatives (Figure 1.5) (Speight, 2014a, 2017). Significant amounts of porphyrin derivatives (which contain nickel, Ni, and vanadium, V) may occur in the nonbasic fraction of the nitrogen compounds. Other

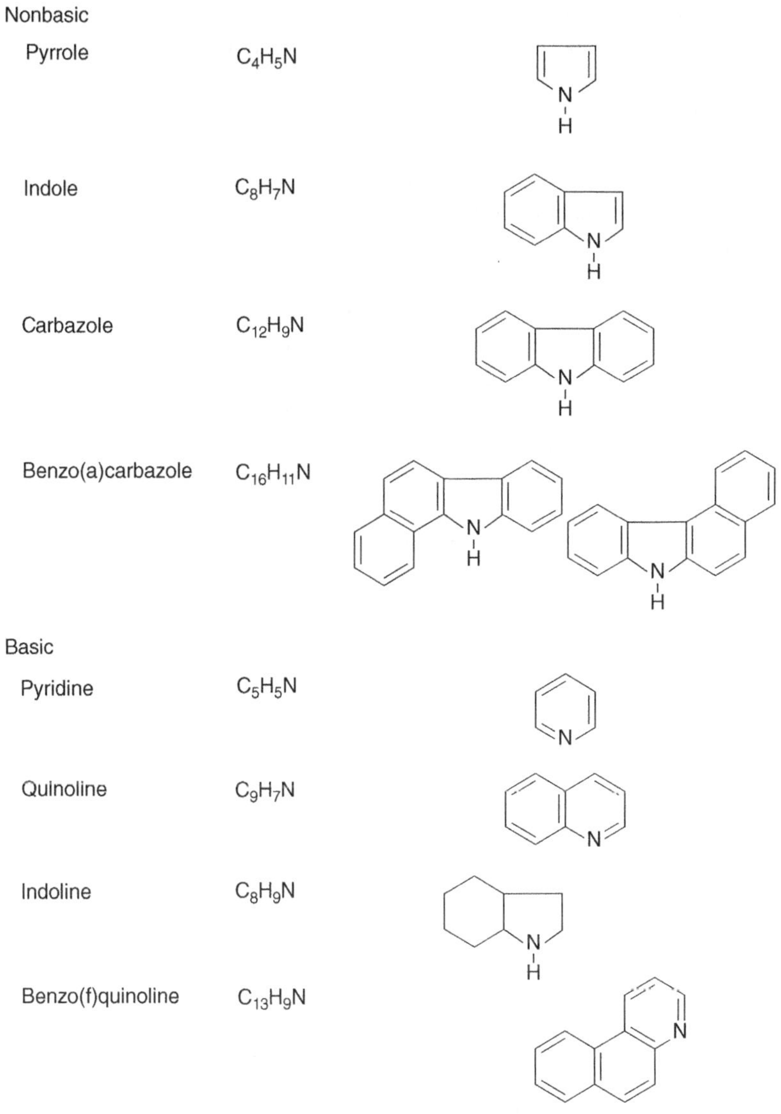

Figure 1.5 Examples of organic nitrogen derivatives that occur in crude oil and the viscous oils.

metal-containing derivatives are generally present in the form of organic salts (R–M+) where R is an organic moiety and M is a metal dissolved in oil-emulsified water. Heavy crude oils often contain a large portion of nickel and vanadium, which form chelates with porphyrins that are responsible for catalyst contamination and corrosion problems (Speight, 2014c).

Oxygenated compounds appear as carboxylic derivatives (RCO_2H, where R is an aliphatic group or aromatic group) and phenol derivatives ($ArCO_2H$, where Ar is an aromatic group), although the presence of ketone derivatives (R^1COR^2, where R^1 and R^2 are either aliphatic or aromatic hydrocarbon groups) and ether derivatives (R^1OR^2, where R^1 and R^2 are either aliphatic or aromatic hydrocarbon groups) have also been identified in some feedstocks (Figure 1.6) (Speight, 2014a, 2017). The content of the oxygen-containing compounds (especially the carboxylic acid derivatives) and the phenol derivatives determines the acidity of the feedstock and, hence, the corrosivity (Speight, 2014c), which is particularly important in the refining processes. These features can occur on the same molecular structure, further increasing the complexity and difficulty of the characterization of the compounds present in the crude oil.

Due to the impossibility of an elemental characterization of crude oils and viscous oils because of their complex nature, a complete characterization has been satisfactorily obtained by fractionation, based on fraction polarity and solubility (Chapter 3). The saturates, aromatics, resins, asphaltene (SARA) analysis is the most widely used method to subdivide refinery feedstocks (and many higher boiling refinery products) into four subfractions, namely (1) a saturates fraction, (2) an aromatics fraction, (3) a resins fraction, and (4) an asphaltene fraction. The method produces the fractions based on molecular weight (by solvent treatment of the

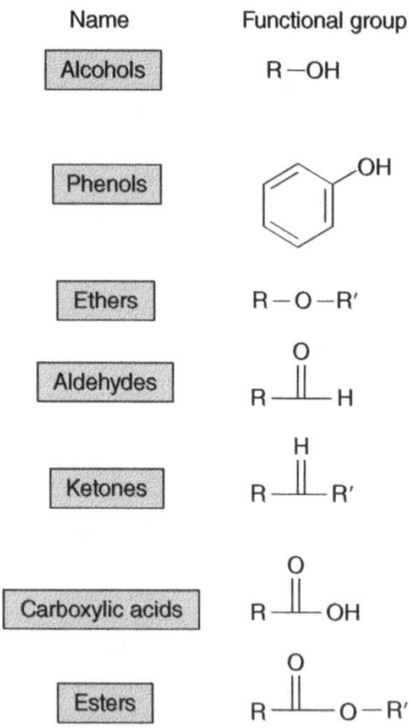

Figure 1.6 Examples of organic oxygen derivatives that occur in crude oil and the viscous oils.

feedstock) and polarity through a chromatographic technique. Thus, by using different solvents (such as *n*-pentane or *n*-heptane), the asphaltene fraction can be separated from the whole feedstock (or a topped feedstock—the feedstock from which the more volatile constituents boiling up to, say, 200 °C, 390 °F have been removed)—while an adsorbent is (or different adsorbents are) employed to separate the saturates, aromatics, and resin fractions (Chapter 3) (Speight, 2014a, 2015a). The method is reproducible and applicable to a wide variety of heavy crude oil, extra heavy crude oil, tar sand bitumen, and crude oil residua.

Finally, there is not one single heavy feedstock upgrading solution that will fit all refineries. Market conditions, existing refinery configuration, and available crude prices all can have a significant effect on the final configuration. Furthermore, a proper evaluation, however, is not a simple undertaking for an existing refinery. The evaluation starts with an accurate understanding of the market for the various products along with corresponding product values at various levels of supply. The next step is to select a set of crude oils that adequately cover the range of crude oils that may be expected to be processed. It is also important to consider new unit capital costs as well as incremental capital costs for revamp opportunities along with the incremental utility, support, and infrastructure costs. The costs, although estimated at the start, can be better assessed once the options have been defined leading to the development of the optimal configuration for refining the incoming feedstocks.

In fact, the limitations of processing these heavy feedstocks depend to a large extent on the amount of higher molecular weight constituents (i.e., asphaltene constituents and resin constituents) that contain the majority of the heteroatom-containing compounds, which are responsible for high yields of thermal and catalytic coke (Parkash, 2003; Gary et al., 2007; Speight, 2014a; Hsu and Robinson, 2017; Speight, 2017). Be that as it may, the essential step required of a modern refinery is the knowledgeable upgrading of the heavy feedstocks through the data acquisition programs (analytical test methods) that provide indication of the behavior of the feedstocks during refining.

Furthermore, there is not one single upgrading option that will accommodate all of the viscous feedstocks and that will fit all refineries and feedstock composition can have a significant effect on the necessary processing configuration. Thus, a proper evaluation that starts with an analytical program that will identify the characteristics of the feedstocks accepted by the refinery as well as afford the luxury of predictability of the behavior of the feedstock constituents during refining.

Thus, the focus of this text is on the test methods that will assist the refiner to identify the appropriate options for the various viscous refinery feedstocks that are currently accepted as feedstocks for refineries. This chapter begins with a presentation of the terminology of the feedstocks.

1.2 Historical Perspectives

The history of any subject is the means by which the subject is studied in the hopes that much can be learned from the events of the past. In the current context, the occurrence and use of crude oil, crude oil derivatives (naphtha), and the viscous feedstocks (such as heavy crude oil, extra heavy crude oil, and tar sand bitumen) are not new. The use of crude oil and the viscous materials that seeped from outcrops was practiced in pre-Christian times and is known largely through historical use in many of the older civilizations, particularly the civilization of the Middle Eastern countries where seepages from outcrops were widely recognized (Henry, 1873; Abraham, 1945; Forbes, 1958a, 1958b; James and Thorpe, 1994; Speight, 2014a).

As a brief introduction to the historical aspects of the subject, crude oil (also known as petroleum) was originally referred to as "rock oil" which was derived from the Latin "petra" which, in turn, was derived from the Greek word "πέτρα" (meaning "rock" or "stone") and "oleum" which originally meant "olive oil" and, in turn, was derived from the ancient Greek word "ἔλαιον" (élaion also meaning "olive oil"). However, the development of a system of terminology is more complex and has evolved from pre-Christian times (years typically designated by the term "BC") to the present time and, with the incursion of the viscous feedstocks into refining processing, is continuing to evolve.

1.2.1 Pre-Christian Era Use of Heavy Oil and Bitumen

The word asphalt is derived from the Akkadian term "asphaltu" or "sphallo," meaning to split, which is indicative of the early use of an asphalt as a mastic between stones.

As an explanatory note, the Akkadian Empire was the first ancient empire of Mesopotamia that existed after the long-lived southern Mesopotamian civilization of Sumer. The empire was centered in the city of Akkad and the immediately surrounding region, and the empire united the Sumerian and Akkadian speakers under one rule and language. Sargon of Akkad (also known as Sargon the Great and sometimes identified as the first person in recorded history to rule over an empire rather than a city-state) was the first ruler of the Akkadian Empire, known for his conquests of the Sumerian city-states in the 24th to 23rd centuries BC.

The term "asphaltu" was later adopted by the Homeric Greeks in the form of the signifying firm, stable, secure, and the corresponding verb ασφαλίζω ίσω meaning to make firm or stable, to secure. On the other hand, the origin of the word bitumen is more difficult to trace and subject to considerable speculation. The word has been proposed to have originated in the Sanskrit, in which language the words "jatu", meaning pitch, and "jatukrit" occur and have been proposed to mean "pitch creating" (Speight, 2011b, 2014a, 2016).

From the Sanskrit, the word "jatu" was incorporated into the Latin language as "gwitu" and is reputed to have eventually morphed into the word "gwitumen" (supposedly meaning "pertaining to pitch"). Another word, "pixtumen" (which has been proposed and assumed to mean "exuding pitch" or "bubbling pitch") is also reputed to have been in the Latin language, although the construction of this Latin word, from which the word "bitumen" was reputedly formed and derived, is certainly suspect (Speight, 2014a, 2016). However, there is the possibility that subsequent derivation of the word led to a shortened version (which eventually became the modern version) "bitûmen" from which the word has passed into the English language by way of the French language.

In this text for convenience and to avoid confusion that can exist from the use of different names, the word "bitumen" is used to indicate the naturally occurring viscous organic material that occurs in various deposits and outcrops. In the modern world, the word "asphalt" is used to indicate the manufactured organic material and does not include the inorganic aggregate that is mixed with the paving as a lay-down for road paving systems (Speight, 2014a, 2016).

More generally, bitumen (often incorrectly referred to as natural asphalt since the word "asphalt" is used to describe a refinery product) was used as mortar from very early times, and sand, gravel, or clay was employed in preparing these mastics. Bitumen-coated tree trunks were often used to reinforce wall corners and joints, for instance in the temple tower of Ninmach in Babylon. In vaults or arches, a mastic-loam composite was used as mortar for the bricks, and the keystone was usually dipped in bitumen before being set in place. The use of bituminous mortar was introduced into the city of Babylon by King Hammurabi, but the use of bituminous mortar was abandoned toward the end of the reign of King Nebuchadnezzar in favor of lime mortar to which varying amounts of bitumen were added. The Assyrians recommended the use of bitumen for medicinal purposes, as well as for building purposes, and perhaps there is some merit in the fact that the Assyrian moral code recommended that bitumen, in the molten state, be poured onto the heads of criminals as a form of punishment.

One of the earliest recorded uses of bitumen was by the pre-Babylonian inhabitants of the Euphrates Valley in southeastern Mesopotamia, present-day Iraq, formerly called Sumer and Akkad and, later, Babylonia. In this region, there are various bitumen deposits, and uses of the material have become evident. For example, King Sargon Akkad (Agade) (at about 2550 BC) was (for reasons that are lost in the annals of time) set adrift by his mother in a basket of bulrushes on the waters of the Euphrates and he was discovered by Akki the husbandman (the irrigator), whom he brought up to serve as gardener in the palace of Kish. Sargon eventually ascended to throne.

On the other hand, the bust of Manishtusu, King of Kish, an early Sumerian ruler (about 2270 BC), was found in the course of excavations at Susa in Persia, and the eyes, composed of white limestone, are held in their sockets with the aid of bitumen. Fragments of a ring composed of bitumen have been unearthed above the flood layer of the Euphrates at the site of the prehistoric city of Ur in southern Babylonia, ascribed to the Sumerians of about 3500 BC.

The Tigris-Euphrates valley, in what is now Iraq, was inhabited as early as 4000 BC by the people known as the Sumerians who established one of the first great cultures of the civilized world. The Sumerians devised the cuneiform script, built the temple-towers known as ziggurats, and had an impressive law, literature, and mythology. As the culture developed, bitumen was used as a sealant (sometimes incorrectly referred to as natural occurring bitumen). In fact, there is evidence that bitumen was used as a sealant for water channels to, and within, several of the ancient cities in the region (Speight, 1978).

Nevertheless, these writings do offer documented examples of the use of crude oil and related materials. For example, in the Epic of Gilgamesh written more than 2,500 years ago, in anticipation of a forthcoming deluge, the principal of the story builds an ark (a very large boat) to accommodate and save his family as well as a host of animals. The wooden boat is caulked with bitumen and pitch (see for example, Kovacs, 1990) and in a related story, also set in Mesopotamia just prior to the deluge (the Biblical Flood), (it is not the intent here to discuss the similarities or relationship or origin of the two stories), a man (Noah) is commanded to build a boat that also includes instructions for caulking the vessel with pitch (Genesis 6:14).

> Make yourself an ark of cypress wood; rooms shalt thou make in the ark, and shalt coat it with pitch inside and out.

The occurrence of slime (bitumen) pits in the Valley of Siddim, a valley at the southern end of the Dead Sea, is also noted (Genesis 14:10). There is also reference to the use of tar as a mortar when the Tower of Babel was under construction (Genesis 11:3).

> And they said one to another, Go to, let us make brick, and burn them thoroughly. And they had brick for stone, and slime had they for mortar.

In the Septuagint, or Greek version of the Bible, this work is translated as "asphaltos" and in the Vulgate or Latin version, as bitumen. In the Bishop's Bible of 1568 and in subsequent translations into English, the word is given as slime. In the Douay translation of 1600, it is "bitume" while in the German version of the Bible (assigned to Martin Luther), it appears as the word "thon"—the word is of linguistically questionable origin and has been suggested to mean "clay." Another example of the use of pitch (and slime) is given in the story of Moses (Exodus 2:3):

> And when she could no longer hide him, she took for him an ark of bulrushes, and daubed it with slime and with pitch, and put the child therein; and she laid it in the flags by the river's brink.

It is convenient (if not very reasonable) to assume that the word "slime" referred to a lower melting bitumen (such as extra heavy crude oil) whereas the word "pitch" referred to a higher melting material—the slime acting as a flux for the pitch.

It is most probable that, in both these cases, the *pitch* and the *slime* were obtained from the seepage of oil to the surface, which was a fairly common occurrence in the area. And during biblical times, bitumen was exported from Canaan to various parts of the countries that surround the Mediterranean (Armstrong, 1997).

In terms of liquid products, there is a noteworthy reference that relates to bringing oil out of flinty rock (Deuteronomy 32:13). The exact nature of the oil is not described nor is the nature of the rock. The use of oil for lamps is also referenced (Matthew 23:3) but whether it was mineral oil (a crude oil derivative such as naphtha or kerosene) or whether it was vegetable oil is not known.

Excavations conducted at Mohenjo-Daro and Harappa (founded and occupied in approximately 2500 BC) in the Indus Valley indicated that a bitumen-type mastic composed of a mixture of bitumen, clay, gypsum, and organic matter was found between two brick walls in a layer about 25 mm thick, probably a waterproofing material. Also unearthed was a bathing pool that contained a layer of mastic on the outside of its walls and beneath its floor.

Also in the Bronze Age (approximately 3300 to 1200 BC), dwellings were constructed on piles in lakes close to the shore to better protect the inhabitants from the ravages of wild animals and attacks from marauders. The wooden piles were preserved from decay by coating with bitumen and there are also references to deposits of bitumen at Hit (the ancient town of Tuttul on the Euphrates River in Mesopotamia) and the bitumen from these deposits was transported to Babylon for use in construction (Herodotus, The Histories, Book I). There is also reference (Herodotus, The Histories, Book IV) to a Carthaginian story in which birds' feathers smeared with pitch are used to recover gold dust from the waters of a lake.

Use of bitumen by the Babylonians (1500 to 538 BC) is also documented and each monarch commemorated their reign and perpetuated their name by construction of building or other monuments. This includes references to deposits of bitumen at Hit (the ancient town of Tuttul on the Euphrates River) and the bitumen from these deposits was transported to Babylon for use in construction (Herodotus, The Histories, Book I). There is also reference (Herodotus, The Histories, Book IV) to a Carthaginian story in which birds' feathers smeared with pitch are used to recover gold dust from the waters of a lake. Also, the use of bitumen mastic as a sealant for water pipes, water cisterns, and in outflow pipes leading from flush toilets in ancient cities such as Babylon, Nineveh, Calah, and Ur has been observed and the bitumen lines are still evident in uncovered systems (Speight, 1978).

Early references to crude oil and its derivatives occur in the Bible, although by the time the various books of the Bible were written, the use of crude oil and bitumen was established. For example, the combustion properties of bitumen (and its fractions) were known in Biblical times. There is the reference to these properties (Isaiah 34:9) when it is stated that:

> And the stream thereof shall be turned into pitch, and the dust thereof into brimstone, and the land thereof shall become burning pitch. And the streams thereof shall be turned into pitch.
>
> It shall not be quenched night nor day; the smoke thereof shall go up forever: from generation to generation it shall lie waste; none shall pass through it.

It can be concluded (based on current knowledge) that the effects of the burning bitumen and sulfur (brimstone) were long lasting and quite devastating.

The Egyptians were the first to adopt the practice of embalming their dead rulers and wrapping the bodies in cloth. Before 1000 BC, bitumen was rarely used in mummification, except to coat the cloth wrappings and thereby protect the body from the elements. From 500 to about 40 BC, bitumen was generally used both to fill the corpse cavities (after the viscera had been removed) as well as to coat the cloth wrappings. The word mûmûia first made its appearance in Arabian and Byzantine literature about 1000 AD, signifying bitumen. In fact, it was through the spread of the Islamic Empire that, it is believed, brought Arabic science and the use of bitumen to Western Europe.

In the Roman world, bitumen was also an important commodity and could also be used for medicinal purposes such as to stop bleeding, heal wounds, drive away snakes, treat cataracts, and straighten out eyelashes which inconvenience the eyes as well as a wide variety of other diseases.

From the same roots as used by the pre-Christian builders, the Anglo Saxon word "cwidu" (mastic, adhesive) was derived. Also, the German word "kitt" (translated as filler, cement, or mastic) which is also found in the old Norse language as being descriptive of the material used to waterproof the long ships and other sea-going vessels. It is just as (perhaps even more than) likely that the word is derived from the Celtic bethe or beithe or "bedw" that was the birch tree that was used as a source of resin (tar). From this, the word appears in Middle English as bithumen (Speight, 2014a, 2016).

In fact, a variety of words exist in ancient languages which, from the use of the words in the relevant contexts, can be proposed as referring to bitumen

Table 1.1 Examples of the Origins of Words Related to Crude Oil and Bitumen

Language	Word	Possible meaning
Sumerian	esir	Petroleum, bitumen
	esir-harsag	Rock asphalt
	esir-ud-du-a	Pitch
	kupru	Slime
	kupru	Pitch
Sanskrit	jatu	Bitumen
	Jatu	Pitch
	śilā-jatu	Rock asphalt
Assyrian/Akkadian	idd	Bitumen
	Ittû	Bitumen
	it-tû-u	Bitumen
	amaru	Bitumen
	sippatu	Pitch
Greek	maltha	Soft asphalt
	asphaltos	Bitumen
	pittasphaltos	Rock asphalt
	pittolium	Rock oil
	pittolium	Rock asphalt
	pissa	Pitch
	pitta	Pitch
Latin	maltha	Soft asphalt
	bitumen liquidum	Soft asphalt
	pix	Pitch

(Table 1.1) (Abraham, 1945). From the Greek language, the word passed into (early medieval) Latin as "asphaltum" or "aspaltum" and from there into French ("asphalte") and Old English ("aspaltoun").

In addition to bitumen, there was also an interest in the distillation product derived from crude oil (nafta, which is presumably the product now referred to as naphtha) which was also derived from the thermal treatment of bitumen. It was discovered that this material could be used as an illuminant and as a supplement to asphalt incendiaries in warfare. For example, there are records of the use of mixtures of pitch and/or naphtha with sulfur as a weapon of war during the Battle of Palatea, Greece, in the year 429 BC (Forbes, 1959).

1.2.2 Post-Christian Era Use of Heavy Oil and Bitumen

Approximately two thousand years ago, Arabian scientists developed methods for the distillation of crude oil, which were introduced into Europe by way of Spain. This represents another documented use of the volatile derivatives of crude oil which led to a continued interest in crude oil and its derivatives as medicinal materials and materials for warfare, in addition to the usual construction materials.

There are references to the use of naft as an incendiary material (Greek fire, a precursor and chemical cousin to napalm) during various battles (James and Thorpe, 1994). Greek fire is also recorded as being used in the period 674–678 AD when the city of Constantinople was saved by the use of the fire against the by an Arab fleet (Davies, 1996). In 717–718 AD, Greek fire was again used to save the city of Constantinople from attack by another Arab fleet; again with deadly effect (Dahmus, 1995).

Many other references to bitumen occur throughout the empires of Greece and Rome and from then to the Middle Ages, early scientists (alchemists) frequently alluded to the use of bitumen for a variety of purposes of which bitumen use as a mastic appears to have been popular. In later times, both Christopher Columbus and Sir Walter Raleigh (depending upon the country of origin of the biographer) have been credited with the discovery of the bitumen deposit on the island of Trinidad and apparently used the material to caulk the ships.

Bitumen (or a bitumen-type material, i.e., a viscous naturally occurring material) was used in prescriptions, as early as the 12th Century, by the Arabian physician Al Magor, for the treatment of contusions and wounds. The 13th Century scientist Al-Kazwînî alluded to the healing properties of mûmûia, and Ibn Al-Baitâr gives an account of its source and composition. The region around Baku (currently, the capital and largest city of Azerbaijan, as well as the largest city on the Caspian Sea and of the Caucasus region) was also reported (by Marco Polo in 1271–1273) as having an established commercial crude oil industry. It is believed that the prime interest was in the kerosene fraction that was then known for its use as an illuminant. By inference, it has to be concluded that the distillation, and perhaps the thermal decomposition, of crude oil were established technologies. If not, Polo's diaries might well have contained a description of the stills or the reactors.

In addition, bitumen was investigated in Europe during the Middle Ages (Bauer, 1546, 1556), and the separation and properties of bituminous products were thoroughly described. Other investigations continued, leading to a good understanding of the sources and use of this material even before the birth of the modern crude oil industry (Forbes, 1958a, 1958b). Also, in Europe, Engelbert Kämpfer (1651–1716) gave a detailed account of the collection of bitumen in his treatise Amoenitates Exoticae (Exotic Pleasantries) as well as the different grades and types and the use of the material as a curative in medicine.

There are also records of the use of crude oil spirit, probably a higher boiling fraction of or than naphtha that closely resembled the modern-day liquid paraffin, for medicinal purposes. In fact, the so-called liquid paraffin has continued to be prescribed up to modern times. The naphtha of that time was obtained from shallow wells or by the destructive distillation of asphalt.

Parenthetically, the destructive distillation operation may be likened to modern thermal cracking or coking operations (Parkash, 2003; Gary et al., 2007; Speight, 2014a; Hsu and Robinson, 2017; Speight, 2017) in which the overall objective is to convert the feedstock into distillates for use as fuels. This particular interest in crude oil and its derivatives continued with an increasing interest in nafta (naphtha) because of its aforementioned used as an illuminant and as a supplement to asphaltic incendiaries for use in warfare.

In summary, the use of crude oil and bitumen has been observed for almost 6000 years (Mallowan and Rose, 1935; Nellensteyn and Brand, 1936; Mallowan, 1954; Marschner et al., 1978). During this time, the use of the raw materials and the products has progressed from the relatively simple use of asphalt from Mesopotamian seepage sites to the present-day refining operations that yield a wide variety of products and petrochemicals (Parkash, 2003; Gary et al., 2007; Speight, 2014a; Hsu and Robinson, 2017; Speight, 2017, 2019a).

1.3 Definitions and Terminology

While the focus of this chapter is predominantly on the various viscous refinery feedstocks, the nonviscous fractions of crude oil are also used for the production of products through the application of thermal cracking processes and catalytic cracking processes. These are (1) crude oil and (2) other assorted crude oils such as opportunity crude oil, high-acid crude oil, and foamy oil.

These feedstocks are produced from reservoirs and are part of the crude oil family but are, nevertheless, worthy of mention as potential feedstocks for comparison with the viscous feedstocks which, when submitted to thermal cracking processes and catalytic cracking processes, lead to a range of products that also includes petrochemical products (Parkash, 2003; Gary et al., 2007; Speight, 2014a; Hsu and Robinson, 2017; Speight, 2017, 2019a, 2021).

Throughout the millennia in which crude oil has been known and used, it is only in the last decade or so that some attempts have been made to standardize the nomenclature and terminology. But confusion may still exist. Therefore, it is the purpose of this chapter to provide some semblance of order into the disordered state that exists in the segment of crude oil technology that is known as terminology.

Terminology is the means by which various subjects are named so that reference can be made in conversations and in writings and so that the meaning is passed on.

Definitions are the means by which scientists and engineers communicate the nature of a material to each other and to the world, through either the spoken or the written word. Thus, the definition of a material can be extremely important and have a profound influence on how the technical community and the public perceive that material.

Because of the need for a thorough understanding of crude oil and the associated viscous feedstocks, it is essential that the definitions and the terminology of crude oil science and technology be given prime consideration (Meyer and De Witt, 1990). This will aid in a better understanding of crude oil, its constituents, and its various fractions. Of the many forms of terminology that have been used, not all have survived, but the more commonly used are illustrated here. Particularly troublesome, and more confusing, are those terms that are applied to the more viscous materials, for example the use of the terms bitumen and asphalt (Speight, 2014a, 2016). This part of the text attempts to alleviate much of the confusion that exists, but it must be remembered that the terminology of crude oil is still open to personal choice and historical usage.

In general, the definition of crude oil has been varied, unsystematic, diverse, and often archaic (Speight, 2014a, 2017, 2021) because the terminology used to describe crude oil is subject to many conventions of which major ones are chemistry-aligned definitions, company-aligned definitions, engineering-aligned definitions, and definitions aligned with the geological sciences. Thus, the long-established use of an expression, however inadequate it may be, is altered with difficulty, and a new term, however precise, is at best adopted only slowly.

Of the many forms of terminology that have been used not all have survived and the more commonly used are illustrated here but, as a note, particularly troublesome, and more confusing, are those terms that are applied to the more viscous materials, for example the use of the terms bitumen and asphalt. This part of the text attempts to alleviate much of the confusion that exists, but it must be remembered that the terminology of crude oil is still open to personal choice and historical usage.

By way of introduction, when crude oil occurs in a reservoir that allows the crude material to be recovered by pumping operations as a free-flowing dark to light colored liquid, it is often referred to as conventional crude oil. Heavy crude oils are categorized as types of crude oil that are different from conventional crude oil insofar as they are much more difficult to recover (but can be recovered) from the subsurface reservoir (Speight, 2014a). There have been arbitrary attempts to rationalize the definition of heavy crude oil (and other naturally occurring viscous materials such as extra heavy crude oil and tar sand bitumen) based upon viscosity, API gravity, and density when it is preferable to use a more practical method (that is less prone to misinterpretation and confusion) such as the method of recovery of the material (Speight, 2014a, 2017, 2021).

Viscous feedstocks are refinery feedstocks such as crude oil residua, (1) heavy crude oil, (2) extra heavy crude oil, (3) tar sand bitumen, and (4) distillation residua that have a low volatility, high density, and high viscosity. Although other feedstocks such as naphtha, kerosene, and gas oil are also used as feedstocks for thermal processes and catalytic processes in the refinery (Parkash, 2003; Gary et al., 2007; Speight, 2014a; Hsu and Robinson, 2017; Speight, 2017), the focus of this book remains on the viscous feedstocks.

1.3.1 Nonviscous Feedstocks

By way of introduction, when crude oil occurs in a reservoir that allows the crude material to be recovered by pumping operations as a free-flowing dark to light colored liquid, it is often referred to as conventional crude oil. A heavy crude oil is categorized as a type of crude oil that is different from conventional crude oil insofar as the heavy crude oil is typically recovered using a thermal-based (i.e., a steam-based) process and is much more difficult to recover from the subsurface reservoir (Speight, 2014a).

In spite of the different recovery methods that are applied to heavy oil, the definition of heavy crude oil is often arbitrarily (and incorrectly) based on the API gravity or viscosity, although there have been arbitrary attempts to rationalize the definition based upon viscosity, API gravity, and density when it is preferable to use the method of recovery of the material (Speight, 2014a, 2017, 2021).

1.3.1.1 Crude Oil

The definition of crude oil has been varied, unsystematic, diverse, and often archaic. Furthermore, the terminology of crude oil is a product of many years of growth. Thus, the long-established use of an expression, however inadequate it may be, is altered with difficulty, and a new term, however precise, is at best adopted only slowly.

Crude oil is a naturally occurring mixture of hydrocarbons, generally in a liquid state, which may also include compounds of sulfur nitrogen oxygen metals and other elements (ASTM D4175, 2021). Crude oil has also been defined (ITAA, 1936) as: (1) any naturally occurring hydrocarbon, whether in a liquid, gaseous, or solid state; (2) any naturally occurring mixture of hydrocarbons, whether in a liquid, gaseous, or solid state; and (3) any naturally occurring mixture of one or more hydrocarbons, whether in a liquid, gaseous, or solid state and one or more of the following, that is to say, hydrogen sulfide, helium, and carbon dioxide. The definition also includes any crude oil as defined by section (1), (2), or (3) that has been returned to a natural reservoir.

Table 1.2 Boiling Fractions of Conventional Crude Oil

	Boiling range	
Fraction	**°C**	**°F[a]**
Low boiling range naphtha	30–100	30–210
Medium boiling range naphtha	100–150	210–300
High boiling range naphtha	150–180	300–400
Fuel oil (six classes)	175–600	345–1,110
Middle distillates[b]	180–290	400–555
Kerosene	180–260	355–500
Light gas oil	205–250	400–480
Heavy gas oil	250–290	480–550
Atmospheric gas oil	260–315	500–800
Vacuum gas oil	425–600	800–1,100
Light vacuum gas oil	315–425	600–800
Heavy vacuum gas oil	425–600	425–1,110
Residuum	>510	>950

[a] For convenience, boiling ranges which can vary from refinery to refinery are approximate and, for convenience, are converted to the nearest 5°. Also, there is overlap between some of the distillates because of the lack of a standard terminology that has been applied to these materials insofar as the terminology employed to name the fractions is either company dependent or refinery dependent.
[b] Obtained in the "middle" boiling range which is on the order of 180–290 °C (355–550 °F) during the distillation process. The middle distillates are so named because the fractions are removed at mid-height of the distillation tower during the multistage process of thermal separation.

Thus, the term "crude oil" and the equivalent term "petroleum" cover a wide assortment of materials of different boiling range (Table 1.2) consisting of mixtures of hydrocarbon derivatives and other constituents that contain variable amounts of sulfur, nitrogen, and oxygen, which may vary widely in volatility, specific gravity, and viscosity (Table 1.3).

As a special note, and to avoid any apparent confusion that may arise from an inspection of the boiling ranges, the boiling ranges of the various distillate fraction from refinery distillation can vary from refinery to refinery and, as presented here (Table 1.2) are approximate and, for convenience, are converted to the nearest 5°. There is also an overlap between some of the distillates because of the lack of a standard terminology that has been applied to these materials insofar as the terminology employed for the fractions is either company dependent or refinery

Table 1.3 Hydrocarbon and Heteroatom Types in Crude Oil, Heavy Oil, and Tar Sand Bitumen

Class	Compound types
Saturated hydrocarbons	*n*-Paraffins
	iso-Paraffins and other branched paraffins
	Cycloparaffins (naphthenes)
	Condensed cycloparaffins (including steranes, hopanes)
	Alkyl side chains on ring systems
Unsaturated hydrocarbons	Olefins: nonindigenous
	Present in products of thermal reactions
Aromatic hydrocarbons	Benzene systems
	Condensed aromatic systems
	Condensed naphthene-aromatic systems
	Alkyl side chains on ring systems
Saturated heteroatomic systems	Alkyl sulfides
	Cycloalkyl sulfides
Sulfides	Alkyl side chains on ring systems
Aromatic heteroatomic systems	Furans (single-ring and multi-ring systems)
	Thiophenes (single-ring and multi-ring systems)
	Pyrroles (single-ring and multi-ring systems)
	Pyridines (single-ring and multi-ring systems)
	Mixed heteroatomic systems
	Amphoteric (acid–base systems)
	Alkyl side chains on ring systems

dependent. The middle distillates are those distillates and those fractions that occur in the "middle" boiling range which is on the order of 180–260 °C (355–500 °F) during the crude oil distillation process. The middle distillates are so named because the fractions are removed at mid-height of the distillation tower during the multistage process of thermal separation.

Metal-containing constituents, notably those compounds that contain vanadium and nickel, usually occur in the more viscous crude oils in amounts up to several thousand parts per million and can have serious consequences during processing of these feedstocks (Gruse and Stevens, 1960; Speight, 1984). Because crude oil is a mixture of widely varying constituents and proportions, its physical properties also vary widely and the color varies from colorless to black (Parkash, 2003; Gary et al., 2007; Speight, 2012a, 2014a; Hsu and Robinson, 2017; Speight, 2017).

Crude oil occurs in underground formations that are at various pressures depending on the depth and, because of the pressure, conventional crude oil typically contains considerable natural gas in solution. Crude oil underground is much more fluid than it is on the surface and is generally mobile under reservoir conditions because the elevated temperatures (the geothermal gradient) in subterranean formations decrease the viscosity. The geothermal gradient depths vary from site to site and from formation to formation but it is generally accepted to be on the order of 15 °F/1,000 ft or 120 °C/1,000 ft, i.e., 0.015 °C per foot of depth or 0.012 °C per foot of depth.

In the raw (unrefined) state, crude oil has minimal value, but when refined, it provides high-value liquid fuels, solvents, lubricants, and many other products (Purdy, 1957). The fuels derived from crude oil contribute approximately one-third to one-half of the total world energy supply and are used not only for transportation fuels (i.e., gasoline, diesel fuel, and aviation fuel, among others) but also to heat buildings. Crude oil products have a wide variety of uses that vary from gaseous and liquid fuels to near-solid machinery lubricants. In addition, the residue of many refinery processes, asphalt—a once-maligned by-product—is now a premium value product for highway surfaces, roofing materials, and miscellaneous waterproofing uses.

Crude oil is a mixture of compounds boiling at different temperatures that can be separated into a variety of different generic fractions by distillation. And the terminology of these fractions has been bound by utility and often bears little relationship to composition.

Since there is a wide variation in the properties of crude oil (Table 1.4), the proportions in which the different constituents occur vary with origin (Parkash, 2003; Gary et al., 2007; Speight, 2014a; Hsu and Robinson, 2017; Speight, 2017). Thus, some crude oils have higher proportions of the lower boiling components and while other oils (such as heavy crude oil, extra heavy crude oil, and tar sand bitumen) have higher proportions of higher boiling components (usually referred to as the asphaltic components and the residuum) (Speight, 2014a, 2015a, 2017).

There is another type of crude oil that is recovered from shale formation and is generally referred to as "oil from tight shale." This is a relatively new term to the crude oil lexicon and should not be confused with "shale oil," which is oil produced by the thermal treatment of oil shale and the decomposition of the kerogen contained in the shale formation (Speight, 2012b).

The properties of oils from tight shale formations are significantly different to those of typical crude oils. Unlike most crude oils, shale oils are highly volatile, sweet (low-sulfur) liquids, with a high content of paraffin derivatives (alkanes)

Table 1.4 Examples of the Variation in the Properties of Crude Oil

Crude oil source	Specific gravity	API gravity	Residuum %v/v >535 °C (>1,000 °F)
United States			
California	0.858	33.4	23.0
Oklahoma	0.816	41.9	20.0
Pennsylvania	0.800	45.4	2.0
Texas	0.827	39.6	15.0
Other countries			
Bahrain	0.861	32.8	26.4
Iran	0.836	37.8	20.8
Iraq	0.844	36.2	23.8
Kuwait	0.860	33.0	31.9
Saudi Arabia	0.840	37.0	27.5
Venezuela	0.950	17.4	33.6

and low acidity. They also have minimal asphaltene (and resin) content phase and varying contents of filterable solids, hydrogen sulfide (H_2S), and mercaptans. There are significant differences in the sulfur content and the filterable solids loading. Solids loading (and the potential for inorganic fouling) of samples from a single producing region can be highly variable and associated with the stage of fracturing and production from which the oil is produced.

Finally, crude oil from tight shale formations may contain much higher levels of iron (Fe) than conventional crude oils (Benoit and Zurlo, 2014; De Graaf et al., 2014). In addition, other contaminant metals that promote the detrimental effects of iron (such as calcium, sodium, and potassium) are often present at elevated levels. Feedstocks with high iron content create a number of problems for refining and iron poisoning of fluid catalytic cracking catalyst occurs readily leading to: (1) an incase in slurry yield increases and (2) emission of sulfur oxides often increases. In the case of iron poisoning, activity and surface area tests are subject to artifacts that do not reveal that the catalyst is rendered inactive in the unit when the contact time with feed is short (contrary to the activity test unit).

1.3.1.2 Opportunity Crude Oil

While the term nonviscous feedstocks may suggest a highly suitable feedstock for a refinery, it may not be the case. In addition to conventional crude oil—often considered to be the prime refinery feedstocks—there are other types of crude ones that are not always suitable for the refinery. Such feedstocks, which are occurring more often as refinery feedstocks, are (1) opportunity crude oil, (2) high-acid crude oil, and (3) foamy oil, all three of which can also be accommodated under the umbrella term "heavy crude oil."

Thus, there is also the need for a refinery to be ready to accommodate opportunity crude oils and/or high-acid crude oils and/or foamy oils which, for many purposes, are often shipped to a refinery under the banner of heavy crude oil.

Opportunity crude oils are often dirty and need cleaning before refining by removal of undesirable constituents such as high-sulfur, high-nitrogen, and high-aromatics (such as polynuclear aromatic) components (Speight, 2014a, 2014b; Yeung, 2014).

More specifically, opportunity crude oils are either (1) new-to-the-refinery crude oils with unknown or poorly understood properties relating to processing issues or are (2) existing crude oils with well-known properties and processing concerns (Ohmes, 2014). Opportunity crude oils are often, but not always, heavy crude oils that are more difficult to process due to a high level of solids (and other contaminants) as well as a high level of acidity and a high viscosity. Opportunity crude oils (while cheaper to purchase because of the contaminants) may also (because of the heavy crude oil character) be incompatible with other (lighter crude oils) in the refinery feedstock blend and cause excessive equipment fouling when processed either in a blend or separately (Speight, 2014a, 2015b).

During the refinery process, fouling can occur in a variety of processes, usually inadvertently when the separation is detrimental to the process outcome (Speight, 2015b, 2017). Thus, separation of solids occurs whenever the solvent characteristics of the liquid phase are no longer adequate to maintain polar and/or high molecular weight constituents in solution.

Examples of such occurrences are: (1) separation of asphaltene constituents, which occurs when the paraffin nature of the liquid medium increases, (2) wax separation which occurs when there is a drop in temperature or the aromaticity of the liquid medium increases, and (3) sludge/sediment formation in a reactor which occurs (1) when the solvent characteristics of the liquid medium change so that asphaltic or wax materials separate or (2) when there is coke formation which occurs at high temperatures and commences when the solvent power of the liquid phase is not sufficient to maintain the coke precursors in solution, and (3) when there is sludge/sediment formation in fuel products which occurs because of the interplay of several chemical and physical factors.

In general, the foulant consists of organic material and/or inorganic material that is deposited by the feedstock at the time of (or shortly thereafter if there is an induction period) instability or incompatibility of the feedstock (one crude oil) with another during and shortly after a blending operation (Speight, 2014a, 2015b). However, fouling can also be a consequence of corrosion in a unit when deposits of inorganic solids become evident (Speight, 2015b). With the influx of opportunity crudes, high-acid crudes, heavier crude oils, extra heavy crude oils, and tar sand bitumen into refineries (Chapter 1) fouling phenomena are more common and diverse (Speight, 2014a, 2014b, 2015b).

Fouling can be classified into two broad categories: (1) microfouling and (2) microfouling. Common types of microfouling are: (1) biofouling, which is caused by microorganisms, (2) chemical reaction fouling, (3) precipitation fouling, (4) corrosion fouling, and (5) composite fouling, which is caused by more than one fouling mechanism or foulant. The occurrence of marine fouling typically occurs during the time when seaweed, bacteria, and other living organisms in the water adhere to immersed surfaces such as a ship hull which results in the formation of a layer that covers the surface thereby acting as a base for more material. In any case, the extent and severity of fouling is dependent on variables such as process parameter and the immediate environment.

On the other hand, macrofouling is caused by matter (or constituents) of either inorganic or organic origin, such as animals and plants. An example is the occurrence of fouling that occurs in heat transfer components in heat exchangers which can cause blockages or fretting damage. By way of explanation, *fretting* refers to wear damage as well as corrosion damage at the uneven (or rough) areas of metal surfaces and such damage is induced under load and in the presence of repeated relative surface motion, as induced for example by vibration which, as a result of contact with another surface, can cause mechanical wear and material transfer at the surface. This is often followed by the oxidation of both the metallic debris and the freshly exposed metallic surfaces. Typically, in many cases, the oxidized debris is much harder (and abrasive) than the surfaces from which it originated and often acts as an abrasive agent that increases the rate of both fretting and brinelling.

Briefly, fretting refers to wear and sometimes corrosion damage of loaded surfaces in contact while they encounter small oscillatory movements tangential to the surface. Fretting is caused by adhesion of contact surface asperities, which are subsequently broken again by the small movement. In addition, brinelling refers to mechanical wear which involves the permanent indentation of a hard surface.

Whatever the cause, fouling is a serious problem in the crude oil industry and is dependent on the properties of the feedstock (Table 1.5) and the properties are indicative of refinery performance and profitability (Ohmes, 2014). Therefore, understanding the properties and contaminants of various crude oils as well as the intermediate streams and final products is critical to selecting the crude slate for the refinery. In fact, the occurrence of fouling in reactors during processing has become more common with the influx of heavier feedstocks (such as heavy oil, extra heavy oil, and tar sand bitumen—tar sand bitumen is an exception insofar as it is not classed as a member of the crude oil family as defined by the United States Department of Energy) and the requirement of more complex processing units to convert such feedstocks into saleable products (Speight, 2014a).

Table 1.5 Feedstock Properties and Their Respective Impacts on Refining

Property	Comment	Impact
API	Low API gravity	High potential for coke formation Carbon deposition on catalyst Catalyst fouling and deactivation
Sulfur	Requires hydrogen for removal as hydrogen sulfide	Corrosion Catalyst fouling and deactivation
Nitrogen	Requires hydrogen for removal as ammonia	Corrosion Catalyst fouling and deactivation
Metals Ni/V/Fe	Typically associated with asphaltene fraction Require guard bed catalysts	Catalyst fouling and deactivation
Metals Na/Ca/Mg	Typically associated with high-acid crudes Require guard bed catalysts	Catalyst fouling and deactivation Corrosion Catalyst fouling and deactivation
Coke precursors	Require carbon rejection process	Catalyst fouling and deactivation
Asphaltene fraction	Increases potential for fouling	Phase separation during process Fouling Catalyst fouling and deactivation
Naphthenic acids	High levels cause corrosion	Corrosion and fouling
Compatibility	Certain crude blends are incompatible	Affects allowable blend behavior Phase separation Fouling
Chlorides	Typically associated with alkali metals	Corrosion Fouling
Viscosity	High viscosity	High potential for coke formation Phase separation during process Fouling

In addition to taking the necessary steps to process these feedstocks without serious fouling and any ensuing corrosion of the processing equipment, refiners need to develop programs for detailed feedstock evaluation in order to determine the quality of a feedstock so that the processing steps can be planned meticulously. For example, the compatibility of an opportunity crude oil not only with the viscous oils but also with other opportunity crude oils is an important aspect that should be considered prior to the onset of processing and to mitigate the possibility of any processing difficulties and the potential for equipment fouling (Speight, 2014a, 2015a).

Thus, before refining, there is the need for comprehensive evaluations of opportunity crude oils in order to make informed decisions regarding the suitability of an opportunity crude oil for a refinery.

1.3.1.3 High-Acid Crude Oil

High-acid crude oils are crude oils that contain considerable proportions of organic acid derivatives—commonly referred to as naphthenic acids—and, in many cases (but not all cases), high-acid crude oils are more viscous than the conventional crude oils (Speight, 2014a, 2014b). For convenience, the high-acid crude oils are presented in this section rather than within the section on "viscous feedstocks."

By definition, a naphthenic acid is a monobasic carboxyl group attached to a saturated cycloaliphatic structure and the term naphthenic acid refers to a complex mixture of low molecular weight to high molecular weight acid derivatives as well as other acidic species (such as phenol derivatives). In fact, a major portion of the acidic constituents in a high acid feedstock may not be carboxylic acids and some of the constituents may be polycyclic and may have unsaturated bonds, aromatic rings, and hydroxyl groups. Some naphthenic acids may even contain heteroatoms other than the two oxygen atoms ($-CO_2H$) of the carboxylic acid group leading to a broad distribution of species.

High-acid crude oils cause corrosion in the refinery—corrosion is predominant at temperatures in excess of 180 °C (355 °F) (Ghoshal and Sainik, 2013)—and occur particularly in the atmospheric distillation unit (the first point of entry of the high-acid crude oil) and also in the vacuum distillation units (Speight, 2014b, 2014c). In addition, overhead corrosion is caused by the mineral salts, magnesium, calcium, and sodium chloride which are hydrolyzed to produce volatile hydrochloric acid, causing a highly corrosive condition in the overhead exchangers which can (and do, if ignored) present a significant contamination in opportunity crude oils. Other contaminants in the opportunity crude oils which are shown to accelerate the hydrolysis reactions include inorganic clay minerals and organic acid derivatives (Speight, 2014a, 2017, 2021).

In addition to taking preventative measure for the refinery to process these feedstocks without serious deleterious effects on the equipment, refiners will need to develop programs for detailed and immediate feedstock evaluation so that they can understand the qualities of a crude oil very quickly and processing of such feedstocks can be planned meticulously (Speight, 2014a, 2017, 2021).

High-acid crude oils cause corrosion in the refinery—corrosion is predominant at temperatures in excess of 180 °C (355 °F) (Ghoshal and Sainik, 2013; Speight, 2014c)—and occurs particularly in the atmospheric distillation unit (the first point of entry of the high-acid crude oil) and also in the vacuum distillation units. In addition, overhead corrosion is caused by the mineral salts, magnesium, calcium, and sodium chloride which are hydrolyzed to produce volatile hydrochloric acid, causing a highly corrosive condition in the overhead exchangers. Therefore, these salts present a significant contamination in opportunity crude oils. Other contaminants in opportunity crude oils include inorganic clay minerals and organic acid derivatives which have the ability to accelerate the hydrolysis reactions.

In addition to taking preventative measure for the refinery to process these feedstocks without serious deleterious effects on the equipment, there is the need to develop programs that can be applied to feedstock evaluation so that the feedstock can be evaluated in terms of identification of the applicable necessary refinery operations (Speight, 2014a, 2015b, 2017).

1.3.1.4 Foamy Oil

Foamy oil is an oil-continuous foam that contains dispersed gas bubbles produced at the well head from heavy oil reservoirs under solution gas drive. The manner in which the gaseous constituents are dispersed in the oil distinguishes foamy oil behavior from conventional heavy crude oil. When the gas emerges from the solution in the reservoir, it does not coalesce into large gas bubbles or into a continuous flowing gas phase. Instead, it remains as small bubbles entrained in the oil, keeping the effective oil viscosity low while providing expansive energy that helps drive the oil toward the production. In addition, the crude oil known as "foamy oil" can contribute to unusually high production in heavy crude oil reservoirs and is considered to be solution-gas drive (Sun et al., 2013).

Thus, foamy oil is formed in a solution gas drive reservoir when gas is released from the oil due to a decrease in the reservoir pressure (Speight, 2014a, 2015b, 2017). In such cases, the oil appears at the wellhead of these heavy crude oil reservoirs as a foam, hence the term foamy oil. The gas initially exists in the form of small bubbles within individual pores in the rock but, as time passes and the pressure continues to decline, the bubbles grow to fill the pores. As the reservoir pressure continues to decrease, the gas bubbles created within the reservoir become sufficiently large to form a continuous gas phase. Once this occurs (i.e., when gas saturation exceeds the critical level—the minimum saturation at which a continuous gas phase exists in porous media), traditional two-phase (oil and gas) flow with classical relative permeability occurs. As a result, the GOR increases rapidly after the critical gas saturation has been exceeded.

1.3.2 Viscous Feedstocks

Also, the viscous feedstocks are richer in polynuclear aromatic derivatives (Figure 1.7)—of which the peri-condensed system (Figure 1.7a) (in keeping with the structures of the precursor that eventually evolve to the crude oil derivatives) predominate over the kata-condensed systems (Figure 1.7b).

While the terminology used for the various refinery feedstocks has been described in detail elsewhere (Chapter 1) (Speight, 2021), for the sake of completeness and to mitigate the potential for any confusion or misunderstanding, a short description of the various viscous feedstocks is worthy of inclusion at this point.

As a general introduction, in the context of this book, a viscous feedstock typically has a relatively low proportions of volatile constituents (low molecular weight constituents) and high proportions of low volatility constituents (high molecular weight constituents). Thus, a viscous feedstock is comprised of complex assortment of different molecular and chemical types with high melting points and high pour points that greatly contribute to the poor fluid properties of the viscous feedstock and, hence, to low mobility (compared to conventional crude oil). The same is true for extra heavy oil and tar sand bitumen (Speight,

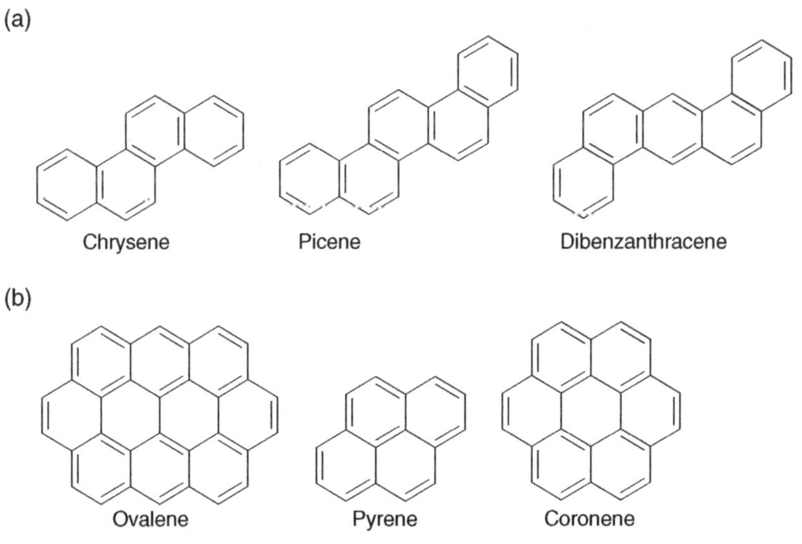

Figure 1.7 Examples of (a) peri-condensed and (b) kata-condensed aromatic systems.

2013b, 2013c, 2014a, 2019b). More generally a viscous feedstock typically has moderate-to-high levels of asphaltene constituents. The asphaltene constituents are not necessarily the primary cause for the high specific gravity (low API gravity) of the oil nor are they always the prime cause for production problems.

Briefly the asphaltene constituents are those constituents of oil that are insoluble in *n*-heptane or *n*-pentane while the resin constituents are those constituents of oil that are soluble in *n*-heptane or *n*-pentane but are adsorbed from these hydrocarbon solutions on to an adsorbent such as a clay mineral or alumina (Al_2O_3). In addition to the asphaltene constituents, it is also essential to consider the resin constituents and the aromatic constituents (especially the aromatic constituents that contain heteroatoms).

In the fractionation procedure (Figure 1.8), the hydrocarbon (*n*-pentane or *n*-heptane) must be specified since each liquid hydrocarbon gives different yield of the asphaltene fraction. Both the resin constituents and the asphaltene constituents produce coke in thermal processes (Speight, 2014a, 2017) and, therefore, are capable of interfering with refinery processes by means of the formation of sediment—a phenomenon known as instability or incompatibility (Chapter 10).

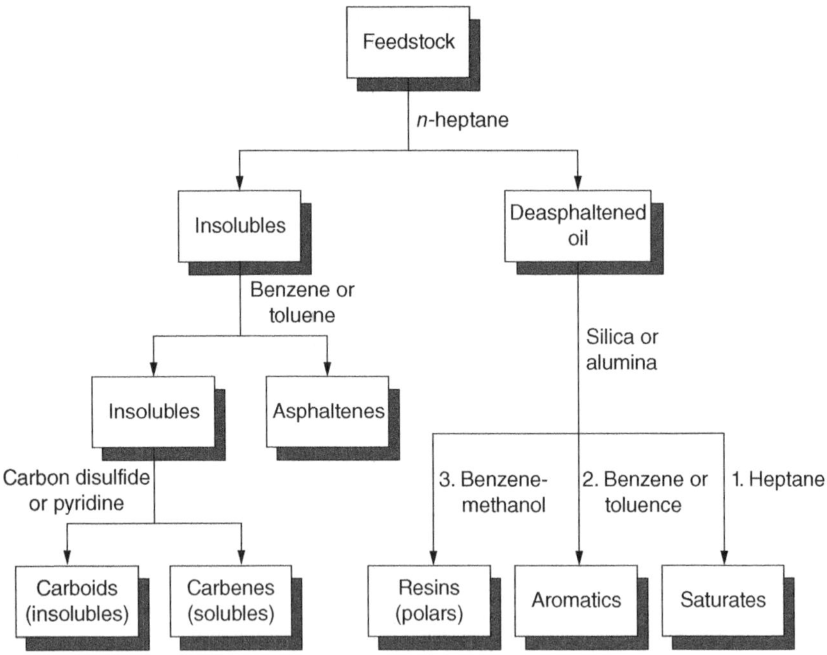

Figure 1.8 Schematic of the separation of oil into various bulk fractions.

1.3.2.1 Gas Oil

In the strictest sense, gas oil (Table 1.6) is not a member of the heavy oil family but the properties of this material often classify the gas oil as a viscous material from which lubricating oil and grease and cracked products can be manufactured (Parkash, 2003; Gary et al., 2007; Speight, 2014a; Hsu and Robinson, 2017; Speight, 2017, 2021). In this respect, gas oil (especially the vacuum gas oil fraction) deserves to be given attention here.

Vacuum gas oil (often identified by the acronym VGO) is a for fluid catalytic cracker units and is used to produce distillates that may be employed as solvents or as blend stock for transportation fuels. On the basis of the product, the vacuum gas oil can be subdivided into (1) light vacuum gas oil, often identified by the acronym LVGO, and (2) heavy vacuum gas oil, often identified by the acronym HVGO.

Light vacuum gas oil is the lower boiling fraction of the vacuum gas oil while heavy vacuum gas oil is produced in the vacuum distillation unit (as the higher boiling fraction of the vacuum gas oil) through the distillation of the residuum (i.e., the nonvolatile fraction) from the atmospheric distillation column. The higher boiling fraction of the vacuum gas oil (the HVGO) is a heavy a complex

Table 1.6 Example of the Properties of Vacuum Gas Oil

Specific gravity at 60/60 °F	0.8739
Pour point (°C)	5
Viscosity at 37.8 °C, SSU	51.22
Viscosity at 98.8 °C, SSU	33.54
Aniline point (°C)	70
Molecular weight	278
Refractive index at 20 °C	1.4875
K_{uop}-factor	11.70
K_{W}-factor	11.67
ASTM distillation(D-86) (°C)	
IBP	265
10%	281
30%	304
50%	316
70%	334
90%	381
FBP	400

viscous fraction that is produced by vacuum distillation unit in the refinery as a black semisolid material that often has an odor similar to asphalt.

More generally, the heavy vacuum gas oil (HVGO) is that is produced as the lower side product from the vacuum distillation unit in a refinery. Vacuum gas oil has a boiling range on the order of (for example) 350–565 °C (660–1,050 °F) although this temperature range is subject to the refinery distillation parameters. More specifically, because the composition of the gas oil is dependent upon the source of the gas oil and there is no "typical composition" of a gas oil fraction, the maximum temperature at which cracking occurs is different for each gas oil and cannot be determined reliably. Nevertheless, vacuum gas oil is prone to thermal decomposition (cracking) at temperatures higher than 360 °C (680 °F) (Song et al., 2013; Wang et al., 2016). In many cases, the gas oil is mixed with hydrogen to suppress the cracking reactions that occur above the cracking temperature.

Finally, as a point of reference, there are two main uses of the vacuum gas oil which are (1) naphtha production, which is used as a gasoline blend stock—the heavy vacuum gas oil is sent to the fluidized catalytic cracking unit (often represented by the acronym FCU or FCCU) for conversion in the presence of a catalyst, to produce the desired product, and (2) kerosene production, which is used for production of diesel fuel—the heavy vacuum gas oil is sent to a hydrocracking unit where it is cracked to produce a mixture of products rich in middle distillates.

Briefly, the fluid catalytic cracking unit is one of the processes that is a type of secondary unit operation in the refinery plant. The technology is used for a high molecular weight hydrocarbon (such as the heavy vacuum gas oil) to convert the feedstock to valuable products such as naphtha (the prime product) and olefin gases using a high temperature and catalyst. In the process, the feedstock and the catalyst (in the form of a fluidized power) react in a reactor, and catalyst is regenerated and recycled (regenerator) (Parkash, 2003; Gary et al., 2007; Speight, 2014a; Hsu and Robinson, 2017; Speight, 2017).

On the other hand, the vacuum gas oil may be sent to a hydrotreating unit which is used to remove impurities by using hydrogen to bind with sulfur (hydrodesulfurization, HDS), nitrogen (hydrodenitrogenation, HDN), and oxygen (hydrodeoxygenation, HDO). Thus, hydrotreating is a refining process for treating crude oil fractions from an atmospheric distillation unit or from a vacuum distillation unit (such as the heavy gas oil) in the presence of a catalyst and substantial quantities of hydrogen.

1.3.2.2 Heavy Crude Oil

The term heavy oil has also been arbitrarily used to describe both the heavy oil that require thermal stimulation of recovery from the reservoir and the bitumen in tar sand formation (i.e., bituminous sand formation) from which the bitumen is recovered by, for example, a mining operation (Speight, 2014a, 2019b).

When crude oil occurs in a reservoir that allows the crude material to be recovered by pumping operations as a free-flowing dark to light colored liquid, it is often referred to as conventional crude oil (Table 1.7). Heavy crude oil is a type of crude oil that is different from the conventional crude oil insofar as they are much more difficult to recover from the subsurface reservoir (Speight, 2014a, 2019b). A heavy oil has a higher viscosity (and lower API gravity) than conventional crude oil, and primary recovery of the heavy oil requires thermal stimulation of the reservoir (Speight, 2017, 2019b).

Heavy crude oils are more difficult to recover from the subsurface reservoir than conventional crude oil. The definition of heavy crude oil has been (and continues to be) typically based on the API gravity or viscosity but the form of definition is arbitrary although there have been attempts to rationalize the definition based upon viscosity and API gravity (Speight, 2014a).

For many years, crude oil and heavy oil were very generally defined in terms of physical properties. For example, heavy oils were considered to be those crude oils that had gravity somewhat less than 20° API with the heavy oils falling into the API gravity range 10°–15°. For example, Cold Lake heavy crude oil has an API gravity equal to 12° and extra heavy oils, such as tar sand bitumen, usually have an API gravity in the range 5°–10° (Athabasca bitumen = 8° API). Residua would

Table 1.7 Example of the Properties of Heavy Crude Oil

		Heavy crude oil[a]
Whole oil	API gravity	12.0
	Carbon (% w/w)	84.5
	Hydrogen (% w/w)	12.1
	H/C atomic ratio	1.72
	Carbon (% w/w)	83.6
	Hydrogen (% w/w)	10.5
	Sulfur (% w/w)	3.84
	Nitrogen (% w/w)	0.31
	Oxygen	0.2
	Vanadium (ppm)	190
	Nickel (ppm)	70
	C5 asphaltenes	15.3
	C7 asphaltenes	7.8
	Conradson carbon residue (% w/w)	10.3

[a] Cold Lake, Alberta, Canada.

vary depending upon the temperature at which distillation was terminated but usually vacuum residua are a range on the order of 2–8° API (Speight, 2000, 2014a).

Heavy crude oils have a much higher viscosity (and lower API gravity) than conventional crude oil, and primary recovery of these crude oil types usually requires thermal stimulation of the reservoir (Speight, 2014a, 2019b). The generic term "heavy crude oil" is often applied to a crude oil that has less than 20° API or less than 22° API or, in some cases, less than 25° API. Obviously, the use of the density-API descriptor is subject to variation depending on the company or the country or the limits of accuracy of the test method (by which the API was determined) and there is always the question that refers to the differences between a heavy crude oil having an API gravity of 19.5 and a heavy crude oil having an API gravity of 20.5 (Speight, 2014a, 2015a, 2019b). In addition, and to add further confusion to the so-called definition, heavy crude oil is often (a heavy crude oil having an API gravity of 19.5) described as having a sulfur content higher than 2% w/w (Speight, 2000, 2014a, 2017) and, in contrast to conventional crude oil, heavy crude oil is darker in color and may even be black.

Heavy crude oil is typically recognized as a type of crude oil but is different to the other types of crude oil insofar as heavy crude oil differs conventional crude oil (often referred to as light oil, i.e., low boiling low density) insofar as the heavy crude oil is more difficult to recover from the subsurface reservoir. The heavy crude oil has a higher viscosity (and lower API gravity) than conventional crude oil and recovery of heavy crude oil usually requires thermal stimulation of the reservoir leading to application of various thermal methods (such as coking) in the refinery for suitable conversion to low-boiling distillates.

There has been the continuing tendency to define heavy crude oil using the API gravity or viscosity, and the definition is quite arbitrary although there have been attempts to rationalize the definition based upon viscosity, API gravity, and density (Speight, 2014a, 2015a, 2019b, 2021). For example, heavy crude oil is considered to be the type of crude oils that had gravity somewhat less than 20° API with many heavy crude oils falling into the API gravity range 10°–15°. For example, Cold Lake heavy crude oil (Alberta, Canada) has an API gravity equal on the order of 12° and tar sand bitumen (Athabasca, Fort McMurray, Alberta, Canada) usually have an API gravity on the order of 5°–10° (Athabasca bitumen = 8° API). Using this scale, crude oil residua vary depending upon the temperature at which distillation was terminated but atmospheric residua are typically in the range 10°–15° API while vacuum residua are in the range 2°–8° API (Parkash, 2003; Gary et al., 2007; Speight, 2014a; Hsu and Robinson, 2017; Speight, 2017, 2021).

In the simplest terms, heavy crude oil is a type of crude oil which is very viscous and does not flow easily. The common characteristic properties (relative to conventional crude oil) are high specific gravity, low hydrogen to carbon ratios, high carbon residues, and high contents of asphaltenes, heavy metals (i.e., metallic chemical elements that have a relatively high density and are toxic or poisonous at low concentrations; examples of heavy metals include [alphabetically]: arsenic, As, cadmium, Cd, chromium, Cr, mercury, Hg, lead, Pb, and thallium, Tl), sulfur, and nitrogen. Specialized refining processes are required to produce more useful fractions, such as: naphtha, kerosene, gas oil, and wax.

Thus, the term heavy crude oil has also been arbitrarily used to describe the viscous crude oil that requires a thermal stimulation method to recover the oil from the reservoir. It must also be remembered that the term "heavy crude oil" may also include opportunity crude oil, high-acid crude oil, and foamy oil and, thus, there is the need for a refinery to be ready to accommodate opportunity crude oils and/or high-acid crude oils and/or foamy oils which, for many purposes, may have been shipped to a refinery under the banner of "heavy crude oil."

1.3.2.3 Extra Heavy Crude Oil

Extra heavy oil is a nondescript term (related to viscosity) of little scientific meaning, which is usually applied to tar sand bitumen, which is generally incapable of free flow under reservoir conditions. Thus, the term extra heavy oil is used to describe materials that occur in the solid or near-solid state in the deposit or reservoir and are generally incapable of free flow under ambient conditions. Whether or not such a material exists in the near-solid or solid state in the reservoir can be determined from the pour point and the reservoir temperature (Tables 1.8 and 1.9).

The general difference is that extra heavy oil, which may have properties similar to tar sand bitumen in the laboratory under ambient conditions of temperature and pressure (e.g., the API gravity is on the order of 8, the sulfur content is on the order of 4.8% w/w, and the C5 asphaltene content is on the order or 17% w/w as well as other properties that are similar to the properties of tars and bitumen) but unlike tar sand bitumen in the deposit, the extra heavy crude oil exhibits mobility in the reservoir or deposit (Speight, 2014a, 2019b, 2021). Thus, extra heavy oil can flow at the reservoir temperature and can be produced without additional viscosity-reduction techniques through variants of conventional processes such as long horizontal wells, or multilateral arrangement of the production wells.

This is the case, for instance, in the Orinoco basin (Venezuela) or in offshore reservoirs of the coast of Brazil but, once outside of the influence of the high reservoir temperature, these oils are too viscous at surface to be transported through conventional pipelines and require heated pipelines for transportation. Alternatively, the oil must be partially upgraded or fully upgraded or diluted with a low boiling (low density) hydrocarbon (such as aromatic naphtha) to create a mix that is suitable for transportation (Speight, 2014a, 2017).

Thus, extra heavy oil is a material that occurs as in fluid at the high temperature of the reservoir-deposit and, thus, has mobility under temperature conditions in the reservoir-deposit insofar as the temperature of the reservoir-deposit is higher

Table 1.8 Simplified Differentiation Between Conventional Crude Oil, Tight Oil, Heavy Crude Oil, Extra Heavy Crude Oil, and Tar Sand Bitumen[a]

Conventional Crude Oil
Mobile in the reservoir
API gravity: >20°
Primary recovery methods
Secondary recovery methods
Heavy Crude Oil
More viscous than conventional crude oil
API gravity: 10–20°
Mobile in the reservoir
Secondary recovery methods
Tertiary recovery (enhanced oil recovery—EOR, e.g., steam stimulation)
Extra Heavy Crude Oil
Similar properties to the properties of tar sand bitumen
API gravity: <10°
Temperature of the reservoir/deposit higher than pour point of the oil
Mobile in the reservoir/deposit
Secondary recovery methods
Tertiary recovery methods (enhanced oil recovery—EOR, e.g., steam stimulation)
Tar Sand Bitumen
API gravity: <10°
Temperature of the reservoir/deposit lower than pour point of the oil
Immobile in the deposit
Mining (often preceded by explosive fracturing)
Extreme heating methods
Innovative methods[b]
Residuum
Semi-mobile to immobile at ambient conditions
Typically, API gravity: 5–15°; sulfur: 2.0–5.0% w/w; metals: 100–1,000 ppm
Does not occur naturally; manufactured product produced in a refinery
Atmospheric residuum: source of heavy gas oil under vacuum distillation
Vacuum residuum: nonvolatile under vacuum distillation conditions
Typically immobile under ambient conditions

[a] This list is not intended for use as a means of classification.
[b] Innovative methods exclude tertiary recovery methods and methods such as steam-assisted gravity drainage (SAGD) and vapor-assisted extraction (VAPEX) as well as variants or hybrids thereof (Speight, 2016).

Table 1.9 Use of Pour Point and Reservoir/Deposit Temperature to Differentiate Between Heavy Crude Oil, Extra Heavy Crude Oil, and Tar Sand Bitumen

Oil	Location	Temperature	Effect on oil
Heavy crude oil	Reservoir or deposit	Higher than oil pour point	Fluid
			Mobile
	Surface/ambient	Higher than oil pour point	Fluid
			Mobile
Extra heavy crude oil	Reservoir or deposit	Higher than oil pour point	Fluid
			Mobile
	Surface/ambient	Lower than oil pour point	Solid to near-solid
			Fluidity much reduced
			Immobile
Tar sand bitumen	Reservoir or deposit	Lower than oil pour point	Solid to near-solid
			Not fluid
			Immobile
	Surface/ambient	Lower than oil pour point	Solid to near-solid
			Not fluid
			Immobile

than the pour point of the oil. While this type of oil may resemble tar sand bitumen and does not flow easily, extra heavy oil is generally recognized as having mobility in the reservoir compared to tar sand bitumen, which is typically incapable of mobility (free flow) under reservoir conditions. For example, the tar sand bitumen located in Alberta, Canada, is not mobile in the deposit and requires extreme methods of recovery to recover the bitumen. On the other hand, much of the extra heavy oil located in the Orinoco basin of Venezuela requires recovery methods that are less extreme because of the mobility of the material in the reservoir. Whether the mobility of extra heavy oil is due to a high reservoir temperature (i.e., the reservoir temperature is higher than the pour point of the extra heavy oil) or subject to other factors due to localized conditions in the reservoir. This may also be reflected in the choice of suitable conversion process for extra heavy crude oil or for tar sand bitumen in the refinery.

In the context of this book, the methods outlined in this book for refining viscous feedstocks focus on heavy oil with an API gravity of less than 20 with a variable sulfur content (which includes atmospheric residua and vacuum residua) (Parkash, 2003; Gary et al., 2007; Speight, 2014a, 2017). However, it must be recognized that some of the heavy oil are sufficiently liquid to be recovered by pumping operations and are already being recovered by this method. Refining depends on the characteristics (properties) since these heavy oils fall into a range of high viscosity and the viscosity is subject to temperature effects (Speight, 2001, 2014a, 2015a), which is the reason for the application of thermal conditions or dilution with a suitable solvent (such as aromatic naphtha) to enable heavy oil to flow in pipeline or within the refinery system.

1.3.2.4 Tar Sand Bitumen

In addition to heavy crude oil and the so-called extra heavy crude oil, there remains an even more viscous material (Table 1.10) that offers some relief to the potential shortfalls in supply (Meyer and De Witt, 1990; Speight, 2014a, 2016). This is the bituminous material that is found in tar sand (oil sand) deposits (referred to as oil sand deposits in Canada). However, many of these reserves are only available with some difficulty due to the highly viscous nature of the bitumen (because of the substantial differences in character between conventional crude oil and tar sand bitumen) and optional refinery scenarios are necessary for conversion of the

Table 1.10 Examples of the Properties of Tar Sand Bitumen

Property	Athabasca (Alberta, Canada)	Sunnyside (Utah, USA)
Specific gravity	1.03	1.015
API gravity	6.5	7.9
Viscosity (cP)		
38 °C/100 °F	750,000	1,500,000
Pour point (°F)	50	75
Elemental analysis (%, w/w)		
Carbon	83.0	86.8
Hydrogen	10.6	10.8
Nitrogen	0.5	1.1
Sulfur	4.9	0.7
Oxygen (by difference)	1.0	0.6
Composition (%, w/w)		
C5 Asphaltenes	17.0	23.7
Carbon residue (%, w/w)		
Conradson	14.0	15.0

bitumen to liquid products (Speight, 2000, 2014a, 2016, 2021). Tar sand deposits, also variously called oil sand deposits or bituminous sand deposits, vary from a loose sandstone formation to a consolidated sandstone formation or a porous carbonate rock formation, that is, impregnated with bitumen in the form of a viscous asphaltic oil that has an extremely high viscosity under reservoir conditions.

The term "tar sand bitumen" (also, on occasion, incorrectly referred to as native asphalt) includes a wide variety of near-black to black materials that occur in the form of near-solid to solid in the deposit either (1) with no mineral impurity or (2) with a mineral matter content that may exceed 50% w/w. Bitumen is frequently found filling pores and crevices of sandstone, limestone, or argillaceous sediments, in which case the organic and associated mineral matrix is known as rock asphalt. Even though the term tar sand is incorrect since tar is a product of the thermal processing of coal (Speight, 2013a), the term has remained in use. Also, it is incorrect to refer to native bituminous materials as tar or pitch even though the word "tar" is descriptive of the black, nonvolatile bituminous material, it is best to avoid its use with respect to natural materials and to restrict the meaning to the volatile or near-volatile products produced in the destructive distillation of such organic substances as coal (Speight, 2013a). In the simplest sense, pitch is the distillation residue of the various types of tar. Thus, alternative names, such as bituminous sand or oil sand, are gradually finding usage, with the former name (bituminous sands) more technically correct. The term oil sand is also used in the same way as the term tar sand, and these terms are used interchangeably throughout this text.

The permeability of a tar sand deposit is low and passage of fluids through the deposit can only be achieved by prior application of fracturing techniques. Alternatively, bitumen recovery can be achieved by conversion of the bitumen to a product in situ (often referred to as in situ upgrading) followed by product recovery from the deposit (Speight, 2013b, 2013c, 2014a, 2016, 2019b). The bitumen is a high-boiling material with little, if any, material boiling below 350 °C (660 °F) and the boiling range approximates the boiling range of an atmospheric residuum (Speight, 2014a, 2015a, 2019b).

There have been many attempts to define tar sand deposits and the bitumen contained therein. In order to define conventional crude oil, heavy oil, and bitumen, the use of a single physical parameter such as viscosity is not sufficient. Other properties such as API gravity, elemental analysis, composition, and, most of all, the properties of the bulk deposit must also be included in any definition of these materials. In fact, the most appropriate definition of tar sands is found in the writings of the United States government (Speight, 2014a, 2016):

> Tar sands are the several rock types that contain an extremely viscous hydrocarbon which is not recoverable in its natural state by conventional oil well production methods including currently used enhanced recovery techniques.

This definition refers to the character of the bitumen through the method of recovery. Mining methods match the requirements of this definition (since mining is not one of the specified recovery methods) and the bitumen can be recovered by alteration of its natural state such as thermal conversion to a product that is then recovered. In this sense, changing the natural state (the chemical composition) as occurs during several thermal processes (such as some in-situ combustion processes) also matches the requirements of the definition.

By inference and by omission, conventional crude oil and heavy crude oil are also included in this definition. Crude oil is the material that can be recovered by conventional oil well production methods whereas heavy oil is the material that can be recovered by enhanced recovery methods. Tar sand is currently recovered by a mining process followed by separation of the bitumen by the hot water process. The bitumen is then used to produce hydrocarbon derivatives by a conversion process.

The only commercial operations for the recovery of bitumen from tar sand and its subsequent conversion to liquid fuels exist in the Canadian Province of Alberta where Suncor (initially as Great Canadian Oil Sands before the name change) went onstream in 1967 and Syncrude (a consortium of several companies) went onstream in 1977. Thus, throughout this text, frequent reference is made to tar sand bitumen, but because commercial operations have been in place for approximately 50 years, it is not surprising that more is known about the Alberta (Canada) tar sand reserves than any other reserves in the world. Therefore, when discussion is made of tar sand deposits, reference is made to the relevant deposit, but when the information is not available, the Alberta material is used for the purposes of the discussion.

Refining heavy crude oil, extra heavy crude oil, tar sand bitumen, and residua depends to a large degree on the composition of the feedstock and there are wide differences in properties between, say, tar sand bitumen and conventional crude oil (Table 1.11) (Parkash, 2003; Gary et al., 2007; Speight, 2014a, 2017). Generally, the bitumen found in tar sand deposits is an extremely viscous material that is immobile and cannot be refined by the application of conventional refinery. On the other hand, extra heavy oil, which is often likened to tar sand bitumen because of similarities in the properties of the two, has a degree of mobility under reservoir or deposit conditions but typically suffers from the same drawbacks as the immobile bitumen in refinery operations.

Thus, alternative names, such as bituminous sand or oil sand, are gradually finding usage, with the former name (bituminous sands) more technically correct. The term oil sand is also used in the same way as the term tar sand, and these terms are used interchangeably throughout this text.

Bituminous rock and bituminous sand are those formations in which the bituminous material is found as a filling in veins and fissures in fractured rocks or impregnating relatively shallow sand, sandstone, and limestone strata. The deposits contain as much as 20% bituminous material, and if the organic material in the

Table 1.11 Comparison of the Properties of Conventional Crude Oil and Tar Sand Bitumen

Property	Bitumen (Athabasca)	Crude oil
Specific gravity	1.03	0.85–0.90
Viscosity (cP)		
38 °C/100 °F	750,000	<200
100 °C/212 °F	11,300	
Pour point (°F)	>50	<32
Elemental analysis (% w/w)		
Carbon	83.0	86.0
Hydrogen	10.6	13.5
Nitrogen	0.5	0.2
Oxygen	0.9	ca. 0.5
Sulfur	4.9	ca. 2.0
Ash	0.8	0.0
Nickel (ppm)	250	ca. 10.0
Vanadium (ppm)	100	ca. 10.0
Composition (% w/w)		
Asphaltenes (pentane)	17.0	<10.0
Resins	34.0	<20.0
Aromatics	34.0	ca. 30.0
Saturates	15.0	ca. 30.0
Carbon residue (% w/w)		
Conradson	14.0	<10.0

rock matrix is bitumen, it is usual (although chemically incorrect) to refer to the deposit as rock asphalt to distinguish it from bitumen that is relatively mineral free. A standard test method (ASTM D4, 2021) is available for determining the bitumen content of various mixtures with inorganic materials, although the use of word bitumen as applied in this test might be questioned and it might be more appropriate to use the term organic residues to include tar and pitch. If the material is the asphaltite-type or asphaltoid-type, the corresponding terms should be used: rock asphaltite or rock asphaltoid.

Bituminous rocks generally have a coarse, porous structure, with the bituminous material in the voids. A much more common situation is that in which the organic material is present as an inherent part of the rock composition insofar as it is a diagenetic residue of the organic material detritus that was deposited with the sediment. The organic components of such rocks are usually refractory and are only slightly affected by most organic solvents.

Tar sand deposits occur throughout the world and the largest deposits occur in Alberta, Canada (the Athabasca, Wabasca, Cold Lake, and Peace River areas), and in Venezuela. Smaller deposits occur in the United States, with the larger individual deposits in Utah, California, New Mexico, and Kentucky. The term tar sand, also known as oil sand (in Canada), or bituminous sand, commonly describes sandstones or friable sand (quartz) impregnated with a viscous organic material known as bitumen (a hydrocarbonaceous material that is soluble in carbon disulfide). Significant amounts of fine material, usually largely or completely clay, are also present. The degree of porosity in tar sand varies from deposit to deposit and is an important characteristic in terms of recovery processes for the bitumen which is the desirable fraction of the tar sand and from which a variety of distillable products can be produced (Speight, 2014a, 2017). However, the bitumen is usually not recoverable by conventional crude oil production techniques (Speight, 2013b, 2013c, 2014a, 2016).

The properties and composition of the tar sand deposits and the bitumen contained therein significantly influence the selection of recovery and the bitumen conversion processes and vary among the bitumen from different deposits. In the so-called wet sands or water-wet sands of the Athabasca deposit, a layer of water surrounds the sand grain, and the bitumen partially fills the voids between the wet grains. The Utah tar sand deposits typically lack the water layer and the bitumen in is direct contact with the sand grains and, without any intervening water; such tar sands are typically sometimes referred to as oil-wet sands.

Typically, more than 99% w/w of mineral matter is composed of quartz and clay minerals—the latter are detrimental to refining operations. The typical composition of the P.R. Spring (also written as Peor Spring) tar sand area has a porosity on the order of 8.4% v/v with the solid/liquid fraction typically being on the order of 90.5% w/w sand, 1.5% w/w fines, 7.5% w/w bitumen, and 0.5% w/w water. The high concentration of heteroatoms (nitrogen, oxygen, sulfur, and metals) contribute to (1) an increase the viscosity of the bitumen, (2) an increase in the bonding of bitumen with the minerals, (3) a reduced yield of recovered bitumen, and (4) difficulties encountered during processing of the bitumen.

Finally, as an addition to this subsection and for information purposes, synthetic crude oil is produced by upgrading bitumen to reduce its density and viscosity for transport by pipeline (Speight, 2014a, 2017). A disadvantage to this concept is that supplies of synthetic crude oils are limited by the availability of upgraders at the source of the extraction of the bitumen from the tar sand and that a mixture on the order of 50% v/v and 50% v/v synthetic crude oil is required to obtain the desired density and viscosity.

1.3.2.5 Residuum

In contrast to the naturally occurring heavy crude oil, extra heavy crude oil, and tar sand bitumen, a residuum (pl. residua, also shortened to resid, pl. resids) is a manufactured material (rather than a naturally occurring material) insofar as a residuum is produced as a result of crude oil refining (Parkash, 2003; Gary et al., 2007; Speight, 2014a, 2017). Thus:

Feedstock	Atmospheric distillation
	Naphtha
	Kerosene
	Middle distillates
	Atmospheric gas oil
	Atmospheric residuum
Atmospheric residuum	Vacuum distillation
	Gas oil
	Light gas oil
	Heavy gas oil
	Vacuum residuum

In addition, the properties are variable depending upon the cut point (i.e., either atmospheric residua or vacuum residua) (Table 1.12) (Parkash, 2003; Gary et al., 2007; Speight, 2014a, 2017). Many residua are subject to further refining after production and are worthy of mention here as optional refinery viscous feedstocks either further refined as single feedstocks or as blends with other feedstocks, such as gas oil, as well as conventional whole crude oil, heavy crude oil, extra heavy crude oil, and tar sand bitumen.

Typically, a residuum is the residue obtained from crude oil after nondestructive distillation has removed all the volatile materials. The temperature of the distillation is usually maintained below 350 °C (660 °F) since the rate of thermal decomposition of crude oil constituents is minimal below this temperature but the rate of thermal decomposition of crude oil constituents is substantial above 350 °C (660 °F). If the temperature of the distillation unit rises above 350 °C (660 °F) as happens in certain units where temperatures up to 395 °C (740 °F) are known to occur, cracking can be controlled by adjustment of the residence time.

Residua (pl: residuum, often shortened to "resid") are black, viscous materials that are obtained by distillation of a crude oil under atmospheric pressure (atmospheric residuum) or under reduced pressure (vacuum residuum) and the properties of the residuum do vary with the initial boiling point of the residuum (Table 1.13) (Parkash, 2003; Gary et al., 2007; Speight, 2014a; Hsu and

Table 1.12 Properties of Residua Showing the Differences Between Atmospheric Residua (650$^+$ Resids) and Vacuum Residua (1,050$^+$ Resids)

Feedstock	Gravity API	Sulfur (wt%)	Nitrogen (wt%)	Nickel (ppm)	Vanadium (ppm)	Asphaltenes (heptane) (wt%)	Carbon residue (Conradson) (wt%)
Arabian light, >650 °F	17.7	3.0	0.2	10.0	26.0	1.8	7.5
Arabian light, >1,050 °F	8.5	4.4	0.5	24.0	66.0	4.3	14.2
Arabian heavy, >650 °F	11.9	4.4	0.3	27.0	103.0	8.0	14.0
Arabian heavy, >1,050 °F	7.3	5.1	0.3	40.0	174.0	10.0	19.0
Alaska, North Slope, >650 °F	15.2	1.6	0.4	18.0	30.0	2.0	8.5
Alaska, North Slope, >1,050 °F	8.2	2.2	0.6	47.0	82.0	4.0	18.0
Lloydminster (Canada), >650 °F	10.3	4.1	0.3	65.0	141.0	14.0	12.1
Lloydminster (Canada), >1,050 °F	8.5	4.4	0.6	115.0	252.0	18.0	21.4
Kuwait, >650 °F	13.9	4.4	0.3	14.0	50.0	2.4	12.2
Kuwait, >1,050 °F	5.5	5.5	0.4	32.0	102.0	7.1	23.1
Tia Juana, >650 °F	17.3	1.8	0.3	25.0	185.0		9.3
Tia Juana, >1,050 °F	7.1	2.6	0.6	64.0	450.0		21.6
Taching, >650 °F	27.3	0.2	0.2	5.0	1.0	4.4	3.8
Taching, >1,050 °F	21.5	0.3	0.4	9.0	2.0	7.6	7.9

Note: >650 °F$^+$ to >1,050 °F$^+$ is accompanied by: (1) decrease in API gravity, (2) increase in sulfur content, (3) increase in nitrogen content, (4) increase in nickel content, (5) increase in vanadium content, (6) increase in asphaltene content, and (7) increase in carbon residue.

Table 1.13 Properties of Tia Juana Crude Oil and the Different Residua

		Residua		
	Whole crude	**650 °F+**	**950 °F+**	**1,050 °F+**
Yield (vol%)	100.0	48.9	23.8	17.9
Sulfur (wt%)	1.08	1.78	2.35	2.59
Nitrogen (wt%)		0.33	0.52	0.60
API gravity	31.6	17.3	9.9	7.1
Carbon residue (wt%) Conradson		9.3	17.2	21.6
Metals (ppm)				
Vanadium		185		450
Nickel		25		64
Viscosity				
Kinematic				
At 100 °F	10.2	890		
At 210 °F		35	1,010	7,959
Furol				
At 122 °F		172		
At 210 °F			484	3,760
Pour point (°F)	−5	45	95	120

Robinson, 2017; Speight, 2017). More typically, residua are black, viscous materials that may be near liquid at room temperature (typically, an atmospheric residuum) or solid (typically, a vacuum residuum) depending upon the properties of the crude oil source.

When a residuum is obtained from a feedstock and thermal decomposition has commenced, it is more usual to refer to this product as cracked residuum (sometimes referred to as pitch, a name derived from the coal industry). The differences between conventional crude oil and the related residua are due to the relative amounts of various constituents present, which are removed or retained by according to the relative volatility of the constituents.

The vacuum residuum (also commonly referred to as "vacuum bottoms," typically 950 °F$^+$ or 1050 °F$^+$) is the most complex of crude oil and, in many cases, may even resemble heavy oil or extra heavy oil or tar sand bitumen in composition. Vacuum residua contain the majority of the heteroatoms originally in the crude oil and molecular weight of the constituents range up to several thousand (as near as can be determined but subject to method dependence). The fraction is so complex that the characterization of individual species is virtually impossible, no matter what claims have been made or will be made. Separation of vacuum residua by group type can be difficult because of the multi-substitution of aromatic and naphthenic species as well as by the presence of multiple functionalities in single molecules (Chapters 2 and 3) (Speight, 2014a, 2021).

Typically, the precipitation of the asphaltene fraction *n*-pentane or *n*-heptane is used as the initial step for the characterization of a viscous feedstock, such as a vacuum residuum (Chapter 3) (Speight, 2014a, 2015a). The insoluble fraction, the pentane–asphaltene fraction or the heptane–asphaltene fractions, may be as much as 50% by weight of a vacuum residuum. The pentane-soluble or the heptane-soluble portion (often referred to as the maltene fraction) of the residuum is then fractionated by adsorption chromatographically into several solubility or adsorption classes for characterization (Chapter 3). However, in spite of claims to the contrary, the method is not a separation by chemical type. However, the separation of the asphaltene fraction does, however, provide a simple way to remove some of the highest molecular weight and most polar components but the asphaltene fraction is so complex that compositional detail based on average parameters is of questionable value.

For the 565 °C$^+$ (1050 °F$^+$) fractions of crude oil, the levels of nitrogen and oxygen may begin to approach the concentration of sulfur. These elements consistently concentrate in the most polar fractions to the extent that every molecule contains more than one heteroatom. At this point, structural identification is somewhat fruitless and characterization techniques are used to confirm the presence of the functionalities found in lower boiling fractions such as, for example, acid derivatives, phenol derivatives, nonbasic (carbazole-type) nitrogen derivatives, and basic (quinoline-type) nitrogen derivatives (Speight, 2014a).

The nickel and vanadium that are concentrated into the vacuum residuum appear to occur in two forms: (1) porphyrins and (2) non-porphyrins. Because the metalloporphyrins can provide insights into crude oil maturation processes, they have been studied extensively and several families of related structures have been identified. On the other hand, the non-porphyrin metals remain not clearly identified although some studies suggest that some of the metals in these compounds still exist in a tetra-pyrrole (porphyrin-type) environment.

It is more than likely that, in a specific residuum molecule, the heteroatoms are arranged in different functional types thereby contributing to an extremely complex fraction. Thus, considering the potential (and the reality) for the number of different combinations that are possible, the chances of determining every structure in a residuum are minimal. Because of this seemingly insurmountable task, it may be better to determine ways of utilizing the residuum rather attempting to determine (at best questionable) molecular structures.

In summary, the chemical composition of a residuum is complex. Physical methods of fractionation usually indicate high proportions of asphaltene constituents and resin constituents, even in amounts up to 50% (or higher) of the residuum (Parkash, 2003; Gary et al., 2007; Speight, 2014a, 2017). In addition, the presence of ash-forming metallic constituents, including such organometallic compounds as those of vanadium and nickel, is also a distinguishing feature of residua and the heavier oils. Furthermore, the deeper the cut into the crude oil, the greater is the concentration of sulfur and metals in the residuum and the greater the deterioration in physical properties.

1.3.2.6 Asphalt

Although not a naturally occurring material but manufactured from crude oil and any of the viscous feedstocks as a black or brown material that has a consistency varying from a viscous liquid to a glassy solid, a description of the production of asphalt in the refinery is warranted to dismiss any potential confusion between a residuum and asphalt (Parkash, 2003; Gary et al., 2007; Speight, 2014a, 2016; Hsu and Robinson, 2017; Speight, 2017).

Put simple, asphalt is produced simply by distillation of a feedstock to produce a nonvolatile product which can be referred to as residual asphalt or straight run asphalt—straight run asphalt, like any other straight run product—refers to the product that emerges from the refinery distillation column.

Feedstock [distillation]	Residuum
Residuum + [aerial oxidation]	Asphalt[a]
Asphalt [+ solvent]	Cutback asphalt

[a] Often referred simply to as "oxidized asphalt" or "blown asphalt."

However, in addition to this simplified manufacturing outline, if the asphalt is prepared by the solvent extraction method in which a residuum is subject to extraction by a low boiling liquid hydrocarbon (such as propane) or if blown or otherwise treated, the term should be modified accordingly to qualify the product (e.g., "propane asphalt," "blown asphalt").

In very general terms, asphalt may often resemble bitumen, hence the tendency to refer to bitumen (incorrectly) as native asphalt. It is recommended that there is a differentiation between asphalt (manufactured) and bitumen (naturally occurring) other than by use of the qualifying terms crude oil and native bitumen (i.e., the native or naturally occurring material) and asphalt (i.e., the manufactured material) in a way that reflects the origins of each of the materials which is also reflected in the chemical properties and physical properties of the two types of materials.

Furthermore, it advantages (if not necessary) to distinguish between the asphalt which originates from crude oil by refining and the product in which the source of the asphalt is a material other than crude oil, such as wurtzilite asphalt or gilsonite asphalt (Bland and Davidson, 1967; Speight, 2014a, 2016). In the absence of a qualifying word, it should be assumed that the term asphalt refers the product manufactured from crude oil and has satisfied the properties as specified in the standard test methods, some of which are applicable to compare the properties of asphalt that has been spilled into an ecosystem on a before-and-after basis (Table 1.14) (ASTM, 2021).

Table 1.14 Examples of ASTM Standard Test Methods for Asphalt and Pitch

Number	Title
ASTM D4	Standard Test Method for Bitumen Content
ASTM D5	Standard Test Method for Penetration of Bituminous Materials
ASTM D6	Standard Test Method for Loss on Heating of Oil and Asphaltic Compounds
ASTM D20	Standard Test Method for Distillation of Road Tars
ASTM D36	Standard Test Method for Softening Point of Bitumen (Ring-and-Ball Apparatus)
ASTM D56	Standard Test Method for Flash Point by Tag Closed Cup Tester
ASTM D61	Standard Test Method for Softening Point of Pitches (Cube-in-Water Method)
ASTM D70	Standard Test Method for Density of Semi-Solid Bituminous Materials (Pycnometer Method)
ASTM D88	Standard Test Method for Saybolt Viscosity
ASTM D92	Standard Test Method for Flash and Fire Points by Cleveland Open Cup Tester
ASTM D93	Standard Test Methods for Flash Point by Pensky-Martens Closed Cup Tester
ASTM D94	Standard Test Methods for Saponification Number of Petroleum Products
ASTM D95	Standard Test Method for Water in Petroleum Products and Bituminous Materials by Distillation
ASTM D113	Standard Test Method for Ductility of Bituminous Materials
ASTM D129	Standard Test Method for Sulfur in Petroleum Products (General High Pressure Decomposition Device Method)
ASTM D139	Standard Test Method for Float Test for Bituminous Materials
ASTM D140	Standard Practice for Sampling Bituminous Materials
ASTM D189	Standard Test Method for Conradson Carbon Residue of Petroleum Products
ASTM D244	Standard Test Methods and Practices for Emulsified Asphalts

Table 1.14 (Continued)

Number	Title
ASTM D524	Standard Test Method for Ramsbottom Carbon Residue of Petroleum Products
ASTM D1370	Standard Test Method for Contact Compatibility Between Asphaltic Materials (Oliensis Test)
ASTM D1552	Standard Test Method for Sulfur in Petroleum Products (High Temperature Method)
ASTM D1669	Standard Practice for Preparation of Test Panels for Accelerated and Outdoor Weathering of Bituminous Coatings
ASTM D1754	Standard Test Method for Effects of Heat and Air on Asphaltic Materials (Thin-Film Oven Test)
ASTM D2042	Standard Test Method for Solubility of Asphalt Materials in Trichloroethylene
ASTM D2170	Standard Test Method for Kinematic Viscosity of Asphalts (Bitumens)
ASTM D2171	Standard Test Method for Viscosity of Asphalts by Vacuum Capillary Viscometer
ASTM D2319	Standard Test Method for Softening Point of Pitch (Cube-in-Air Method)
ASTM D2397	Standard Specification for Cationic Emulsified Asphalt
ASTM D2622	Standard Test Method for Sulfur in Petroleum Products by Wavelength Dispersive X-ray Fluorescence Spectrometry
ASTM D2872	Standard Test Method for Effect of Heat and Air on a Moving Film of Asphalt (Rolling Thin-Film Oven Test)
ASTM D3104	Standard Test Method for Softening Point of Pitches (Mettler Softening Point Method)
ASTM D3142	Standard Test Method for Specific Gravity, API Gravity, or Density of Cutback Asphalts by Hydrometer Method
ASTM D3143	Standard Test Method for Flash Point of Cutback Asphalt with Tag Open-Cup Apparatus
ASTM D3279	Standard Test Method for *n*-Heptane Insolubles
ASTM D3461	Standard Test Method for Softening Point of Asphalt and Pitch (Mettler Cup-and-Ball Method)
ASTM D4045	Standard Test Method for Sulfur in Petroleum Products by Hydrogenolysis and Rateometric Colorimetry
ASTM D4055	Standard Test Method for Pentane Insolubles by Membrane Filtration
ASTM D4072	Standard Test Method for Toluene-Insoluble (TI) Content of Tar and Pitch

(Continued)

Table 1.14 (Continued)

Number	Title
ASTM D4124	Standard Test Method for Separation of Asphalt into Four Fractions
ASTM D4287	Standard Test Method for High-Shear Viscosity Using a Cone/Plate Viscometer
ASTM D4294	Standard Test Method for Sulfur in Petroleum and Petroleum Products by Energy Dispersive X-ray Fluorescence Spectrometry
ASTM D4312	Standard Test Method for Toluene-Insoluble (TI) Content of Tar and Pitch (Short Method)
ASTM D4402	Standard Test Method for Viscosity Determination of Asphalt at Elevated Temperatures Using a Rotational Viscometer
ASTM D4469	Standard Practice for Calculating Percent Asphalt Absorption by the Aggregate in an Asphalt Pavement Mixture
ASTM D4530	Standard Test Method for Determination of Carbon Residue (Micro Method)
ASTM D4715	Standard Test Method for Coking Value of Tar and Pitch (Alcan)
ASTM D4893	Standard Test Method for Determination of Pitch Volatility
ASTM D6560	Standard Test Method for Determination of Asphaltenes (Heptane Insolubles) in Crude Petroleum and Petroleum Products
ASTM E102	Standard Test Method for Saybolt Furol Viscosity of Bituminous Materials at High Temperatures

Annual Book of Standards. ASTM International, West Conshohocken, Pennsylvania.

Asphalt softens when heated and is elastic under certain conditions. The mechanical properties of asphalt are of particular significance when it is used as a binder or adhesive (Speight, 2014a, 2016, 2017). The principal application of asphalt is in road construction, which may be done in a variety of ways (Speight, 2014a, 2016), as well as, for example (as there are other uses), water-proofing for the roof of buildings and water-proofing for the foundations of building walls.

An asphalt emulsion is a mixture of asphalt and an anionic agent such as the sodium or potassium salt of a fatty acid. The fatty acid is usually a mixture and may contain palmitic, stearic, linoleic, and abietic acids and/or high molecular weight phenols. Sodium lignate (a product of the wood pulping industry) is often added to alkaline emulsions to effect better emulsion stability. Nonionic cellulose derivatives are also used to increase the viscosity of the emulsion if needed. The acid number of asphalt is an indicator of its emulsification capability and reflects the presence of high molecular weight asphaltogenic acids or naphthenic acids. Diamines, frequently used as cationic agents, are made from the reaction of tallow acid amines with acrylonitrile, followed by hydrogenation. The properties of asphalt emulsions (ASTM D977, 2021; ASTM D2397, 2021) are adaptable to a variety of uses. As with other crude oil products, sampling is an important precursor to asphalt analysis and a standard method (ASTM D140, 2021) is available that provides guidance for the sampling of asphalt material (both liquid asphalt and semisolid asphalt) at point of manufacture, storage, or delivery.

Other important applications of asphalt include canal and reservoir linings, dam facings, and sea works. The asphalt so used may be a thin, sprayed membrane, covered with earth for protection against weathering and mechanical damage, or thicker surfaces, often including riprap (also known as rip rap, rip-rap, shot rock, rock armor, or rubble, is human-placed rock or other material used to protect shoreline structures against scour and water, wave, or ice erosion). Asphalt is also used for roofs, coatings, floor tiles, soundproofing, waterproofing, and other building-construction elements and in a number of industrial products, such as batteries. In some applications, an asphalt emulsion is used—in the emulsion, fine globules of the asphalt are suspended in water (Speight, 2014a, 2016).

When a residuum is obtained from a crude oil and thermal decomposition has commenced, it is more usual to refer to this product (typically a solid product) as "pitch." Although there are occasions when refinery personally have referred to the non-cracked vacuum residuum as pitch or vacuum pitch.

The differences between a parent crude oil and the residua are due to the relative amounts of various constituents present, which are removed or remain by virtue of their relative volatility.

1.3.2.7 Tar and Pitch

Tar is a product of the destructive distillation of many bituminous substances or other organic materials and is a brown to black, oily, viscous liquid to semisolid material.

Tar is most commonly produced from bituminous coal and is generally understood to refer to the product from coal, although it is advisable to specify coal tar if there is the possibility of ambiguity (Speight, 2013a). Furthermore, an important (or key) factor in determining the yield and character of the coal tar is the carbonizing temperature. Three general temperature ranges are recognized, and the products have acquired the designations: low-temperature tar (approximately 450–700 °C; 540–1,290 °F), mid-temperature tar (approximately 700–900 °C; 1,290–1,650 °F), and high-temperature tar (approximately 900–1,200 °C; 1,650–2,190 °F). Tar released during the early stages of the decomposition of the organic material is called primary tar since it represents a product that has been recovered without the secondary alteration that results from prolonged residence of the vapor in the heated zone (Speight, 2013a).

Treatment of the distillate (boiling up to 250 °C, 480 °F) of the tar with caustic soda causes separation of a fraction known as tar acids; acid treatment of the distillate produces a variety of organic nitrogen compounds known as tar bases. The residue left following removal of the high boiling material, or distillate, is typically referred to as pitch, which is a black, hard, and highly ductile material. In the chemical-process industries, pitch is the black or dark brown residue obtained by distilling coal tar, wood tar, fats, fatty acids, or fatty oils.

Coal tar pitch is typically a hard and brittle substance that typically contains aromatic resinous compounds along with aromatic and other hydrocarbons and their derivatives; it is used chiefly as road tar, in waterproofing roofs and other structures, and to make electrodes (Speight, 2013a). On the other hand, wood tar pitch is a bright, lustrous substance containing resin acids; it is used chiefly in the manufacture of a variety of products such as plastics, insulating materials, and caulking material. Pitch derived from fats, fatty acids, or fatty oils by distillation are usually soft substances containing polymers and decomposition products; they are used chiefly in varnishes and paints and in floor coverings.

Pitch lake is the name that has been applied to large surface deposit of bitumen. Guanoco Lake in Venezuela covers more than 1,100 acres and contains an estimated 35,000,000 bbl of bitumen. It was used as a commercial source of asphalt from 1891 to 1935. Smaller deposits occur commonly where Tertiary marine sediments outcrop on the surface; an example is the tar pits at Rancho La Brea in Los Angeles (brea and tar have been used synonymously with bitumen). Although most pitch lakes the remains of are formerly active seeps, some pitch lakes, such as the Pitch Lake on the island of Trinidad (covers 115 acres and contains an estimated 40,000,000 bbl of bitumen) continue to be supplied with fresh crude oil seeping from a subterranean source.

1.3.2.8 Sludge

In respect of completion of the potential viscous feedstocks, the material arbitrarily referred to as sludge should also be given some consideration.

Sludge can vary from a semisolid material to a solid material and can be generated from many of the refining processes, crude oil handling operations, as well as wastewater treatment (Parkash, 2003; Gary et al., 2007; Speight, 2014a; Hsu and Robinson, 2017; Speight, 2017). Sludge is often also referred to as a refinery waste and each component of the sludge can have its own impact on refinery operations as well as on the environment and there is always the potential for associated reactions between the sludge components that change the potential for the environmental impact—usually to a higher potential. Thus, knowledge of the composition of the sludge can be a necessary piece of information that will assist all aspects of the refinery operations. Application of the methods used to fractionate the viscous feedstocks may be applicable to defining the composition of the sludge into material that can be further refined or materials that are designated as refinery waste which should be disposed accordingly.

A significant portion of the sludge may be transported off-site and sold as by-products (Asia et al., 2006). These outputs include sulfur, acetic acid, phosphoric acid, and recovered metals. Metals from catalysts and from the crude oil that have deposited on the catalyst during the production often are recovered by third-party recovery facilities.

Storage tanks are used throughout the refining process to store crude oil and intermediate process feeds for cooling and further processing. Finished crude oil products are also kept in storage tanks before transport off site. Storage tank bottoms are mixtures of sludge (typically emulsified oil and incompatible organic constituents as well as iron rust from corrosion) which accumulates at the bottom of tanks. Liquid tank bottoms are periodically drawn off to prevent their continued build up—sludge is also removed during periodic cleaning of tanks for inspection. Tank bottoms may contain amounts of tetraethyl or tetramethyl lead (although this is increasingly rare due to the phase-out of leaded products), other metals, and phenols.

Disposal of the sludge can be achieved by installation of a gasifier and, in fact, a wide variety of feedstocks can be considered for gasification, ranging from solids to liquids to gaseous streams. Although when the feed is a gas or liquid, the operation is frequently referred to as partial oxidation (POX). From a process perspective, partial oxidation of gases and liquids is very similar to the gasification of solids.

The major requirement for a suitable feedstock is that it contains a significant content of carbon and hydrogen. Solid feedstocks include solid waste, residua, visbreaker bottoms, crude oil coke, and even biomass—the streams most commonly employed are generally low-value by-products or waste streams generated by other processes (Parkash, 2003; Gary et al., 2007; Speight, 2014a; Hsu and Robinson, 2017; Speight, 2017).

1.4 Classification

As already noted, the generic term heavy oil is often applied to crude oil that has an API gravity of less than 20° as well as those highly viscous naturally occurring materials (typically referred to as bitumen) having an API gravity less than 10° that have been referred to as bitumen (Speight, 2014a, 2015a). Considering the potential range of values from the standard test methods, the differentiation between an API gravity of 19.5° and 20.5° (for a top value of 20°) or 21.5° and 22.5° (for a top value of 22°) is difficult to determine as is the difference in the overall character of feedstocks having a viscosity of 9,500 and 10,500 cP.

Also, there has been an attempt to classify crude oil, heavy crude oil, and tar sand bitumen using the viscosity scale, and 10,000 cP being the fine line of demarcation between heavy oil and tar sand bitumen. Use of such a system leads to confusion when having to differentiate between a material having a viscosity of

9,950 cP and one having a viscosity of 10,050 cP as well as taking into account the limits of accuracy of the method of viscosity determination (Speight, 2014a, 2015a). Such definitions based on a single property should be discounted as being of little scientific value. Whether the limits are the usual laboratory experimental difference (typically ±3% of the acquired value by the relevant test method) or more likely the limits of accuracy of the method (±5% to ±10%), there is the question of viability and reliability of the definition. In fact, the limits of experimental difference of the method of measuring viscosity also increase the potential for misclassification using this (or any) single property for classification purposes.

Thus, in order to classify crude oil, heavy crude oil, and tar sand bitumen, the use of a single parameter such as API gravity or viscosity is not enough for classification purposes and any attempt to classify these feedstocks on the basis of a single property is no longer sufficient to define the nature and properties of crude oil and crude oil-related materials. The general classification method used for conventional crude oil, heavy crude oil, extra heavy crude oil, and tar sand bitumen involves not only an inspection of several properties but also some acknowledgment of the method of recovery.

Technically, heavy crude oil, extra heavy crude oil, and tar sand bitumen (relative to the more conventional crude oil fluids) have a high specific gravity, high viscosity, low atomic hydrogen-to-carbon (H/C) ratio as well as a high content of asphaltene constituents, resin constituents, sulfur, nitrogen, heavy metals and are often black in color (McCain, 1990; Meyer and De Witt, 1990; Meyer and Attanasi, 2003; Speight, 2014a).

Furthermore, to repeat and emphasize, one of the most appropriate means of classification of crude oil and the viscous feedstocks comes from the definition of tar sand bitumen offered by the United States Government for tar sands (Speight, 2014a, 2016):

> Tar sands are the several rock types that contain an extremely viscous hydrocarbon which is not recoverable in its natural state by conventional oil well production methods including currently used enhanced recovery techniques.

This definition allows a method by which refinery feedstocks can be categorized further reference to (1) the existence of the feedstock in the reservoir or deposit and (2) the method of (Table 1.8).

Heavy crude oil, extra heavy crude oil, and tar sand bitumen are all unconventional oil resources. The general classification of these feedstocks is related to the ease of flow of the material (Table 1.8) which indicates (1) the methods by which the material exits in the reservoir/deposit, (2) the methods of transportation, and (3) the methods of refining. Other pertinent properties used commercially are density and viscosity.

Other definitions (or attempted classification) of the viscous feedstocks are often based on the API gravity (which is a scale based on the oil relative density) as the criterion for classification. The API degree range that has been arbitrarily selected to define and classify refinery feedstocks oils has not been standardized. For example, the World Petroleum Conference has classified heavy crude oil as crude oil having an API gravity less than 22.3° API while the American Petroleum Institute suggests that heavy crude oil has an API gravity less and 20°. On the other hand, Petrobras suggests that heavy oil has an API gravity that falls in the range 10–19° API (Santos et al., 2014). There have also been attempts to classify crude oil and other refinery feedstocks on properties such as viscosity and sulfur content but classification based on a single property is fraught with errors and speculation.

Generally, heavy oil and tar bitumen are characterized by high viscosity (i.e., resistance to flow measured in centipoises or cP) and high density compared to conventional oil. Attempts have been made to define heavy oil as an oil with a gas-free viscosity between 100 and 10,000 cP at reservoir temperature (tar sand bitumen is incorrectly defined to have a viscosity greater than 10,000 cP, and may be as high as 10,000,000 cP). Attempting to draw a finite line in order to define heavy oil on the one side of the line and bitumen on the other side of the line is meaningless and incorrect and offers very little in the way of scientific integrity since the standard test methods that are used to determine viscosity suffer from reproducibility and accuracy (Speight, 2015a).

On the other hand, heavy oil is slightly less dense than water with a density on the order of 0.973 which represents an API gravity of 14° (water has a density of 1.00 which represents an API gravity of 10°). In addition, heavy oil can flow in some reservoirs at downhole temperatures that are above the pour point of the oil and/or with in situ solution gas, but at the surface, it is a thick, black, viscous fluid, which typically has an API gravity <10°. It should also be noted that the term oil sands as used in Canada was created for incentive tax purposes for those heavy crude oils found above a certain latitude where the infrastructure was almost nonexistent.

For correctness and to avoid confusion, the terms heavy oil, extra heavy oil, and bitumen should not be used interchangeably.

1.5 Feedstock Evaluation

The evaluation of any refinery feedstock involves application of a suite of analytical test methods and the data are used to determine the most appropriate processes to be employed in the refinery to convert the feedstock to a suite of useful products. As brief examples, the following properties are commonly applied (but are only a small selection of the analytical methods) to an evaluation of a refinery feedstock: (1) density, which is used to assess the quality of the feedstock as well as in weight and volume conversion, (2) kinematic viscosity, which is an indicator of fluidity and is used in flow rate calculations and an indicator of fluid behavior, (3) pour point, which is the minimum temperature at which the feedstock will flow, (4) sulfur content which is used in quality control and feedstock evaluation, especially in terms of catalyst poisoning and equipment corrosion, and (5) component analysis, which is typically carried out by fractionation methods and the data are used to determine the processes that should be applied in the refinery.

Thus, the evaluation of refinery feedstocks is a necessary series of actions because of the variety of material that are accepted as feedstocks for the modern refinery. Also, the time when one crude oil as the sole feedstock for a refinery is limited (if not nonexistent) in the modern refinery because of the use of a blend of two or more crude oils as the refinery feedstock. Thus, evaluation of the blend (in fact, each of the components of the blend) is necessary to allow an assessment of the product slate and the processes that are necessary to create the saleable products. Thus, feedstock evaluation is the necessary application of a series of test methods in order to assess the chemical and physical properties of each of the constituents of the blend as well as the blend itself against a specific set of refinery-developed standards. A comprehensive feedstock evaluation is based on a detailed analysis necessary to determine the value of these crude oils to the refinery.

Accordingly, this section presents an outline of the tests that may be applied to crude oil, heavy crude oil, extra heavy crude oil, tar sand bitumen, and residua as well as biomass and their respective products as a pointer to the chemical properties and physical properties from which a feedstock or product can be evaluated.

The series of test methods are often referred to as the "feedstock assay" which consists of determination of (1) the feedstock assay, (2) the physical properties, (3) the thermal properties, (4) the chromatographic properties, and, when deemed appropriate, (5) the spectroscopic properties which offer insights into the chemical constituents in the feedstock and the potential for reaction of these chemical constituents with other constituents of the blend and with catalysts used in catalytic processes.

Thus, analyses are performed (by a series of standard test methods) to determine whether each batch of crude oil or the viscous feedstock received at the refinery is suitable not only for refining purposes but also to determine the preferential process options and the order in which the processes should be applied. To obtain the necessary information, two different analytical schemes are commonly used and these are: (1) an inspection assay and (2) a comprehensive assay (Table 1.15). The test applied to the feedstock to obtain the assay data are, for the purposes of this text, the standard methods as described by the ASTM International (formerly the American Society for Testing and Materials) (ASTM, 2021).

Table 1.15 Recommended Inspection Data Required for Crude Oil and the Viscous Feedstocks[a]

Crude oil	Viscous feedstocks
Density, specific gravity	Density, specific gravity
API gravity	API gravity
Carbon (wt%)	Carbon (wt%)
Hydrogen (wt%)	Hydrogen (wt%)
Nitrogen (wt%)	Nitrogen (wt%)
Sulfur (wt%)	Sulfur (wt%)
	Nickel (ppm)
	Vanadium (ppm)
	Iron (ppm)
Pour point	Pour point
Wax content	
Wax appearance temperature	
Viscosity (various temperatures)	Viscosity (various temperatures)
Carbon residue of residuum	Carbon residue[a]
	Ash (wt%)
Distillation profile:	Fractional composition:
All fractions plus vacuum residue	Asphaltenes (% w/w)
	Resin constituents (% w/w)
	Aromatics (% w/w)
	Saturates (% w/w)

[a] Includes heavy crude oil, extra heavy crude oil, tar sand bitumen, and residua.

An inspection assay (which is routinely performed on all feedstocks received at a refinery) typically involves determination of several bulk properties of the feedstock (such as API gravity, sulfur content, pour point, distillation range, viscosity, salt content, water and sediment content, and trace metals) as a means to determine if any major changes in characteristics have occurred since the last comprehensive assay was performed within the refinery. Other properties that are determined on an as needed basis include, but are not limited to, the following: (1) vapor pressure (Reid method) and the (2) total acid number. The results from these tests with the archived data from a comprehensive assay provide an estimate of any changes that have occurred in the crude oil that may be critical to refinery operations. The aforementioned inspection assay test methods are not exhaustive but are the ones most commonly used and provide data on the impurities present as well as a general idea of the products that may be recoverable (Speight 2015a).

On the other hand, a full assay (also known as a comprehensive assay) is more complex and is usually performed when the feedstocks from a new field (in the case of crude oil and heavy crude oil) or a new deposit (in the case of extra heavy oil and tar sand bitumen) come onstream, or when the inspection assay indicates that significant changes in the composition of the feedstock have occurred. Except for these circumstances, a comprehensive assay of a particular feedstock may not (unfortunately) be updated for several years.

The data derived from the test assay can be used to assess the quality (and potential behavior) of the feedstock in the refinery and, therefore, indicate acquiring a degree of predictability of feedstocks performance during refining as well as the potential for changes to any material spilled into the environment. However, knowledge of the basic concepts of refining will help the analyst understand the production and, to a large extent, the anticipated properties of the product, which in turn is related to storage, sampling, and handling the products.

1.6 Modern Analytical Perspectives

The importance of chemical and physical properties is dependent upon the purity of the feedstock. In the strictest sense, refinery feedstocks are complex chemical mixtures which contain hydrocarbon derivatives and non-hydrocarbon derivatives that confer properties on the mixture that may not be reflected in the composition. Therefore, it is necessary to apply various standard test methods (as presented in the following chapters) to the feedstocks to determine if the material is suitable for processing and (in the case of the products) for sale with a designated use in mind. The more common tests are introduced in the following chapters and the test methods are presented in alphabetical order with no preference given to any particular test method.

Typically there are many standard test methods that have been developed for determining the properties of conventional crude oil and, therefore, some of these methods require modification for application to viscous refinery feedstocks while there is also the need to develop test methods that are particularly applicable to the viscous feedstocks not only as refinery feedstocks (or products) but also when the material is spilled into the environment. In fact, since the 1960s (especially after the so-called oil shocks of the 1970s) the development of viscous oil resources (particularly tar sand resources) became a major focus of many companies (Speight, 2013b, 2013c). These methods allowed refiners to have access to a literature database and to assess the capabilities and limitations of the application of the analytical methods to the viscous feedstocks in order to accept these materials for refining as well as the identification of possible product slates.

However, before submitting a sample of a viscous feedstock for analysis, there is the need to ensure that the sample is representative. Typically, such methods also involve cleaning the sample by, for example, centrifugation and filtration, however, after dilution of the sample with a suitable (and easily removable) solvent, and the method should be acknowledged in a description of the method and the results (Wallace, 1988; Speight, 2015a). In addition, the viscous feedstocks typically contain substantial quantities of asphaltene constituents which not only may be a cause of instability and/or incompatibility but are also a precursor to coke formation as well as an indicator of the choice of upgrading options (Chapters 3 and 10) (Speight, 2015a, 2017).

More generally, and relative to conventional (light) crude oils, heavy crude oil, extra heavy crude oil, tar sand bitumen, and distillation residua are characterized by (1) low atomic H/C ratio; (2) high specific gravity; (3) high viscosity; (4) high contents of asphaltene constituents, resin constituents, and other constituents containing nitrogen, oxygen, sulfur, and heavy metals; and (5) often exhibit a near-black to black color (Parkash, 2003; Gary et al., 2007; Speight, 2014a, 2017).

Analytical methods used to describe conventional oils must be modified for the characterization of the viscous feedstocks because of the variation in composition and properties of the viscous feedstocks when compared to conventional crude oil. For analytical properties that might be of minor importance in the conventional crude oil refining industry, there can be a quantum shift in importance of these same properties when used for refining the viscous feedstocks and there needs to be continued development of an acceptable analytical method that often begins with identification of the property that has to be determined.

As a result, research into methods for the measurement of properties of interest is ongoing in many laboratories that leads to the development of a variety of methods or modification of current methods. After this important step, there needs to be a reviewing process, deficiencies are identified, and promising methods are improved and acceptable.

In some cases, there have been questions about the application of methods developed for conventional crude oil to the viscous feedstocks. For example, (1) several methods are often available for the measurement of a given property but not all methods may be suitable for application to the viscous feedstocks, (2) some methods are more appropriate than others, (3) not all methods provide equivalent data, (4) different laboratories using the same method do not always report equivalent data for a given sample, and (5) in some analyses, the error in sampling and subsampling is greater than the error in the measurement.

Thus, a single standard method for conventional oil can therefore be the forerunner of several similar but distinct methods for a viscous feedstock. Moreover, in order to meet the objective of accepting a new method of analysis, the following goals help the investigator (or the investigating organization) to meet this objective: (1) evaluation of existing methods by identifying the limitations and applicability of the method, (2) improvement of existing analytical methods, (3) development of new analytical methods, (4) standardization of analytical methods and adoption of recommended methods where applicable or desirable, and (5) development of recommended methods for sampling, sample storage, and sample preparation (Wallace, 1988).

References

Abraham, H. 1945. *Asphalts and Allied Substances*. Van Nostrand Publishers, New York.

Armstrong, K. 1997. *Jerusalem: One City, Three Faiths*. Ballantine Books, New York. p. 3.

Asia, I.O., Enweani, I.B., and Eguavoen, I.O. 2006. Characterization and Treatment of Sludge from the Petroleum Industry. *African Journal of Biotechnology*, 5(5): 461–466.

ASTM 2021. *Annual Book of Standards*. ASTM International, West Conshohocken, Pennsylvania.

ASTM D4. 2021. Standard Test Method for Bitumen Content. Annual Book of Standards. ASTM International, West Conshohocken, Pennsylvania.

ASTM D140. 2021. Standard Practice for Sampling Bituminous Materials. Annual Book of Standards. ASTM International, West Conshohocken, Pennsylvania.

ASTM D977. 2021. Standard Specification for Emulsified Asphalt. Annual Book of Standards. ASTM International, West Conshohocken, Pennsylvania.

ASTM D2397. 2021. Standard Specification for Cationic Emulsified Asphalt. Annual Book of Standards. ASTM International, West Conshohocken, Pennsylvania.

ASTM D4175. 2021. Standard Terminology Relating to Petroleum, Petroleum Products, and Lubricants. Annual Book of Standards. ASTM International, West Conshohocken, Pennsylvania.

Bauer, G. (Georgius Agricola) 1546. *De Natura Fossilium. Book IV*, Basel, Switzerland.

Bauer, G. (Georgius Agrícola) 1556. *De Re Metallica. Book XII*, Basel, Switzerland.

Benoit, B., and Zurlo, J. 2014. Overcoming the Challenges of Tight/Shale Oil Refining. *Petroleum Technology Quarterly*, Q1: 37–44.

Bland, W.F., and Davidson, R.L. 1967. *Petroleum Processing Handbook*. McGraw-Hill, New York.

Dahmus, J. 1995. *A History of the Middle Ages*. Barnes and Noble, New York. p. 135 and 136; previously published with the same title by Doubleday Book Co., 1968

Davies, N. 1996. *Europe: A History*. Oxford University Press, Oxford, United Kingdom. p. 245 and 250

De Graaf, B., Tang, T., Oberlin, J., and Diddams, P. 2014. Shale Crudes and FCC: A Mismatch from Heaven? Digital Refining.

Forbes, R.J. 1958a. *A History of Technology*. Oxford University Press, Oxford, United Kingdom.

Forbes, R.J. 1958b. *Studies in Early Petroleum Chemistry*. E.J. Brill, Leiden, Netherlands.

Forbes, R.J. 1959. *More Studies in Early Petroleum Chemistry*. E.J. Brill, Leiden, Netherlands.

Gary, J.G., Handwerk, G.E., and Kaiser, M.J. 2007. *Petroleum Refining: Technology and Economics*. 5th Edition. CRC Press, Taylor & Francis Group, Boca Raton, Florida.

Ghoshal, S., and Sainik, V. 2013. Monitor and Minimize Corrosion in High-TAN Crude Processing. *Hydrocarbon Processing*, 92(3): 35–38.

Gruse, W.A., and Stevens, D.R. 1960. *The Chemical Technology of Petroleum*. McGraw-Hill, New York.

Henry, J.T. 1873. *The Early and Later History of Petroleum*. Volume I and II. APRP Co., Philadelphia, PA.

Hsu, C.S., and Robinson, P.R. (Editors). 2017. *Handbook of Petroleum Technology*. Springer International Publishing AG, Cham, Switzerland.

ITAA. 1936. Income Tax Assessment Act. Government of the Commonwealth of Australia.

James, P., and Thorpe, N. 1994. *Ancient Inventions*. Ballantine Books, New York.

Khan, M.R., and Patmore, D.J. 1997. Heavy Oil Upgrading Processes. In: *Petroleum Chemistry and Refining*. J.G. Speight (Editor). Taylor & Francis, Washington, DC. Chapter 6

Kovacs, M.G. 1990. *The Epic of Gilgamesh*. Stanford University Press, Stanford, California. Tablet XI

Mallowan, M.E.L. 1954. The Excavations at Nimrud (Kahlu), 1953. *Iraq*, 16: 59–114, and 16: 115–163

Mallowan, M.E.L., and Rose, J.C. 1935. Excavations at Tell Arpachiyah, 1933. *Iraq*, 2(1): 1–178.

Marschner, R.F., Duffy, L.J., and Wright, H. 1978. Asphalts from ancient towns in Southwestern Iran. *Paléorient*, 4: 97–112.

McCain, W.D. 1990. *The Properties of Petroleum Fluids*. PennWell Pub. Co., Tulsa, Oklahoma.

Meyer, R.F., and Attanasi, E.D. 2003. Heavy Oil and Natural Bitumen: Strategic Petroleum Resources. Fact Sheet No. FS0070-03, 2p. United States Geological Survey, Reston, Virginia.

Meyer, R.F., and De Witt, W. Jr. 1990. Definition and World Resources of Natural Bitumens. Bulletin No. 1944. US Geological Survey, Reston, Virginia.

Motaghi, M., Shree, K., and Krishnamurthy, S. 2010a. Consider New Methods for Bottom of the Barrel Processing – Part 1. *Hydrocarbon Processing*, 89(2): 35–40.

Motaghi, M., Shree, K., and Krishnamurthy, S. 2010b. Consider New Methods for Bottom of the Barrel Processing – Part 2. *Hydrocarbon Processing*, 89(2): 55–88.

Nellensteyn, F.J., and Brand, J. 1936. Chem. Weekblad, 261.

Ohmes, R. 2014. Characterizing and Tracking Contaminants in Opportunity Crudes. Digital Refining.

Parkash, S. 2003. *Refining Processes Handbook*. Gulf Professional Publishing, Elsevier, Amsterdam, Netherlands.

Purdy, G.A. 1957. *Petroleum: Prehistoric to Petrochemicals*. Copp Clark Publishing, Toronto.

Rana, M.S., Sámano, V., Ancheyta, J., and Diaz, J.A.I. 2007. A Review of Recent Advances on Process Technologies for Upgrading of Heavy Oils and Residua. *Fuel*, 86: 1216–1231.

Rispoli, G., Sanfilippo, D., and Amoroso, A. 2009. Advanced Hydrocracking Technology Upgrades Extra Heavy Oil. *Hydrocarbon Processing*, 88(12): 39–46.

Santos, R.G., Loh, W., Bannwart, A.C., and Trevisan, O.V. 2014. An Overview of Heavy Oil Properties and Its Recovery and Transportation Methods. *Brazilian Journal of Chemical Engineering*, 31(3): 571–590.

Song, H., Jiao, G., Zhu, X., and Zhou, X. 2013. Hydrocarbon Composition of Different VGO Feedstocks and Its Correlation with FCC Product Distribution. *China Petroleum Processing & Petrochemical Technology*, 15(1): 32–39.

Speight, J.G. 1978. Personal Observations at Archeological Digs at the Cities of Babylon, Calah, Nineveh, and Ur, Iraq. Report submitted to the Director of Antiquities, Baghdad, Iraq.

Speight, J.G. 1984. Upgrading Heavy Oils and Residua: The Nature of the Problem. In: *Catalysis on the Energy Scene*. S. Kaliaguine and A. Mahay (Editors). Elsevier, Amsterdam. p. 515.

Speight, J.G. 2000. *The Desulfurization of Heavy Oils and Residua*. 2nd Edition. Marcel Dekker, New York.

Speight, J.G. 2001. *Handbook of Petroleum Analysis*. John Wiley & Sons Inc., Hoboken, New Jersey.

Speight, J.G. 2011a. *The Refinery of the Future*. Gulf Professional Publishing, Elsevier, Oxford, United Kingdom.

Speight, J.G. 2011b. *An Introduction to Petroleum Technology, Economics, and Politics*. Scrivener Publishing, Salem, Massachusetts.

Speight, J.G. 2012a. *Crude Oil Assay Database*. Knovel, New York. http://www.knovel.com/web/portal/browse/display?_EXT_KNOVEL_DISPLAY_bookid=5485&VerticalID=0

Speight, J.G. 2012b. *Shale Oil Production Processes*. Gulf Professional Publishing, Elsevier, Oxford, United Kingdom.

Speight, J.G. 2013a. *The Chemistry and Technology of Coal*. 3rd Edition. CRC Press, Taylor & Francis Group, Boca Raton, Florida.

Speight, J.G. 2013b. *Heavy Oil Production Processes*. Gulf Professional Publishing Company, Elsevier, Oxford, United Kingdom.

Speight, J.G. 2013c. *Heavy and Extra Heavy Oil Upgrading Technologies*. Gulf Professional Publishing Company, Elsevier, Oxford, United Kingdom.

Speight, J.G. 2014a. *The Chemistry and Technology of Petroleum*. 5th Edition. CRC Press, Taylor & Francis Group, Boca Raton, Florida.

Speight, J.G. 2014b. *High Acid Crudes*. Gulf Professional Publishing Company, Elsevier, Oxford, United Kingdom.

Speight, J.G. 2014c. *Oil and Gas Corrosion Prevention*. Gulf Professional Publishing Company, Elsevier, Oxford, United Kingdom.

Speight, J.G. 2015a. *Handbook of Petroleum Product Analysis*. 2nd Edition. John Wiley & Sons Inc., Hoboken, New Jersey.

Speight, J.G. 2015b. *Fouling in Refineries*. Gulf Professional Publishing Company, Elsevier, Oxford, United Kingdom.

Speight, J.G. 2016. *Asphalt Materials Science and Technology*. Butterworth-Heinemann, Elsevier, Kidlington, Oxford, United Kingdom and Waltham, Massachusetts.

Speight, J.G. 2017. *Handbook of Petroleum Refining*. CRC Press, Taylor & Francis Group, Boca Raton, Florida.

Speight, J.G. 2019a. *Handbook of Petrochemical Processes*. CRC Press, Taylor & Francis Group, Boca Raton, Florida.

Speight, J.G. 2019b. *Heavy Oil Recovery and Upgrading*. Gulf Publishing Company, Elsevier, Cambridge, Massachusetts.

Speight, J.G. 2021. *Petroleum Refining Technology Series Volume 1: Refinery Feedstocks*. CRC Press, Taylor & Francis Group, Boca Raton, Florida.

Stratiev, D., and Petkov, K. 2009. Residue Upgrading: Challenges and Perspectives. *Hydrocarbon Processing*, 88(9): 93–96.

Stratiev, D., Tzingov, T., Shishkova, I., and Dermatova, P. 2009. Hydrotreating Units Chemical Hydrogen Consumption Analysis: A Tool For Improving Refinery Hydrogen Management. Proceedings. 44th International Petroleum Conference, Bratislava, Slovak Republic. September 21-22.

Sun, X., Zhang, Y., Li, X., Cui, G., and Gu, J. 2013. A Case Study on Foamy Oil Characteristics of the Orinoco Belt, Venezuela. *Advances in Petroleum Exploration and Development*, 5(1): 37–41.

Wallace, D. (Editor). 1988. *A Review of Analytical Methods for Bitumens and Heavy Oils*, AOSTRA Technical Publication Series No. 5. Alberta Oil Sand Technology and Research Authority, Edmonton, Alberta, Canada.

Wang, W., Liu, Y., Liu, Z., and Tian, S. 2016. Detailed Chemical Composition of Straight-Run Vacuum Gas Oil and Its Distillates as a Function of the Atmospheric Equivalent Boiling Point. *Energy & Fuels*, 30(2): 968–974.

Yeung, T.W. 2014. Evaluating Opportunity Crude Processing. Digital Refining.

2

Sampling and Measurement

2.1 Introduction

Crude oil and the viscous feedstocks exhibit a wide range of physical properties and several relationships can be made between various physical properties (Chapter 1)(Parkash, 2003; Gary et al., 2007; Speight, 2014; Hsu and Robinson, 2017; Speight, 2017). Whereas the properties such as viscosity, density, boiling point, and color of crude oil may vary widely, the elemental analysis (ultimate analysis) varies, as already noted, over a narrow range for a large number of crude oil samples (Speight, 2014). The carbon content is relatively constant (i.e., on the order of $85 \pm 2\%$ w/w), while the hydrogen content and the heteroatom (nitrogen plus oxygen plus sulfur plus metals) content are responsible for the major differences between the various refinery feedstocks. Coupled with the changes brought about to the feedstock constituents by refinery operations, it is not surprising that feedstock characterization is a monumental task.

The chemical and physical properties of crude oil and the viscous feedstocks and the yields and properties of products or factions produced therefrom vary considerably and are dependent on the concentration of the hydrocarbon constituents and non-hydrocarbon constituents present in the crude oil. Some types of crude oils have pronounced advantages as sources of fuels and lubricants because they require less specialized processing than that needed for production of the same products from other types of crude oil. Other crude oils may contain unusually low concentrations of constituents that are desirable fuel or lubricant constituents or precursors to these products, and the production of these products from such crude oils may require considerable refining.

Evaluation of crude oil for use as a refinery feedstock usually involves an examination of one or more of the physical properties of the material (Speight, 2001, 2014). By this means, a set of basic characteristics can be obtained that can be correlated with utility. To satisfy specific needs with regard to the type of crude oil to be

Handbook of Heavy Oil Properties and Analysis, First Edition. James G. Speight.

processed, as well as to the nature of the product, various standards organizations, such as the American Society for Testing and Materials (ASTM, 2021) in North America and the Energy Institute in the United Kingdom (formerly the Institute of Crude oil), have devoted considerable time and effort to the correlation and standardization of methods for the inspection and evaluation of crude oil and crude oil products. Many other countries—too numerous to mention here—also have similar organizations for developing standard test methods or industrial products. For the purposes of this text, it is appropriate that in any discussion of the physical properties of crude oil and crude oil products, reference be made to the corresponding test methods (designated by the letters ASTM along with a letter, typically for fossil fuel materials "D" and a number), and accordingly, the various test numbers have been included in the text.

In North America, the test methods designed and recommended by ASTM International (formerly the American Society for Testing and Materials) are the are prime assets for the evaluation and assessment of the physical, mechanical, rheological, thermal, and chemical properties of crude oils, viscous feedstocks as well as their respective products and other naturally occurring energy resources used for various industrial applications.

In addition, while the focus of this chapter is on the sampling and preparation of the viscous feedstocks for refining, it is worthwhile (because of the similarity of the methods) to consider the application of the sampling and measurement methods to situations where the feedstock has been spilled into the environment. In fact, the spills of crude oil and viscous feedstocks can occur anywhere that the material is being stored, transported, transferred from one tank to another, or refined.

Thus, methods that are applicable to both refining and environmental issues are cited in this chapter. This will save unnecessary repetition of the methods at other parts of the book.

2.2 Sampling

Sampling is a critical part in any analytical methods and if the sampling is performed without the use of adequate techniques and according to specific objectives, the result of the analysis can have serious errors. Moreover, the sampling strategy should reflect the composition of the bulk material and the sample should be preserved in order to prevent any form of contamination. Furthermore, the design of the sampling plan should be linked to the purposes of the sampling and to the specific information that is required at the culmination of the application of the analytical method.

Thus, sampling is the process in analytical science by which a predetermined number of observations is taken by selection of sample from the bulk material.

However, the method of sampling depends on the type of analysis being performed but can include (1) simple random sampling, (2) systematic sampling, (3) stratified sampling, and (4) clustered sampling.

Simple sampling (or random sampling) is the means by which a sample taken is entirely by chance from the bulk material each member of the population has an equal chance, or probability, of being selected. On the other hand, systematic sampling is a probability sampling method where the samples are taken from the bulk material at a regular interval such as from specific areas of the bulk material. Another technique, known as stratified sampling, may also be applied to the bulk material, such as a pile of a solid substance or a near-solid substance such as when tar sand bitumen is stored in large tanks in which flow is limited (if not prevented) by the properties of the material. Finally, clustered sampling is the process by which the bulk material is divided into smaller groups (referred to as clusters) and the samples to be taken are randomly selected among these clusters to form a sample.

Thus, the reliability of a sampling method is the degree of perfection with which the identical composition and properties of the entire whole crude oil or crude oil product are obtained (preserved completely without alteration) in the sample—such a sample is referred to as a "representative sample."

By definition, a sample from a bulk material that accurately represents the characteristics of a material and the results from the representative sample must accurately reflect the properties of the bulk material.

In contrast to conventional (liquid low density/high API gravity) crude oil, when dealing with the viscous feedstocks which, because of the properties of the material and/or the ambient conditions, may have no flow properties, it is essential to decide on the most appropriate method of sampling and the process must be well documented. It is important to recognize that the outcome of any analytical method in terms of reproducibility and accuracy depends on the reliability of the sampling method.

The reliability and the value of any analysis are judged by the characteristics of the sample as determined by laboratory tests. As a word of caution, the high viscosity of the viscous feedstocks can (and often does) impede the mobility of the feedstock thereby introducing difficulties in sampling prior to application of the test methods and emphasizes the need for a well-designed sampling plan insofar as poorly designed (haphazard) sampling plan may make the analytical results unreliable or even impossible to interpret. This consideration is especially important when addressing the problem of sampling bulk materials such as crude oil or the viscous feedstocks that may not only be difficult to sample but also be highly heterogeneous.

When dealing with materials as heterogeneous as the viscous feedstocks, the sampling protocol must be designed to produce a sample that is of the bulk material. If this is not the case, the analytical data will not be representative of the material and will be, to be blunt, incorrect no matter how accurate or precise the test method. In addition, the type and cleanliness of sample containers are important; if the container is contaminated or is made of material that either reacts with the product or is a catalyst, the test results may be wrong.

However, before the actual sampling begins, there should be a sampling plan in which not only the nature of the material to be sampled but also the size, number, and location of the sample increments should be defined. In fact, it is recommended that before the work commences, the plan should be recorded (in writing) as a detailed protocol which should include (1) the procedures for all steps, from sampling through sample treatment, (2) the measurement, and (3) the data evaluation which should be revised as new information is obtained. This not only assures consistency in the use of the analytical methods but also can be used (if necessary) if there are any legal disputes about the use of the data (Speight, 2009, 2015).

The ability to collect and preserve a sample that is representative of the material under investigation is a critically important step (Dean, 1998, 2003; Patnaik, 2004). Obtaining a representative environmental sample (especially in analysis of a material that has been spilled into the environment) is always a challenge due to the heterogeneity of different sample matrices. Additional difficulties are encountered with feedstock constituents due to the wide range in volatility, solubility, biodegradation, and adsorption potential of the individual constituents. Also, the procedures used for sample collection and preparation must be legally defensible.

In order to accomplish such a goal, the design of bulk sampling programs involves the following criteria: (1) identification of the population from which the sample is to be obtained, (2) selection and withdrawal of valid samples of this population, (3) reduction of each bulk sample to a laboratory sample, and (4) suitable for the analytical techniques to be used.

The sampling methods used for feedstock constituents are generally considered to be methods for determination of the total hydrocarbon derivatives in the crude oil or viscous feedstock. In addition, because of the overall complexity of the constituents of the feedstock, the potential of the constituents for impact on human health and on the environment are not well defined.

Although most site investigations to determine the assessment of contamination of an ecosystem by crude oil constituents and viscous feedstock constituents are regulated by local or by regional (state) governments, sample collection and preservation recommendations follow strict guidelines (Table 2.1) thereby reducing the potential for sample compromise. Therefore, before a sample is collected, the particular sample collection and preservation requirements must be investigated. And, because of holding (storage) time considerations, the laboratory must be selected and notified prior to the collection of the samples.

Table 2.1 Example of the Content and Makeup of a Sampling Log

Parameter	Comment
Location	The precise geographic (or other precise) location from which the sample was obtained
	The site or refinery from which the sample was obtained
	The refinery process by which the sample was produced
	The process engineer
The sample	The character of the bulk material (solid, liquid, or gas) at the time of sampling
	The means by which the sample was obtained
	The means and protocols that were used to obtain the sample
	The date and the amount of sample that was originally placed into storage
Storage	The manager of the storage area
	The type of container used for sample storage
	The location of the storage area
	The location of the sample within the storage area
	The temperature of the storage area
Analysis	Any chemical analyses that have been determined to date
	Any physical analyses that have been determined to date
	The date of any such analyses
	The methods used for the analyses
	The analysts who carried out the work
Log sheet	A list of the names of the persons who removed sample
	The reason for the removal of an aliquot
	The date of removal of an aliquot
	The amount of the aliquot removed
	The fate (or means of disposal) of any unused aliquot

Furthermore, the reliability of the storage procedure is the degree to which the sample remains unchanged thereby guaranteeing the accuracy and usefulness of the analytical data. Thus, the composition of the sample should be preserved during storage and transport and even when in the laboratory just prior to application of the various test methods. These precautions are necessary to (1) to retard oxidation, (2) to retard hydrolysis of any of the constituents, (3) to reduce any potential for evaporation, (4) to reduce adsorption, and (5) to avoid photodegradation.

In addition, sampling devices and containers have to be chosen depending on the compound to be analyzed to avoid adsorption or contamination.

At this point, a review of the sampling methods applied to crude oil and the viscous feedstocks products is worth of inclusion.

Thus, the importance of the correct sampling of any sample destined for analysis should always be overemphasized. In fact, the sampling protocol—the procedure used to select samples from the bulk material to be measured—is to select samples that are representative of the study bulk material. An incorrect sampling protocol can lead to erroneous analytical data from which decisions about processing issues or, in the case of an environmental sample, regulatory issues cannot be made with any degree of confidence. In addition, adequate records of the circumstances and conditions during sampling have to be made, for example, in sampling from storage tanks, the temperatures and pressures of the separation plant and the atmospheric temperature would be noted.

At the high-boiling end of the volatility scale, any samples that contain, or are composed of, high molecular weight paraffin hydrocarbon derivatives (commonly referred to as wax) that are also in a solid state, may require judicious heating (to dissolve the wax) and agitation (homogenized, to ensure thorough mixing) before sampling. Also, if the sampling is performed at ambient temperature, homogenization of the feedstock is necessary to ensure correct sampling because there is cause for the wax to precipitate as a separate (solid) phase.

The first task in any analysis is to separate the analytes from the bulk of the sample. The traditional liquid extraction is the most common means employed but alternate methods are also available (Dean, 1998, 2003). A portion of the sample is mixed with an organic solvent into which the analyte is preferentially partitioned. The idea of partitioning is key to the success of this procedure. No organic analyte is completely removed from a sample by a single washing with an organic solvent.

Representative samples are prerequisite for the laboratory evaluation of any type of environmental sample and many precautions are required in obtaining and handling representative samples (ASTM D270, 2021; ASTM D1265, 2021). The precautions depend upon the sampling procedure, the characteristics (low boiling or high boiling constituents) of the product being sampled, and the storage tank, container, or tank carrier from which the sample is obtained. In addition, the sample container must be clean and the type to be used depends not only on the product but also on the data to be produced.

However, obtaining a representative sample is not just a matter of relying on partition coefficients. Partitioning is thermodynamically controlled and to achieve equilibrium requires a certain amount of intimate contact time between the sample matrix, the analyte molecules, and the organic solvent. Swirling an aqueous sample with the solvent is not always an effective method for achieving a thermodynamic equilibrium. Mixing the two phases together as thoroughly as possible and then allowing the phases to separate achieves the desired equilibrium. The procedure should be repeated with additional portions of solvent. If equilibrium is not established, the efficiency of the extraction procedure is low and the analytical data will be subject to error and indefensible (Dean, 1998, 2003).

The basic objective of each procedure is to obtain a representative sample of the material under investigation or, more often, a composite of several samples that can be considered to be a representative sample. In some cases, because of the size of the storage tank and the lack of suitable methods of agitation, several samples may be taken from different locations in the tank (especially in the case of a large storage tank because of the tendency of some materials to undergo stratification in the tank). The sample must be acquired in such a manner that the samples represent the properties of the bulk material and, thus, the composite sample must be a true representative sample of the entire lot being sampled.

This procedure allows for differences in sample that might be due to the stratification of the bulk material due to tank size or temperature at the different levels of the storage tank. Solid samples require a different protocol that might involve melting (liquefying) the bulk material (assuming that thermal decomposition is not induced) followed by homogenization. On the other hand, the protocol used for coal sampling (ASTM D346, 2021) might also be applied to sampling crude oil products (such as coke) which are solid and for which accurate analysis is required before sales.

Once the sampling procedure has been accomplished, the sample container should be accurately labeled immediately after taking the sample. The labeling should include information such as (1) the product, (2) time of sampling, and (3) location of the sampling point as well as any other information necessary for the sample identification. Furthermore, if the sample is a composite sample and contains individual samples that were taken from different levels of the storage tank, the levels from which the samples were taken and the amounts taken and mixed into the composite should be specified as part of the sample documentation.

In addition, the sampling records for any procedure must be complete and should include, but is not restricted to, items relating to (1) the origin of the sample, (2) the methods of storage, (3) analytical tests performed, (4) the test methods used, and (5) the analyst or analysts who performed the tests (Table 2.1). It is essential that there is a means to identify the sample history as carefully as possible so that each sample is tracked and defined in terms of source and activity. Thus, the accuracy of the data from any subsequent procedures and tests for which the sample is used will be placed beyond a reasonable doubt.

Since the value of any product is judged by the characteristics of the sample as determined by laboratory tests, the ability to collect and preserve a sample that is representative of the site is a critically important step. Furthermore, obtaining a representative sample of crude oil or a crude oil product is always a challenge due to the heterogeneity of the materials. Additional difficulties are encountered due to the ranges in volatility, solubility, and adsorption potential of individual constituents. And the procedures used for sample collection and preparation must be analytically reproducible with a high degree of accuracy (Dean, 1998; Weisman, 1998; Speight, 2001; Dean, 2003; Patnaik, 2004; Speight, 2014).

Moreover, it is essential that the sample to be used for the test methods must be representative of the bulk material or data will be produced that are not a true representation of the material and will, to be blunt, incorrect no matter how accurate or precise the test method and not matter how much *fine-tuning* is applied by the submitter of the sample. In addition, the type and cleanliness of sample containers are important if the container is contaminated or is made of material that either reacts with the product or is a catalyst, the test results may be wrong.

Finally, if the sample is to be acquired from a spill site, the ability to collect and preserve a sample that is representative of the site is also a critically important step. Obtaining representative environmental samples is always a challenge due to the heterogeneity of different sample matrices. Additional difficulties are encountered with sample of a viscous feedstock due to the wide range of potential interactions between the constituents of the spilled material and the adsorption potential of individual constituents.

As a final observation for this subsection, the procedure used for collection of a sample and the procedures for sample transportation and storage must be beyond reproach and also be legally defensible (Speight, 2005).

2.2.1 Sampling Protocol

The importance of the correct sampling of any product destined for analysis should always be overemphasized. Incorrect sampling protocols can lead to erroneous analytical data from which performance of the product in service cannot be accurately deduced. For example, properties such as specific gravity, distillation yield, vapor pressure, hydrogen sulfide content, and octane number of the gasoline are affected by the content of low boiling hydrocarbon derivatives so that suitable cooling and/or pressure sampling methods have to be used and care taken during the subsequent handling of the sample in order to avoid the loss of any volatile constituents. In addition, adequate records of the circumstances and conditions during sampling have to be made, for example, in sampling from storage tanks, the temperatures and pressures of the separation plant, and the atmospheric temperature would be noted.

At the other end of the volatility scale, products that contain, or are composed of, high molecular weight paraffin hydrocarbon derivatives (wax) that are also in a solid state, may require judicious heating (to dissolve the wax) and agitation (homogenized, to ensure thorough mixing) before sampling. If room temperature sampling is the *modus operandi* and product cooling causes wax to precipitate, homogenization to ensure correct sampling is also necessary.

Representative samples are prerequisite for the laboratory evaluation of any type of product and many precautions are required in obtaining and handling representative samples (ASTM D1265, 2021). The precautions depend upon the sampling procedure, the characteristics (low boiling or high boiling constituents) of the product being sampled, and the storage tank, container, or tank carrier from which the sample is obtained. In addition, the sample container must be clean and the type to be used depends not only on the product but also on the data to be produced.

2.2.1.1 Sampling Semi-volatile and Nonvolatile Compounds

In almost all cases of crude oil analysis (especially if the feedstock has been spilled into the environment and has been subject to biodegradation), some attention will have to be paid to the presence of semi-volatile constituents and nonvolatile constituents.

As a general definition for the purposes of this text, semi-volatile organic compounds (SVOCs) are a subgroup of volatile organic compounds (VOCs) that have boiling points on the order of 260–400 °C (500–750 °F) which is the general range for the gas oil fractions. On the other hand, the nonvolatile organic compounds (NVOCs) are, as is evident from the name, are nonvolatile (initial boiling point 400 °C 750 °F) under the general handling conditions.

Thus, the collection, handling of samples, and their ultimate preparation for analysis is entirely different from that used for VOCs. In general, it is not necessary to take such rigorous procedures to prevent loss of any of the sample following collection, although the procedures should still be verified using appropriate quality control (QC) measures (Dean, 1998, 2003; Stratiev et al., 2010; Speight, 2015).

Before analysis of semi-volatile constituents or nonvolatile constituents can proceed, it may be necessary for these constituents be brought into solution. In a sample from a contaminated site, semi- and nonvolatile molecules may exist in the soil pores in the free form within the pore spaces, but are far more likely to be adsorbed by organic matter attached to the soil. Indeed, the probability of such adsorption increases with increasing hydrophobicity of the molecules.

For most analyses, it is necessary to separate the analytes of interest from the matrix (i.e., soil, sediment, and water). Extraction of analytes can be performed using one or more of the following methods: (1) extracting the analytes into a solvent, (2) heating the sample, as may be necessary to remove the solvent and for the analysis of volatile compounds, and (3) purging the sample with an inert gas, as is also used in the analyses of volatile compounds.

In the environmental arena, procedures such as (1) Soxhlet extraction using the Soxhlet apparatus, (2) sonication, (3) supercritical fluid extraction, (4) subcritical fluid extraction, (5) accelerated solvent extraction, and (6) purge and trap extraction have been introduced into a variety of methods for the extraction of organic materials from contaminated soil.

Also, shaking/vortexing is adequate for the extraction of hydrocarbon derivatives in most environmental samples. For these extraction methods, the ability to extract the hydrocarbon derivatives from soil and water samples depends on the solvent and the sample matrix. Surrogate compounds (i.e., compounds of known identity and quantity) are frequently added to monitor extraction efficiency. Environmental laboratories also generally perform matrix spikes (i.e., the addition of target analytes) as control compounds to determine if the soil mix or the water matrix retains or discharges the analytes.

Thus, solvents have different extraction efficiencies and, thus, extracting the same sample in the same manner by two different solvents may result in different concentrations. The choice of solvents is determined by many factors such as cost, spectral qualities, method regulations, extraction efficiency, toxicity, and availability. Methylene chloride has been the solvent of choice for many semi-volatile analyses due to its high extraction efficiency. Chlorofluorocarbon solvents such as trichlorotrifluoroethane (Freon 113) have been used in the past for oil and grease analyses because of their spectral qualities (they do not absorb in the 2,930 cm^{-1} infrared measurement wavelength) and low human toxicity. The use of chlorofluorocarbons is to be questioned because not all of the crude oil constituents are soluble in such solvents. Furthermore, the use of chlorofluorocarbon solvents is being reduced, even phased out of analytical methods, because of their detrimental effects on stratospheric ozone. Tetrachloroethylene and carbon tetrachloride are possible replacements but caution is advised since these solvents may be sensitive to light and leave residual chlorine in the sample. Methanol is the most common solvent used to preserve and extract volatiles such as benzene, toluene, ethylbenzene, and xylene(s) in soil. But methanol is unsuitable for many constituents of crude oil and crude oil products. In short, there is no one solvents that will satisfy all of the criteria necessary of a complete and full extraction or solubility of crude oil and its products.

If the release has contaminated a water system, there are several methods that can be employed for sample separation (Patnaik, 2004). Volatile compounds (gasoline, solvents) in water are generally separated from the aqueous matrix by purging with an inert gas and trapping the compounds on a sorbent. The sorbent is later heated to release the volatile compounds, and a carrier gas sweeps the compounds into a gas chromatograph. Headspace analysis is recommended as a screening method, although it performs well in particular situations, especially field analysis. In this method, the water sample is placed in a closed vessel with a headspace and heated to drive volatiles into the gas phase and instrument contamination is minimized because only volatile compounds are introduced into the instrument. Addition of salts or acids may enhance this process.

Briefly, headspace analysis is a technique for sampling and examining any volatile constituents associated with a liquid sample or a solid sample. The actual headspace itself is the volume of vapor or gas above the sample and for most headspace analysis purposes, the sample and its associated headspace are held within an enclosed container.

Samples containing viscous materials, along with the volatile constituents, can severely contaminate purge and trap instrumentation and caution is advised when interpreting the data. For such sample, it may be advisable to use a separatory funnel for the water extraction method for semi-volatiles involves extraction using the method in which the sample is poured into a funnel-shaped piece of glassware, solvent is added, and the mixture is shaken vigorously. After layer separation, the extract (i.e., the solvent layer) is removed, filtered, dried with a desiccant, and concentrated.

Another commonly used water extraction method for semi-volatile materials involves continuous liquid–liquid extraction. In this method, the sample (rather than being shaken with the solvent) is treated with a continuously heated solvent that is nebulized (broken into small droplets) and sprayed on top of the water. Liquid–liquid extraction is excellent for samples containing emulsion-forming solids, but it is more time-consuming than separatory funnel extractions. Nevertheless, time-consuming or not, the method must produce reliable data that can be used without question for monitoring or regulatory purposes.

Solid phase extraction also can be used for extraction and concentration of semi-volatile material. The technique involves passing the water sample through a cartridge or disk containing an adsorbent such as silica or alumina. The adsorbent is often coated with compounds that impart selectivity for particular products or analytes such as polynuclear aromatic hydrocarbon derivatives (known by the acronyms PAHs or PNAs). After extraction, the analytes are separated from the solid phase by elution with a small amount of organic solvent. A variant of solid phase extraction involves dipping a sorbent-coated fiber into the water (solid phase micro-extraction). Adsorbed analytes are thermally desorbed directly into a heated chromatographic injection port. Generally, the solid phase extraction method requires much less solvent and glassware than separatory funnel and liquid–liquid extraction.

For the separation of samples from contaminated soil, there are also several possible methods depending upon whether the contaminant is volatile or semi-volatile. Volatile compounds (such as BTEX, benzene, toluene, ethyl benzene, xylene, and gasoline) may be extracted from soil using, for example, methanol. In the method, the extraction is usually accomplished by mechanically shaking the soil with methanol. A portion of the methanol extract is added to a purge vessel and diluted in reagent grade water. The extract is then purged similar to a water sample.

The application of headspace analysis is suitable for the analysis of the constituents of a feedstock that have entered the soil. In the test method, the soil is placed in a headspace vial and heated to drive out the volatiles from the sample into the area above (the headspace) the sample in a container. Similar to water headspace analysis, the soil headspace technique is useful when the viscous feedstocks and high analyte concentrations of the analyte(s) are present which can severely contaminate the instrumentation used of purge and trap analysis. Moreover, the detection limits are generally higher for headspace analysis than for purge and trap analysis.

The simplest method to separate semi-volatile compounds from soil is to shake or vortex (vigorous mechanical stirring) the soil with a solvent. Adding a desiccant to the soil/solvent mixture can help to break up soil and increase the surface area. The word of caution here is to ensure that the drying agent does not adsorb the solute. Assuming that no such adsorption occurs, the extract can be analyzed directly. Simple shaking is quick and easy, making it an excellent field extraction technique. However, extraction efficiency will vary depending on soil type and whether or not clay minerals (excellent adsorbents for many organic compounds) are present.

Soxhlet extraction is a very efficient extraction process that is commonly used for semi-volatile crude oil constituents. In the method, the solvent is heated and refluxed (recirculated continuously) through the soil sample for 16 hr. This method generates a relatively large volume of extract that needs to be concentrated. Thus, it is more appropriate for semi-volatile constituents than for volatile constituents. Sonication extraction (which is a faster technique than Soxhlet extraction and typically requires less solvent) can also be used for semi-volatile compounds and, as the name suggests, involves the use of sound waves to enhance the transfer of soluble constituents from the sample to the solvent.

Supercritical fluid extraction is applicable to the extraction of semi-volatile constituents. Supercritical fluid extraction involves heating and pressuring a mobile phase to supercritical conditions (where the solvent has the properties of a gas and a liquid). The supercritical fluid is passed through the soil sample, and the analytes are concentrated on a sorbent or trapped cryogenically. The analytes are eluted with a solvent and analyzed with conventional techniques. Carbon dioxide is the most popular mobile phase.

In some cases, when crude oil and/or crude oil products are released to the environment, a free phase is formed and sample(s) of the hydrocarbon material can be collected directly for characterization. The ability to analyze free product greatly aids the determination of product type and potential source. The samples may be diluted prior to analysis but caution is advised since, as part of the initial sample collection procedure, water and sediment may be inadvertently included in the sample. There are several protocols involved in initial isolation and cleaning of the sample that must be recognized. In fact, considerable importance attaches to the presence of water or sediment in crude oil (ASTM D1796, 2021; ASTM D4007, 2021) for they lead to difficulties in other analyses.

Sediment usually consists of finely divided solids that may be dispersed in the feedstock or carried in water droplets. The solids may be acquired by the sample during refining and transportation and can consist of inorganic material (especially chloride derivatives that arise from the evaporation of brine droplets in the oil). In any event, the sediment can lead to serious plugging of the equipment, corrosion due to chloride decomposition that can result in a lowering of the gas oil or residuum (Speight, 2005, 2014, 2017).

Water may be found in the crude either in an emulsified form or in large droplets. The quantity is generally limited by pipeline companies and by refiners, and steps are normally taken at the wellhead to reduce the water content as low as possible. However, after a spill, water can be introduced by climatic conditions and the relevant tests (ASTM D1796, 2021) are regarded as important in crude oil analyses. Prior to analyses, it is often necessary to separate the water from a crude oil sample and this is usually carried out by one of the procedures described in the preliminary distillation of crude oil.

2.2.1.2 Solids

While the viscous feedstocks may be considered to be semi-solid (i.e., tar-like, which is a misnomer), the occurrence of any of these feedstocks as a solid may require application of techniques for sampling solid samples include (1) accelerated solvent extraction, (2) supercritical-fluid extraction, (3) microwave-assisted extraction, or (4) thermal extraction. In solid–liquid extraction, the sample is present in a closed container and a solvent that is able to dissolve the analyte(s) is added. After evaporation of the solvent, the sample can be prepared for analysis. However, there is also the potential for the sample to be acquired from a site where the feedstock has been spilled into the environment. Such an occurrence requires a change in the sampling method before the sample can be submitted for analysis with a high degree of certainty that the sample is truly representative of the bulk material.

For homogeneous materials, sampling protocols are relatively simple and straightforward, although caution is always advised lest overconfidence causes errors in the method of sampling as well as introduces extraneous material. On the other hand, the heterogeneous nature of soil and contaminated soil complicate the sampling procedures. If the soils and the sample are visibly heterogeneous, there is a very strong emphasis on the need to obtain representative samples for testing and analysis.

Thus, the variable composition of contaminated soil as well as solid sample, such as petroleum coke, offers many challenges to environmental analysts who must ensure that the sample under investigation is representative of the contaminated site or the coke. Furthermore, sample transportation can initiate (due to movement of the sample) processes that result in size and density segregation, in a manner analogous to variations in coal quality from sample-to-sample (ASTM D346, 2021; ASTM D2234, 2021; ASTM D6543, 2021; ASTM D6883, 2021; ASTM D7430, 2021).

Therefore, the challenge in sampling solids for environmental analysis is to collect a relatively small portion of the sample that accurately represents the composition of the whole. This requires that sample increments must be collected in such a manner that no piece, regardless of position (or size) relative to the sampling position and implement, is selectively collected or rejected. Optimization of solids sampling is a function of the many variable constituents of coal and is reflected in the methods by which an unbiased sample can be obtained, as is required by coal sampling.

Thus, in order to evaluate any particular environmental solid sample, there are two criteria that must be followed and these are: (1) obtain a sample of the solid and (2) ensure that the sample is a true representative of the bulk material, and (last but by no means least) (3) to ensure that the sample does not undergo any chemical or physical changes after completion of the sampling procedure and during storage prior to analysis. In short, the reliability of a sampling method is the degree of perfection with which the identical composition and properties of the entire whole coal are obtained in the sample. The reliability of the storage procedure is the degree to which the coal sample remains unchanged thereby guaranteeing the accuracy and usefulness of the analytical data. At this point, a review of the sampling methods applied to coal, and which, under favorable conditions can be applied to environmental solids, is worth of inclusion.

The sampling procedures (ASTM D346, 2021; ASTM D2234, 2021; ASTM D6543, 2021; ASTM D6883, 2021; ASTM D7430, 2021) are designed to give a precision such that if gross samples are taken repeatedly from a lot or consignment and prepared according to the standard test methods and one ash determination is made on the analysis sample from each gross sample, the majority (usually specified as 95 out of 100) of these determinations will fall within ±10% of the average of all of the determinations. When other precision limits are required or when other constituents are used to specify precision, defined special-purpose sampling procedures may need to be employed.

Thus, when a property of the sample (which exists as a large volume of material) is to be measured, there may be differences (depending upon the reliability or consistency of the sampling method) between the analytical data derived from application of the test methods to a "gross lot" or "gross consignment" and the data from the "sample lot." This difference (the "sampling error") has a frequency distribution with a mean value and a variance. "Variance" is a statistical term defined as the mean square of errors; the square root of the variance is more generally known as the "standard deviation" or the "standard error of sampling."

Recognition of the issues involved in obtaining representative samples of coal and minimization of the sampling error has resulted in the designation of methods that dictate the correct manner for coal sampling (ASTM D346, 2021; ASTM D2234, 2021; ASTM D6543, 2021; ASTM D6883, 2021; ASTM D7430, 2021).

Finally, and this applied to all samples where separation of the solute or released material from the matrix is necessary, it must be recognized that the organic compounds, for example, in soil or in a liquid matrix (water or an organic solvent), have an affinity for the sample matrix to one degree or another. It is this affinity that allows the sample to be retained by the matrix. The affinity may be due to adsorption on the surface of sample particles or solvation in water or other solvent medium.

Therefore, it is necessary in an extraction process to recognize the potential solute–matrix interactions in order to overcome such interactions. A suitable choice of solvent is necessary, making sure that the solvent itself does not react with the solid or with the matrix. If the first extraction solvent (e.g., pyridine or carbon disulfide) of choice is too powerful, switching to a solvent of lesser ability, such as pentane, hexane, or crude oil ether, remembering that if it is a high-boiling crude oil sample that is being extracted, these solvents of lesser ability will leave some of the sample unextracted. In other cases, particularly when the sample is a hydrocarbon material, a polar solvent (e.g., methanol) can be used to extract polar organic materials like the phenols from the matrix.

Although widely used, solvent extraction procedures have been demonstrated as sensitive to such variables as (1) the content of humic matter, which consists of organic compounds that are the major organic fraction of soil, and (2) the moisture within the samples. Supercritical fluid extraction appears to be a more robust procedure. Thermal extraction procedures are sensitive to the size of the soil sample and in some cases since the technique can result in cracking higher molecular weight constituents that do not volatilize out of the thermal zone. In the case of solvent extraction procedures, it is necessary to concentrate and also to clean up the samples. With complex mixtures of semi-volatile hydrocarbon derivatives, it is generally advisable to separate the aliphatic and aromatic fractions.

2.2.1.3 Extract Concentration

Extract concentration is the one area in the isolation procedure that has the greatest potential for loss of analytes. The operative physical concept during concentration is vapor pressure, not boiling point. However, boiling is uncontrolled insofar as all of the molecules in the sample are attempting to convert from the liquid state to the gaseous state. Therefore, equating sample concentration to boiling point leads to erroneous conclusions about choice of an appropriate method. On the other hand, vapor pressure is a continuous function that relates to the rate of evaporation, and evaporation is what is desired during sample concentration. Thus, the proper technique is to control the rate of solvent evaporation while minimizing analyte loss. A completely successful sample extraction can be performed that is completely negated by an inappropriate method for sample extract concentration.

In many cases, sample extracts are generally filtered, dried with desiccant, and concentrated before analysis. Concentration of the extract may allow for lower sample detection limits. Frequently, sample extracts must be concentrated to obtain detection limits low enough to meet regulatory action limits. Concentration may be achieved by:

For volatile materials	For semi-volatile materials
Sorbent trapping	Snyder column
Cryogenic trapping	Kuderna–Danish concentrator
Nitrogen evaporation	Vacuum

The trapping step in a purge and trap analysis is essentially a concentration step. Analytes are purged from the matrix into a gas stream and captured on a sorbent trap. The analytes are released by heating the trap. Cryogenic trapping is also used in place of sorbent trapping. In cryogenic trapping, a very cold material (such as liquid nitrogen) surrounds a sample loop and as analytes are purged and swept through the sample loop, they freeze in the sample loop. The analytes are released when the trap is heated.

A Snyder column (a distillation column that utilizes separate chambers to promote multiple condensations and vaporizations during the distillation process) is designed to allow highly volatile solvents to escape while retaining semi-volatile analytes of interest. Snyder columns are generally fitted onto the tops of flasks containing extracts and column design permits solvent to escape as the flask is heated. The analytes of interest condense from a gas to a liquid phase and fall back down into the solvent reservoir. The Kuderna–Danish concentrator is a Snyder column with a removable collection tube attached to the bottom. As solvent is evaporated, the extract is collected in the collection tube.

As an alternative to a Snyder column, the sample extracts may be concentrated with nitrogen evaporation by directing a slow stream of the gas over the extract surface at room temperature, resulting in minor loss of volatiles. Placing the extract container in warm water helps to speed the process, but then some loss of volatiles can occur. Concentration by evaporating excess solvent with a vacuum is not very common in environmental laboratories. Many semi-volatile analytes are lost in the procedure. Additionally, evaporating as a means of concentrating the sample cannot be used if the goal is to detect volatile analytes.

Cleanup steps are an important component of infrared (IR)-based methods and gravimetric methods because these methods are very sensitive to non-crude oil hydrocarbon interferences. Cleanup steps are not always a part of the crude oil analytical process, but when they are necessary, the goals of extract cleanup steps typically include one or more of the following: (1) removal of non-crude oil compounds, (2) isolation of a particular crude oil fraction, and (3) concentration of analytes of interest.

The methods employed to extract the constituents (or analytes) of interest may also include other undesirable constituents that may interfere with the efficiency of the extraction process. Polar compounds such as animal and plant fats, proteins, and small biological molecules may be improperly identified as constituents of the spilled material that may require further treatment to remove the unwanted material. In an ideal situation, only interfering compounds are removed. In reality, some polar crude oil constituents can also be removed.

Briefly, there are two main methods that can be used to clean crude oil extracts which are (1) the undesirable constituents are removed by passing the extract through a glass column filled with sorbent and (2) a method in which the extract is treated with a sorbent material which is then removed by filtration. Other methods involve trapping the interfering compounds on a sorbent column such as alumina that is designed to remove interfering compounds and to fractionate crude oil wastes into aliphatic, aromatic, and polar fractions. The fractions can be analyzed separately or combined for measurement of the total petroleum hydrocarbons. Alternatively, silica gel is commonly used for polynuclear aromatic hydrocarbon derivatives and phenol derivatives. Variations of this technique are used to clean extracts before infrared analysis. In addition, the gel permeation technique has been used for cleanup and works on the principle of size exclusion. Large macromolecules such as lipids, polymers, and proteins are removed from the sample extract. Extracts obtained from soil that have (or have had) high biological activity may be cleaned by this method.

There are two non-column cleanup methods, one of which uses acid partition to separate the base/neutral constituents and the acid constituents by adjusting pH. This method is often used before alumina column cleanup to remove the acid constituents. The other method is used for sulfur removal and uses copper, mercury, and tetrabutylammonium sulfite as desulfurization compounds. Sulfur is a common interfering compound for sample analysis, particularly for samples taken from sediments.

Even though cleanup procedures are advocated before sample analysis, there can be several limitations to various cleanup steps. The reasons for decreased effectiveness of cleanup procedures include (1) sample loading may exceed the capacity of cleanup columns, (2) non-crude oil compounds may have chemical structures similar to crude oil constituents and may behave like a crude oil constituent, (3) analytes of interest may be removed during the cleanup, and (4) no single cleanup technique removes all of the chemical interferences.

2.2.1.4 Sample Cleanup

The use of any sample clean up in environmental analysis is always accompanied by the possibility of analyte loss. Procedures that depend upon polarity interactions between the eluting solvent, the solid adsorbent, and the target analytes to achieve selective isolation, are particularly prone to having the desired compounds ending up in the wrong fraction. Sources of these errors include mistakes in the preparation of the eluting solvent, use of the wrong or a deactivated absorbent, the presence of traces of polar solvents in the sample solution, and the structure of the analyte molecule that will determine the behavior of the analytes during separation (Speight, 2014, 2015). Attention to detail and procedure are required for successful use of sample cleanup techniques and the aspects of the procedure that need to be examined include (1) suitability of the materials to achieve the cleanup that will be sample dependent, (2) introduction of laboratory contamination, and (3) success of the procedure on the individual sample.

The introduction of laboratory contamination is a significant, but often overlooked concern in sample cleanup. Examples are the introduction of extractable materials from plastics used for joining tubes and contaminated solvents are only two of the potential points of laboratory contamination.

2.2.2 Representative Sample

Representative samples are always a prerequisite for the laboratory evaluation of crude oil and crude oil products (Speight, 2001) and precautions are required in (1) obtaining and (2) handling representative samples. The commercial value of any crude oil product (as well as the parent crude oil) is assessed by the characteristics of the sample as determined by standard test methods. In summary, the sample must be representative of the lot or a wrong evaluation will be made, no matter how accurately the sample may be tested. In addition, each type of crude oil or crude oil product (gas, liquid, or solid) will more than likely require application of a different sampling method to assure that the sample is truly a representative sample.

Whether or not a sample is a representative sample depends on the sampling procedure, the type of crude oil or crude oil product being sampled, and the storage tank, container, or tank carrier from which the sample is obtained. The sample container must be clean and the type to be used depends not only on the product but also on the laboratory data desired. Thus, the objective of each procedure is to obtain a sample or a composite of several samples in such manner and from such location in the tank or other container that the sample or composite will be representative of the entire bulk material. The sample container—which must be clean and free of any possible source of sample contamination and should not be made of material which either reacts with the sample or is a catalyst—should be labeled immediately by the sampler, to indicate at least (1) the nature/name of the sample, (2) time of sampling, (3) location of the sampling point, and any other information necessary for the sample identification.

Thus, the basic objective of each sampling procedure is to obtain a truly representative sample or, more often, a composite of several samples that can be considered to be a representative sample. Once the sampling procedure is accomplished, the sample container should be labeled immediately, to indicate the product, time of sampling, location of the sampling point, and any other information necessary for the sample identification. And, for further emphasis, if the samples were taken from different levels of the storage tank, the levels from which the samples were taken and the amounts taken and mixed into the composite, should be indicated on the sample documentation.

While the above text has focused on the acquisitions of samples from storage tanks, sampling records for any procedure must be complete and should include, but is not restricted to items, information such as: (1) the precise (geographic or other) location (or site or refinery or process) from which the sample was obtained, (2) the identification of the location (or site or refinery or process) by name, (3) the character of the bulk material (solid, liquid, or gas) at the time of sampling, (4) the means by which the sample was obtained, (5) the means and protocols that were used to obtain the sample, (6) the date, the amount of sample that was originally placed into storage, (7) any chemical analyses (elemental analyses, fractionation by adsorbents or by liquids, functional type analyses) that have been determined to date, (8) any physical analyses (API gravity, viscosity, and distillation profile) that have been determined to date, (9) the date of any such analyses included in items five and six, (10) the methods used for analyses that were employed in five and six, (11) the analysts who carried out the work in five and six, and (12) a log sheet showing the names of the persons (with the date and the reason for the removal of an aliquot) who removed the samples from storage and the amount of each sample (aliquot) that was removed for testing.

In summary, there must be a means to accurately to track and identify the sample history so that each sample is tracked and defined in terms of source, activity, and the personnel involved in any of the above stages. Thus, the accuracy of the data from any subsequent procedures and tests for which the sample is used will be placed beyond a *reasonable doubt* and would stand the test of time in court should legal issues arise.

2.2.3 Sampling Error

Analytical test methods applied in the field—and such logic can be applied to all sampling protocols—must take into consideration the various site conditions comprising local heterogeneity, as well as the changing ambient conditions. Furthermore, the analysts must take into account the potential for samples to be disturbed during sampling, storage, and pretreatment. Rather than attempt to rationalize the application of corrections because of such disturbances, it is beer that steps to taken to avoid sample disturbance. If corrections must be applied, the manner in which the corrections are applied must be beyond reproach or claims of that data will not stand up (when needed) to legal scrutiny and falsification of the data will be the most likely result (Speight and Foote, 2011).

2.3 Measurement

Measurement is the quantification of attributes of a material which can be used to compare with other objects or events. Put simply, measurement is the process of determining how large or small a physical quantity is as compared to a basic reference quantity of the same kind.

The issues related to the viscous feedstocks that face analysts include the need to provide higher-quality results. In addition, environmental regulations may influence the method of choice. Nevertheless, the method of choice still depends to a large extent on the boiling range (or carbon number) of the sample to be analyzed. For example, there is a large variation in the carbon number range and boiling points (of normal paraffins) for some of the more common crude oil products and thus a variation in the methods that may be applied to these products (Speight, 2001, 2015). In some cases, the predominant methods of measuring the properties of crude oil and crude oil products (ASTM D287, 2021; ASTM D1250, 2021; ASTM D1298, 2021) may be modified for application to the viscous feedstocks.

Testing for suspended water and sediment is used primarily with fuel oils, where appreciable amounts of water and sediment may cause fouling of facilities for handling the oil and give trouble in burner mechanisms. Three standard methods are available for this determination. The centrifuge method gives the total water and sediment content of the sample by volume, the distillation method gives the water only, volumetrically, and the extraction method gives the solid sediment in per cent by weight.

The determination of density of specific gravity (ASTM D287, 2021; ASTM D1298, 2021) in the measurement and calculation of volume of crude oil products is important since gravity is an index of the weight of a measured volume of the product. Two scales are in use in the crude oil industry, specific gravity and API gravity, the determination being made in each case by means of a hydrometer of constant weight displacing a variable volume of oil. The reading obtained depends upon both the gravity and the temperature of the oil.

There are also procedures for determining the temperatures of crude oil and crude oil products when in a liquid state. Temperatures are determined at specified locations in tanks, ships and barges, tank cars, and tank trucks. For a non-pressure tank, a temperature is obtained by lowering a tank thermometer of proper range through the gaging hatch to the specified liquid level. After the entire thermometer assembly has had time to attain the temperature of the product, the thermometer is withdrawn and read quickly. This procedure is also used for low-pressure tanks equipped with gaging hatches or standpipes, and for any pressure tank that has a pressure lock. For tanks equipped with thermometer wells, temperatures are obtained by reading thermometers placed in the wells with their bulbs at the desired tank levels. If more than one temperature is determined, the average temperature of the product is calculated from the observed temperatures. Electrical-resistance thermometers are sometimes used to determine both average and spot temperatures.

As part of the analytical procedure for crude oil and crude oil products, refiners also need to know not only the bulk properties of the material but also the amount of materials held in storage tanks—as part of, say, the refinery inventory. Once the crude oil or crude oil product is sampled, there is the need to calculate the total volume of the material in storage. Following from this, the amount of crude oil or crude oil product involved in commercial transactions can be calculated with a high degree of accuracy.

The complex nature of crude oil and crude oil products requires several procedures for calculating volume or density of liquid products in commercial transactions. In addition, temperature affects the volume and density of crude oil and its products. Measured volumes must be converted to net quantities at 15.6 °C (60 °F), which is the accepted base temperature used widely in commercial transactions. Again, obtaining a representative sample of the material that is held in storage or carrier tanks is essential for converting measured quantities to the standard volume.

To assist with volume calculations, crude oil measurement tables are available for use in the calculation of quantities of crude oil and crude oil products as they exist under designated conditions (ASTM D1250, 2021). The tables provide data that cover typical operating ranges for the reduction of gravity and volume to standard states, for calculation of weight–volume relationships, and interconversion of a wide variety of commercially useful units.

Following from this, the issues that always face analytical laboratory personnel include the need to provide higher-quality results. In addition, environmental regulations may influence the method of choice. Nevertheless, the method of choice still depends to a large extent on the boiling range (or carbon number) of the sample to be analyzed. For example, there is a large variation in the carbon number range and boiling points (of normal paraffins) for some of the more common crude oil products and thus a variation in the methods that may be applied to these products (Speight, 2001).

Testing for suspended water and sediment is used primarily with fuel oils, where appreciable amounts of water and sediment may cause fouling of facilities for handling the oil and give trouble in burner mechanisms. Three standard methods are available for this determination: (1) the centrifuge method, which gives the total water and sediment content of the sample by volume, (2) the distillation method, which gives the volumetric amount of water, and (3) the extraction method, which gives the solid sediment in per cent by weight.

The term "sediment" is typically used as an indicator of the presence of finely divided solids that may be dispersed in the oil or carried in water droplets. The solids may be drilling mud or sand or scale picked up during the transport of the oil, or may consist of chlorides derived from evaporation of brine droplets in the oil. In any event, the presence of sediment can lead to (1) plugging of the equipment, (2) corrosion due to chloride decomposition, and (3) lowering of quality of the residuum.

Water may be found in the crude oil or the viscous feedstock either in an emulsified form or in large droplets. The quantity is generally limited by pipeline companies and by refiners, and steps are normally taken at the wellhead to reduce the water content as low as possible. Prior to analyses, it is often necessary to separate the water from a crude oil sample and this is usually carried out by one of the procedures described in the distillation of crude oil (ASTM D4006, 2021) or by the coulometric Karl Fischer titration method (ASTM D4928, 2021).

Gaging crude oil products involves the use of procedures for determining the liquid contents of tanks, ships and barges, tank cars, and tank trucks. Depth of liquid is determined by gaging through specified hatches, or by reading gage glasses or other devices. There are two basic types of gages, innage and outage. The procedures used depend upon the type of tank to be gaged, its equipment, and the gaging apparatus.

An innage gage is the depth of liquid in a tank measured from the surface of the liquid to the tank bottom or to a datum plate attached to the shell or bottom. The innage gage is used directly with the tank calibration table and temperature of the product to calculate the volume of product (ASTM D1250, 2021). On the other hand, an outage gage is the distance between the surface of the product in the tank and the reference point above the surface, which is usually located in the gaging hatch. The outage gage is used either directly or indirectly with the tank calibration table and the temperature of the product to calculate the volume of product. The amount of any free water and sediment in the bottom of the tank is also gaged so that corrections can be made when calculating the net volume of the crude oil or crude oil product.

In general, the volume received or delivered is calculated from the observed gage readings. Corrections are made for any *free* water and sediment as determined by the gage of the water level in the tank. The resultant volume is then corrected to the equivalent volume at 15.6 °C (60 °F) by use of the observed average temperature and the appropriate volume correction table (ASTM D1250, 2021). When necessary, a further correction is made for any suspended water and sediment that may be present in materials such as crude oil and the more viscous feedstocks.

2.4 Method Validation

Method validation is the process by which an analytical method is demonstrated to be acceptable for its intended purpose. In general, methods for product specifications and regulatory submission must include studies on specificity, linearity, detection limit, and quantitation limit as well as accuracy, precision.

The process of method development and validation covers all aspects of the analytical procedure and the best way to minimize method problems is to perform validation experiments during development. However, method validation cannot mitigate out all of the potential problems, the process of method development and validation should address the most common (anticipated) problems.

In general, methods for product specifications and regulatory submission must include studies on specificity, linearity, accuracy, precision, range, detection limit, and quantitation limit. However, in order to perform validation studies, the approach should be viewed with the understanding that validation requirements are continually changing and vary widely, depending on the type of product under test and compliance with any necessary regulatory group.

The first step in the method development and validation cycle should be to set minimum requirements, which are essentially acceptance specifications for the method. A complete list of criteria should be agreed upon during method development and the end users before the method is developed so that expectations are clear. Once the validation studies are complete, the method developers should be confident in the ability of the method to provide good quantitation in their own laboratories. The remaining studies should provide greater assurance that the method will work well in other laboratories, where different operators, instruments, and reagents are involved and where it will be used over much longer periods of time.

The remaining precision studies comprise much of what is often referred to as "ruggedness." In addition, "intermediate precision" is the precision obtained when an assay is performed by multiple analysts using several instruments on different days in one laboratory. Intermediate precision results are used to identify which of the above factors contribute significant variability to the final result.

The last type of precision study is "reproducibility" which is determined by testing homogeneous samples in multiple laboratories, often as part of interlaboratory crossover studies. Statistical equivalence is often used as a measure of acceptable interlaboratory results. As an alternative, more practical approach is the use of "analytical equivalence" in which a range of acceptable results is chosen prior to the study and used to judge the acceptability of the results obtained from the different laboratories.

Performing a thorough method validation can be a tedious process, but the reliability of the data generated with the method is directly linked to the application of quality assurance (QA) and QC protocols, which must be followed assiduously (Quevauviller, 2002).

Briefly, to assure QA and QC, samples are analyzed using standard analytical procedures. A continuing program of analytical laboratory QC verifies data quality and involves participation in interlaboratory crosschecks, and replicate sampling and analysis. When applicable, it is advisable, even insisted by the Environmental Protection Agency, that analytical labs must be certified to complete the analysis requested.

However, in many cases, time constraints often do not allow for sufficient method validation. Many researchers have experienced the consequences of invalid methods and realized that the amount of time and resources required to solve problems discovered later exceeds what would have been expended initially if the validation studies had been performed properly. One method that is often used for method validation is the addition of a surrogate to the sample matrix and following the track of the surrogate through the separation procedure. It is appropriate that at this point, comments should be made on such procedures.

However, it is the nature of the sampling protocols, and the location of the site from which the sample are taken, that it should be assumed that no two samples are identical. Factors such as (1) depth from which the sample is taken, (2) distance of the sampling point from the initial location of the spill, and (3) ambient conditions play a role in determining sample character. In fact, each sample is a unique combination of matrix–analyte interactions. Adding surrogate compounds to each sample and then determining the recovery of the surrogates is used as the benchmark for gaging the success of the extraction procedure. The best surrogates are those that are most likely the target analytes and example are the isotopically labeled versions of each of the target molecules. Whether or not use of such molecules is justified relates to the data that are required.

As an alternative, other similar chemical compounds can be used but the choice of the surrogate compounds can limit or it can maximize data interpretation. For most analyses, it is possible to choose surrogates that will always generate excellent recoveries, regardless of the complexity of the sample. Choosing surrogate molecules to produce "acceptable quality control" is not realistic. The use of surrogates is to obtain reliable information about the overall strength and weaknesses of the analytical method. In addition, the surrogate(s) should reflect the chemical behavior and physical properties of the analytes. Important chemical behavior and properties include (1) acidic and basic properties of the analytes, (2) the range of polarity of the analytes, (3) reactivity in chemical derivatization procedures, and (4) the sensitivity of the analytes to decomposition caused by extremes of chemical or physical environment.

2.4.1 Requirements

In order to perform validation studies, the approach should be viewed with the understanding that validation requirements are continually changing and vary widely, depending on the type of product under test and compliance with any necessary regulatory group. Furthermore, issues arising from the use of analytical methods increase as additional (new) people unfamiliar with the method, different laboratories, and equipment are used to perform the analytical investigation. For example, a small adjustment (no adjustment is really *small*) can usually be made to make the method work but the flexibility to change or reverse the adjustment on an as-needed basis, it is lost once the method is transferred to other analysts or to other analytical laboratories.

Performing a thorough method validation is a necessary part of any sampling and measurement program because the quality of data generated with the method is directly linked to the quality of this process.

Once the validation studies are complete, there should be a high degree of confidence in the ability of the method to provide good quantitation. Any remaining studies should provide greater assurance that the method will work well in other laboratories, where different operators, instruments, and reagents are involved and where it will be used over much longer periods of time.

2.4.2 Method Detection Limit

The method detection limit (MDL) is the smallest quantity or concentration of a substance that the instrument can measure (Patnaik, 2004). The MDL is related to the instrument detection limit (IDL) that depends on the type of instrument and its sensitivity, and on the physical and chemical properties of the test substance.

The MDL is, in reality, a statistical concept that is applicable only in trace analysis of certain type of substances, such as organic pollutants by gas chromatographic methods. The MDL measures the minimum detection limit of the method and involves all of the analytical steps including (1) sample extraction, (2) sample concentration, and (3) the determination of the result by an analytical instrument.

In the environmental analysis of organic materials, the MDL is the minimum concentration of a substance that can be measured and reported with a 99% confidence level that the analyte concentration is greater than zero and is determined from the analysis of a sample in a given matrix containing the analyte. For determination of the MDL, several replicate analyses are performed at the concentration level of the instrument detection limit or at a level equivalent to two to five times the background noise level. The standard deviation of the replicate tests is found. The MDL is determined by multiplying the standard derivation by the "*t*-factor."

In environmental analysis, however, periodic determination of the MDL (e.g., once per year or with any change in personnel, location, or instrument) is part of the QC requirement.

2.4.3 Accuracy

The "accuracy" of a test is a measure of how close the test result will be to the true value of the property being measured. As such the accuracy can be expressed as the *bias* between the test result and the true value. However, the *absolute accuracy* can only be established if the true value is known.

In the simplest sense, a convenient method to determine a relationship between two measured properties is to compare (often, for example, by using an *x*–*y* plot) one to the other such as (1) the relationship of the API gravity to the sulfur content, (2) the relationship of the API gravity to the carbon residue, (3) the relationship of the API gravity to the asphaltene content, (4) the relationship of the API gravity to the asphaltene content plus the resins content, and (5) the relationship of the API gravity to the viscosity (Speight, 2014, 2015). Such an exercise will provide either a line fit of the points or a spread that may or may not be within the limits of experimental error. The data can then be used to determine the approximate accuracy of one or more points employed in the plot. For example, a point that lies outside the limits of experimental error (typically referred to as a "flyer") will indicate an issue of accuracy with that test and the need for a repeat determination.

However, the graphical approach is not appropriate for finding the absolute accuracy between more than two properties. The well-established statistical technique of regression analysis is more pertinent to determining the accuracy of points derived from one property and any number of other properties. There are many instances in which relationships of this sort enable properties to be predicted from other measured properties with as good precision as they can be measured by a single test. It would be possible to examine in this way the relationships between all the specified properties of a product and to establish certain key properties from which the remainder could be predicted, but this would be a tedious task.

An alternative approach to that of picking out the essential tests in a specification using regression analysis is to take a look at the specification as a whole, and extract the essential features (termed principal components analysis, PCA).

The principal components analysis involves an examination of set of data as points in n-dimensional space (corresponding to n original tests) and determines (first) the direction that accounts for the biggest variability in the data ("first principal component"). The process is repeated until n principal components are evaluated, but not all of these are of practical importance since some may be attributable purely to experimental error. The number of significant principal components shows the number of independent properties being measured by the tests considered.

Following from this, it is necessary to establish the number of independent properties that are necessary to predict product performance in service with the goals of rendering any specification more meaningful and allowing a high degree of predictability of product behavior. On a long-term approach, it might be possible to obtain new tests of a fundamental nature to replace, or certainly to supplement, existing tests. In the short term, selecting the best of the existing tests to define product quality is the most beneficial route to predictability.

2.4.4 Precision

The precision of a test method is the variability between test results obtained on the same material, using a specific test method. The precision of a test is usually unrelated to its accuracy. The results may be precise but not necessarily accurate and in order to be meaningful, the precision study must be performed using the exact sample and standard preparation procedures that will be used in the final method. Precision is expressed as repeatability and reproducibility.

The terms "intra-laboratory precision" or the "within-laboratory precision" refer to the precision of a test method when the results are obtained by the same operator in the same laboratory using the same apparatus. In some cases, the precision is applied to data gathered by a different operator in the same laboratory using the same apparatus. Thus, intra-laboratory precision has an expanded meaning insofar as it can be applied to laboratory precision.

Repeatability or repeatability interval of a test (r) is the maximum permissible difference due to test error between two results obtained on the same material in the same laboratory.

$$r = 2.77 \times \text{standard deviation of test}$$

The repeatability interval (r) is, statistically, the 95% probability level or the differences between two test results are unlikely to exceed this repeatability interval more than five times in a hundred.

The *interlaboratory precision* or the *between-laboratory precision* is defined in terms of the variability between test results obtained on the aliquots of the same homogeneous material in different laboratories using the same test method.

The term *reproducibility* or *reproducibility interval* (R) is analogous to the term repeatability but it is the maximum permissible difference between two results obtained on the same material but now in different laboratories. Therefore, differences between two or more laboratories should not exceed the reproducibility interval more than five times in a hundred.

$$R = 2.77 \times \text{standard deviation of test}$$

The repeatability value and the reproducibility value have important implications for quality (Ralli et al., 2003). As the demand for clear product specifications, and hence control over product consistency grows, it is meaningless to establish product specifications that are more restrictive than the reproducibility/repeatability values of the specification test methods.

2.5 Quality Control and Quality Assurance

QC and QA programs are key parts of all analytical protocols in all areas of analysis, including environmental, pharmaceutical, and forensic testing, among others (Patnaik, 2004). These programs mandate that the laboratories follow a set of well-defined guidelines to achieve valid analytical results to a high degree of reliability and accuracy within an acceptable range. Although such programs may vary depending on the regulatory authority, certain key features of these programs are more or less the same.

However, there is often confusion between the terms *QA* and *QC*, perhaps because there is also considerable overlap between certain aspects of QA and QC programs.

2.5.1 Quality Control

The QC method involves the application of single procedures that are performed in conjunction with the analysis to help assess in a quantitative manner the success of the individual analysis. Examples of the QC method are blanks, calibration, calibration verification, surrogate additions, matrix spikes, laboratory control samples, performance evaluation samples, and determination of detection limits. Moreover, the success of the QC is evaluated against an acceptance limit. The actual generation of the acceptance limit is a function of QA; it would not be termed a QC.

Unlike QA plans that mostly address regulatory requirements involving comprehensive documentation, QC programs are science-based, the components of which may be defined statistically. The two most important components of QC are (1) determination of precision of analysis and (2) determination of accuracy of measurement.

Whereas *precision* is a measure of the reproducibility of data from replicate analyses, the *accuracy* (Section 4.3) of a test estimates how accurate are the data, that is, how close the data would fall to probable true values or how accurate is the analytical procedure to give results that may be close to true values. Both the precision and accuracy are measured on one or more samples selected at random from a given batch of samples for analysis. The precision of analysis is usually determined by running duplicate or replicate tests on one of the samples in a given batch of samples. It is expressed statistically as standard deviation, relative standard deviation (RSD), coefficient of variance (CV), standard error of the mean (M), and the relative percent difference (RPD).

The standard deviation in measurements, however, can vary with the concentrations of the analytes. On the other hand, the RSD, which is expressed as the ratio of standard deviation to the arithmetic mean of replicate analyses and is given as a percent, does not have this problem and is a more rational way of expressing precision:

$$\text{RSD} = -(\text{standard deviation}/\text{arithmetic mean of replicate analysis}) \times 100\%$$

The standard error of the mean, *M*, is the ratio of the standard deviation and the square root of the number of measurements (*n*):

$$M = \text{Standard deviation}/n$$

This scale, too, will vary in the same proportion as standard deviation with the size of the analyte in the sample.

In routine testing, many repeat analyses of a sample aliquot may not be possible. Alternatively, therefore, the precision of a test may be determined from duplicate analyses of sample aliquots and expressed as the relative percent difference:

$$\text{RPD} = (a_1 - a_2)/\left[(a_1 + a_2)/2\right] \times 100\%$$

In another iteration,

$$\text{RPD} = (a_2 - a_1)/\left[(a_1 + a_2)/2\right] \times 100\%$$

In this equation, a_1 and a_2 are the results of duplicate analyses of a sample. Since only two tests are performed on a selected sample, the RPD may not be as accurate a measure of precision as the RSD and since the RSD does not vary with sample size, it should be used whenever possible to estimate precision of analysis from replicate tests.

The accuracy of an analysis can be determined by several procedures. One common method is to analyze a "known sample" such as a standard solution or a QC check standard solution that may be commercially available or a laboratory-prepared standard solution. Such samples must be subjected to all analytical steps, including sample extraction, digestion, or concentration, similar to regular samples.

Alternatively, the accuracy of an analytical method may be estimated from the recovery of a known standard solution "spiked" or added into the sample in which a known amount of the same substance that is to be tested is added to an aliquot of the sample, usually as a solution, prior to the application of the analytical method. The concentration of the analyte in the spiked solution of the sample is then measured. The percent spike recovery is then calculated. A correction for the bias in the analytical procedure can then be made, based on the percent spike recovery. However, in most routine analysis such bias correction is not required. Percent spike recovery then may be calculated as follows:

$$\text{Recovery, \%} = (\text{measured concentration})/(\text{theoretical concentration}) \times 100\%$$

The percent spike recovery to measure the accuracy of analysis may also be determined by the U.S. Environmental Protection Agency (EPA) method often used in environmental analysis (EPA, 1998, 2004):

$$\text{Recovery, \%} = \left[100\left(X_\text{s} - X_\text{u}\right)\right]/K$$

X_s is the measured value for the spiked sample and X_u is the measured value for the unspiked sample adjusted for the dilution of the spike and K is the known value of spike in the sample.

2.5.2 Quality Assurance

Quality assurance is an umbrella term that is correctly applied to everything that the laboratory does to assure product reliability. As the product of a laboratory is information, anything that is done to improve the reliability of the generated information falls under QA.

Quality assurance includes all the QCs, the generation of expectations (acceptance limits) from the QCs, plus a great number of other activities such as (1) analyst training and certification, (2) data review and evaluation, (3) preparation of final reports of analysis, (4) information given to clients about tests that are needed to fulfill regulatory requirements, (5) use of the appropriate tests in the laboratory, (6) obtaining and maintaining laboratory certifications/accreditations, (7) conducting internal and external audits, (8) preparing responses to the audit results, (9) the receipt, storage, and tracking of samples, and (10) tracking the acquisition of standards and reagents.

Thus, the objective of QA (usually in the form of a *QA plan*) is to obtain reliable and accurate analytical results that may be stated with a high level of confidence (statistically), so that such results are legally defensible. The key features of any plan involve essentially documentation and record keeping. In short, QA program involves documentation of sample collection for testing, the receipt of samples in the laboratory, and their transfer to the individuals who perform the analyses. The information is recorded on chain-of-custody forms stating the dates and times along with the names and signatures of individuals who carry out these tasks. Also, other pertinent information is recorded, such as any preservatives added to the sample to prevent degradation of test analytes, the temperature at which the sample is stored, the temperature to which the sample is brought prior to its analysis, the nature of the container (which may affect the stability of the sample), and its holding time prior to testing.

In fact, the performance of QC is just one small aspect of the QA program. The functions of QA are embodied in the terms *analytically valid* and (in these days of perpetual litigation) *legally defensible.*

Analytically valid means that the target analyte has been (1) correctly identified and (2) quantified using fully calibrated tests. In addition, the sensitivity of the test (MDL) has been established. And the analysts have demonstrated that they are capable of performing the test. The accuracy and precision of the test on the particular sample must also have been determined and the possibility of false positive and false negative results has been evaluated through performance of blanks and other test-specific interference procedures.

2.6 Assay and Specifications

Crude oil exhibits wide variations in composition and properties, and these occur not only in crude oil from different fields but also in crude oils recovered through different production depths in the same well. Historically, physical properties such as boiling point, density (gravity), and viscosity have been used to describe crude oil but the needs are even more extensive as increased amounts of viscous feedstocks are introduced into refineries (Table 2.2).

Crude oil analysis involves not only determining the composition of the material under investigations but, more appropriately, determining the suitability of

Table 2.2 Recommended Analytical Inspections[a] for Crude Oil and the Viscous Feedstocks (Heavy Crude Oil, Extra Heavy Crude Oil, Tar Sand Bitumen, and Residua)

Crude oil	Viscous feedstocks
API gravity (density specific gravity)	API gravity (density specific gravity)
Carbon residue of residuum (% w/w)	Carbon residue (% w/w)
Distillation profile	N/A
All distillable fractions	N/A
Atmospheric residuum	Initial boiling point-final boiling point
Vacuum residuum	Initial boiling point
Elemental analysis	Elemental analysis
Carbon (% w/w)	Carbon (% w/w)
Hydrogen (% w/w)	Hydrogen (% w/w)
Nitrogen (% w/w)	Nitrogen (%w/w)
Sulfur (% w/w)	Sulfur (% w/w)
Nickel (ppm)	Nickel (ppm)
Vanadium (ppm)	Vanadium (ppm)
	Iron (ppm)
	Fractional composition
	Asphaltenes (% w/w)
	Resins (% w/w)
	Aromatics (% w/w)
	Saturates (% w/w)
Pour point	Pour point
Wax content	N/A
Wax appearance temperature	N/A
Viscosity (various temperatures)	Viscosity (various temperatures)

[a] Listed alphabetically rather than by order of preference.

the crude oil for refining or the product for use. In this sense, the end product of crude oil analysis or (testing) is a series of data that allow the investigator to *specify* the character and quality of the material under investigation. Thus, a series of specifications are determined for crude oil and its products.

Because of the differences in crude oil composition, the importance of the correct sampling of crude oil that contains low boiling (low density) hydrocarbon derivatives cannot be overestimated. Properties such as specific gravity, distillation profile, vapor pressure, hydrogen sulfide content, and octane number of gasoline are affected by the low boiling (low density) hydrocarbon content so that suitable cooling or pressure sampling methods have to be used and care taken during the subsequent handling of the oil in order to avoid the loss of any volatile constituents. In addition, adequate records of the circumstances and conditions during sampling have to be made. For example, sampling from oilfield separators, the temperatures and pressures of the separation plant, and the atmospheric temperature should be noted.

Hence, the production of data focuses on (1) measurement, (2) accuracy, (3) precision, and (4) method validation. All of which depend upon the sampling protocols that were used to obtain the sample. Without strict sampling protocols, variation and loss of accuracy (or precision) must be anticipated. For example, correct sampling of the product in storage or carrier tanks is important in order to obtain a representative sample for the laboratory tests which are essential in converting measured quantities to the standard volume.

Elemental analyses of crude oil show that it contains mainly carbon and hydrogen. Nitrogen, oxygen, and sulfur (heteroelements) are present in smaller amounts and also trace elements such as vanadium, nickel, and so forth. The mixture of the constituents is highly complex since (using the hydrocarbon constituents as the examples) paraffin derivatives, naphthene derivatives, and aromatic derivatives can occur in the same molecule and the complexity of the constituents increases with boiling range (Parkash, 2003; Gary et al., 2007; Speight, 2014; Hsu and Robinson, 2017; Speight, 2017). The attempted classification of crude oils in terms of these three main structural types has proved inadequate (Speight, 2014).

2.6.1 Assay

An assay, as the term is used in the crude oil industry, is the application of a series of test to crude oil, heavy crude oil, extra heavy crude oil, or tar sand bitumen that present data upon which the character and processability of the feedstock can be estimated (Table 2.2). Since crude oil is a naturally occurring mixture of hydrocarbon derivatives, generally in a liquid state, that may also include compounds of sulfur, nitrogen, oxygen, metals, and other elements (ASTM D4175, 2021). Consequently, it is not surprising that crude oil can vary in composition properties and produce wide variations in refining behavior as well as product properties. Thus, a feedstock assay is a necessary preliminary step before deciding upon the refining sequence necessary to convert the feedstock to saleable products.

The crude oil being processed in refineries is becoming increasingly more viscous (higher amounts of residuum) and has higher sulfur content (Parkash, 2003; Gary et al., 2007; Speight, 2014; Hsu and Robinson, 2017; Speight, 2017). Market demand ("market pull") dictates that the viscous feedstocks must be upgraded to higher value products. In short, the value of crude oil depends upon its quality for refining and whether or not a product slate can be obtained to fit market demand.

Thus, the process units in a refinery require analytical test methods that can adequately evaluate feedstocks (with a high degree of accuracy) in order to monitor the process parameters and the amount and quality of the products (Parkash, 2003; Gary et al., 2007; Speight, 2014; Hsu and Robinson, 2017; Speight, 2017). In addition, the high sulfur content of crude oil and regulations limiting the maximum sulfur content of fuels makes sulfur removal a priority in refinery processing. Here again, analytical methodology is key to the successful determination of the sulfur compound types present and their subsequent removal.

Upgrading residua involves processing (usually conversion) into a more salable, higher valued product. Improved characterization methods are necessary for process design, crude oil evaluation, and operational control. Definition of the boiling range and the hydrocarbon-type distribution in viscous feedstocks (including residua) is increasingly important because there is often the need to provide a quantitative boiling range distribution (that accounts for non-eluting constituents) as well as the distribution of chemical types in the higher boiling materials which is an important aspect of the evaluation feedstocks for further processing.

Desulfurization processes (i.e., sulfur reduction processes) are sensitive to both amount and structure of the sulfur compounds being removed (Parkash, 2003; Gary et al., 2007; Speight, 2014; Hsu and Robinson, 2017; Speight, 2017). Tests that can provide information about both are becoming increasingly important and analytical tests that provide information about other constituents of interest (e.g., nitrogen, organometallic constituents) are also valuable and being used for characterization.

But before emerging into the detailed aspects of crude oil product analysis, it is necessary to understand the nature and character of crude oil as well as the methods used to produce crude oil products. This will present to the reader the background that is necessary to understand crude oil and the processes used to convert it to products. The details of the chemistry are not presented here and can be found elsewhere (Speight, 2014, 2017).

The underlying premise for these methods of definition or classification is uniformity of the molecular nature of the feedstocks. This is not the case. And when blends are employed as refinery feedstocks, the methods do not take into account any potential interactions between the constituents of each member of the blend.

The most adequate definitions of crude oil come from legal documents where crude oil is defined directly or by inference (Speight, 2014). In addition, crude oil varies in physical appearance from a light colored liquid to the more viscous feedstocks. The near-solid or solid bitumen that occurs in tar sand deposits is different to crude oil and the viscous feedstocks, as evidenced by the respective methods of recovery.

Elemental analyses of crude oil shows that the major constituents are carbon and hydrogen with smaller amounts of sulfur (0.1–8% w/w), nitrogen

(0.1–1.0% w/w), and oxygen (0.1–3% w/w) and trace elements such as vanadium, nickel, iron, and copper present at the part-per-million (ppm) level. Of the heteroelement elements (i.e., the non-carbon and non-hydrogen elements), sulfur is the most abundant and often considered the most important by refiners (Parkash, 2003; Gary et al., 2007; Speight, 2014; Hsu and Robinson, 2017; Speight, 2017). However, nitrogen and the trace metals also have deleterious effects on refinery catalysts and should not be discounted because of relative abundance. Process units that have, for example, a capacity of 50,000 bbl/day and which are in operation continuously can soon reflect the presence of the trace elements. The effects of oxygen, which also has an effect on refining catalysts, has received somewhat less study than the other heteroelements but remains equally important in refining.

Crude oil suitability for refining (to produce a slate of predetermined products) is determined by application of a series of analytical methods that provide information that is sufficient to assess the potential quality of the crude oil as a feedstock and also to indicate whether any difficulties might arise in handling, refining, or transportation.

Moreover, the suitability of a crude oil product for a specific use cannot be assumed and the suitability must be determined by a sequence of analytical test data which must be in agreement with (or within the range of) the data presented as part of the specifications. Thus, an efficient assay is derived from a series of test data that give an accurate description of crude oil quality and crude oil product quality and allow an indication of its behavior during refining or service. The first step is, of course, to assure adequate (correct) sampling by use of the prescribed protocols (ASTM D4057, 2021).

The general condition or properties of a refinery feedstock is determined by a feedstock assay which is derived from a series of test data that give an accurate description of feedstock quality and allow an indication of its behavior during refining. The first step is, of course, to assure adequate (correct) sampling by use of the prescribed protocols (ASTM D4057, 2021). In the assay, analyses are performed to determine whether each batch of feedstock received at the refinery is suitable for refining purposes. However, in order to obtain the necessary information, two different analytical schemes are commonly used and these are: (1) an inspection assay and (2) a comprehensive assay (Table 2.1) (Speight, 2007, 2014, 2015).

Inspection assays usually involve determination of several key bulk properties of the feedstock (e.g., API gravity, sulfur content, pour point, and distillation range) as a means of determining if *major* changes in characteristics have occurred since the last comprehensive assay was performed. For example, a more detailed inspection assay might consist of the following tests: API gravity (or density or relative density), sulfur content, pour point, viscosity, salt content, water and sediment content, trace metals (or organic halides). The results from these tests with the archived data from a comprehensive assay provide an estimate of any changes that have occurred in the feedstock that may be critical to refinery operations. Inspection assays are routinely performed on all feedstocks received at a refinery.

On the other hand, the comprehensive (or full) assay is more complex (as well as time-consuming and costly) and is usually only performed only when a new field comes onstream, or when the inspection assay indicates that significant changes in the composition of the feedstock have occurred. Except for these circumstances, a comprehensive assay of a particular feedstock stream may not (unfortunately) be updated for several years. A full feedstock assay may involve at least determinations of (1) carbon residue yield, (2) density or specific gravity, (3) sulfur content, (4) distillation profile or volatility, (5) metallic constituents, (6) viscosity, and (7) pour point as well as any tests designated necessary to understand properties and behavior of the feedstock under examination.

The inspection assay tests discussed above are not exhaustive but are the ones most commonly used and provide data on the impurities present as well as a general idea of the products that may be recoverable. Other properties that are determined on an as needed basis include, but are not limited to, the following: (1) vapor pressure by the Reid method, ASTM D323 (2021), and (2) the total acid number, ASTM D664 (2021).

Using the data derived from the test assay, it is possible to assess feedstock quality acquire a degree of predictability of performance during refining. However, knowledge of the basic concepts of refining will help the analyst understand the production and, to a large extent, the anticipated properties of the product, which in turn is related to storage, sampling, and handling the products.

Thus, analyses are performed to determine whether each batch of crude oil received at the refinery or each batch of the product is suitable for refining purposes or for the designated use. The tests are also applied to determine if there has been any contamination during storage or transportation that may increase the processing difficulty (cost). The information required is generally specific to a particular refinery and is also a function of refinery operations and desired product slate.

To obtain the necessary information, two different analytical schemes are commonly used and these are: (1) an inspection assay and (2) a comprehensive assay (Speight, 2014, 2015).

Inspection assays usually involve determination of several key bulk properties of the sample (such as API gravity, sulfur content, pour point, and distillation range) as a means of determining if *major* changes in characteristics have occurred since the last comprehensive assay was performed. For example, a more detailed inspection assay might consist of the following tests: API gravity (or density or relative density), sulfur content, pour point, viscosity, salt content, water and sediment content, trace metals (or organic halides). The results from these tests—as well as the archived data from a comprehensive assay—provide an estimate of any changes that have occurred in the crude oil that may be critical to refinery operations. Inspection assays are routinely performed on all crude oils received at a refinery.

On the other hand, the comprehensive (or full) assay is more complex (as well as time-consuming and costly) and is usually only performed only when a new field comes onstream, or when the inspection assay indicates that significant changes in the composition of the crude oil have occurred. Except for these circumstances, a comprehensive assay of a particular crude oil stream may not (unfortunately) be updated for several years.

2.6.2 Specifications

A *feedstock specification* or *product specification* is the data that give adequate control of feedstock behavior in a refinery or product quality. More accurately, the specifications are derived from the set of tests and data limits applicable to the crude oil or to a finished product in order to ensure that every batch is of satisfactory and consistent quality at release for sales. The specifications should include all critical parameters in which variations would be likely to affect the safety and in-service use of the product—they are, in fact, part of the assay.

Often, the specifications against which a finished product is tested before release for sale are referred to as the "batch release specifications" which are based on the product data against which the finished product is tested to ensure satisfactory quality. If the product should be in-service before a certain date, the purchases will be advised (by law) through a note on the product of the "expiry specifications" and the product, if tested at any time within the functional life period must comply with the requirements in the expiry specifications.

In terms of the whole crude oil, a specification offers the luxury of *predictability* of feedstock behavior in a refinery or "predictability" of product quality (therefore, product behavior) relative to market demand. Ultimately, feedstock behavior and/or product quality is judged by an assessment of performance and it is the "performance" of a product that is the ultimate criterion of quality. It is therefore necessary to determine those properties, the values of which can be established precisely and relatively simply by inspection tests in a control laboratory that correlate closely with the important performance properties.

The value of any feedstock to a refiner depends upon the quality of the feedstock and whether or not a satisfactory product pattern that matches market demand ("market pull"). In the main, the refiner is not concerned with the actual chemical nature of the material but in methods of analysis which would provide information sufficient to assess the potential quality of the oil, to supply preliminary engineering data, and also to indicate whether any difficulties might arise in handling, refining, or transporting crude oil or its products. Such information may be obtained in one of two ways: (1) inspection data from a preliminary assay and (2) inspection data from a full assay, which involves the derivation of a true boiling point curve and the analysis of fractions and product blends throughout the full range of the crude oil.

The "preliminary assay" provides general data on the oil and is based on simple tests such as distillation range, water content, specific gravity, and sulfur content that enable desirable or undesirable features to be noted. This form of assay requires only a small quantity of sample and is therefore particularly useful for the characterization of oilfield samples produced from cores, drill stem tests, or seepages.

The tests in the preliminary assay are relatively simple and can be completed in a short time and generally on a routine basis. This assay gives a useful general picture of the quality of crude oil but it does not cover the work necessary to provide adequate data, for example, for the design of refinery equipment, nor does it produce a sufficient quantity of the various products from the crude so that they can be examined for quality. A "full assay" of crude oil is based on a true boiling point distillation of the crude, and sufficient data are obtained to assess the yields and properties of the straight run products, covering the naphtha fractions, the middle distillates, the gas oil fractions, and the residuum. Often, both the middle-ground is reached between the preliminary assay and the full assay but the requirements may also be feedstock dependent (Table 2.2).

Sometimes the inspection tests attempt to measure these properties, for example the carbon residue of a feedstock that is an approximation of the amount of the thermal coke that will be formed during refining. In other cases, the behavior must be determined indirectly from a series of test results.

2.6.3 Metallic Constituents

Crude oil, as recovered from the reservoir, contains metallic constituents but also pick up metallic constituents during recovery, transportation, and storage. Even trace amounts of these metals can be deleterious to refining processes, especially processes where catalysts are used. Trace constituents, such as metallic constituents, can also produce adverse effects in refining either (1) by causing corrosion or (2) by affecting the quality of refined products.

Hence, it is important to have test methods that can determine metals, both at trace levels and at major concentrations. Thus, test methods have evolved that are used for the determination of specific metals as well as for the multielement methods determination using techniques such as atomic absorption (AA) spectrometry, inductively coupled plasma (ICP) atomic emission spectrometry, and X-ray fluorescence spectroscopy.

Nickel and vanadium along with iron and sodium (from the brine) are the major metallic constituents of crude oil. These metals can be determined by AA spectrophotometric methods, wave length-dispersive X-ray fluorescence spectrometry (IP 433), and inductively coupled plasma emission spectrometry (ICPAES). Several other analytical methods are available for the routine determination of trace elements in crude oil, some of which allow direct aspiration of the samples (diluted in a solvent) instead of the time-consuming sample preparation procedures such as wet ashing (acid decomposition) or flame or dry ashing (removal of volatile/combustible constituents) (ASTM D5863, 2021). Among the techniques used for trace element determinations are conductivity, flameless, and flame AA spectrophotometry and inductively coupled argon plasma spectrophotometry (ASTM D5863, 2021; ASTM D5708, 2021) (Speight, 2014, 2015).

Inductively coupled argon plasma emission spectrophotometry (ASTM D5708, 2021) has an advantage over AA spectrophotometry (ASTM D4628, 2021; ASTM D5863, 2021) because it can provide more complete elemental composition data than the AA method. Flame emission spectroscopy is often used successfully in conjunction with AA spectrophotometry (ASTM D3605, 2021). X-ray fluorescence spectrophotometry (ASTM D4927, 2021; ASTM D6443, 2021) is also sometimes used but matrix effects can be a problem.

The method to be used for the determination of metallic constituents in crude oil is often a matter of individual preference.

All the standard methods for the determination of the wax involve precipitating the wax from solvents such as methylene chloride or acetone under specified conditions of solvent/oil ratio and temperature. Measurements such as these give comparative results that are often useful in characterizing the wax content of crude oil, or for investigating factors involved in flow problems.

The inspection assay tests discussed above are not exhaustive but are the ones most commonly used and provide data on the impurities present as well as a general idea of the products that may be recoverable.

2.6.4 Water Content

Some of the crude oil and viscous feedstocks sampled may contain substantial amounts of water. Because any process that would separate the feedstock and water would also change the composition of the oil, most properties are often determined on the oils as received. Therefore, for those oils with significant water contents (>5% w/w), many of the properties measured will not represent the properties of the dry feedstock. Generally, water content is determined by the Karl Fischer titration (ASTM D6304, 2021) which is adopted in many official standards as the standard test method for water determination in crude oil and crude oil products.

2.7 Environmental Issues

Despite the nature of the environmental regulations and the precautions taken by the refining industry, the accidental release of nonhazardous chemicals and hazardous chemicals into the environment has occurred and, without being unduly pessimistic, will continue to occur (by all industries—not wishing to select the refining industry as the only industry that suffers accidental release of chemicals into the environment). It is a situation that, to paraphrase *chaos theory*, no matter how well one prepares, the unexpected is always inevitable.

It is at this point that the environmental analyst has to identity the nature of the chemicals and their potential effects on the ecosystem(s) (Smith, 1999). Although crude oil itself and its various products are complex mixtures of many organic chemicals (Chapters 2 and 3), the predominance of one particular chemical or one particular class of chemicals may offer the environmental analyst or scientist an opportunity for predictability of behavior of the chemical(s).

Briefly, for environmental purposes, chemicals are subdivided into two classes: (1) organic chemicals and (2) inorganic chemicals. Furthermore, classification occurs insofar as organic chemicals are classified as VOCs or semi-volatile organic compounds (on occasion, the word chemicals is substituted for the word compounds without affecting the definition).

The first class of organic compounds, the VOCs, is subdivided into regulated compounds and unregulated compounds. Regulated compounds have maximum contaminant levels, but unregulated compounds do not. Regulated generally (but not always) have low boiling points, or low boiling ranges, and some are gases. Many of these chemicals can be detected at extremely low levels by a variety of instrumentation, including the human nose! In the case of the crude oil industry, sources for VOCs typically are crude oil refineries, fuel stations, naphtha (i.e., dry cleaning solvents, paint thinners, and cleaning solvents for auto parts) and, in some cases, refrigerants that are manufactured from petrochemicals.

The second class of organic compounds, the *semi-volatile compounds*, typically have high boiling points, or high boiling ranges, and are not always easily detected by the instrumentation that may be used to detect the VOCs (including the human nose). Some of the common sources of contamination are high boiling crude oil products (e.g., lubricating oils), pesticides, herbicides, fungicides, wood preservatives, and a variety of other chemicals that can be linked to the refining industry.

Regulations are in place that set the maximum contamination concentration levels that are designed to ensure public safety. There are primary and secondary standards for inorganic chemicals. Primary standards are those chemicals that cause neurological damage, cancer, or blood disorders. Secondary standards are developed for other environmental reasons. In some instances, the primary standards are referred to as the *Inorganic Chemical Group*. The secondary standards are referred to as the *General Mineral Group* and *General Physical Testing Group*.

The inorganic chemical group includes aluminum, antimony, arsenic, barium, beryllium, cadmium, chromium, lead, mercury, nickel, selenium, silver, fluoride, nitrate, nitrite, and thallium. The mineral group includes calcium, magnesium, sodium, potassium, bicarbonate, carbonate, chloride, sulfate, pH, alkalinity, hardness, electrical conductivity, total dissolved solids, surfactants, copper, iron, manganese, and zinc. The physical group includes turbidity, color, and odor. Many of these chemicals arise from desalting residues and from other processes where catalysts are used. A high level of any of these three chemicals in the soil or in the water is an indication that one or more specific processes (identified from the chemicals that have been released) or pollution prevention processes are not performing according to operational specifications.

Another source of toxic compounds is combustion (Chapter 4). In fact, some of the greater dangers of fires are from toxic products and by-products of combustion. The most obvious of these is carbon monoxide (CO), which can cause serious illness or death because it forms carboxyhemoglobin with hemoglobin in the blood so that the blood no longer carries oxygen to body tissues. Toxic sulfur dioxide and hydrogen chloride are formed by the combustion of sulfur compounds and organic chlorine compounds, respectively. In addition, a large number of noxious organic compounds such as aldehydes are generated as by-products of combustion. In addition to forming carbon monoxide, combustion under oxygen-deficient conditions produces polynuclear aromatic hydrocarbon derivatives consisting of fused-ring structures. Some of these compounds, such as benzo(a)pyrene are pre-carcinogenic compounds, insofar as they are acted upon by enzymes in the body to yield cancer-producing metabolites.

Most investigations involving crude oil hydrocarbon derivatives are regulated by various agencies that may require methodologies, action levels, and cleanup criteria that are different. Indeed, the complex chemical composition of crude oil and crude oil products can make it extremely difficult to select the most appropriate analytical test methods for evaluating environmental samples and to accurately interpret and use the data.

Accordingly, general methods of environmental analysis (Smith, 1999), that is, analysis for the determination of crude oil or crude oil products that have been released, are available. The data determine whether or not a release of such chemicals will be detrimental to the environment and may lead to regulations governing the use and handling of such chemicals. But first, sample collection, preservation, preparation, and handling protocols must be followed to the latter. This, of course, includes "chain of custody" or "sampling handling protocols" that will be defensible if and when legal issues arise. Thus, an accurate sample handling and storage log should be maintained and should include the basic necessary information (Table 2.1) (Dean, 1998; Weisman, 1998; Dean, 2003). Attention to factors such as these enables standardized comparisons to be made when subsequent samples are taken.

In summary, many of the specific chemicals in crude oil are hazardous because of the (1) chemical reactivity, (2) fire hazard, and (3) toxicity as well as other properties that are typically product dependent. In fact, a simple definition of a hazardous chemical (or hazardous waste) is that it is a chemical substance (or chemical waste) that has been inadvertently released, discarded, abandoned, neglected, or designated as a waste material and has the potential to be detrimental to the environment. Alternatively, a hazardous chemical may be a chemical that may interact with other (chemical) substances to give a product that is hazardous to the environment.

Whatever the case, methods of analysis must be available to determine the nurture of the released chemical (often designated as chemical waste) and from the data predict the potential hazard to the environment.

References

ASTM 2021. *Annual Book of Standards*. ASTM International, West Conshohocken, Pennsylvania.

ASTM D270. 2021. Standard Method of Sampling Petroleum and Petroleum Products. Annual Book of Standards. ASTM International, West Conshohocken, Pennsylvania.

ASTM D287. 2021. Standard Test Method for API Gravity of Crude Petroleum and Petroleum Products (Hydrometer Method). Annual Book of Standards. ASTM International, West Conshohocken, Pennsylvania.

ASTM D323. 2021. Standard Test Method for Vapor Pressure of Petroleum Products (Reid Method). Annual Book of Standards. ASTM International, West Conshohocken, Pennsylvania.

ASTM D346. 2021. Standard Practice for Collection and Preparation of Coke Samples for Laboratory Analysis. Annual Book of Standards. ASTM International, West Conshohocken, Pennsylvania.

ASTM D664. 2021. Standard Test Method for Acid Number of Petroleum Products by Potentiometric Titration. Annual Book of Standards. ASTM International, West Conshohocken, Pennsylvania.

ASTM D1250. 2021. Standard Guide for Use of the Petroleum Measurement Tables. Annual Book of Standards. ASTM International, West Conshohocken, Pennsylvania.

ASTM D1265. 2021. Standard Practice for Sampling Liquefied Petroleum (LP) Gases, Manual Method. Annual Book of Standards. ASTM International, West Conshohocken, Pennsylvania.

ASTM D1298. 2021. Standard Test Method for Density, Relative Density, or API Gravity of Crude Petroleum and Liquid Petroleum Products by Hydrometer Method. Annual Book of Standards. ASTM International, West Conshohocken, Pennsylvania.

ASTM D1796. 2021. Standard Test Method for Water and Sediment in Fuel Oils by the Centrifuge Method (Laboratory Procedure). Annual Book of Standards. ASTM International, West Conshohocken, Pennsylvania.

ASTM D2234. 2021. Standard Practice for Collection of a Gross Sample of Coal. Annual Book of Standards. ASTM International, West Conshohocken, Pennsylvania.

ASTM D3605. 2021. Standard Test Method for Trace Metals in Gas Turbine Fuels by Atomic Absorption and Flame Emission Spectroscopy. Annual Book of Standards. ASTM International, West Conshohocken, Pennsylvania.

ASTM D4006. 2021. Standard Test Method for Water in Crude Oil by Distillation. Annual Book of Standards. ASTM International, West Conshohocken, Pennsylvania.

ASTM D4007. 2021. Standard Test Method for Water and Sediment in Crude Oil by the Centrifuge Method (Laboratory Procedure). Annual Book of Standards. ASTM International, West Conshohocken, Pennsylvania.

ASTM D4057. 2021. Standard Practice for Manual Sampling of Petroleum and Petroleum Products. Annual Book of Standards. ASTM International, West Conshohocken, Pennsylvania.

ASTM D4175. 2021. Standard Terminology Relating to Petroleum, Petroleum Products, and Lubricants. Annual Book of Standards. ASTM International, West Conshohocken, Pennsylvania.

ASTM D4628. 2021. Standard Test Method for Analysis of Barium, Calcium, Magnesium, and Zinc in Unused Lubricating Oils by Atomic Absorption Spectrometry. Annual Book of Standards. ASTM International, West Conshohocken, Pennsylvania.

ASTM D4927. 2021. Standard Test Methods for Elemental Analysis of Lubricant and Additive Components—Barium, Calcium, Phosphorus, Sulfur, and Zinc by Wavelength-Dispersive X-ray Fluorescence Spectroscopy. Annual Book of Standards. ASTM International, West Conshohocken, Pennsylvania.

ASTM D4928. 2021. Standard Test Method for Water in Crude Oils by Coulometric Karl Fischer Titration. Annual Book of Standards. ASTM International, West Conshohocken, Pennsylvania.

ASTM D5708. 2021. Standard Test Methods for Determination of Nickel, Vanadium, and Iron in Crude Oils and Residual Fuels by Inductively Coupled Plasma (ICP) Atomic Emission Spectrometry. Annual Book of Standards. ASTM International, West Conshohocken, Pennsylvania.

ASTM D5863. 2021. Standard Test Methods for Determination of Nickel, Vanadium, Iron, and Sodium in Crude Oils and Residual Fuels by Flame Atomic Absorption Spectrometry. Annual Book of Standards. ASTM International, West Conshohocken, Pennsylvania.

ASTM D6304. 2021. Standard Test Method for Determination of Water in Petroleum Products, Lubricating Oils, and Additives by Coulometric Karl Fischer Titration.

Annual Book of Standards. ASTM International, West Conshohocken, Pennsylvania.
ASTM D6443. 2021. Standard Test Method for Determination of Calcium, Chlorine, Copper, Magnesium, Phosphorus, Sulfur, and Zinc in Unused Lubricating Oils and Additives by Wavelength Dispersive X-ray Fluorescence Spectrometry (Mathematical Correction Procedure). Annual Book of Standards. ASTM International, West Conshohocken, Pennsylvania.
ASTM D6543. 2021. Standard Guide to the Evaluation of Measurements Made by Online Coal Analyzers. Annual Book of Standards. ASTM International, West Conshohocken, Pennsylvania.
ASTM D6883. 2021. Standard Practice for Manual Sampling of Stationary Coal from Railroad Cars, Barges, Trucks, or Stockpiles. Annual Book of Standards. ASTM International, West Conshohocken, Pennsylvania.
ASTM D7430. 2021. Standard Practice for Mechanical Sampling of Coal. Annual Book of Standards. ASTM International, West Conshohocken, Pennsylvania.
Dean, J.R. 1998. *Extraction Methods for Environmental Analysis*. John Wiley & Sons, Inc., New York.
Dean, J.R. 2003. *Methods for Environmental Trace Analysis*. John Wiley & Sons Inc., Hoboken, New Jersey.
EPA 1998. *Test Methods for Evaluating Solid Waste—Physical/Chemical Methods*. EPA/SW-846, 3rd Edition, 1986, Update 1, 1992, Update II, 1994, Update III, 1996, Update IV, 1998. Environmental Protection Agency, Washington, DC.
EPA. 2004. Environmental Protection Agency, Washington, DC. www.epa.gov
Gary, J.G., Handwerk, G.E., and Kaiser, M.J. 2007. *Petroleum Refining: Technology and Economics*. 5th Edition. CRC Press, Taylor & Francis Group, Boca Raton, Florida.
Hsu, C.S., and Robinson, P.R. (Editors). 2017. *Handbook of Petroleum Technology*. Springer International Publishing AG, Cham, Switzerland.
Parkash, S. 2003. *Refining Processes Handbook*. Gulf Professional Publishing, Elsevier, Amsterdam, Netherlands.
Patnaik, P. (Editor). 2004. *Dean's Analytical Chemistry Handbook*. 2nd Edition. McGraw-Hill, New York.
Quevauviller, P. 2002. *Quality Assurance for Water Analysis*. John Wiley & Sons Inc., Hoboken, New Jersey.
Ralli, D.K., Pandey, S.C., Saxena, A.K., and Alamkhan, W.K. 2003. Impact of a Quality Management System on Product Quality at ONGC, Uran. *Journal of Scientific & Industrial Research*, 62: 1001–1007.
Smith, R.K. 1999. *Handbook of Environmental Analysis*. 4th Edition. Genium Publishing, Schenectady, New York.
Speight, J.G. 2001. *Handbook of Petroleum Analysis*. John Wiley & Sons Inc., Hoboken, New Jersey.

Speight, J.G. 2005. *Environmental Analysis and Technology for the Refining Industry*. John Wiley & Sons Inc., Hoboken, New Jersey.

Speight, J.G. 2007. *Natural Gas: A Basic Handbook*. GPC Books, Gulf Publishing Company, Houston, Texas.

Speight, J.G. 2009. *The Scientist or Engineer as an Expert Witness*. CRC Press, Taylor and Francis Group, Boca Raton, Florida.

Speight, J.G. 2014. *The Chemistry and Technology of Petroleum*. 5th Edition. CRC Press, Taylor & Francis Group, Boca Raton, Florida.

Speight, J.G. 2015. *Handbook of Petroleum Product Analysis*. 2nd Edition. John Wiley & Sons Inc., Hoboken, New Jersey.

Speight, J.G. 2017. *Handbook of Petroleum Refining*. CRC Press, Taylor & Francis Group, Boca Raton, Florida.

Speight, J.G., and Foote, R. 2011. *Ethics in Science and Engineering*. Scrivener Publishing, Salem, Massachusetts.

Stratiev, D., Dinkov, D., Petkov, K., and Stanulov, K. 2010. Evaluation of Crude Oil Quality. *Petroleum and Coal*, 52(1): 35–43. https://www.researchgate.net/publication/43968852_Evaluation_of_crude_oil_quality

Weisman, W. 1998. *Analysis of Petroleum Hydrocarbons in Environmental Media*. Amherst Scientific Publishers, Amherst, Massachusetts. Volume 1: Total Petroleum Hydrocarbons Criteria Working Group Series; Volume 2: Composition of Petroleum Mixtures; Volume 3: Selection of Representation Total Petroleum Hydrocarbons Fractions Based on Fate and Transport Considerations; Volume 4: Development of Fraction-Specific; Volume 5: Human Health Risk-Based Evaluation of Petroleum Contaminated Sites—Implementation of the Working Group Approach

3

Chemical Composition

3.1 Introduction

Crude oil and the various viscous feedstocks (heavy crude oil, extra heavy crude oil, and tar sand bitumen—the latter feedstock is referred to as "oil sand bitumen" in Canada) are not always usually (in many cases, rarely) found where the precursors were laid down, but in reservoirs or deposits where accumulation has occurred after the source material (typically in a liquid form or in a semiliquid form) has migrated from the source rocks through geologic strata to a geologic trap where further movement of the material is prohibited by virtue of the surrounding impermeable strata. In addition, the theory that the precursors to the various feedstocks form a mix that is often referred to as "protopetroleum" (and also referred to as the "primordial precursor soup" or "petroleum porridge" or "crude oil porridge") is an acceptable generalization (Speight, 2014, 2017, 2021). In addition, the structure and character of the molecular constituents in any particular fraction are controlled by (1) the chemical nature of the precursors which contributed to the formation of crude oil, (2) the chemical structures of the precursors which contributed to the formation of crude oil, and (3) the physical conditions that are prevalent during the maturation during which the source material undergoes a variety of conversion processes.

In the natural state, crude oil and the viscous feedstocks are not homogeneous materials and the physical characteristics differ depending on where the material was produced. This is due to the fact that any of these feedstocks from different geographical locations will naturally has unique properties. In the natural, unrefined state, crude oil and the viscous feedstocks (or, for that matter refinery feedstock) range in density and consistency from very thin, lightweight, and volatile fluidity to an extremely thick, semisolid. In addition, there is also a very obvious gradation in the color that ranges from a light, golden yellow conventional crude

Handbook of Heavy Oil Properties and Analysis, First Edition. James G. Speight.

oil to the near-black heavy crude oil to the black tar sand bitumen (Speight, 2014, 2017, 2021).

The viscous feedstocks typically exhibit a high content of higher molecular weight hydrocarbon derivatives, especially the condensed aromatic constituents (Table 3.1) and the elevated levels of heteroatom constituents, which are the constituents that contain sulfur and/or nitrogen and/or oxygen and/or metals (Santos et al., 2014; Speight, 2014). Typically, the molecular constituents that are present in heavy crude oil and the more viscous feedstocks have more than 15 carbon atoms in the molecule. Although the amount of compounds containing heteroatoms is relatively small, the effect of these constituents on the oil properties can be anticipated to be substantial. Moreover, the presence of constituents containing sulfur atoms of which in the viscous feedstocks, the most common types are the derivatives, sulfide derivatives, and the cyclic thiophene derivatives (Figure 3.1) that are known to exert and often regarded as harmful to the refining processes (Parkash, 2003; Gary et al., 2007; Speight, 2014; Hsu and Robinson, 2017; Speight, 2017).

Organic nitrogen-containing constituents in refinery feedstocks cause major problems in refineries (though poisoning of the catalysts) especially when the nitrogen levels in the feedstocks exceed 0.5% w/w. The nitrogen compounds are generally basic, formed by pyridine and its homologs. However, nitrogen compounds can also occur in nonbasic forms, formed by species including pyrrole derivatives, indole derivatives, and carbazole derivatives (Figure 3.2).

Oxygenated compounds appear as carboxylic acid derivatives (RCO_2H, where R is an aliphatic or aromatic moiety) and phenol derivatives ($R_{Ar}OH$, where R_{Ar} is an aromatic moiety), although the presence of ketone derivatives (RCOR, where R is an aliphatic or aromatic moiety), ether derivatives (ROR, where R is an aliphatic or aromatic moiety), and anhydride derivatives (RCOOCOR, where R is an aliphatic moiety or an aromatic moiety) has also been identified (Figure 3.3). The amount of these constituents in a feedstock determines the acidity of the feedstock, which is particularly important in the refining process (Parkash, 2003; Gary et al., 2007; Speight, 2014; Hsu and Robinson, 2017; Speight, 2017).

Thus, crude oil and the viscous feedstocks are not (even within the individual types of viscous feedstocks—that is, within the feedstocks grouped together as heavy crude oil, extra heavy crude oil, and tar sand bitumen) uniform materials and the chemical and physical (fractional) composition of each of these refinery feedstocks can vary not only with the location and age of the reservoir or deposit but also with the depth of the individual well within the reservoir or deposit (Speight, 2014, 2017, 2021). On a molecular basis, the feedstocks are complex mixtures containing (depending upon the type of feedstock) derivatives of hydrocarbons that also contain varying amounts of hydrocarbonaceous constituents (i.e., hydrocarbon derivatives that contain sulfur, oxygen, and nitrogen as well as

Table 3.1 Hydrocarbon Derivatives and Heteroatom Derivatives that Occur in Crude Oil, Heavy Crude Oil, Tar Sand Bitumen, and Residua

Class	Compound types
Hydrocarbon systems	Saturated
	n-Paraffins
	iso-Paraffins plus other branched paraffins
	Cycloparaffins (naphthenes)
	Condensed cycloparaffins (including steranes, hopanes)
	Alkyl side chains on ring systems
	Aromatic
	Benzene systems
	Condensed aromatic systems
	Condensed naphthene-aromatic systems
	Unsaturated[a]
	Olefins
Heteroatomic systems	Saturated (nonaromatic systems)
	Alkyl sulfides
	Cycloalkyl sulfides (naphthene sulfides)
	Aromatic
	Furan derivatives (single-ring and multi-ring systems)
	Phenol derivatives
	Thiophene derivatives (single ring and multi-ring systems)
	Pyrrole derivatives (single-ring and multi-ring systems)
	Pyridine derivatives (single-ring and multi-ring systems)
	Amphoteric
	Acid–base systems in the same molecule

[a] Olefins do not occur naturally in crude oil but are present in the products of thermal reactions.

constituents containing metallic constituents, particularly those constituents that containing vanadium, nickel, iron, and copper). For example, in a conventional (light) crude oil, the hydrocarbon content may be as high as 97% w/w while in a tar sand bitumen, the hydrocarbon content may not exceed 50% w/w (Speight, 2014, 2017, 2021).

Residua (as a result of the concentration effect of distillation) contain more heteroatomic species and less hydrocarbon constituents than conventional crude oil. Thus, there have been different approaches to refining the heavier feedstocks as

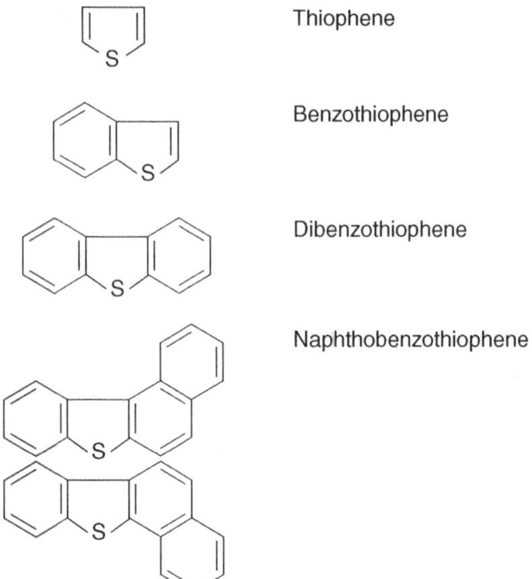

Figure 3.1 Nomenclature and types of thiophene compounds.

well as the recognition that knowledge of the constituents of these higher boiling feedstocks is also of some importance. The problems encountered in processing the viscous feedstocks can be equated to the chemical character and the amount of complex, higher boiling constituents in the feedstock. Refining these materials is not just a matter of applying know-how derived from refining conventional crude oil but requires knowledge of the chemical structure and chemical behavior of these more complex constituents.

As part of the evaluation of the various feedstocks, it is also necessary to acquire data related to the chemical composition of the feedstock. In fact, the production of this type of data is an important aspect of refining since the product distribution is dependent on (1) the chemical composition of feedstock, (2) the primary reactions of the feedstock constituents, (3) the secondary reactions of the feedstock constituents, (4) the interactions of the feedstock constituents with each other, and (5) the secondary reactions of the primary products (Parkash, 2003; Gary et al., 2007; Speight, 2014; Hsu and Robinson, 2017; Speight, 2017).

Heavy crude oil, extra heavy crude oil, and tar sand bitumen (and residua—as a result of the concentration effect of distillation) contain more heteroatomic species and less hydrocarbon constituents than conventional crude oil. Thus, to obtain as much naphtha and other distillate products as possible, there have been different approaches to refining the viscous feedstocks as well as the recognition that knowledge of the constituents of these higher boiling feedstocks is also of some importance. The problems encountered in processing the heavier feedstocks

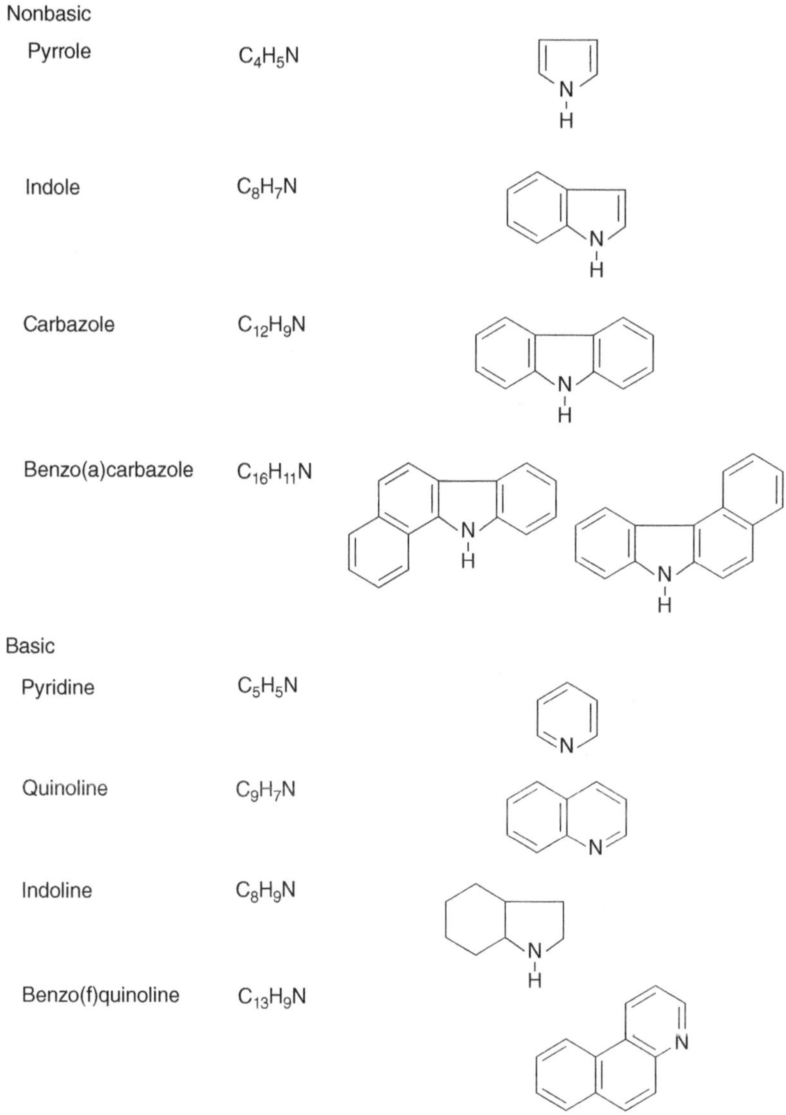

Figure 3.2 Nomenclature and types of organic nitrogen compounds.

can be equated to the chemical character and the amount of complex, higher boiling constituents in the feedstock. Refining these materials is not just a matter of applying know-how derived from refining conventional crude oils but requires knowledge of the chemical structure and chemical behavior of these more complex constituents.

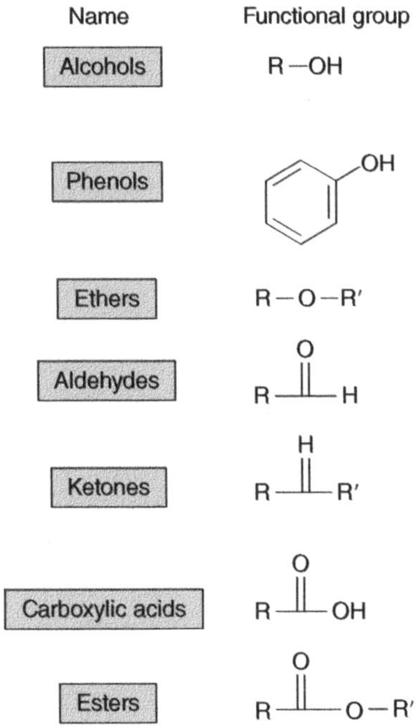

Figure 3.3 Examples of the common organic-containing functional groups.

It is the purpose of this chapter to present a brief overview of the types of constituents that are found in crude oil, heavy crude oil, extra heavy crude oil, and tar sand bitumen.

3.2 Elemental Composition

As part of the determination of the composition of refinery feedstocks, the first step is the determination of the elemental composition. For all feedstocks, the higher the atomic hydrogen–carbon (H/C) ratio, the higher is the value of the feedstock to the refinery because of the lower hydrogen requirements for upgrading. Similarly, the lower the heteroatom content, the lower the hydrogen requirements for upgrading. Thus, inspection of the elemental composition of feedstocks is an initial indication of the quality of the feedstock and, with the molecular weight, indicates the molar hydrogen requirements for upgrading (Speight, 2014, 2015, 2017).

Crude oils are naturally occurring complex and diverse mixtures and contain a multitude of thousands of different organic compounds that belong to several compound classes. The main classes are (1) hydrocarbon derivatives, which include nonaromatic and aromatic compounds and (2) heteroatom derivatives, which include a variety of sulfur, nitrogen, oxygen derivatives, and metal-containing compounds which are collectively grouped together as the heteroatom compounds. In general, all crude oils contain the same types of chemical structures, but these compounds can be in highly variable proportions in crude oils and are based on the (1) the composition of the source material, (2) the geologic conditions that were present during the maturation (conversion) of the source material to crude oil, and (3) the mineralogical composition of the pathway from the source rock to the reservoir rock which can (physically and chemically) influence the composition, and (4) the mineralogical composition of the reservoir rock which can also influence (physically and chemically) the composition.

However, it has become apparent, with the introduction of the heavier feedstocks (such as heavy crude oil, extra heavy crude oil, and tar sand bitumen) into refinery operations, that the amounts of the elements and the elemental ratios are only a start to feedstock evaluation and much more is required in order to predict the behavior of the feedstock during the refining operations. In fact, although the atomic ratios may be used on a comparative basis between feedstocks, there is no guarantee that a particular feedstock will behave as predicted from these data. Additional knowledge such as defining the various chemical reactions of the constituents as well as the reactions of these constituents with each other also plays a role in determining the processability of a feedstock. Moreover, insofar as the viscous feedstocks contain higher molecular weight material as well as more polar material makes these feedstocks more difficult to handle. The chemical reactions and the associated physical reaction of these feedstocks are not the same as these two types of reactions of crude oil.

As another note that is relevant to the spillage of crude oil into the environment is that both the various constituents and the respective quantities of these constituents can change rapidly once the crude oil is spilled into the (land-based or water-based) environment and become the target of oxidation processes (through the influence of the aerial oxygen) (Speight and Arjoon, 2012; Overton et al., 2016; Arjoon and Speight, 2023). And, moreover, there are no simple chemical reactions or physical reactions making the circumstances associated with every spill a unique chemical and physical event.

In general, the lower molecular weight constituents of crude oil (relevant to the constituents for the viscous, higher molecular weight feedstocks) are more susceptible to processes such as (1) evaporation, (2) dissolution, and (3) biodegradation, which is the degradation of the spilled material by any one of several classes of microorganisms (Speight, and Arjoon, 2012; Arjoon and Speight, 2023), while the higher molecular weight (and more dense) and more hydrophobic compounds tend to adhere to living organisms or particulates and persist in the environment.

The presence of certain compounds, such as polynuclear aromatic hydrocarbon derivatives (PNAs, also referred to as polyaromatic hydrocarbon derivatives, PAHs), also determines the acute and chronic toxicity of the spilled oil. Natural processes can degrade virtually all compounds in crude oils, with aerobic oxidation proceeding much faster than anaerobic degradation, although not all crude oil components are degraded with the same speed. This is especially true for the more dense viscous feedstocks where the interaction of the feedstock constituents and the microorganism becomes a diffusion-controlled process which can, to a great extent, have an adverse effect on the activity of the microorganisms as they attempt to "feed" on the feedstock constituents. In addition, the dependency of the analytical data from the various standard test methods may be questioned (justifiably or unjustifiably) and can cause further delay in the processing of the feedstock or in the analysis of the spilled material as it relates to cleanup efforts.

The analysis of feedstocks for the percentages by weight of carbon, hydrogen, nitrogen, oxygen, sulfur, and metals (elemental composition, ultimate composition) is perhaps the first method used to examine the general nature, and perform an evaluation, of a feedstock (Speight, 2001). The atomic ratios of the various elements to carbon (i.e., H/C, N/C, O/C, and S/C) are frequently used for indications of the overall character of the feedstock. For example, since carbon and hydrogen are the two major elements in crude oil and the viscous feedstocks, the ratio of hydrogen to carbon (i.e., the H/C atomic ratio) has been and continued to be an indicator of the amount of hydrogen that will be required to upgrade the feedstock to a variety of products. In addition, the elemental analysis of a feedstock (Table 3.2) is also of value to determine the amounts of trace elements, such as vanadium, nickel, and any other metals since these materials can have serious deleterious effects on catalyst performance during refining by catalytic processes (Parkash, 2003; Gary et al., 2007; Speight, 2014; Hsu and Robinson, 2017; Speight, 2017).

For example, the carbon content of a feedstock can be determined by the method designated for coal and coke (ASTM D3178, 2015) or by the method designated for municipal solid waste (ASTM E777, 2015). There are also methods designated for: (1) hydrogen content—ASTM D1018 (2015), ASTM D3343

Table 3.2 Standard Test Methods Designated for Elemental Analysis

Analysis	Test methods
Carbon and hydrogen	ASTM D1018, ASTM D3178, ASTM D3343, ASTM D3701, ASTM D5291, ASTM E777
Nitrogen	ASTM D3179, ASTM D3228, ASTM E258, ASTM D5291, ASTM E778
Oxygen	ASTM E385
Sulfur	ASTM D129, ASTM D139, ASTM D1266, ASTM D1552, ASTM D1757, ASTM D2622, ASTM D3120, ASTM D3177, ASTM D4045, ASTM D4294
Metals content	ASTM C1109, ASTM C1111, ASTM D482, ASTM D1318, ASTM D3340, ASTM D3341, ASTM D3605

(2015), ASTM D3701 (2015), and ASTM E777 (2015); (2) nitrogen content—ASTM D3179 (2015), ASTM D3228 (2015), ASTM E258 (2015), and ASTM E778 (2015); (3) oxygen content—ASTM E385 (2015); and (4) sulfur content—ASTM D1266 (2015), ASTM D1552 (2015), ASTM D1757 (2015), ASTM D2622 (2015), ASTM D3177 (2015), ASTM D4045 (2015), and ASTM D4294 (2015). On particular note, the viscous feedstocks can contain up to 5% w/w sulfur in the form of mercaptan derivatives and sulfide derivatives. Removal of these sulfur-containing constituents is key to meet new ultralow specifications for products. Test methods ASTM D2622 and D4294 (both are X-ray methods) are applicable to the viscous feedstocks but are matrix dependent. Test method ASTM D1552 (the oxidative microcoulometry LECO method) is also applicable to viscous feedstocks.

In fact, for all refinery feedstocks, the higher the H/C atomic ratio, the higher the value as a refinery feedstock because of the lower amounts of hydrogen that are required for upgrading. Following this statement, the lower the heteroatom content of the feedstock, the lower the amount of hydrogen required for upgrading. Thus, inspection of the elemental composition of feedstocks is an initial indication of the quality of the feedstock and, with the molecular weight, indicates the molar hydrogen requirements for upgrading (Parkash, 2003; Gary et al., 2007; Speight, 2014; Hsu and Robinson, 2017; Speight, 2017).

However, it has become apparent with the introduction of the viscous feedstocks into refinery operations that these ratios are not the only requirement (and are not always satisfactory) for accurate prediction of the character and behavior of the feedstock prior to the commencement of (and during) the refining operations. The use of more viscous (and chemically—in terms of the heteroatom content—and structurally complex) feedstocks (in terms of molecularly complex composition) has added a new dimension to refining operations. Thus, although atomic ratios, as determined by elemental analyses, may be used on a comparative basis between feedstocks, there is now no guarantee that a particular feedstock will behave as predicted from these data (Speight, 2014, 2017, 2021). Product slates cannot be predicted accurately, if at all, from these ratios. Additional knowledge such as defining the various chemical reactions of the constituents as well as the reactions of these constituents with each other also play a role in determining the processability of a feedstock.

The elemental analysis of tar sand bitumen has also been widely reported but the data suffer from the disadvantage that identification of the source is too general (i.e., Athabasca bitumen which covers several deposits in the area) and is often not site specific (Speight, 1990, 2009, 2014, 2017). In addition, the analysis is quoted for separated bitumen, which may have been obtained by any one of several procedures and may therefore not be representative of the total bitumen on the sand. However, recent efforts have focused on a program to produce sound, reproducible data from samples for which the origin is carefully identified (Wallace et al., 1988).

In summary, crude oil contains carbon, hydrogen, nitrogen, oxygen, sulfur, and metals (particularly nickel and vanadium) and the amounts of these elements in a whole series of feedstocks including the conventional crude oils and the viscous feedstocks vary over fairly narrow limits (Speight, 2014, 2017, 2021):

Element	Viscous feedstocks (% w/w)	Conventional crude oil (% w/w)
Carbon	83.0–87.0	83.0–87.0
Hydrogen	10.0–14.0	10.0–14.0
Nitrogen	0.2–2.0	0.1–2.0
Oxygen	0.5–1.5	0.05–1.5
Sulfur	4.5–6.0	0.05–3.0
Metals (ppm)		
(Ni + V)	>1,000	<1,000

The major exception to these narrow limits is the oxygen content that can vary from as little as 0.2% to as high as 4.5% which is to be expected since the oxygen content (which often is estimated by difference) is subject to the accumulation of all of the errors in the other elemental data. In addition, the viscous feedstocks are susceptible to aerial oxygen and the oxygen content is also very dependent on sample history.

This narrow range is an apparent contradiction when the wide variation in physical properties is considered and covers feedstocks from the more conventional crude oils at one end of the range which are the lighter, more mobile crude oils, to the more viscous feedstocks at the other extreme (Charbonnier et al., 1969; Draper et al., 1977; Speight, 2012, 2014, 2017). In addition, because of the narrow range of carbon and hydrogen content, it is not possible to classify crude oil on the basis of carbon content as coal is classified since the carbon content of the various types of coal can vary from a value on the order of as low as 75% w/w in lignite to a value on the order of 95% w/w in anthracite (Speight, 2013). Of course, other subdivisions are possible within the various carbon range of the various types of coals but crude oil and the viscous feedstocks are restricted to a much narrower range of elemental composition (Speight, 2014, 2017, 2021).

Furthermore, heteroatoms in feedstocks affect every aspect of refining. The occurrence of sulfur in feedstocks as organic or elemental sulfur or in produced gas as compounds of oxygen (SO_x) and hydrogen (H_2S) is an expensive aspect of refining and must be removed at some stage of the upgrading and refining process (Speight, 2014, 2017). Although the sulfur content of many conventional crude oils is on the order of 1% w/w, the sulfur content of the viscous feedstocks (such as tar sand bitumen) can exceed 5% w/w or even approach 10% w/w. Of all of the heteroelements, sulfur is usually the easiest to remove and many commercial catalysts are available that routinely remove 90% of the sulfur from a feedstock (Speight, 2000, 2014, 2017).

The nitrogen content is usually (at maximum) on the order of 2% w/w and the presence of this element complicates refining by poisoning the catalysts employed in the various processes. Nitrogen is more difficult to remove from feedstocks than sulfur, and there are fewer catalysts that are specific for nitrogen removal (Speight, 2014, 2017). Furthermore, if the nitrogen is not removed, the potential for the production of nitrogen oxides (NO_x) during processing and use become real.

Metals (particularly vanadium and nickel) are found in most feedstocks (conventional and viscous) and can occur either in the form of salts or as organometallic constituents (such as the metallo-porphyrins), which are extremely difficult to remove from the feedstock. Indeed, the nature of the distillation process by which the residua are produced (and the volatility of the constituents) dictates that all of the metals in the original refinery feedstock are concentrated in the residuum (Speight, 2014, 2017, 2021). The metallic constituents that may actually be entrained in a high boiling distillate under the conditions of the distillation process and appear in the higher boiling distillates are the exceptions to the "concentration effect."

Metal constituents of feedstocks cause problems by poisoning the catalysts used for sulfur and nitrogen removal as well as the catalysts used in other processes such as catalytic cracking. Thus, serious attempts are continuously underway to develop catalysts that can tolerate a high concentration of metals without any serious loss of catalyst activity or catalyst life (Speight, 2014, 2017).

A variety of tests have been designated for the determination of metals in refinery feedstocks and products (ASTM D1318, 2015; ASTM D3340, 2015; ASTM D3341, 2015; ASTM D3605, 2015) and the tests for the products may, with some modification, be adapted for use with whole crude oil and even for application to the viscous feedstocks. Determination of metals in whole feedstocks can be accomplished by combustion of the sample so that only inorganic ash remains. The ash can then be digested with an acid and the solution examined for metal species by atomic absorption (AA) spectroscopy or by inductively coupled argon plasma (ICP) spectrometry.

In addition to metals, many refinery feedstocks (including the viscous feedstocks) may contain chloride-containing impurities, including some organic components (usually referred to as "organic chloride"). Owing to the high dissociation energy of the carbon–chloride bond (the C—Cl bond) and unlike the inorganic chloride salts, the organic chlorides are difficult (if not impossible) to remove during the desalting process. Failure to remove the organic chloride derivatives can have a significant (and highly adverse effect) on refinery equipment which can lead to serious corrosion of the refinery equipment. Therefore, the necessity of identifying chloride derivatives in refinery feedstocks cannot be overemphasized (ASTM D4929, 2021; ASTM D6470, 2021).

In the ASTM D4929 (2021) test method, the procedure covers the determination of organic chloride either by using (1) distillation and sodium biphenyl reduction, or (2) distillation and microcoulometry, or (3) distillation and X-ray fluorescence (XRF) spectrometry. This involves the distillation of the feedstock to obtain a naphtha fraction prior to chloride determination, which is not always a viable method for viscous feedstocks. Such a method might involve dissolution/extraction of the viscous material with a suitable nonreactive solvent from which the chloride concentration could be determined by extrapolation of the data back to the original feedstock.

In order to combat the presence of these chloride derivatives, a guard bed column (typically comprised of metal oxide-based components that operate at a high temperature) can be employed for the removal of the organic chlorides.

3.3 Chemical Composition

Crude oil, heavy crude oil, extra heavy crude oil, and tar sand bitumen contain an extreme range of organic functionality and molecular size. Furthermore, the variety is so great that it is difficult, if not impossible (despite claims to the contrary by some workers), that a complete compound-by-compound description for even a single crude oil would not be possible. As already noted, the composition of any crude oil or viscous feedstock can vary with the location and age of the field in addition to any variations that occur with the depth of the individual well. In fact, two adjacent wells are more than likely to produce crude oil with very different characteristics.

In very general terms (and as observed from elemental analyses), crude oil and the viscous feedstocks (including residua) are a complex composition (listed alphabetically rather than by occurrence) of (1) hydrocarbon derivatives, (2) nitrogen-containing derivatives, (3) oxygen-containing derivatives, (4) sulfur-containing derivatives, and (5) metal-containing derivatives. However, this general definition is not adequate to describe the composition of any feedstock as it relates to the behavior of these feedstocks during refining. Indeed, the consideration of H/C atomic ratio, sulfur content, and API gravity are no longer adequate to the task of determining the refining behavior of the viscous feedstocks.

Furthermore, the composition of a feedstock can be described in terms of three classes of compounds which are (1) saturated compounds, which include normal alkane derivatives, branched alkane derivatives, and cycloalkane derivatives but alkene derivatives (i.e., olefin derivatives, $R^1CH{=}CHR^2$, where R^1 and R^2 are aliphatic or aromatic moieties) are rare to the extent of being considered an oddity in naturally occurring oils and are typically found in refinery feedstocks as a result of thermal reactions. The distribution and characteristics of these molecular species account for the rich variety of crude oils and viscous feedstocks; (2) aromatic compounds, which include monoaromatic constituents that range from benzene derivatives to the multiple fused ring analogs, such as naphthalene derivatives and phenanthrene derivatives; and (3) heteroatom compounds, which include compounds containing nitrogen and/or oxygen and/or sulfur.

Processability is not only a matter of determining the elemental composition of a feedstock but it is also a matter of understanding the bulk properties as they relate to the chemical behavior or the physical behavior of the feedstock as it passes through the various refinery options. For example, while it is difficult to understand the process chemistry of a feedstock from the elemental composition alone, it might be surmised that the major difference between a heavy crude oil and a more conventional material is the H/C atomic ratio alone. This property indicates that a heavy crude oil (having a lower H/C atomic ratio and being more aromatic in character) would require more hydrogen for upgrading to an aliphatic product which may be true, but much more information is necessary to understand and rationalize feedstock processability.

With the necessity of processing crude oil residua, heavy oil, and tar sand bitumen to obtain more naphtha and other distillate products, there has been the recognition that knowledge of the constituents of these higher boiling feedstocks is also of some importance. Indeed, the problems encountered in processing the heavier feedstocks can be equated to the chemical character and the amount of complex, higher boiling constituents in the feedstock. Refining these materials is not just a matter of applying know-how derived from refining conventional crude oils but requires knowledge of the chemical structure and chemical behavior of these more complex constituents.

Thus, it is better to address feedstock behavior in the refinery by means of a thorough consideration and understanding of the molecular composition of the feedstock. As an example, the occurrence of amphoteric species (i.e., compounds having a mixed acid/base nature) is rarely addressed nor is the phenomenon of molecular size or the occurrence of specific functional types which can play a major role in the interactions between the constituents of a feedstock resulting in unforeseen (and unpredictable) chemical reactions that can detract from the production of the desired product slate.

All of these items are important in determining feedstock behavior during refining operations and an understanding of the chemical types (or the chemical composition) of any feedstock can lead to an understanding of the chemical aspects of processing the feedstock.

To complicate matters even further, within each class, there are several families of related compounds. For example, in the viscous feedstocks, the most likely forms of sulfur are the thiophene derivatives (Figure 3.1). In terms of the nitrogen types in feedstocks, the nitrogen is more likely to be found in polar forms, such as pyridine derivatives and pyrrole derivatives (Figure 3.2). In the constituents that contain oxygen, the oxygen in these constituents is more likely to exit as phenol derivatives, carboxylic acid derivatives, amide derivatives, and ether derivatives.

It cannot be overstated that an understanding of the chemical types (or composition) that occur in any refinery feedstock can lead to an understanding of the chemical aspects of processing the feedstock (Speight, 2014, 2017, 2021). Processability is not only a matter of knowing the elemental composition of a feedstock but it is also a matter of understanding the bulk properties as they relate to the chemical or physical composition of the material.

For example, it is difficult (in some cases impossible) to understand the process chemistry of various feedstocks from the elemental composition alone because of the narrowness of the range of the various elements in the feedstock as presented above (Speight, 2014, 2017, 2021). From such data, it might be surmised that the major difference between a heavy crude oil and a more conventional crude oil is the H/C atomic ratio alone. This property indicates that a heavy crude oil (having a lower H/C atomic ratio and being more aromatic in character) would require more hydrogen for upgrading to liquid fuels. This is, indeed, true but much more information is necessary to understand the processability of the feedstock.

3.3.1 Hydrocarbon Constituents

The hydrocarbon components of crude oil and the viscous feedstocks are composed of paraffin-type hydrocarbon derivatives, naphthene-type hydrocarbon derivatives, and aromatic-type hydrocarbon derivatives. Unsaturated hydrocarbon derivatives—such as olefin-type hydrocarbon derivatives—do not typically occur in crude oils or in the viscous feedstock, unless the feedstock is the result of a thermal process in which cracking has occurred and converted paraffin-type hydrocarbon derivatives to the olefin counterparts that can be presented simply by the equation:

$$R^1CH_2CH_2R^2 \rightarrow R^1CH{=}CHR^2 + H_2$$

It is therefore convenient to divide the hydrocarbon components of crude oil into the following three classes: (1) paraffin derivatives, (2) naphthene derivatives, and (3) aromatic derivatives (Table 3.3).

Table 3.3 Hydrocarbon Derivatives Present in Feedstocks

Hydrocarbon	Description
Paraffin derivative	Saturated hydrocarbon derivatives
	Straight (linear) or branched chains
	No ring structure
Naphthene derivatives	More correctly known as alicyclic hydrocarbon derivatives
	Saturated hydrocarbon derivatives
	Containing one or more saturated ring systems
	Each ring may have one or more paraffinic side chains
	If linked up with aromatic ring systems—such as occurs in tetralin—known as naphthene-aromatics
Aromatic derivatives	Hydrocarbon derivatives containing one or more aromatic nuclei, such as benzene, naphthalene, and phenanthrene ring systems
	May be linked up with (substituted) naphthene rings and/or paraffin side chains

3.3.1.1 Paraffin Hydrocarbon Derivatives

The proportion of paraffin derivatives in crude oil varies with the type of crude, but within any one crude oil, the proportion of paraffinic hydrocarbon derivatives usually decreases with increasing molecular weight and there is a concomitant increase in aromaticity and the relative proportions of the nitrogen-, oxygen-, and sulfur-containing species (Speight, 2014, 2017).

The normal-paraffin hydrocarbon derivatives (*n*-paraffins, straight-chain paraffins) occur in varying proportions in most crude oils. In fact, paraffinic crude oil may contain up to 20–50% w/w *n*-paraffins in the gas oil fraction. However, naphthenic or asphaltic crude oils sometimes contain only very small amounts of normal paraffins.

Considerable quantities of *iso*-paraffins have been noted to be present in the crude oil of which the 2-methyl and 3-methyl derivatives are the most abundant. The *iso*-paraffins occur throughout the boiling range of crude oil fractions but the proportion tends to decrease with increasing boiling point. The *iso*-paraffins are present in the gas oil fraction of crude oil.

As the molecular weight (or boiling point) of the crude oil fraction increases, there is a concomitant decrease in the amount of free paraffins in the fraction and in some crude oils, there may be no paraffins at all in the vacuum gas oil (VGO) fraction (Speight, 1994, 2014).

3.3.1.2 Cycloparaffin Hydrocarbon Derivatives

Cycloparaffin derivatives (naphthene derivatives) are represented in all fractions in which the constituent molecules contain more than five carbon atoms. Several series of cycloparaffin derivatives, usually containing five- or six-membered rings or their combinations, occur as polycyclic structures up to and including the gas oil fractions.

The principal structural variation of naphthene derivatives is the number of rings present in the molecule. The mono- and bicyclic naphthenes are generally the major types of cycloparaffin derivatives in the lower boiling factions of crude oil, with boiling point or molecular weight increased by the presence of alkyl chains. The higher boiling point fractions, such as the gas oil fractions, may contain two to six rings per molecule. In the asphaltic (naphthenic) crude oils, the gas oil fraction can contain considerable amounts of naphthenic ring systems that increase even more in consideration of the molecular types in the asphaltene constituents. However, as the molecular weight of the fraction increases, the occurrence of condensed naphthene ring systems and alkyl-substituted naphthene ring systems increases.

3.3.1.3 Aromatic Hydrocarbon Derivatives

Crude oil is a mixture of compounds, and aromatic compounds are common to all crude oil and it is the difference in extent that becomes evident upon examination of a series of crude oil. By far the majority of these aromatics contain paraffinic chains, naphthene rings, and aromatic rings side by side.

There is a general increase in the proportion of aromatic hydrocarbon derivatives with increasing molecular weight of the fraction. In the higher molecular weight fractions (such as the atmospheric and VGO fractions), the rings are usually condensed together. Thus, components with two aromatic rings are presumed to be naphthalene derivatives and those with three aromatic rings may be phenanthrene derivatives. Currently, and because of the consideration of the natural product origins of crude oil, phenanthrene derivatives are favored over anthracene derivatives.

It should also be emphasized that in the higher boiling point crude oil fractions, many polycyclic structures occur in naphthene-aromatic systems. The naphthene-aromatic hydrocarbon derivatives, together with the naphthenic hydrocarbon series, form the major content of higher boiling point crude oil fractions. Usually the different naphthene-aromatic components are classified according to the number of aromatic rings in their molecules.

3.3.1.4 Unsaturated Hydrocarbon Derivatives

The presence of olefins ($R^1CH{=}CHR^2$) in crude oil is usually restricted to thermally produced fractions. The presence of diene derivatives ($RCH{=}CH{=}CHR'$) and acetylene derivatives ($RC{=}CR'$) is considered to be extremely unlikely.

In summary, a variety of hydrocarbon compounds occur throughout crude oil. Although the amount of any particular hydrocarbon varies from one crude oil to another, the family from which that hydrocarbon arises is well represented.

3.3.2 Non-hydrocarbon Constituents

Crude oils and especially the viscous feedstocks contain appreciable amounts of organic non-hydrocarbon constituents, mainly sulfur-, nitrogen-, and oxygen-containing compounds and, in smaller amounts, organometallic compounds. These elements are collectively referred to as heteroelements and appear throughout the entire boiling range of conventional crude oil but tend to concentrate mainly in the higher boiling fractions and in the nonvolatile residua. However, the term "non-hydrocarbon constituents" is a misnomer insofar as the molecular environment in which these non-carbon atoms and non-hydrogen atoms occur is, in reality, a hydrocarbon environment (Figures 3.1–3.3).

Although the concentration of the heteroelements in certain fractions may be quite small, the influence of the heteroelements on the properties of the various feedstocks is of utmost importance. For example, the decomposition of inorganic salts suspended in the crude can cause serious breakdowns in refinery operations; the thermal decomposition of deposited inorganic chlorides with evolution of free hydrochloric acid can give rise to serious corrosion problems in the distillation equipment. The presence of organic acid components, such as mercaptan derivatives and acid derivatives, can also promote metallic corrosion. In catalytic operations, passivation and/or poisoning of the catalyst can be caused by deposition of traces of metals (vanadium and nickel) or by chemisorption of nitrogen-containing compounds on the catalyst, thus necessitating the frequent regeneration of the catalyst or its expensive replacement.

3.3.2.1 Sulfur Compounds

Sulfur compounds (Figure 3.1) are among the most important heteroatomic constituents of crude oil, and although there are many varieties of sulfur compounds (Speight, 2000, 2014, 2017), the prevailing conditions during the formation, maturation, and even in situ alteration may dictate that only preferred types exist in any particular crude oil. In general, the higher the density of the crude oil for the lower API gravity of the crude oil, the higher the sulfur content and the total sulfur in the crude oil can vary from approximately 0.04% w/w for a low-density crude oil (light crude oil) to approximately 5.0% for heavy crude oil, extra heavy crude oil, and tar sand bitumen. However, the sulfur content of crude oils produced from broad geographic regions varies with time, depending on the composition of newly discovered fields, particularly those in different geological environments.

Various thiophene derivatives have also been isolated from a variety of crude oils; benzothiophene derivatives are usually present in the higher boiling crude oil fractions (Table 3.1). On the other hand, disulfides are not regarded as true constituents of crude oil but are generally formed by oxidation of thiols during transportation, storage, or processing. Thus:

$$2R\text{ - }SH + [O] \rightarrow R\text{-}S\text{-}S\text{-}R + H_2O$$

3.3.2.2 Nitrogen Compounds

Nitrogen in crude oil may be classified arbitrarily as basic and nonbasic (Figure 3.2) (Speight, 2014, 2017). The basic nitrogen compounds which are composed mainly of pyridine homologs and occur throughout the boiling ranges have a tendency to exist in the higher boiling fractions and residua. The nonbasic nitrogen compounds, which are usually of the pyrrole, indole, and carbazole types, also occur in the higher boiling fractions and residua.

In general, the nitrogen content of crude oil is low and is generally within the range 0.1–0.9% w/w, although early work indicates that some higher boiling fractions and the viscous feedstocks crude oil may contain up to 2% nitrogen. The presence of nitrogen in crude oil is of much greater significance in refinery operations than might be expected from the small amounts present. Nitrogen compounds can be responsible for the poisoning of cracking catalysts, and they also contribute to gum formation.

3.3.2.3 Oxygen Compounds

Oxygen in organic compounds can occur in a variety of forms (ROH, ArOH, ROR′, RCO_2H, $ArCO_2H$, RCO_2R, $ArCO_2R$, and $R_2C{=}O$) (Figure 3.3) as well as the cyclic furan derivatives, where R and R′ are alkyl groups and Ar is an aromatic group in nature; so, it is not surprising that the more common oxygen-containing compounds occur in crude oil (Figure 3.3) (Speight, 2014). The total oxygen content of crude oil is usually less than 2% w/w, although larger amounts have been reported, but when the oxygen content is phenomenally high, it may be that the oil has suffered prolonged exposure to the atmosphere either during or after production. However, the oxygen content of crude oil increases with the boiling point of the fractions examined; in fact, the nonvolatile residua may have oxygen contents up to 8% w/w which is similar to the oxygen content of some viscous feedstocks.

In addition to the carboxylic acids and phenolic compounds in feedstocks, the presence of ketone derivatives, ester derivatives, ether derivatives, and anhydride derivatives has been claimed for a variety of crude oils. However, the precise identification of these compounds is difficult because most of them occur in the higher molecular weight nonvolatile residua. They are claimed to be products of the air blowing of the residua, and their existence in virgin crude oil may yet need to be substantiated.

In addition to the carboxylic acids and phenolic compounds (ArOH, where Ar is an aromatic moiety), the presence of ketones (>C=O), esters [>C(=O)–OR], ethers (R—O—R), and anhydrides [>C(=O)—O—(O=)C<] has been claimed for a variety of crude oils. However, the precise identification of these compounds is difficult because most of them occur in the higher molecular weight nonvolatile residua. They are claimed to be products of the air blowing of the residua, and their existence in conventional crude oil, heavy crude oil, extra heavy crude oil, or tar sand bitumen is still open to discussion.

3.3.3 Metallic Constituents

The occurrence of metals is a common feature of crude oils and the viscous feedstocks and the metals are typically found in the nonvolatile portion of the feedstock unless "volatilized" by entrainment during the distillation process (Reynolds, 1998; Speight, 2014, 2017).

The metallic constituents of a refinery feedstock can (an do) affect many upgrading processes and cause particular problems through catalyst poisoning. The trace metals nickel (Ni) and vanadium (V) are typically one or more orders of magnitude higher than other metals in refinery feedstocks, except when the feedstock is contaminated with coproduced brine salts that contain a variety of metals (such as sodium, Na, magnesium, Mg, and calcium, Ca) or other metals (such as iron, Fe) that contaminate the feedstock during storage and/or transportation.

The occurrence of metallic constituents in refinery feedstocks is of considerable interest to the refining industry than might be expected from the small amounts present. Even minute amounts of metals—particularly nickel and vanadium—in the feedstocks for catalytic cracking affect the activity of the catalyst and result in increased gas and coke formation and reduced yields of gasoline. In high-temperature power generators, such as oil-fired turbines, the presence of metallic constituents, particularly vanadium in the fuel, may lead to ash deposits on the turbine rotors, thus disturbing the balance of the rotors by reducing the clearances. More particularly, damage by corrosion may be very severe. The ash resulting from the combustion of fuels containing sodium and especially vanadium reacts with refractory furnace linings to lower their fusion points and so cause their deterioration.

Thus, the ash residue left after burning of a crude oil is due to the presence of these metallic constituents, part of which occur as inorganic water-soluble salts (mainly chlorides and sulfates of sodium, potassium, magnesium, and calcium) in the water phase of crude oil emulsions. These are removed in the desalting operations, either by evaporation of the water and subsequent water washing or by breaking the emulsion, thereby causing the original mineral content of the crude to be substantially reduced. Other metals are present in the form of oil-soluble organometallic compounds as complexes, metallic soaps, or in the form of colloidal suspensions, and the total ash from desalted crude oils is of the order of 0.1–100 mg/l.

Two groups of elements appear in significant concentrations in the original crude oil associated with well-defined types of compounds. Zinc, titanium, calcium, and magnesium appear in the form of organometallic soaps with surface-active properties adsorbed in the water/oil interfaces and act as emulsion stabilizers. However, vanadium, copper, nickel, and part of the iron found in crude oils seem to be in a different class and are present as oil-soluble compounds. These metals are capable of complexing with pyrrole pigment compounds derived from chlorophyll and hemoglobin and are almost certain to have been present in plant and animal source materials. It is easy to surmise that the metals in question are present in such form, ending in the ash content. Evidence for the presence of several other metals in oil-soluble form has been produced, and thus zinc, titanium, calcium, and magnesium compounds have been identified in addition to vanadium, nickel, iron, and copper. Examination of the analyses of a number of crude oil for iron, nickel, vanadium, and copper indicates a relatively high vanadium content, which usually exceeds that of nickel, although the reverse can also occur.

Distillation concentrates the metallic constituents in the residues (Reynolds, 1998), although some can appear in the higher boiling distillates, but the latter may be due in part to entrainment. Nevertheless, there is evidence that a portion of the metallic constituents may occur in the distillates by volatilization of the organometallic compounds present in the crude oil. In fact, as the percentage of overhead obtained by vacuum distillation of a reduced crude is increased, the amount of metallic constituents in the overhead oil is also increased. The majority of the vanadium, nickel, iron, and copper in residual stocks may be precipitated along with the asphaltene constituents by hydrocarbon solvents. Thus, removal of the asphaltene constituents with *n*-pentane reduces the vanadium content of the oil by up to 95% with substantial reductions in the amounts of iron and nickel.

3.3.4 Porphyrins

Finally, there is one class of feedstocks constituents that falls into both the nitrogen-containing constituents and the metal-containing constituents and that is the porphyrin derivatives. These chemical derivatives typically occur in a concentration on the order of parts-per-million, but this low concentration should not be ignored. These compounds can have serious deleterious effects on catalysts if they are not removed.

Significant amounts of porphyrin derivatives may occur in the nonbasic fraction of the nitrogen compounds. Metals are generally present in the form of organic salts dissolved in oil-emulsified water. Heavy crude oil often contains a high amount (in excess of several hundred parts-per-million) of nickel and vanadium, each of which forms chelates with porphyrins (Speight, 2014). These metallic constituents are responsible for catalyst contamination and corrosion problems (Parkash, 2003; Gary et al., 2007; Speight, 2014; Hsu and Robinson, 2017; Speight, 2017).

Porphyrins are a naturally occurring chemical species that exist in crude oil and usually occur in the nonbasic portion of the nitrogen-containing concentrate (Bonnett, 1978; Reynolds, 1998; Speight, 2014, 2017, 2021). Typically, the porphyrin constituents are not usually considered among the usual nitrogen-containing constituents of crude oil, nor are they considered a metallo-containing organic material that also occurs in some crude oils. As a result of these early investigations, there arose the concept of porphyrins as biomarkers that could establish a link between compounds found in the geosphere and their corresponding biological precursors.

Chemically, porphyrins are derivatives of porphine that consists of four pyrrole molecules joined by methine (—CH=) bridges (Figure 3.4). The methine bridges occur as conjugated carbon–carbon links between the pyrrole nuclei thereby contributing to a more extended resonance system. Although the resulting structure retains much of the character of the pyrrole components, the larger conjugated system gives increased conjugated (or aromatic-type) character to the porphine molecule. Furthermore, the imine functions (i.e., the —NH— function) within the porphine system allow metals such as nickel (as well as vanadium and iron) to be included into the molecule through chelation (Figure 3.5).

Almost all crude oils and (especially) the viscous feedstocks contain detectable amounts of vanadium and nickel porphyrin derivatives. More mature, lighter crude oils usually contain only small amounts of these

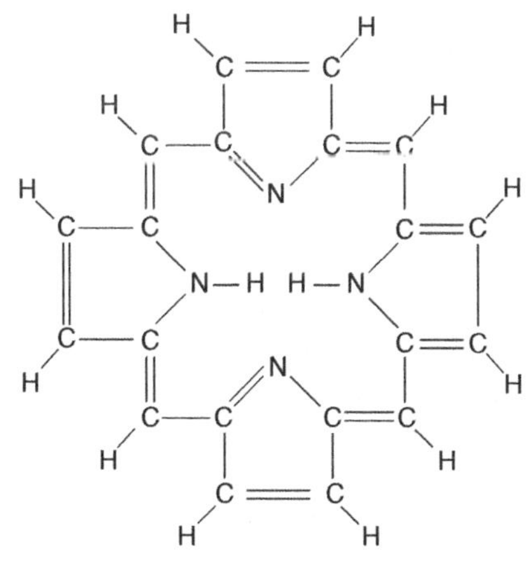

Figure 3.4 The structure of porphine.

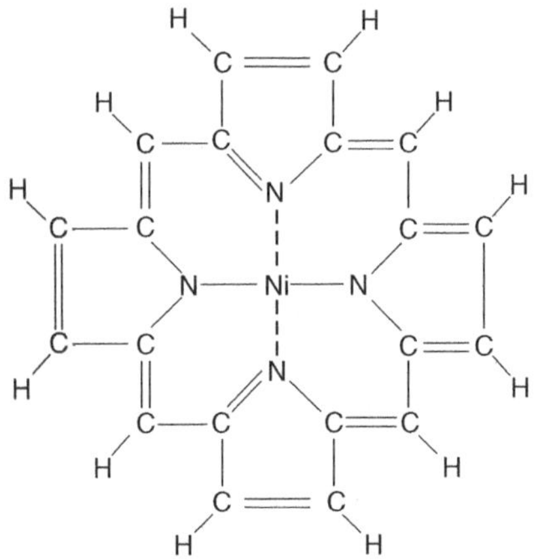

Figure 3.5 The nickel chelate of porphine.

compounds. Heavy crude oil (and the other viscous feedstocks) typically contain high amounts of nickel and vanadium porphyrin derivatives. It can be assumed (but should be checked by a suitable analytical method) that a concentration of vanadium on the order of (or in excess of) 1,000 ppm and a substantial amount of the vanadium in is in the form of a porphyrin chelate (Reynolds et al., 1987; Vogh and Reynolds, 1988; Reynolds, 1998; Speight 2014). Also, in high-sulfur crude oil of marine origin (i.e., formed from marine precursors and not necessarily recovered from offshore sources), vanadium porphyrin derivatives are often more abundant than nickel porphyrin derivatives. Low-sulfur crude oils of lacustrine origin (i.e., onshore lake systems) often contain more nickel porphyrin derivatives than vanadium porphyrin derivatives.

Of all the metals in the periodic table, only nickel and vanadium have been proven definitely to exist as chelates in significant amounts in a large number of crude oils and tar sand bitumen (Hodgson et al., 1967; Baker, 1969; Baker and Palmer, 1978; Baker and Louda, 1986; Filby and Van Berkel, 1987; Quirke, 1987; Speight, 2014, 2021).

If the vanadium and nickel contents of crude oils are measured and compared with porphyrin concentrations, it is usually found that not all the metal content can be accounted for as porphyrin constituents (Reynolds, 1998). In some crude oils, as little as 10% w/w of total metals appears to be chelated with porphyrins and it is rarely that all of the measured nickel and vanadium in a crude oil can be accounted for as porphyrin-type metals. Currently, some investigators believe that part of the vanadium and nickel in crude oils is chelated with ligands that are not porphyrins and are referred to as non-porphyrin metal chelates or non-porphyrin metal complexes (Reynolds et al., 1987; Vogh and Reynolds, 1988; Reynolds, 1998).

Finally, during the fractionation of crude oil, the metallic constituents (metalloporphyrins and non-porphyrin metal chelates) are concentrated in the asphaltene fraction (Speight, 1994, 2014, 2021). On the other hand, and in contrast to the asphaltene fraction, the deasphaltened oil contains lower concentrations of porphyrins than the parent materials and usually only small concentrations of non-porphyrin metals (Reynolds, 1998; Speight, 2021).

3.4 Chemical Composition by Distillation

The distillation process is particularly amenable to the separation of conventional crude oil into distillable fractions that can be identified; the viscous feedstocks are not so amenable to the process (Parkash, 2003; Gary et al., 2007; Speight, 2014; Hsu and Robinson, 2017; Speight, 2017, 2023). Distillation is a means of separating chemical compounds (usually liquids) through differences in the respective vapor pressures. In the mixture, the components evaporate, such that the vapor has a composition determined by the chemical properties of the mixture. Distillation of a given component is possible, if the vapor has a higher proportion of the given component than the mixture which is due to the given component having a higher vapor pressure and a lower boiling point than the other components.

By the nature of the process, it is theoretically impossible to completely separate and purify the individual components of a feedstock when the possible number of isomers are considered for the individual carbon numbers that occur within the paraffin family (Speight, 2014, 2023). When other types of compounds are included, such as the aromatic derivatives and heteroatom derivatives, even though the maturation process might limit the possible number of isomeric permutations (Tissot and Welte, 1978), the potential number of compounds in crude oil is still (in a sense) astronomical.

However, crude oil can be separated into a variety of fractions on the basis of the boiling points of the crude oil constituents. Such fractions are primarily identified by their respective boiling ranges and, to a lesser extent, by chemical composition. However, it is often obvious that as the boiling ranges increase, the nature of the constituents remains closely similar and it is the number of substituents that caused the increase in boiling point. It is through the recognition of such phenomena that the molecular design of the higher boiling constituents can be achieved (Speight, 2014, 2017). Invoking the existence of structurally different constituents in the nonvolatile fractions from those identifiable constituents in the lower boiling fractions is unnecessary and (considering the nature of the precursors and maturation paths) irrational (Speight, 1994, 2014). For example, the predominant types of condensed aromatic systems in crude oil are derivatives of phenanthrene and it is anticipated that the higher peri-condensed homologs shall be present in resin constituents and asphaltene constituents rather than the derivatives of kata-condensed polynuclear aromatic system (Speight, 2014).

The distillation process is the most important separation process in refineries insofar as the distillation characteristics of the feedstock provide distillation cutes (fractions) that contribute to knowledge of the composition of the feedstock (Speight, 2014, 2017, 2023). In terms of the viscous feedstocks, distillation methods (ASTM D2892, 2021; ASTM D5236, 2021) can be applied using temperatures up to 538 °C (1,000 °F) while spinning band distillation can be used at temperatures up to 450 °C (840 °F) and short path distillation can be used up to 700 °C (1,290 °F).

The former test method (ASTM D2892, 2021) presents details of the procedure for a viscous feedstock (in the method defined as a "residuum of standardized quality") on which analytical data can be obtained, and the determination of yields of the above fractions by both mass and volume. This allows a graph of temperature vs. % w/w distilled that can be produced. In addition, the latter test method (ASTM D5236, 2021) covers the procedure for the distillation of "heavy hydrocarbon mixtures" that have an initial boiling point greater than 150 °C (300 °F), such as heavy crude oil, tar sand bitumen, and crude oil residua. In order to accomplish the distillation, the method employs a pot still (a type of distillation apparatus or still used to distill liquids) with a low-pressure drop entrainment separator operated under total takeoff conditions.

Other methods include ASTM D2887 (2021), which is applicable to fractions with a final boiling point up to 538 °C (1,000 °F); ASTM D5307 (2021) for viscous feedstocks, which provides for sample recovery up to 538 °C (1,000 °F); ASTM D6352 (2021), which is applicable to fractions with a final boiling point up to 720 °C (1,330 °F); and ASTM D7169 (2021), which is applicable to viscous feedstocks and provides for sample recovery to 720 °C (1,330 °F).

The ASTM D2887 (2021) test method covers the determination of the boiling range distribution of crude oils and crude oil products and is applicable to petroleum products and fractions having a final boiling point of 538 °C (1,000 °F) or lower atmospheric pressure as measured by this test method, but the method is limited to feedstock samples having a vapor pressure and a sufficiently low amount of sediment to permit sampling at ambient temperature.

The ASTM D5307 (2021) method covers the determination of the boiling range distribution of water-free crude petroleum through 538 °C (1,000 °F). Material boiling above 538 °C (1,000 °F) is reported as a residue. This test method is applicable to samples that can be solubilized in a solvent to permit sampling by means of a micro-syringe.

The ASTM D6352 (2021) method covers the determination of the boiling range distribution of petroleum distillate fractions. The test method is applicable to petroleum distillate fractions having an initial boiling point greater than 174 °C (345 °F) and a final boiling point of less than 700 °C (1,292 °F) (C10–C90) at atmospheric pressure as measured by this test method.

While physical distillation is still considered the preferred method for distillation, the simulated distillation (SIMDIST) method that employs gas chromatography (GC) offers advantages over the physical procedure. Analysis by GC typically has better precision, higher throughput, less hands-on time, and lower cost per sample and the SIMDIST method requires considerably less sample.

Knowledge of the boiling point distribution of stabilized crude oils is important for the marketing, scheduling, and processing of crude oil in the petroleum industry. ASTM methods D7900 and ASTM D7169 can be combined to determine the boiling point distribution of various feedstocks. For feedstocks that contain a significant amount of volatile constituents ("light ends"), this method is recommended and is frequently employed for the determination of the boiling point distribution of feedstocks with a final boiling point on the order of 720 °C (1,330 °F). However, the method may not give a complete result for the separation of lower boiling (C4–C9) constituents, due to the thin film column in the presence of large amounts of carbon disulfide. However, combining the results of two methods (ASTM D7169, 2021; ASTM D7900, 2021) will assist in the analysis of any low boiling fraction of a feedstock up to, and including the nonane (C9) constituents.

Although distillation is of lesser value when used to identify the composition of the viscous feedstocks, it is appropriate to mention distillation here insofar as it is a method by which the constituents of crude oil can be separated and identified. Distillation is a means of separating chemical compounds (usually liquids) through differences in the respective vapor pressures. In the mixture, the components evaporate, such that the vapor has a composition determined by the chemical properties of the mixture. Distillation of a given component is possible, if the vapor has a higher proportion of the given component than the mixture. This is caused by the given component having a higher vapor pressure and a lower boiling point than the other components.

By the nature of the process, it is theoretically impossible to completely separate and purify the individual components of crude oil when the possible number of isomers is considered for the individual carbon numbers that occur within the paraffin family (Table 3.1). When other types of compounds are included, such as the aromatic derivatives and heteroatom derivatives, even though the maturation process might limit the possible number of isomeric permutations, the potential number of compounds in crude oil is still (in a sense) very high (Tissot and Welte, 1978; Tissot, 1984).

However, as the boiling ranges of the distillates increase, the nature and properties of the constituents remain closely similar as it is through the recognition of such phenomena that the molecular design of the higher boiling constituents can be achieved (Parkash, 2003; Gary et al., 2007; Speight, 2014; Hsu and Robinson, 2017; Speight, 2017). Moreover, declaring that the existence of structurally different constituents in the nonvolatile fractions from the identifiable constituents in the lower boiling fractions is unnecessary and (considering the nature of the precursors and maturation paths) is not always a rational conclusion (Speight, 1994, 2014, 2023). For example, the predominant types of condensed aromatic systems in crude oil are derivatives of phenanthrene and it is anticipated that the higher peri-condensed homologs (Figure 3.6a) are present in the resins fraction and in the asphaltene fraction rather than the derivatives of kata-condensed polynuclear aromatic system (Figure 3.6b). However, this type of fractionation is very limited when applied to the relatively nonvolatile viscous feedstocks, although it may be possible to produce a VGO fraction from the viscous feedstocks.

For the purposes of this chapter and in deference to the viscous feedstocks, the lower boiling fraction isolated by distillation from conventional crude oil (i.e., the gases, naphtha, kerosene, other middle distillates, and the atmospheric gas oil) is not included here. Thus, the focus is on the VGO and the associated residua since both fractions, by virtue of their respective properties, is included in the "viscous feedstocks" category.

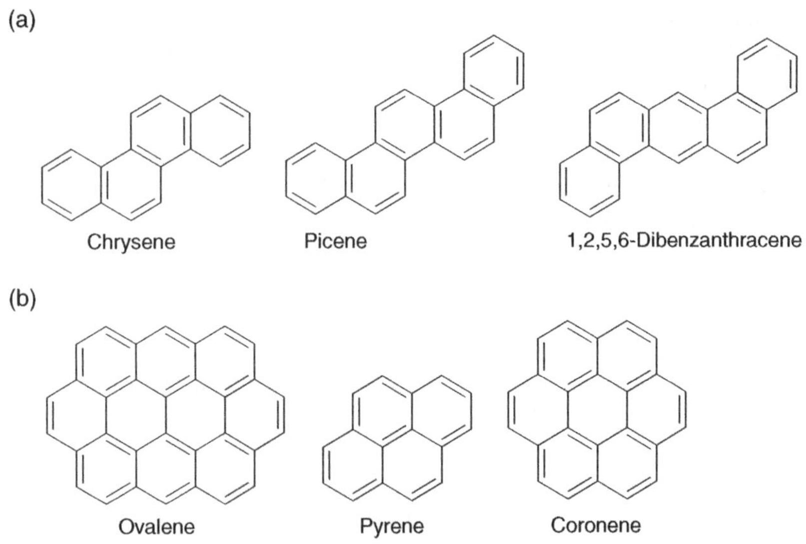

Figure 3.6 (a) Examples of peri-condensed aromatic systems. (b) Examples of kata-condensed (cata-condensed) aromatic systems.

3.4.1 Vacuum Gas Oil

The saturate constituents contribute less to the VGO than the aromatic constituents but more than the polar constituents that are now present at percentage rather than trace levels.

Within the VGO, saturates, distribution of paraffins, *iso*-paraffins, and naphthenes is highly dependent upon the crude oil source. Generally, the naphthene constituents account for approximately two-thirds (60%) of the saturate constituents but the overall range of variation is from <20% to >80%. In most samples, the *n*-paraffins from C20 to C44 are still present in sufficient quantity to be detected as distinct peaks in gas chromatographic analysis.

The bulk of the saturated constituents in VGO consist of *iso*-paraffins and especially naphthene species, although isoprenoid compounds, such as squalane (C30) and lycopane (C40), have been detected. Analytical techniques show that the naphthene derivatives contain from one to more than six fused rings accompanied by alkyl substitution of the available ring hydrogen atoms. For mono-aromatic derivative and di-aromatic derivatives, the alkyl substitution typically involves incorporation of several methyl and ethyl substituents in place of ring hydrogen atoms. Hopane derivatives and sterane derivatives have also been identified and are also used as internal markers for estimating biodegradation of crude oils during bioremediation processes (Prince et al., 1994).

The aromatic constituents in the VGO fraction may (and often do) contain one to six fused aromatic rings that may have attachments such as naphthene rings and alkyl substituents in keeping with the boiling range. Mono-aromatic derivatives and di-aromatic derivatives can account for an amount on the order of 50% w/w of the aromatic derivatives and can be four fused naphthenic rings on some aromatic compounds. This is consistent with the suggestion that these species originate from the aromatization of steroids. Although present at lower concentration, alkyl benzene derivatives and naphthalene derivatives may carry one long alkyl side chain and multiple short alkyl side chains.

The fused ring aromatic derivatives (having three or more rings) in crude oil include phenanthrene, chrysene, and picene as well as fluoranthene, pyrene, benzo(a)pyrene, and benzo(ghi)perylene. The most abundantly reported individual phenanthrene compounds appear to be the three-derivatives. In addition, phenanthrene derivatives outnumber anthracene derivatives by as much as 100:1. In addition, chrysene derivatives are favored over pyrene derivatives.

Heterocyclic constituents are significant contributors to the VGO fraction. In terms of sulfur compounds, thiophene sulfur tends to predominate over sulfide sulfur (Figure 3.1). However, some constituents may contain more than one sulfur atom. The benzothiophene derivatives and the dibenzothiophene derivatives are the prevalent thiophene forms of sulfur as thiophene derivatives. In the VGO range, the nitrogen-containing compounds include higher molecular weight pyridine derivatives, quinoline derivatives, benzoquinoline derivatives, amide derivatives, indole derivatives, carbazole derivatives, and in some cases constituents that contain two nitrogen atoms in the molecule as well as three and four aromatic rings may be prevalent (Green et al., 1989; Speight, 2014).

Typically, approximately one-third of the compounds are basic, that is, pyridine and its benzo-derivatives while the remainder are present as neutral species (amide derivatives and carbazole derivatives). Although benzo- and dibenzoquinoline derivatives found in crude oil are rich in sterically hindered structures, hindered and unhindered structures have been found to be present at equivalent concentrations in source rocks. This has been rationalized as geo-chromatography in which the less polar (hindered) structures moved more readily to the reservoir and are not adsorbed on any intervening rocks structures.

Oxygen levels in the VGO parallel the nitrogen content. Thus, the most commonly identified oxygen compounds are the carboxylic acids and phenols, collectively called naphthenic acids (Seifert and Teeter, 1970).

3.4.2 Vacuum Residua

The vacuum residuum (vacuum bottoms, typically 950 °F$^+$ or 1,050 °F$^+$) is the most complex of crude oil and, in many cases, may even resemble heavy oil or extra heavy oil or tar sand bitumen in composition. Vacuum residua contain the majority of the heteroatoms originally in the crude oil and molecular weight of the constituents range up to several thousand (as near as can be determined but subject to method dependence). The fraction is so complex that the characterization of individual species is virtually impossible, no matter what claims have been made or will be made. Separation of vacuum residua into fractions by group type methods (Chapter 3) can become difficult and confused because of the multi-substitution of aromatic and naphthenic species as well as by the presence of multiple functionalities in single molecules.

For the 565 °C$^+$ (1,050 °F$^+$) fractions of crude oil, the levels of nitrogen and oxygen may begin to approach the concentration of sulfur. These elements consistently concentrate in the most polar fractions to the extent that every molecule contains more than one heteroatom. At this point, structural identification is somewhat fruitless and characterization techniques are used to confirm the presence of the functionalities found in lower boiling fractions such as, for example, acids, phenols, nonbasic nitrogen constituents (constituents containing carbazole-type) and basic nitrogen constituents (constituents containing quinoline-type) nitrogen.

The nickel and vanadium constituents that are concentrated into the vacuum residuum appear to occur in two forms: (1) porphyrins and (2) non-porphyrins (Reynolds, 1998). Because the metalloporphyrins can provide insights into crude oil maturation processes, they have been studied extensively and several families of related structures have been identified. On the other hand, the non-porphyrin metals remain not clearly identified although some studies suggest that some of the metals in these compounds still exist in a tetra-pyrrole (porphyrin-type) environment (Pearson and Green, 1993).

It is more than likely that, in a specific residuum molecule, the heteroatoms are arranged in different functionalities, thereby contributing to extremely complex constituents (Speight, 2014, 2017, 2021). Furthermore, when the many different combinations that are possible are considered, the chances of determining every structure in a viscous feedstock (including residua) is on a low-to-zero level. Because of this seemingly insurmountable task, it may be better to determine ways of utilizing the viscous feedstocks rather attempting to determine (at best questionable) molecular structures, especially attempts to assign a so-called "average structure" for the feedstock by means of the method known as "structural group analysis" which may be limited in accuracy and useful application (Chapter 9).

Typically, precipitation of an insoluble fraction by the addition of *n*-pentane or *n*-heptane to the residuum is used as the initial step for the characterization of vacuum residuum (ASTM D3279, 2015; ASTM D4124, 2015). The insoluble fraction, the pentane- or heptane-asphaltenes, may be as much as 50% w/w of a vacuum residuum. The pentane- or heptane-soluble portion (maltene) constituents of the residuum are then fractionated chromatographically into several solubility or adsorption classes for characterization. However, in spite of claims to the contrary, the method is not a separation by chemical type but it is a separation based on solubility. However, the separation of the asphaltene constituents does, however, provide a simple way to remove some of the highest molecular weight and most polar components of the residuum but the asphaltene fraction is so complex that compositional detail based on average parameters is of questionable value.

3.5 Chemical Composition by Spectroscopy

Analytical spectroscopy-based methods can be used to investigate the reactivity, structures, and associations of petroleum constituents in the environment. Because analytical environmental chemistry deals with very complex naturally occurring mixtures, new spectroscopic approaches must be utilized or further developed to fully understand global environmental processes involved in contamination by petroleum constituents and their subsequent remediation. In addition, it is worthy of note that the majority of the spectroscopic methods were designed for volatile materials (and are referenced here) but it is worthy of note that modification of the methods can produce modified methods that can be applied to nonvolatile materials.

Moreover, continually evolving environmental legislation and the introduction of stringent environmental quality standards relating to environmental contaminants continues to provide significant analytical challenges to traditional methods of analysis. Advancements in spectroscopic techniques and associated chromatographic instrumentation have become indispensable in the environmental analysis laboratory. Without some of the recent advancements in mass spectrometry, many of the applications (discussed below) would have not been possible without significant additional effort (Kanda and Glendinning, 2011).

Thus, spectroscopic studies continue to play an important role in the evaluation of petroleum and of petroleum products and many of the methods are now used as standard methods of analysis of petroleum and its products before and after a spill. Application of these methods to petroleum and its products is a natural consequence for the process scientist and engineer.

3.5.1 Infrared Spectroscopy

Conventional infrared spectroscopy yields information about the functional features of various petroleum constituents. For example, infrared spectroscopy will aid in the identification of N—H and O—H functions, the nature of polymethylene chains, the C—H out-of-place bending frequencies, and the nature of any polynuclear aromatic systems (Speight, 2014, 2015).

Infrared spectroscopy is used for the determination of benzene in motor and/or aviation gasoline (ASTM D4053, 2021) while ultraviolet spectroscopy is employed for the evaluation of mineral oils (ASTM D2269, 2021) and for determining the naphthalene content of aviation turbine fuels (ASTM D1840, 2021). With the recent progress of Fourier transform infrared (FTIR) spectroscopy, which allows the analysis of a relevant amount of compositional and structural information concerning environmental samples, quantitative estimates of the various functional groups can also be made (Mecozzi et al., 2009). This is particularly important for application to the higher molecular weight solid constituents of petroleum (i.e., the asphaltene fraction).

3.5.2 Nuclear Magnetic Resonance Spectroscopy

Because nuclear magnetic resonance is one of the premier techniques for studying intermolecular interactions, it has also been used extensively to investigate the environmental chemistry of humic substances through their interactions with organic contaminants and toxic metals (Nanny et al., 1997; Cardoza et al., 2004).

Nuclear magnetic resonance has frequently been employed for general studies and for the structural studies of petroleum constituents (Bouquet and Bailleul, 1982; Hasan et al., 1989). In fact, proton magnetic resonance (PMR) studies (along with infrared spectroscopic studies) were, perhaps, the first studies of the modern era that allowed structural inferences to be made about the polynuclear aromatic systems that occur in the high molecular weight constituents of petroleum.

The nuclear magnetic resonance spectroscopy technique has been developed as a standard method for the determination of hydrogen types in various fuels (ASTM D3701, 2015) and can, with the necessary modifications applied, also be extended to other products. X-ray fluorescence spectrometry has been applied to the determination of lead in gasoline as well as to the determination of sulfur in various petroleum products (ASTM D2622, 2015; ASTM D4294, 2015).

Carbon-13 magnetic resonance (^{13}CMR) can play a useful role since carbon magnetic resonance deals with analyzing the carbon distribution types. The obvious structural parameter to be determined is the aromaticity, f_a. In fact, the direct determination from the various environments of the carbon types is one of the better methods for the determination of aromaticity. Thus, through a combination of proton and carbon magnetic resonance techniques, refinements can be made on the structural parameters and, for the solid-state high-resolution carbon magnetic resonance technique, additional structural parameters can be obtained.

3.5.3 Mass Spectrometry

Mass spectrometry can play a key role in the identification of the constituents of feedstocks and products (Aczel, 1989). The principal advantages of mass spectrometric methods are (1) high reproducibility of quantitative analyses, (2) the potential for obtaining detailed data on the individual components and/or carbon number homologs in complex mixtures, and (3) a minimal sample size is required for analysis. The ability of mass spectrometry to identify individual components in complex mixtures is unmatched by any modern analytical technique, perhaps the exception is GC. The methods include the use of *mass spectrometry* to determine the (1) hydrocarbon types in middle distillates (ASTM D2425, 2021), (2) hydrocarbon types of gas oil saturate fractions (ASTM D2786, 2021), (3) hydrocarbon types in low-olefin gasoline (ASTM D2789, 2021), and (d) aromatic types of gas oil aromatic fractions (ASTM D3239, 2021).

However, there are disadvantages arising from the use of mass spectrometry and these are: (1) the limitation of the method to organic materials that are volatile and stable at temperatures up to 300 °C (570 °F) and (2) the difficulty of separating isomers for absolute identification. The sample is usually destroyed, but this is seldom a disadvantage.

Mass spectrometry coupled to GC and liquid chromatography are ideal techniques for the determination of organic contaminants in wastewater due to the separation capability of the gas chromatographic and the liquid chromatographic method coupled with the selectivity and sensitivity of the mass spectrometric technique. In fact, coupling a chromatographic method with mass selective detection by mass spectrometry allows the routine determination of many contaminants in parts-per-trillion (ppt) or even lower concentrations. In recent years, developments in chromatography such as the introduction of ultra-high-performance liquid chromatography, which utilizes the same separation methodology as conventional high-performance liquid chromatography, but uses columns packed with smaller particles (generally about 2 μm particle size or less) has further enhanced the applicability of liquid chromatography-mass spectrometry in wastewater analysis.

Gas chromatography-mass spectrometry has become the technique of choice for volatile and semi-volatile, nonpolar compounds including polynuclear aromatic hydrocarbons (PNAs), polychlorinated biphenyls (PCB), pesticides, for example, organo-chlorine insecticides, dioxins, and other volatile priority pollutants. Constituents containing polar functional groups are difficult to analyze by GC because of either (1) the constituents are not sufficiently volatile, (2) the constituents fail badly even on polar analytical columns, and (3) the constituents are thermally unstable and decompose in the GC or mass spectrometry interface source.

Chemical derivatization of samples prior to analysis can be used to overcome some of these problems. Derivatization allows polar compounds to be modified to produce new compounds that have properties which are suitable for analysis using gas chromatography-mass spectrometry. Common chemical derivatization methods include: (1) silylation: used to volatilize compounds, (2) alkylation: used as the first step to further derivatization or as a method of protection of certain active hydrogen atoms, and (3) acylation: used to add fluorinated groups.

GC or mass spectrometry is most often used in positive ion electron ionization mode in either (1) a full scan mode, which is useful in determining unknown compounds in a sample or confirming compounds present or in (2) the selected ion monitoring mode in which only certain ion fragments specific to the analytes are monitored but allows more scans to take place on a per second basis. Ion fragmentation patterns in electron ionization are dependent upon the electron energy applied, typically 70 eV is used to facilitate comparison of generated spectra with library spectra.

Chemical ionization is sometimes used for compounds chemically similar to those analyzed by electron ionization to enhance the abundance of the molecular ion resulting in less fragmentation. In positive ion chemical ionization, a chemical reagent gas (for example, methane, ammonia, isobutene) is used. The reagent gas is present at a much higher concentration than the analyte and is ionized by electron ionization to produce primary reagent ions and the collision of these ions with neutral reagent molecules leads to the formation of stable secondary ions, which are the reactant species that then ionize analyte molecules by ion–molecule reactions.

Negative chemical ionization can be performed when an analyte contains electron-capturing moieties (for example, halogenated compounds). In the negative ion chemical ionization technique, there are two primary mechanisms whereby negative ions are produced: electron capture and reactant ion chemical ionization. Under chemical ionization conditions, electronegative molecules can capture electrons to generate negative ions. True negative ion chemical ionization occurs by reaction of an analyte compound with negatively charged reactant ions. Several types of ion–molecule reactions can occur, the most common being proton abstraction of the analyte.

3.5.4 Other Techniques

Other techniques include the use of flame emission spectroscopy for determining trace metals in gas turbine fuels (ASTM D3605, 2015) and the use of absorption spectrophotometry for the determination of the alkyl nitrate content of diesel fuel. Atomic absorption has been employed as a means of measuring the lead content of gasoline (ASTM D3237, 2021) and also for the manganese content of gasoline (ASTM D3831, 2021) as well as for determining the barium, calcium, magnesium, and zinc content of viscous oils such as lubricating oils (ASTM D4628, 2021). Flame photometry has been employed as a means of measuring the lithium/sodium content of lubricating greases (ASTM D3340, 2015) and the sodium content of residual fuel oil (ASTM D1318, 2015).

Liquid chromatography is a fundamental analytical technique in environmental analysis used for the determination of a wide range of compounds, including the high molecular weight compounds that are common to the viscous feedstocks as well as thermally labile constituents of feedstocks that are not suitable for application of the gas chromatographic method. In fact, liquid chromatography-mass spectrometry method is an ideal technique for the routine analysis of organic compounds in environmental analysis. For most compounds, a mass spectrometer is more sensitive and has greater specificity than other liquid chromatographic detectors.

Many of the advancements in liquid chromatography-mass spectrometry have been in the development of ionization techniques (interface/ion source). Earlier liquid chromatography-mass spectrometry systems used interfaces that did not separate the mobile phase from analyte molecules (direct liquid inlet, thermospray) and were ineffective before the sample ionized. The analyte molecules are ionized in the mass spectrometer under vacuum, often by traditional electron ionization.

The introduction of atmospheric pressure ionization techniques greatly expanded the applicability of liquid chromatography-mass spectrometry. In atmospheric pressure ionization liquid chromatography-mass spectrometry, the analyte molecules are ionized at atmospheric pressure. The ionized constituents are then mechanically and electrostatically separated from neutral constituents.

Atmospheric pressure ionization techniques are: (1) electrospray ionization, (2) atmospheric pressure chemical ionization, and (3) atmospheric pressure photoionization.

Electrospray ionization is used for the analysis of a wide range of compounds including large biomolecules. Atmospheric pressure chemical ionization is used for molecules less than 1,500 µm and is less well-suited than electrospray for analysis of large biomolecules and thermally unstable compounds. Atmospheric pressure chemical ionization is used with normal-phase chromatography more often than electrospray because the analytes are usually nonpolar. Atmospheric pressure photoionization is applicable to many of the compounds that are suitable for analysis by atmospheric pressure chemical ionization. Atmospheric pressure photoionization is particularly suitable for the analysis of highly nonpolar compounds.

References

Aczel, T. 1989. *Preprints. Div. Petrol. Chem. Am. Chem. Soc.*, 34(2): 318.

Arjoon, K.K., and Speight, J.G. 2023. *Petroleum Biodegradation and Oil Spill Bioremediation*. CRC Press, Taylor & Francis Group, Boca Raton, Florida.

ASTM D1018. 2015. Standard Test Method for Hydrogen in Petroleum Fractions. Annual Book of Standards. ASTM International, West Conshohocken, Pennsylvania.

ASTM D1266. 2015. Standard Test Method for Sulfur in Petroleum Products (Lamp Method). Annual Book of Standards. ASTM International, West Conshohocken, Pennsylvania.

ASTM D1318. 2015. Standard Test Method for Sodium in Residual Fuel Oil (Flame Photometric Method). Annual Book of Standards. ASTM International, West Conshohocken, Pennsylvania.

ASTM D1552. 2015. Standard Test Method for Sulfur in Petroleum Products (High-Temperature Method). Annual Book of Standards. ASTM International, West Conshohocken, Pennsylvania.

ASTM D1757. 2015. Standard Test Method for Sulfur in Ash from Coal and Coke. Annual Book of Standards. ASTM International, West Conshohocken, Pennsylvania.

ASTM D1840. 2021. Standard Test Method for Naphthalene Hydrocarbons in Aviation Turbine Fuels by Ultraviolet Spectrophotometry. Annual Book of Standards. ASTM International, West Conshohocken, Pennsylvania.

ASTM D2269. 2021. Standard Test Method for Evaluation of White Mineral Oils by Ultraviolet Absorption. Annual Book of Standards. ASTM International, West Conshohocken, Pennsylvania.

ASTM D2425. 2021. Standard Test Method for Hydrocarbon Types in Middle Distillates by Mass Spectrometry. Annual Book of Standards. ASTM International, West Conshohocken, Pennsylvania.

ASTM D2622. 2015. Standard Test Method for Sulfur in Petroleum Products by Wavelength Dispersive X-ray Fluorescence Spectrometry. Annual Book of Standards. ASTM International, West Conshohocken, Pennsylvania.

ASTM D2786. 2021. Standard Test Method for Hydrocarbon Types Analysis of Gas-Oil Saturates Fractions by High Ionizing Voltage Mass Spectrometry. Annual Book of Standards. ASTM International, West Conshohocken, Pennsylvania.

ASTM D2789. 2021. Standard Test Method for Hydrocarbon Types in Low Olefinic Gasoline by Mass Spectrometry. Annual Book of Standards. ASTM International, West Conshohocken, Pennsylvania.

ASTM D2887. 2021. Standard Test Method for Boiling Range Distribution of Petroleum Fractions by Gas Chromatography. Annual Book of Standards. ASTM International, West Conshohocken, Pennsylvania.

ASTM D2892. 2021. Standard Test Method for Distillation of Crude Petroleum (15-Theoretical Plate Column). Annual Book of Standards. ASTM International, West Conshohocken, Pennsylvania.

ASTM D3177. 2015. Standard Test Methods for Total Sulfur in the Analysis Sample of Coal and Coke. Annual Book of Standards. ASTM International, West Conshohocken, Pennsylvania.

ASTM D3178. 2015. Standard Test Methods for Carbon and Hydrogen in the Analysis Sample of Coal and Coke. Annual Book of Standards. ASTM International, West Conshohocken, Pennsylvania.

ASTM D3179. 2015. Standard Test Methods for Nitrogen in the Analysis Sample of Coal and Coke. Annual Book of Standards. ASTM International, West Conshohocken, Pennsylvania.

ASTM D3228. 2015. Standard Test Method for Total Nitrogen in Lubricating Oils and Fuel Oils by Modified Kjeldahl Method. Annual Book of Standards. ASTM International, West Conshohocken, Pennsylvania.

ASTM D3237. 2021. Standard Test Method for Lead in Gasoline by Atomic Absorption Spectroscopy. Annual Book of Standards. ASTM International, West Conshohocken, Pennsylvania.

ASTM D3239. 2021. Standard Test Method for Aromatic Types Analysis of Gas-Oil Aromatic Fractions by High Ionizing Voltage Mass Spectrometry. Annual Book of Standards. ASTM International, West Conshohocken, Pennsylvania.

ASTM D3279. 2015. Standard Test Method for *n*-Heptane Insolubles. Annual Book of Standards. ASTM International, West Conshohocken, Pennsylvania.

ASTM D3340. 2015. Standard Test Method for Lithium and Sodium in Lubricating Greases by Flame Photometer. Annual Book of Standards. ASTM International, West Conshohocken, Pennsylvania.

ASTM D3341. 2015. Standard Test Method for Lead in Gasoline—Iodine Monochloride Method. Annual Book of Standards. ASTM International, West Conshohocken, Pennsylvania.

ASTM D3343. 2015. Standard Test Method for Estimation of Hydrogen Content of Aviation Fuels. Annual Book of Standards. ASTM International, West Conshohocken, Pennsylvania.

ASTM D3605. 2015. Standard Test Method for Trace Metals in Gas Turbine Fuels by Atomic Absorption and Flame Emission Spectroscopy. Annual Book of Standards. ASTM International, West Conshohocken, Pennsylvania.

ASTM D3701. 2015. Standard Test Method for Hydrogen Content of Aviation Turbine Fuels by Low Resolution Nuclear Magnetic Resonance Spectrometry. Annual Book of Standards. ASTM International, West Conshohocken, Pennsylvania.

ASTM D3831. 2021. Standard Test Method for Manganese in Gasoline by Atomic Absorption Spectroscopy. Annual Book of Standards. ASTM International, West Conshohocken, Pennsylvania.

ASTM D4045. 2015. Standard Test Method for Sulfur in Petroleum Products by Hydrogenolysis and Rateometric Colorimetry. Annual Book of Standards. ASTM International, West Conshohocken, Pennsylvania.

ASTM D4053. 2021. Standard Test Method for Benzene in Motor and Aviation Gasoline by Infrared Spectroscopy. Annual Book of Standards. ASTM International, West Conshohocken, Pennsylvania.

ASTM D4124. 2015. Standard Test Method for Separation of Asphalt into Four Fractions. Annual Book of Standards. ASTM International, West Conshohocken, Pennsylvania.

ASTM D4294. 2015. Standard Test Method for Sulfur in Petroleum and Petroleum Products by Energy Dispersive X-ray Fluorescence Spectrometry. Annual Book of Standards. ASTM International, West Conshohocken, Pennsylvania.

ASTM D4628. 2021. Standard Test Method for Analysis of Barium, Calcium, Magnesium, and Zinc in Unused Lubricating Oils by Atomic Absorption Spectrometry. Annual Book of Standards. ASTM International, West Conshohocken, Pennsylvania.

ASTM D4929. 2021. Standard Test Method for Determination of Organic Chloride Content in Crude Oil. Annual Book of Standards. ASTM International, West Conshohocken, Pennsylvania.

ASTM D5236. 2021. Standard Test Method for Distillation of Heavy Hydrocarbon Mixtures (Vacuum Potstill Method). Annual Book of Standards. ASTM International, West Conshohocken, Pennsylvania.

ASTM D5307. 2021. Standard Test Method for Determination of Boiling Range Distribution of Crude Petroleum by Gas Chromatography. Annual Book of Standards. ASTM International, West Conshohocken, Pennsylvania.

ASTM D6352. 2021. Standard Test Method for Boiling Range Distribution of Petroleum Distillates in Boiling Range from 174 °C to 700 °C by Gas Chromatography. Annual Book of Standards. ASTM International, West Conshohocken, Pennsylvania.

ASTM D6470. 2021. Standard Test Method for Salt in Crude Oils (Potentiometric Method). Annual Book of Standards. ASTM International, West Conshohocken, Pennsylvania.

ASTM D7169. 2021. Standard Test Method for Boiling Point Distribution of Samples with Residues Such as Crude Oils and Atmospheric and Vacuum Residues by High Temperature Gas Chromatography. Annual Book of Standards. ASTM International, West Conshohocken, Pennsylvania.

ASTM D7900. 2021. Standard Test Method for Determination of Light Hydrocarbons in Stabilized Crude Oils by Gas Chromatography. Annual Book of Standards. ASTM International, West Conshohocken, Pennsylvania.

ASTM E258. 2015. Standard Test Method for Total Nitrogen in Organic Materials by Modified Kjeldahl Method. Annual Book of Standards. ASTM International, West Conshohocken, Pennsylvania.

ASTM E385. 2015. Standard Test Method for Oxygen Content Using a 14-MeV Neutron Activation and Direct-Counting Technique. Annual Book of Standards. ASTM International, West Conshohocken, Pennsylvania.

ASTM E777. 2015. Standard Test Method for Carbon and Hydrogen in the Analysis Sample of Refuse-Derived Fuel. Annual Book of Standards. ASTM International, West Conshohocken, Pennsylvania.

ASTM E778. 2015. Standard Test Methods for Nitrogen in Refuse-Derived Fuel Analysis Samples. Annual Book of Standards. ASTM International, West Conshohocken, Pennsylvania.

Baker, E.W. 1969. In: *Organic Geochemistry*. G. Eglinton and M.T.J. Murphy (Editors). Springer-Verlag, New York.

Baker, E.W., and Louda, J.W. 1986. In: *Biological Markers in the Sedimentary Record*. R.B. Johns (Editor). Elsevier, Amsterdam, Netherlands.

Baker, E.W., and Palmer, S.E. 1978. In: *The Porphyrins. Volume I. Structure and Synthesis. Part A*. D. Dolphin (Editor). Academic Press, New York.

Bonnett, R. 1978. In: *In The Porphyrins. Volume I. Structure and Synthesis. Part A*. D. Dolphin (Editor). Academic Press, New York.

Bouquet, M., and Bailleul, A. 1982. In: *Petroananlysis '81. Advances in Analytical Chemistry in the Petroleum Industry 1975–1982*. G.B. Crump (Editor). John Wiley & Sons, Chichester, England.

Cardoza, L., Korir, A., Otto, W., Wurrey, C., and Larive, C. 2004. Applications of NMR Spectroscopy in Environmental Science. *Progress in Nuclear Magnetic Resonance Spectroscopy*, 45(3-4): 209–238.

Charbonnier, R.P., Draper, R.G., Harper, W.H., and Yates, A. 1969. Analyses and Characteristics of Oil Samples from Alberta. Information Circular IC 232. Department of Energy Mines and Resources, Mines Branch, Ottawa, Canada.

Draper, R.G., Kowalchuk, E., and Noel, G. 1977. Analyses and Characteristics of Crude Oil Samples Performed Between 1969 and 1976. Report ERP/ERL 77-59 (TR). Energy, Mines, and Resources, Ottawa, Canada.

Filby, R.H., and Van Berkel, G.J. 1987. In: *Metal Complexes in Fossil Fuels*. R.H. Filby and J.F. Branthaver (Editors), Symposium Series No. 344. American Chemical Society, Washington, DC. p. 2

Gary, J.G., Handwerk, G.E., and Kaiser, M.J. 2007. *Petroleum Refining: Technology and Economics*. 5th Edition. CRC Press, Taylor & Francis Group, Boca Raton, Florida.

Green, J.A., Green, J.B., Grigsby, R.D., Pearson, C.D., Reynolds, J.W., Sbay, I.Y., Sturm, O.P. Jr., Thomson, J.S., Vogh, J.W., Vrana, R.P., Yu, S.K.Y., Diem, B.H., Grizzle, P.L., Hirsch, D.E., Hornung, K.W., Tang, S.Y., Carbognani, L., Hazos, M., and Sanchez, V. 1989. Analysis of Heavy Oils: Method Development and Application to Cerro Negro Heavy Petroleum. NIPER-452 (DE90000200). Volumes I and II. Research Institute, National Institute for Petroleum and Energy Research (NIPER), Bartlesville, Oklahoma.

Hasan, M., Ali, M.F., and Arab, M. 1989. Structural Characterization of Saudi Arabian Extra Light and Light Crudes by ^{1}H and ^{13}C NMR Spectroscopy. *Fuel*, 68: 801–803.

Hodgson, G.W., Baker, B.L., and Peake, E. 1967. In: *Fundamental Aspects of Petroleum Geochemistry*. B. Nagy and U. Columbo (Editors). Elsevier, Amsterdam, Netherlands. Chapter 5

Hsu, C.S., and Robinson, P.R. (Editors). 2017. *Handbook of Petroleum Technology*. Springer International Publishing AG, Cham, Switzerland.

Kanda, R., and Glendinning, R. 2011. Mass Spectrometry for Environmental and Wastewater Monitoring. *Spectroscopy Europe*, 23(5): 14–27.

Mecozzi, M., Moscato, F., Pietroletti, M., Quarto, F., Oteri, F., and Cicero, A.M. 2009. Applications of FTIR Spectroscopy in Environmental Studies Supported by Two Dimensional Correlation Analysis. *Global NEST Journal*, 11(4): 593–600.

Nanny, M.A., Minear, R.A., and Leenheer, J.A. (Editors). 1997. *Nuclear Magnetic Resonance Spectroscopy in Environmental Chemistry*. Oxford University Press, Oxford, United Kingdom.

Overton, E.B., Wade, T.L., Radović, J.R., Meyer, B.M., Miles, M.S., and Larter, S.R. 2016. Chemical Composition of Macondo and Other Crude Oils and Compositional Alterations During Oil Spills. *Oceanography*, 29(3): 50–63.

Parkash, S. 2003. *Refining Processes Handbook*. Gulf Professional Publishing, Elsevier, Amsterdam, Netherlands.

Pearson, C.D., and Green, J.B. 1993. Vanadium and Nickel Complexes in Petroleum Resid Acid, Base, and Neutral Fractions. *Energy & Fuels*, 7: 338–346.

Prince, R.O., Elmendoff, D.L., Lute, B.R., Hsu, C.S., Hath, C., Sunnis, B.P., Decherd, G., Douglas, D., and Butler, E. 1994. 17.Alpha.(H)-21.Beta.(H)-Hopane as a Conserved Internal Marker for Estimating the Biodegradation of Crude Oil. *Environmental Science and Technology*, 28(1): 142.

Quirke, J.M.E. 1987. In: *Metal Complexes in Fossil Fuels*. R.H. Filby and J.F. Branthaver (Editors), Symposium Series No. 344. American Chemical Society, Washington, DC. p. 74

Reynolds, J.G. 1998. Metals and Heteroatoms in Heavy Crude Oils. In: *Petroleum Chemistry and Refining*. J.G. Speight (Editor). Taylor & Francis Publishers, Washington, DC. Chapter 3

Reynolds, J.G., Biggs, W.E., and Bezman, S.A. 1987. In: *Metal Complexes in Fossil Fuels*. R.H. Filby and J.F. Branthaver (Editors), Symposium Series No. 344. American Chemical Society, Washington, DC. p. 205

Santos, R.G., Loh, W., Bannwart, A.C., and Trevisan, O.V. 2014. An Overview of Heavy Oil Properties and Its Recovery and Transportation Methods. *Brazilian Journal of Chemical Engineering*, 31(3): 571–590.

Seifert, W.K., and Teeter, R.M. 1970. Identification of Polycyclic Aromatic and Heterocyclic Crude Oil Carboxylic Acids. *Analytical Chemistry*, 42: 750–758.

Speight, J.G. 1990. Tar Sands. In: *Fuel Science and Technology Handbook*. J.G. Speight (Editor). Marcel Dekker Inc., New York. Chapter 12

Speight, J.G. 1994. Chemical and Physical Studies of Petroleum Asphaltenes. In: *Asphaltenes and Asphalts. I. Developments in Petroleum Science, 40*. T.F. Yen and G.V. Chilingarian (Editors). Elsevier, Amsterdam, Netherlands. Chapter 2

Speight, J.G. 2000. *The Desulfurization of Heavy Oils and Residua*. 2nd Edition. Marcel Dekker, New York.

Speight, J.G. 2001. *Handbook of Petroleum Analysis*. John Wiley & Sons Inc., Hoboken, New Jersey.

Speight, J.G. 2009. *Enhanced Recovery Methods for Heavy Oil and Tar Sands*. Gulf Publishing Company, Houston, Texas.

Speight, J.G. 2012. *Crude Oil Assay Database*. Knovel, Taylor & Francis Group, New York. http://www.knovel.com/web/portal/browse/display?_EXT_KNOVEL_DISPLAY_bookid=5485&VerticalID=0

Speight, J.G. 2013. *The Chemistry and Technology of Coal*. 3rd Edition. CRC Press, Taylor & Francis Group, Boca Raton, Florida.

Speight, J.G. 2014. *The Chemistry and Technology of Petroleum*. 5th Edition. CRC Press, Taylor & Francis Group, Boca Raton, Florida. Marcel Dekker, New York

Speight, J.G. 2015. *Handbook of Petroleum Product Analysis*. 2nd Edition. John Wiley & Sons Inc., Hoboken, New Jersey.

Speight, J.G. 2017. *Handbook of Petroleum Refining*. CRC Press, Taylor & Francis Group, Boca Raton, Florida.

Speight, J.G. 2021. *Petroleum Refining Technology Series Volume 1: Refinery Feedstocks*. CRC Press, Taylor & Francis Group, Boca Raton, Florida.

Speight, J.G. 2023. *Petroleum Refining Technology Series Volume 2: Refinery Dewatering. Desalting, and Distillation in Petroleum Refining*. CRC Press, Taylor & Francis Group, Boca Raton, Florida.

Speight, J.G., and Arjoon, K.K. 2012. *Bioremediation of Petroleum and Petroleum Products*. Scrivener Publishing, Beverly, Massachusetts.

Tissot, B.P. (Editor). 1984. *Characterization of Heavy Crude Oils and Petroleum Residues*. Editions Technip, Paris, France.

Tissot, B.P., and Welte, D.H. 1978. *Petroleum Formation and Occurrence*. Springer-Verlag, New York.

Vogh, J.W., and Reynolds, J.W. 1988. Report No. DE88 001242. Contract FC22-83FE60149. United States Department of Energy, Washington, DC.

Wallace, D., Starr, J., Thomas, K.P., and Dorrence, S.M. 1988. *Characterization of Oil Sands Resources*. Alberta Oil Sands Technology and Research Authority, Edmonton, Alberta, Canada.

4

Fractional Composition

4.1 Introduction

The determination of the composition of crude oil and the viscous feedstocks has always presented a challenge because of the complexity and high molecular weight of the various molecular constituents. The methods employed can be conveniently arranged into a number of categories which are (1) fractionation by precipitation; (2) fractionation by distillation; (3) separation by chromatographic techniques; (4) chemical analysis using spectrophotometric techniques such as infrared spectroscopy, ultraviolet spectroscopy, nuclear magnetic resonance spectroscopy, X-ray fluorescence spectroscopy, emission spectroscopy, neutron activation spectroscopy as well as titrimetric techniques, gravimetric techniques, and elemental analysis; and (5) molecular weight analysis by mass spectrometry, vapor pressure osmometry, and size exclusion chromatography.

Refining crude oil involves subjecting the feedstock to a series of physical and chemical processes as a result of which a variety of products are generated. In some of the processes, for example, distillation, the constituents of the feedstock are isolated unchanged, whereas in other processes, for example, cracking, considerable changes are brought about to the constituents (Parkash, 2003; Gary et al., 2007; Speight, 2014; Hsu and Robinson, 2017; Speight, 2017). Recognition that refinery behavior is related to the composition of the feedstock which is a mixture of various organic compounds that includes a wide range of molecular weights and molecular types that exist in balance with each other. Thus, investigations of the character of refinery feedstocks need to be focused on the influence of feedstock composition and properties on refining operations and the nature of the products that will be produced. One means by which the character of crude oil has been studied is through its fractional composition.

However, the fractional composition of crude oil varies markedly with the method of isolation or separation, thereby leading to potential complications

Handbook of Heavy Oil Properties and Analysis, First Edition. James G. Speight.

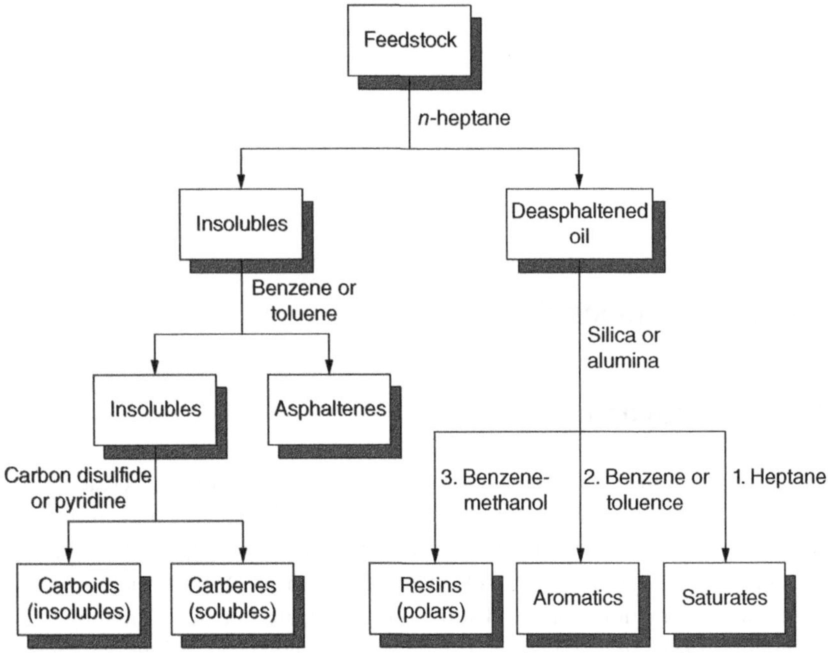

Figure 4.1 General fractionation scheme for crude oil and the viscous feedstocks.

(especially in the case of the heavier feedstocks) in the choice of suitable processing schemes for these feedstocks. Crude oil can be fractionated into four general fractions: (1) the asphaltene fraction, (2) the resin fraction, (3) the aromatics fraction, and (4) the saturates fraction (Figure 4.1). From the data obtained, it is possible to apply the concept of predictability to the various refining sequences and the potential products. The fractionation methods that are available to the crude oil industry (especially the solvent-based methods) allow a reasonably effective degree of separation of the feedstock components (such as the viscous feedstocks, where distillation is not always a viable option) into fractions without alteration of the molecular structure. Thus, the general procedure is to employ techniques that segregate the constituents according to molecular size and/or according to molecular type.

The choice of any (of the several) fractionation methods used is determined not only by the nature and/or composition of the crude oil but also by the effectiveness of the selected method and the compatibility of that method with the other separation procedures to be employed. Thus, although there are wide variations in the nature of crude oil, there have been many attempts to devise standard methods of crude oil fractionation (ASTM, 2021). Recognition that no one particular method may satisfy all the requirements of a refinery feedstock is the first step in any fractionation study which is due to the complexity of refinery feedstocks (especially the viscous feedstocks) because of the distribution of the molecular species which also include the various molecular species including the heteroatom (nitrogen, oxygen, sulfur, and metals) constituents but also because of the wide range of molecular sizes of the constituents in the feedstocks.

It is the purpose of this chapter to present an overview of those methods that can be applied to the separation of refinery feedstocks crude oil which leads not only to an understanding of the subdivision of the feedstock but also to an understanding of the character of the feedstock. Both factors bear a very strong relationship to the processability of the feedstock within the refinery system (Parkash, 2003; Gary et al., 2007; Speight, 2014; Hsu and Robinson, 2017; Speight, 2017).

4.2 Distillation

Distillation is a means of separating chemical compounds (usually liquids) through differences in their vapor pressures (Halvorsen and Skogestad, 2000; Petyluk, 2004; Lei et al., 2005; Speight, 2023). The simple distillation process is very effective for the separation of a volatile liquid from a nonvolatile substance or when separating two liquids that have significant differences in the respective boiling points. However, the application of distillation to the viscous feedstocks is limited in terms of the data produced. The viscous feedstocks have little to no constituents boiling below 300 °C (below 570 °F) and will more than likely be applied to these feedstocks if a valuable gas oil fraction can be produced. That being the case, it might be appropriate to subject the feedstock to flash distillation, where either the temperature is rapidly increased or pressure reduced, and vapor and liquid fractions are thus obtained, which may be processed as such.

Alternatively, distillation at reduced pressure may be an economic option. The higher boiling constituents of a feedstock undergo thermal decomposition at temperatures above 350 °C (660 °F), which results in molecular fragmentation leading to volatile products (that were not indigenous to the crude oil) and to coke. To avoid these thermal reactions, it is necessary to reduce the pressure at which distillation is performed, and since the vapor pressure and temperature are related, the lowering of the pressure is accompanied by a corresponding decrease in the boiling points of the individual constituents and the potential for the occurrence of thermal decomposition is markedly reduced, if not eliminated.

Briefly, the vacuum distillation unit (vacuum pipe still) is a distillation unit that contains a smaller number of theoretical plates than the atmospheric distillation unit and, again, the side streams are taken off at different heights up the distillation tower. The feedstock can be fractionated to produce overhead fractions such as vacuum gas oils and the nonvolatile constituents remain as a vacuum residuum with an initial boiling point (typically) on the order of 565 °C (1,050 °F).

The amounts of the various fractions obtained by distillation depend on the nature of the feedstock and the amount and character of the volatile constituents are important aspects of refinery operations.

4.3 Solvent Treatment

Checking the suitability of a refinery feedstock involves subjecting the feedstock to a series of integrated physical and chemical unit processes (Parkash, 2003; Gary et al., 2007; Speight, 2014; Hsu ad Robinson, 2017; Speight, 2017, 2021) as a result of which a variety of products can be planned and are generated. In some of the processes, for example, distillation, the constituents of the feedstock are isolated unchanged, whereas in other processes (such as cracking) considerable changes occur in the constituents.

Feedstocks can be defined (on a relative or standard basis) in terms of three or four general fractions: asphaltene constituents, resin constituents, saturates, and aromatics (Figure 4.1). Thus, it is possible to compare interlaboratory investigations and thence to apply the concept of predictability to refining sequences and potential products. Recognition that refinery behavior is related to the composition of the feedstock has led to a multiplicity of attempts to establish crude oil and its fractions as compositions of matter. As a result, various analytical techniques have been developed for the identification and quantification of *every molecule* in the lower boiling fractions of crude oil. It is now generally recognized that the name *crude oil* does not describe a composition of matter but rather a mixture of various organic compounds that includes a wide range of molecular weights and molecular types that exist in balance with each other (Speight, 1994; Long and Speight, 1998). There must also be some questions of the advisability (perhaps *futility* is a better word) of attempting to describe *every molecule* in crude oil. The true focus should be to what ends these molecules can be used.

The fractionation methods available to the crude oil industry allow an effective degree of separation of hydrocarbon mixtures (Speight, 2014, 2015, 2021). However, the problems are separating the crude oil constituents without alteration of their molecular structure and obtaining these constituents in a substantially pure state. Thus, the general procedure is to employ techniques that segregate the constituents according to molecular size and molecular type.

It is more generally true, however, that the success of any attempted fractionation procedure involves not only the application of one particular technique but also the utilization of several integrated techniques, especially those techniques involving the use of chemical and physical properties to differentiate among the various constituents. For example, the standard processes of physical fractionation used in the crude oil industry are those of distillation and solvent treatment, as well as adsorption by surface-active materials. Chemical procedures depend on specific reactions, such as the interaction of olefins with sulfuric acid or the various classes of adduct formation. Chemical fractionation is often but not always successful because of the complex nature of crude oil. This may result in unprovoked chemical reactions that have an adverse effect on the fractionation and the resulting data. Indeed, caution is advised when using methods that involve chemical separation of the constituents.

The order in which the several fractionation methods are used is determined not only by the nature and/or composition of the crude oil but also by the effectiveness of a particular process and its compatibility with the other separation procedures to be employed. Thus, although there are wide variations in the nature of refinery feedstocks, there have been many attempts to devise standard methods of crude oil fractionation. However, the various laboratories are inclined to adhere firmly to, and promote, their own particular methods. Recognition that no one particular method may satisfy all the requirements of crude oil fractionation is the first step in any fractionation study. This is due, in the main part, to the complexity of crude oil not only from the distribution of the hydrocarbon species but also from the distribution of the heteroatom (nitrogen, oxygen, and sulfur) species.

Solvent methods have also been applied to feedstock fractionation on the basis of molecular weight. The use of a solvent-based method invokes the concept of the solubility (or insolubility) of a solute in the chosen solvent. The solubility of a solute is the maximum quantity of solute that can dissolve in a specific quantity of solvent at a specified temperature. The main factors that have an effect on solubility of feedstock constituents are (1) the nature of the solute, (2) the nature of the solvent, (3) the temperature, (4) the pressure, and last but by no means least (5) the contact time of the solvent with the feedstock (Speight, 2014, 2015).

The common method for fractional separation of various feedstocks (and refinery products) has been the basis for most composition analysis and the separation methods are used to produce operationally defined fractions (Figure 4.1). Three types of separation procedures are now in use: (1) chemical precipitation in which *n*-pentane separation of an asphaltene fraction is followed by chemical precipitation of other fractions with sulfuric acid of increasing concentration but this method is unsuitable for the separation of the unchanged constituents that have reacted with the sulfuric acid, (2) adsorption chromatography using a clay-gel procedure where, after removal of the asphaltene fraction, the remaining constituents are separated by selective adsorption/desorption on an adsorbent (ASTM D2007, 2021; ASTM D4124, 2021), and (3) size exclusion chromatography in which gel permeation chromatographic (GPC) separation of constituents occurs based on their associated sizes in dilute solutions (ASTM D3593, 2021).

The fractions obtained in these schemes are defined operationally or procedurally. The amount and type of the asphaltene constituents are, for instance, defined by the solvent used for precipitating them. Fractional separation does not provide well-defined chemical components and the separated fractions should only be defined in terms of the particular test procedure (Speight, 2014, 2015). This is analogous to the definition of total petroleum hydrocarbons in which the composition is defined by the method of extraction (Chapter 7). However, these fractions are generated by thermal degradation or by oxidative degradation and are not considered to be naturally occurring constituents. The test method for determining the toluene-insoluble constituents of tar and pitch (ASTM D4072, 2021; ASTM D4312, 2021) is an effective method that can be used to determine the amount of carbenes and carboids in resids and asphalt.

Typically in the test methods, a sample is digested at 95 °C (203 °F) for 25 min and then extracted with hot toluene in an alundum thimble—a thimble composed of hardened alumina which is preferable to paper thimbles insofar as the alundum thimble may be used an indefinite number of times and permit the ignition of the extracted residue without removing the material from the thimble. The extraction time is on the order of 3 hr (ASTM D4312, 2021) or 18 hr (ASTM D4072, 2021) after which the insoluble material is dried and weighed. Combustion will then show if the material is truly carbonaceous or if it is inorganic ash from the metallic constituents (ASTM D482, 2021; ASTM D2415, 2021; ASTM D4628, 2021; ASTM D4927, 2021; ASTM D5185, 2021; ASTM D6443, 2021).

Another method (ASTM D893, 2021) covers the determination of pentane- and toluene-insoluble constituents in used lubricating oils and can also be applied. Pentane-insoluble constituents include oil-insoluble materials and toluene-insoluble constituents can come from external contamination and highly carbonized materials from degradation. A significant change in pentane- or toluene-insoluble constituents indicates a change in properties that could lead to problems in further processing (for resids) or service (for asphalt).

There are two test methods used: Procedure A covers the determination of insoluble constituents without the use of coagulant in the pentane and provides an indication of the materials that can be readily separated from the diluted material by centrifugation. Procedure B covers the determination of insoluble constituents that contain additives and employs a coagulant. In addition to the materials separated by using Procedure A, this coagulation procedure separates some finely divided materials that may be suspended in the resid or asphalt. The results obtained by Procedures A and B should not be compared since they usually give different values. The same procedure should be applied when comparing results obtained periodically and when comparing results determined in different laboratories.

In Procedure A, a sample is mixed with pentane and centrifuged after which the resid or asphalt solution is decanted, and the precipitate washed repeatedly (usually at least twice) with pentane, dried, and weighed. In order to determine the amount of toluene-insoluble constituents, a separate sample of the feedstock is mixed with pentane and centrifuged. The precipitate is washed again (e.g., twice) with pentane, then once with toluene-alcohol solution, after which the precipitate is washed once with toluene and then the washed precipitate is dried and weighed. In Procedure B, Procedure A is followed except that instead of pentane, a pentane-coagulant solution is used.

Many investigations of relationships between composition and properties take into account only the concentration of the asphaltene constituents, independently of any quality criterion. However, a distinction should be made between the asphaltene constituents that occur in straight-run resids and those which occur in blown asphalts. Remembering that asphaltene constituents are a solubility class rather than a distinct chemical class, means that vast differences occur in the makeup of this fraction when it is produced by different processes.

In the current context of feedstock analysis, the separation of a refinery feedstock into a variety of fractions by the use of solvents is of value to understanding the solubility properties of the feedstock constituents.

The simplest application of solvent extraction consists of mixing the feedstock with a solvent which results in the formation of two phases which are (1) the dissolved portion of the feedstock, which is referred to as the extract and (2) the non-dissolved portion of the feedstock, which is referred to as the raffinate.

The ratio of the concentration of any particular component in the two phases is known as the distribution coefficient K:

$$K = C_1 / C_2$$

C_1 and C_2 are the concentrations in the various phases. The distribution coefficient is usually constant and may vary only slightly, if at all, with the concentration of the other components. In fact, the distribution coefficients may differ for the various components of the mixture to such an extent that the ratio of the concentrations of the various components in the solvent phase differs from that in the original crude oil; this is the basis for solvent extraction procedures.

The list of compounds that have been suggested as a selective solvent for the fractionation of a feedstock is considerable, but before any extraction process is attempted, it is necessary to consider the following criteria: (1) the differences in the solubility of the crude oil constituents in the solvent should be substantial, (2) the solvent should be significantly less or more dense than the crude oil (product) to be separated to allow easier countercurrent flow of the two phases, and (3) separation of the solvent from the extracted material should be relatively easy.

It may also be advantageous to consider other properties, such as viscosity and surface tension as well as the optimal temperature for the extraction process.

Thus, aromatic derivatives can be separated from naphthene derivatives and paraffin derivatives by the use of selective solvents. Furthermore, aromatic derivatives with differing numbers of aromatic rings (as might occur in many viscous feedstocks) that may exist in various narrow boiling fractions can also be effectively separated by solvent treatment.

4.3.1 Asphaltene Separation

Put simply, the separation of an asphaltene fraction from hydrocarbonaceous feedstocks involves the following criteria: the use of (1) pentane or *n*-heptane as the precipitating medium, (2) 40 ml per g or per ml of a sample, (3) 2–8 hr contact time, and (4) a dispersant such as toluene or easy-to-evaporate aromatic naphtha—although toluene is preferred—that has a sufficiently low boiling range for the viscous feedstocks. In the latter case (i.e., use of toluene of naphtha), the effect of the dispersant must be countered by the use of a further amount of *n*-pentane or *n*-heptane that is at least equal to 40 volumes of the amount of dispersant used. In some cases, that is, when the feedstock is extremely viscous, it may be advisable to reflux the sample in *n*-heptane (Speight, 2014, 2015).

The asphaltene fraction of a crude oil or a viscous feedstock is the highest molecular weight and most complex fraction in petroleum (ASTM D893, 2021; ASTM D2007, 2021; ASTM D3279, 2021; ASTM D4124, 2021; ASTM D6560, 2021) and is the fraction that produces the highest yields of coke that can be expected during exposure to thermal conditions (Speight, 2001, 2014, 2015, 2017).

In any of the methods for the determination of the asphaltene content (Speight et al., 1984; Speight, 2014, 2015), the feedstock is mixed with a large excess (typically >30 volumes hydrocarbon per volume of sample) low-boiling hydrocarbon such as *n*-pentane or *n*-heptane. If the precipitating hydrocarbon liquid is not present in a large excess, the yields of the asphaltene fraction will vary and will be erroneous (Speight, 2014, 2015).

For an extremely viscous sample, a solvent such as toluene may be added to dissolve the feedstocks prior to the addition of the low-boiling hydrocarbon. This allows the precipitating liquid to penetrate to viscous mass. However, an additional amount of the precipitating hydrocarbon liquid must be added to compensate for the presence of the solvent—typically a minimum of 30 volumes hydrocarbon per volume of, for example, tol solvent. After a specified time (typically 12 hr but most researchers prefer to 16 hr, i.e., overnight) (Table 4.1), the insoluble material (the asphaltene fraction) is separated (by filtration) and dried.

The yield is reported as percentage (% w/w) of the original sample. Furthermore, different hydrocarbon liquids (such as *n*-pentane or *n*-heptane) give different yields of the asphaltene fraction and if the presence of the solvent is not compensated by use of additional hydrocarbon (equivalent to >30 volumes of the added solvent), the yield will be erroneous.

Table 4.1 Definition of the Parameters for the Separation of the Asphaltene Fraction

Parameter	Description
Volume ratio	An excess of the liquid hydrocarbon (>30 ml—usually 40 ml of the liquid hydrocarbon) per ml of feedstock
Liquid hydrocarbon	Volatility constraints (in relation to the use of *n*-pentane) and consistency of the asphaltene fraction favor the use of *n*-heptane over *n*-pentane
Contact time	8–16 hr is preferable; the time is feedstock dependent
Purification	A dissolution–precipitation sequence to remove any non-asphaltene constituents (such as the resin constituents) that separate with the asphaltene constituents

Another method, not specifically described as an asphaltene separation method, is designed to remove pentane-insoluble constituents by membrane filtration (ASTM D4055, 2021). In the method, a sample of oil is mixed with pentane in a volumetric flask, and the oil solution is filtered through a 0.8-μm membrane filter. The flask, funnel, and the filter are washed with pentane to completely transfer any particulates onto the filter after which the filter (with particulates) is dried and weighed to give the pentane-insoluble constituents as a percent by weight of the sample.

The precipitation number is considered to be the same as the yield of the asphaltene fraction as a percent (by weight) of the feedstocks but there are several issues that remain obvious in the rejection of the yield for this purpose. For example, the method to determine the precipitation number advocates the use of naphtha for use with "black oil" or "lubricating oil" and the amount of insoluble material (as a % v/v of the sample) is the precipitating number. In the test method, 10 ml of sample is mixed with 90 ml of ASTM precipitation naphtha (that may or may not have a constant chemical composition) in a graduated centrifuge cone and centrifuged for 10 min at 600–700 rpm. The volume of material on the bottom of the centrifuge cone is noted until repeat centrifugation gives a value within 0.1 ml (the precipitation number). Obviously, this can be substantially different to the asphaltene content and is dependent on the composition of the naphtha which can vary from a liquid that is predominantly to a liquid that is predominantly aromatic.

If the viscous feedstock is the result of a thermal process (such as visbreaking), it may also be necessary to determine if toluene-insoluble material is present by the methods, or modifications thereof, used to determine the toluene insoluble of tar and pitch (ASTM D4072, 2021; ASTM D4312, 2021). In these two methods, a sample is digested at 95 °C (203 °F) for 25 min and then extracted with hot toluene in an alundum thimble. The extraction time is 18 hr (ASTM D4072, 2021) or 3 hr (ASTM D4312, 2021). The insoluble matter is dried and weighed.

4.3.1.1 Influence of Solvent Type

On the basis of the solubility in a variety of solvents, it has become possible to distinguish among the various constituents of crude oil and the viscous feedstocks (Figure 4.1). The feedstocks under consideration in this text do not contain carboids and carbenes that are, in this context, considered to be the products of thermal processes. For example, a residuum from cracking distillation of cracking processes may contain 2% w/w (or more) of the fractions designated as carbenes and carboids.

The separation of a refinery feedstock into two fractions which are (1) the insoluble asphaltene fraction and (2) the soluble maltene fraction is conveniently brought about by means of low molecular weight paraffin hydrocarbon liquids (such as *n*-pentane or *n*-heptane). Addition of 40 volumes of *n*-pentane or *n*-heptane (relative to the single volume of the feedstock) is the method generally preferred (Speight et al., 1984; Speight, 1994, 2001, 2014, 2015). It is no doubt a separation of the chemical components with the most complex structures from the mixture, and this fraction, which should correctly be called *n*-pentane asphaltenes or *n*-heptane asphaltenes is qualitatively and quantitatively reproducible (Figure 4.1).

If the precipitation method (deasphalting) involves the use of a solvent and a residuum and is essentially a leaching of the soluble constituents from the insoluble residue, this process may be referred to as extraction. However, under the prevailing conditions now in laboratory use, the term precipitation is perhaps more correct and descriptive of the method.

Also, these methods were developed for use with more conventional feedstocks rather than for viscous feedstocks. Therefore, adjustments in the methods may be necessary to ensure efficient separation. In the case of a viscous feedstock in which the penetration of the liquid hydrocarbon into the feedstock is diffusion controlled (or diffusion limited), it is recommended that a volume (equivalent to the volume of the feedstock) of an aromatic solvent (such as toluene or aromatic naphtha) be added to the feedstock in which case the amount of the *n*-pentane or *n*-heptane should be double in order to counteract the effect of the aromatic solvent.

Variation in the solvent type also causes significant changes in asphaltene yield (Speight, 1994, 2001, 2014, 2015). The solvent power of the solvents (i.e., the ability of the solvent to dissolve asphaltenes) increases in the order

2-Methyl paraffin (*iso*-paraffin) < *n*-paraffin < terminal olefin

Cycloparaffin solvents (naphthene solvents) have a remarkable effect on asphaltene yield and give results totally unrelated to those from any other nonaromatic solvent (Mitchell and Speight, 1973). For example, when cyclopentane, cyclohexane, or their methyl derivatives are employed as precipitating media, only about 1% of the material remains insoluble.

These differences can be explained by considering the solvent power of the precipitating liquid, which can be related to molecular properties (Hildebrand et al., 1970). Thus, the solvent power of nonpolar solvents has been expressed as a solubility parameter, δ, and equated to the internal pressure of the solvent, that is, the ratio between the surface tension γ and the cubic root of the molar volume V:

$$\delta_1 = \gamma^3\sqrt{V}$$

Alternatively, the solubility parameter of nonpolar solvents can be related to the energy of vaporization ΔR^{V} and the molar volume,

$$\delta_2 = (\Delta E^{\mathrm{V}}/V)^{1/2}$$

or

$$\delta_2 = (\Delta H^{\mathrm{V}} - RT/V)^{1/2}$$

where ΔH^{V} is the heat of vaporization, R is the gas constant, and T is the absolute temperature.

Consideration of this approach shows that there is indeed a relationship between the solubility parameters for a variety of solvents and the amount of precipitate (Mitchell and Speight, 1973). The introduction of a polar group (heteroatom function) into the molecule of the solvent has significant effects on the quantity of precipitate. For example, treatment of a residuum with a variety of ethers or treatment of asphaltenes with a variety of solvents illustrates this point (Speight, 1979). In the latter instance, it was not possible to obtain data from addition of the solvent to the whole feedstock *per se* since the majority of the nonhydrocarbon materials were not miscible with the feedstock. It is nevertheless interesting that, as with the hydrocarbons, the amount of precipitate, or asphaltene solubility, can be related to the solubility parameter.

The solubility parameter allows an explanation of certain apparent anomalies, for example, the insolubility of asphaltenes in pentane and the near-complete solubility of the materials in cyclopentane. Moreover, the solvent power of various solvents is in agreement with the derivation of the solubility parameter; for any one series of solvents, the relationship between the amount of precipitate (or asphaltene solubility) and the solubility parameter δ is quite regular.

In any method used to isolate the asphaltene fraction, standardization of the technique is essential. The use of both *n*-pentane and *n*-heptane has been widely advocated, and although n-heptane is becoming the deasphalting liquid of choice, but this is by no means a hard-and-fast rule. In addition, it must be recognized that large volumes of precipitating liquid are required to effect a reproducible and consistent yield of the fraction. It is also preferable that the solvents be of sufficiently low boiling point that complete removal of the solvent from the fraction can be effected and, most important, the solvent must not react with the feedstock. Hence, there has been a preference for hydrocarbon liquids. Although several standard methods are available, they are not unanimous in the particular hydrocarbon liquid or in ratio of hydrocarbon liquid to feedstock.

Method	Deasphalting liquid	Volume (ml/g)[a]
ASTM D893	n-Pentane	10
ASTM D2007	n-Pentane	10
ASTM D3279	n-Heptane	100
ASTM D4124	n-Heptane	100
ASTM D6560	n-Heptane	100

[a] The ratio can also be ml/ml since the density of a viscous feedstock is usually on the order of 1.0.

Typically, n-pentane and n-heptane are the solvents of choice in the laboratory (other solvents can be used) (Mitchell and Speight, 1973; Speight, 1979, 2001, 2014, 2015) and cause the separation of asphaltenes as brown-to-black powdery materials. In the refinery, supercritical low molecular weight hydrocarbons (e.g., liquid propane, liquid butane, or mixtures of both) are the solvents of choice and the product is a semisolid (tacky) to solid asphalt. The amount of asphalt that settles out of the paraffin/residuum mixture depends on the size of the paraffin, the temperature, and the paraffin-to-feedstock ratio (Girdler, 1965; Corbett and Petrossi, 1978; Speight et al., 1984; Parkash, 2003; Gary et al., 2007; Speight, 2014, 2015; Hsu and Robinson, 2017; Speight, 2017).

4.3.1.2 Influence of the Degree of Dilution

At constant temperature, the quantity of precipitate first increases with increasing ratio of solvent to feedstock and then reaches a maximum. In fact, there are indications that when the proportion of solvent in the mix is <35%, little or no asphaltenes are precipitated (Mitchell and Speight, 1973).

4.3.1.3 Influence of Temperature

When pentane and the lower molecular weight hydrocarbon solvents are used in large excess, the quantity of precipitate and the composition of the precipitate changes with increasing temperature (Mitchell and Speight, 1973).

One particular example is the separation of asphaltenes from using n-pentane. At ambient temperatures (21 °C, 70 °F) the yield of asphaltenes is 17% w/w, but at 35 °C (95 °F), 22.5% by weight asphaltenes are produced using the same feedstock–pentane ratio. This latter precipitate is in fact asphaltenes plus resins; similar effects have been noted with other hydrocarbon solvents at temperatures up to 70 °C (160 °F). These results are self-explanatory when it is realized that the heat of vaporization ΔH^{V} and the surface tension γ, from which the solubility parameters are derived, both decrease with increasing temperature.

4.3.1.4 Influence of Contact Time

Contact time between the hydrocarbon and the feedstock (especially feedstocks such as heavy oil, residua, and tar sand bitumen) also plays an important role in asphaltene separation (Speight et al., 1984). Yields of the asphaltenes reach a maximum after approximately 8 hr, which may be ascribed to the time required for the asphaltene particles to agglomerate into particles of a filterable size as well as the diffusion-controlled nature of the process since the heavier feedstocks also need time for the hydrocarbon to penetrate their mass.

4.3.2 Fractionation

After removal of the asphaltene fraction, further fractionation of crude oil is also possible by variation of the hydrocarbon solvent. For example, liquefied gases, such as propane and butane, precipitate as much as 50% by weight of a viscous feedstock. The can vary from a black, tacky, semisolid material to an amorphous solid that can vary in color from dark brown to black. However, caution is advised when using color as a monitor of the efficiency of the process. In some cases, a black solid can be an indication that resin constituents are adhering to the asphaltene solids. Also, treatment of the propane precipitated fraction with pentane yields the insoluble brown, amorphous asphaltene fractions and soluble, near-black, semisolid resins, which are, as near as can be determined, equivalent to the resins isolated by adsorption techniques.

There are also claims that solvent treatment at low temperatures (−4 to −20 °C, −4 to 25 °F) brings about fractionation of the maltenes. The hydrocarbon solvents pentane and hexane have been claimed adequate for this purpose but may not be successful with maltenes from tar sand bitumen and residua. The author has had considerable success using acetone at −4 °C (25 °F) for the fractionation of the maltenes fraction from viscous feedstocks, other than cracked residua.

Other miscellaneous fractionation procedures involving the use of solvents are available. One method includes a procedure using polar solvents for separation into five fractions. These fractions are: (1) asphaltene constituents, which are insoluble in hexane; (2) hard resin constituents, which are insoluble in 80:20

iso-butyl alcohol-cyclohexane mixture; (3) wax constituents, which are insoluble in a 1:2 mixture of acetone and methylene chloride at 0 °C (−18 °F); (4) soft resin constituents, which are insoluble in *iso*-butyl alcohol; and (5) oil constituents, which is the balance of the sample that remains soluble in *iso*-butyl alcohol. The only drawback to this particular scheme appears to be the use of hexane as the precipitating medium instead of the recommended *n*-pentane and *n*-heptane, as per the ASTM standard test methods. Use of hexane may not always completely precipitate the asphaltene constituents from a sample (as does *n*-pentane) and it is more than likely that the fraction designated as "hard resins" will contain considerable portions of asphaltene constituents.

A fractionation procedure was also devised using *n*-butanol and acetone as the solvents and consists of (1) separation of an asphaltic fraction by *n*-butanol and (2) separation of the butanol-soluble portion into a paraffinic fraction and a fraction containing a high proportion of ring-containing organic constituents fraction by chilling an acetone solution of the two, but it has the unfortunate result that all three fractions obtained may contain asphaltenes. It is difficult to compare the method with any other or even establish any correlation with previous experience. Even precipitation of the asphaltenes with *n*-pentane and subtracting the asphaltenes from the asphaltic, arriving at a fourth fraction (asphaltic resins), did not appear to correct the faults of the method since the asphaltic fraction does not contain all the asphaltenes of the specimen.

Another method of fractionation consisted of stepwise separation into the following fractions:

1) Asphaltene constituents, precipitated by *n*-pentane
2) Resin constituents, precipitated with propane and subdivided by fractionation with aniline into soft resins and hard resins
3) Wax constituents, precipitated with methyl *iso*-butyl ketone
4) Oil constituents, remaining fraction separated with acetone into paraffinic oils and naphthenic oils.

A strong feature of this method is the subdivision of the resins fraction by solubility in aniline and the subdivision of the paraffinic fraction into the three components, wax, paraffinic oils, and naphthenic oils.

It is unfortunate that, with the exception of asphaltene precipitation, no standard method exists for the fractionation of crude oil, residua, bitumen, by solvent treatment. The procedures described here each have their own individual merits, but there has not been any serious effort to apply these methods to a wide variety of carbonaceous liquids to assess their general applicability. Fractionation by means of solvents alone (perhaps with the exception of propane, which requires pressure equipment) would be convenient indeed, provided that facile separation of the solvent and the products could be achieved at a later stage.

Another all-solvent procedure involves the use of acetone, which discharges a resin fraction, and dimethylformamide, which then discharges a saturates fraction.

The use of all-solvent methods for the separation of crude oil allows the fractionation of feedstocks to be achieved without loss of material (on the adsorbent) and produces fractions of varying polarity (Speight, 1989). The obvious benefit of the all-solvent techniques is the complete recovery of material, thereby allowing a more quantitative and qualitative examination of the feedstocks.

The disadvantages of an all-solvent separation technique are that, first, in some instances, low temperatures (e.g., 0 to −10 °C 32 to 50 °F, and the like) are advocated as a means of effecting the fractionation of various feedstocks by means of solvents (Rostler, 1965; Speight, 1979, 2014, 2015). Such requirements may cause inconvenience in a typical laboratory operation by requiring a permanently cool temperature during the separation. Second, it must be recognized that large volumes of solvent may be required to effect a reproducible separation in the same manner as the amounts required for consistent asphaltene separation (ASTM D893, 2021; ASTM D2007, 2021; ASTM D4124, 2021). Finally, it is also essential that the solvent be of sufficiently low boiling point so that complete removal of the solvent from the product fraction can be effected. Although not specifically included in the three main disadvantages of the all-solvent approach, it should also be recognized that the solvent must not react with the feedstock constituents. In addition, caution is still required to ensure that there is no interaction between the solvent and the solute.

4.3.3 Carbenes and Carboids

By definition, the carbene and carboids are the fractions that can be isolated from feedstocks that have been treated thermally and represent (typically if defined as a composite fraction) the material that is insoluble in toluene or carbon disulfide which can then be subdivided into the carbene subfraction and the carboid subfraction using pyridine or carbon disulfide (Figure 4.1).

The fraction that has been called the "carbene" fraction contains molecular species that are soluble in carbon disulfide and soluble in pyridine but which are insoluble in toluene. The subfraction that has the highest carbon content is termed the "carboid" and consists of species that are insoluble in carbon disulfide or in pyridine.

The carbene and carboid fractions are not considered to be naturally occurring constituents of any of the viscous feedstock (unless the feedstock is a residuum that has been subject to cracking during the distillation process or to subsequent to heat treatment). The carbene or carboids can also be generated by oxidative reaction of a residuum (as in the case of excessive blowing of asphalt during the production stage). The test method for determining the toluene-insoluble constituents of tar and pitch (ASTM D4072, 2021; ASTM D4312, 2021) can be used to determine the amount of carbenes and carboids thermally altered viscous feedstocks or oxidative treatment of viscous feedstocks.

4.4 Adsorption

Adsorption is the bonding of molecules or particles to a surface. On the other hand, absorption is the filling of pores in a solid. The bonding to the surface is usually (but not always) weak and reversible. Compounds that contain functional groups are very often strongly adsorbed on activated carbon.

Separation by adsorption chromatography commences with the preparation of a porous bed of finely divided solid, the adsorbent which is typically contained in an open tube (hence, referred to as "column chromatography"). The sample is introduced at one end of the adsorbent bed (the top end of the column) and induced to flow through the bed by means of a suitable solvent to flow from the bottom of the column. As the sample moves through the adsorbent, some of the constituents of the sample are held (adsorbed) depending on the chemical nature of the constituent. The constituents that are strongly adsorbed (usually near to the top of the column) rather than in the moving (solvent) phase remain on the adsorbent, but components that are slightly adsorbed move through the bed comparatively rapidly to be adsorbed in the lower regions of the column.

The most common industrial adsorbents are clay minerals, activated carbon, silica gel (SiO_2), and alumina (Al_2O_3) because each one presents a high surface area per unit weight of solid. By way of explanation, clay is a naturally occurring mineral. Roasting organic material to decompose it to granules of carbon produces activated carbon. Coconut shell, wood, and bone are common sources of activated carbon. Silica gel is a matrix of hydrated silicon dioxide. Alumina is mined or precipitated aluminum oxide and hydroxide. The effects of temperature on the adsorption process can be substantial and, because of this possibility, measurements are usually (or should be) made at a constant temperature. Graphs of the data are called isotherms. Most steps using adsorbents have little variation in temperature.

As a general note about the effectiveness of the method, it is advisable that, before application of the adsorption step, the asphaltene constituents first be completely removed from the feedstock by, for example, any of the methods outlined in the previous section. The prior removal of the asphaltene constituents is essential insofar as they are usually difficult to remove from the adsorbent and may actually be irreversibly adsorbed on the adsorbent which can interfere with the effectiveness of the adsorbent (in terms of the fractionation of the non-asphaltene constituents) by blocking any effective adsorption sites on the adsorbent.

4.4.1 Chemical Factors

As already stated (Chapter 8), crude oil is a complex mixture of paraffin, naphthene, and aromatic hydrocarbons as well as nitrogen-containing, oxygen-containing, and sulfur-containing compounds and traces of a variety of metal-containing compounds and the amounts of non-hydrocarbon compounds increase with molecular weight. By definition, the saturate fraction consists (or should consist) of paraffin derivatives and cycloparaffin derivatives (naphthene derivatives). The single-ring naphthene derivatives present in various feedstocks oil are primarily alkyl-substituted cyclopentane derivatives and cyclohexane derivative. The alkyl groups are usually short (i.e., four or less carbon atoms per alkyl group) with methyl, ethyl, and (on occasion) isopropyl groups the predominant substituents. However, as the molecular weight of the naphthene derivatives increases, the fraction contains more condensed rings with six-membered naphthene rings predominating over the five-membered ring systems that are still present in the complex higher molecular weight constituents.

The aromatics fraction consists of those compounds containing an aromatic ring and vary from monoaromatics (containing one benzene ring in a molecule) to diaromatics (substituted naphthalene) to triaromatic derivatives (substituted phenanthrene derivatives). Higher condensed ring systems (tetra-aromatic derivatives, penta-aromatic derivatives) are also known but are somewhat less prevalent than the lower-number ring systems, and each aromatic type will have increasing amounts of condensed ring naphthene derivative attached to the aromatic ring as molecular weight is increased.

However, depending upon the adsorbent employed for the separation, a compound having an aromatic ring (i.e., six aromatic carbon atoms) carrying side chains consisting in toto of more than six carbon atoms (i.e., more than six non-aromatic carbon atoms) will appear in the aromatic fraction.

The typical nitrogen compounds found in crude oil are generally divided into two groups, basic and nonbasic (Chapters 8 and 99), each of which has alkyl chains and other ring systems. The basic nitrogen compounds cause difficulty with the many acid-catalyzed processes used in crude oil refining. For example, in catalytic cracking, the basic nitrogen adsorbs on the catalytic acid sites and reduces the cracking activity of the catalyst. Furthermore, in reactions catalyzed by liquid acids, the presence of basic nitrogen compounds in the feed increases acid consumption and, thus, the cost of the process. Typical of such processes are alkylation, isomerization, and olefin absorption.

Another type of nitrogen compound is located in the porphyrin derivatives (Chapter 8), which consist of four pyrrole rings connected together with methylene bridges at the carbons next to the nitrogen atoms. They are normally found in trace quantities in the high molecular weight fractions of crude oil as metal complexes.

The oxygen compounds found in crude oil fractions are often products of exposure to air. However, some naturally occurring oxygenated compounds do exist in crude oil; these are typically phenols, naphthenic acids, and esters (Chapter 8). They increase in quantity and complexity as molecular weight increases, just as sulfur and nitrogen compounds do.

A variety of sulfur compounds occur in crude oil and the viscous feedstocks, and include mercaptan derivatives (—SH, also called thiol derivatives), sulfide derivatives (—S—), and disulfide derivatives (—SS—). They are much less thermally stable than the thiophene derivatives and often lose hydrogen sulfide on heating. They can also react thermally to form more stable sulfur compounds. In the chromatographic separation, these sulfur compounds are found in the polar aromatics fraction, even though they may not all be aromatic. Thiophene derivatives tend to exhibit aromatic behavior and are collected with the aromatics when adsorption separation is used on crude oil fractions.

Sulfur can contaminate catalysts and cause corrosion in refinery equipment. Therefore, extensive sulfur removal procedures must be employed so that the viscous feedstocks (including residua) meet refinery requirements. Emissions of sulfur oxides (SO_x, where $x = 2$ or 3) along with nitrogen oxides (NO_x, where $x = 1$ or 2) from refineries can contribute significantly to the formation of acid rain.

$$SO_2 + H_2O \rightarrow \underset{\text{sulfurous acid}}{H_2SO_3}$$

$$2SO_2 + O_2 \rightarrow 2SO_3$$

$$SO_3 + H_2O \rightarrow \underset{\text{sulfuric acid}}{H_2SO_4}$$

$$NO + H_2O \rightarrow \underset{\text{nitrous acid}}{HNO_2}$$

$$2NO + O_2 \rightarrow NO_2$$

$$NO_2 + H_2O \rightarrow \underset{\text{nitric acid}}{HNO_3}$$

Thus, meeting the various environmental regulations requires an accurate measurement of the sulfur content of the feedstock.

4.4.2 Fractionation Methods

After separation of an asphaltene fraction, it is common (and often necessary) practice to determine the composition of the feedstock in terms of other fractions (Figure 4.1).

4.4.2.1 General Methods

Separation of crude oil, heavy crude oil, extra heavy crude oil, tar sand bitumen, and residua by adsorption chromatography essentially commences with the preparation of a porous bed of finely divided solid adsorbent (Hoiberg, 1964; Speight, 2014, 2015).

In the method, the adsorbent is usually contained in a vertical open tube (often referred to as column chromatography); the sample is introduced at the top end of the adsorbent bed and induced to flow through the bed by means of a suitable solvent. As the sample moves through the bed, the various components are held (adsorbed) to a greater or lesser extent depending on the chemical nature of the component. Thus, those molecules that are strongly adsorbed are held on the adsorbent surface longer than the less-absorbable components that are in the moving (solvent) phase.

Numerous factors randomly affect the process of migration through a bed, and in fact, the total distance traveled in a given time by different molecules of the same material is not constant. Nevertheless, the suitable choice of a bed and a moving (solvent) phase allows adequate separation of even multicomponent mixtures to be achieved. Thus, the overall adsorption chromatographic process may be viewed as migration of different compounds along the bed, which varies with compound structure as well as a range of migrations by different molecules of the same compound.

The fractionation of crude oil and residua by adsorption on such materials as fuller's earth, animal charcoal, and various types of clay dates back to the beginning of the 20th century (Hoiberg, 1964). These materials effect an arbitrary separation of the material into a number of fractions that have variously been described as oil, resins, hard resins, and soft resins, to mention only the more commonly used terms.

It is essential that, before application of the adsorption technique to the crude oil, the asphaltenes first be completely removed, for example, by any of the methods outlined in the previous section. The prior removal of the asphaltenes is essential insofar as they are usually difficult to remove from the earth or clay and may actually be irreversibly adsorbed on the adsorbent.

Nevertheless, careful monitoring of the experimental procedures and the nature of the adsorbent have been responsible for the successes achieved with this particular technique. Early procedures consisted of warming solutions of the crude oil fraction with the adsorbent and subsequent filtration. This procedure has continued to the present day, and separation by adsorption is used commercially in plant operations in the form of clay treatment of crude oil fractions and products.

In the laboratory, very little use is made of the technique of warming a solution of the sample with the adsorbent. Rather, a chromatographic technique is employed in which the sample is washed through a column of the adsorbent using various solvents.

Early investigations involved filtration of crude oils through a column of fuller's earth and it was observed that the gasoline components appeared in the initial part of the filtrate. Subsequent investigations showed that if light oil is drawn upward by a pump through a column of earth, the light aliphatic hydrocarbons accumulate in the top section. The aromatic constituents did not rise as high and the nitrogen- and sulfur-containing compounds were largely retained on the adsorbent.

A later chromatographic method involved separation into four principal fractions, asphaltenes, asphaltic resins, dark oils, and water white oils, and a fifth fraction constituting the balance of the specimen. The method consists of precipitation of asphaltenes with 40 volumes of *n*-pentane (or *n*-heptane) and elution of the *n*-pentane (or *n*-heptane) soluble fraction from a chromatographic column of fuller's earth. The elution technique gave a series of fractions which are (1) an "oils" fraction, which is eluted with *n*-pentane, or *n*-heptane as the case may be; (2) a "dark oils" fraction, which is eluted with methylene chloride; (3) a "resins" fraction, which is eluted with methyl ethyl ketone; and (4) a "hard resins" fraction, which is desorbed with an acetone–chloroform mixture.

The proportions of each fraction are subject to the ratio of fuller's earth to *n*-pentane soluble materials. For example, a change in the ratio from 10:1 to 25:1 causes a decrease in the percentage of water white oils by a factor of 3:2, and the percentage of dark oils and asphaltic resins increases by about the same factor. The method, like all chromatographic procedures proposed for the fractionation of crude oils, is dominated by equilibrium conditions, and does not give fractions that are different components but only blends of the same components in different proportions.

Other methods of fractionation by the use of adsorbents include separation of the maltene fraction by elution with *n*-heptane from silica gel into two fractions named aromatics and nonaromatics and is, in fact, a separation into the two broad groups called resins and oils in other methods. The silica gel method may also be modified to produce three fractions: (1) nonaromatics eluted with *n*-heptane, (2) aromatics eluted with benzene, and (3) compounds that contain oxygen as well as sulfur and nitrogen, eluted with pyridine. Prior separation of the asphaltenes renders the procedure especially suitable and convenient for use with the viscous feedstocks. Other modifications to the methods include successive elution with *n*-pentane, benzene, carbon tetrachloride, and ethanol or any other suitably designated solvent. As always, the key to the use of other designated solvent is reproducibility.

Alumina has also been used as an adsorbent and involves (1) precipitation of asphaltenes with normal pentane, (2) elution of oils from alumina with pentane, and (3) elution of resins from alumina with a methanol–benzene mixture. In fact, the choice of the adsorbent appears to be arbitrary, as does the choice of the various solvents or solvent blend. The use of ill-defined adsorbents, such as earths or clays is a disadvantage in that certain components of the crude oil may undergo changes (e.g., polymerization) caused by the catalytic nature of the adsorbent and can no longer be extracted quantitatively. Furthermore, extraction of the adsorbed components may require the use of solvents of comparatively high solvent power, such as chloroform or pyridine, which may be difficult to remove from the product fractions.

It is also advisable, once a procedure using an earth or clay has been established, that the same type of adsorbent be employed for future fractionation since the ratio of the product fractions varies from adsorbent to adsorbent. It is also very necessary that the procedure be used with caution and that the method not only be reproducible but quantitative recoveries be guaranteed; reproducibility with only, say, 85% of the material recoverable is not a criterion of success.

There are two procedures that have received considerable attention over the years and these are (1) the United States Bureau of Mines-American Crude oil Institute (USBM-API) method and (2) the saturates-aromatics-resins-asphaltenes (SARA) method (Figure 4.1) (Speight, 2014, 2015). This latter method is often also called the saturates-aromatics-polars-asphaltenes (SAPA) method. These two methods are used as representing the standard methods of crude oil fractionation. Other methods are also noted, especially when the method has added further meaningful knowledge to compositional studies.

The USBM-API method employs ion-exchange chromatography and coordination chromatography with adsorption chromatography to separate heavy oils and residua into seven broad fractions: acids, bases, neutral nitrogen compounds, saturates, and mono-, di-, and polyaromatic compounds. The acid and base fractions are isolated by ion-exchange chromatography, the neutral nitrogen compounds by complexation chromatography using ferric chloride, and the saturates and aromatics by adsorption chromatography using activated alumina or using a combined alumina-silica column (Hirsch et al., 1972; Jewell et al., 1972).

The feedstock sample can be separated into chemically significant fractions that are suitable for analysis according to compound type (McKay et al., 1975). Although originally conceived for the separation of distillates, this method has been successfully applied to determining the composition of heavy oil and bitumen (McKay et al., 1976). Originally the method required distillation of a feedstock into narrow boiling point fractions, but whole feedstocks, such as the bitumen from Utah or Athabasca tar sands have been separated into classes of compounds without previous distillation or removal of asphaltenes. The latter finding is supported by the separation of asphaltenes using ion-exchange materials (McKay et al., 1977; Francisco and Speight, 1984).

The SARA method (Jewell et al. 1972) is essentially an extension of the API method that allows more rapid separations by placing the two ion-exchange resins and the $FeCl_3$-clay-anion-exchange resin packing into a single column. The adsorption chromatography of the nonpolar part of the same is still performed in a separation operation. Since the asphaltene content of crude oil (and synthetic fuel) feedstocks is often an important aspect of processability, an important feature of the SARA method is that the asphaltenes are separated as a group. Perhaps more important is that the method is reproducible and applicable to a large variety of the most difficult feedstocks, such as residue tar sand bitumen, shale oil, and coal liquids.

Due to the impossibility of an elemental characterization of refinery feedstocks because of the complex nature of these materials, a meaningful characterization has been satisfactorily obtained by fractionation, based on fraction polarity and solubility (Speight, 2014, 2015). In fact, the SARA analysis is the most widely used method to describe feedstock fractions that employs a precipitation-chromatographic technique that divides the feedstocks into four main fractions, which provide the name for the analysis: (1) saturates, (2) aromatics, (3) resins, and (4) asphaltenes. Feedstock fractionation by this method is performed with regard to the polarity of these fractions by using different solvents and adsorbents and, moreover, the method is reproducible and applicable to a wide variety of feedstocks (Speight, 2014, 2015).

The "saturates" fraction is composed of nonpolar hydrocarbons with linear or branched chains, as well as aliphatic cyclic paraffins. The "aromatics" fraction contains compounds with one or more aromatic rings linked to aliphatic chains. The "resins" fraction and the "asphaltene" fraction are composed of high molecular weight compounds, which contain heteroatom derivatives. The resin constituents are insoluble in liquid propane but soluble in pentane and higher hydrocarbons, whose density is approximately equal to $1\,g\,cm^{-3}$ with a molecular weight on the order of 500–2,000. The asphaltene fraction is composed of constituents that are insoluble in liquid *n*-alkane derivatives (such as *n*-pentane and *n*-heptane, as defined by the test method) but are soluble in benzene or toluene (Speight, 1994, 2014, 2015) and are in a class of petroleum macromolecules composed of polycondensed aromatic rings and lateral aliphatic chains, presenting a smaller proportion of acidic and basic functional groups.

Both the USBM-API and SARA methods require some caution if the asphaltenes are first isolated as a separate fraction. For example, the asphaltene yield varies with the hydrocarbon used for the separation and with other factors (Girdler, 1965; Mitchell and Speight, 1973; Speight et al., 1984). An inconsistent separation technique can give rise to problems resulting from residual asphaltenes in the deasphaltened oil undergoing irreversible adsorption on the solid adsorbent.

The USBM-API and SARA methods are widely used separation schemes for studying the composition of heavy crude oil fractions and other fossil fuels, but several other schemes have also been used successfully and have found common usage in investigations of feedstock composition. For example, a simple alternative to the SARA sequence is the chromatographic pre-separation of a deasphaltened sample on deactivated silica or alumina with pentane (or hexane) into saturated materials followed by elution with benzene for aromatic materials and with benzene-methanol for polar materials (resins). This allows further chromatography into narrower (more similar) fractions without mutual interference on the adsorbent.

The selection of any separation procedure depends primarily on the information desired about the feedstock. For example, separation into multiple fractions to examine the minute details of feedstock composition requires a complex sequence of steps. An example of such a separation scheme involves the fractionation of Athabasca tar sand bitumen into four gross fractions and subfractionation of these four fractions, which allows further study of the distribution of the functional types within the feedstock.

A combination chromatography using alumina and silica gel is also suitable for deasphaltened oils. There is also the direct fractionation of hydrocarbon and hetero compounds from deasphaltened residua on a dual alumina-silica column with subsequent treatment of the polar fraction with cation- and anion-exchange resins into basic, acid, and neutral materials. The method also includes chromatography of the asphaltenes, but only with highly polar asphaltene samples are the basic and acid compounds first removed with ion-exchange resins. The remainder of the feedstock is separated into saturates, aromatics, and hetero compounds using alumina-silica adsorption. Separation of the saturates fraction into *n*-alkanes and *iso*-alkanes plus cycloalkanes is achieved by the use of urea ($H_2NC{=}ONH_2$) and thiourea ($H_2NC{=}SNH_2$).

One of the problems of such a fractionation scheme is the initial separation of the feedstocks into two ill-defined fractions (colloids and dispersant) without first removing the asphaltenes. As already noted, asphaltenes are specifically defined by the method of separation. They are less well defined using such liquids as ethyl acetate in place of the more often used hydrocarbons, such as pentane and heptane (Speight, 1979). The use of ethyl acetate undoubtedly leads to asphaltene material in the dispersoids and non-asphaltene material in the colloids. Application of a more standard deasphalting technique would undoubtedly improve this method and provide an excellent insight into feedstock composition.

One of the common issues related to the use of any adsorption-based fractionation scheme is the nature of the adsorbent. In the early reports of crude oil fractionation (Pfeiffer, 1950), clays often appeared as an adsorbent to effect the separation of the feedstock into various constituent fractions. However, clay (fuller's earth, Attapulgus clay, and the like) is often difficult to define with any degree of precision from one batch to another. Variations in the nature and properties of the clay can, and will, cause differences not only in the yields of composite fractions but also in the distribution of the compound types in those fractions. In addition, irreversible adsorption of the more polar constituent to the clay can be a serious problem when further investigations of the constituent fractions are planned.

One option for resolving this problem has been the use of more standard adsorbents, such as alumina and silica. These materials are easier to define and are often accompanied by guarantees of composition and type by various manufacturers. They also tend to irreversibly adsorb less of the feedstock than a clay. Once the nature of the adsorbent is guaranteed, reproducibility becomes a reality. Without reproducibility the analytic method does not have credibility.

4.4.2.2 ASTM Methods

There are three ASTM methods that provide for the separation of a feedstock into four or five constituent fractions and it is interesting to note that as the methods have evolved, there has been a change from the use of *n*-pentane as the asphaltene precipitant (ASTM D2007, 2021) to *n*-heptane as the asphaltene precipitant (ASTM D4124, 2021) (Speight, 2014, 2015). This is, in fact, in keeping with the production of a more consistent fraction that represents these higher molecular weight, more complex constituents of crude oil (Girdler, 1965; Speight et al., 1984). Two of the standard test methods use adsorbents to fractionate the deasphaltened oil (ASTM D2007, 2021; ASTM D4124, 2021).

Obviously, there are precautions that must be taken when attempting to separate viscous feedstocks or polar feedstocks into constituent fractions. The disadvantages in using ill-defined adsorbents are that adsorbent performance differs with the same feed and, in certain instances, may even cause chemical and physical modification of the feed constituents. The use of a chemical reactant like sulfuric acid should only be advocated with caution since feeds react differently and may even cause irreversible chemical changes and/or emulsion formation. These advantages may be of little consequence when it is not, for various reasons, the intention to recover the various product fractions in toto or in the original state, but in terms of the compositional evaluation of different feedstocks, the disadvantages are very real.

In summary, the terminology used for the identification of the various methods might differ. However, in general terms, group-type analysis of crude oil is often identified by the acronyms for the names: PONA (paraffins, olefins, naphthenes, and aromatics), PIONA (paraffins, *iso*-paraffins, olefins, naphthenes, and aromatics), PNA (paraffins, naphthenes, and aromatics), PINA (paraffins, *iso*-paraffins, naphthenes, and aromatics), or SARA (saturates, aromatics, resins, and asphaltenes). However, it must be recognized that the fractions produced by the use of different adsorbents will differ in content and will also be different from fractions produced by solvent separation techniques.

The variety of fractions isolated by these methods and the potential for the differences in composition of the fractions makes it even more essential that the method is described accurately and that it is reproducible not only in any one laboratory but also between various laboratories.

4.5 Chemical Methods

The methods of fractionation that involve the use of chemical reactants are different in nature from the methods described in the preceding sections. Although several methods that employ chemical reactants have been applied to fractionation (which can advertently or inadvertently according to the wishes of the analyst) change the chemical nature of the feedstock constituents and, hence, methods such as adsorption and solvent treatment are often preferred (Speight, 2014).

The method of chemical separation commonly applied to separate crude oil into various fractions is treatment with sulfuric acid and, since this method has also been applied in the refinery but with limited success in the fractionation of heavy oil and/or bitumen due to the formation of complex sulfates and difficult-to-break emulsions (Speight, 2014, 2015). Obviously, the success of this fractionation method is feedstock dependent and, in conclusion, it would appear that the test be left more as a method of product cleaning for which it was originally designed rather than a method of separation of the various fractions.

The most common methods that have been used for the separation procedure, other than the use of ferric chloride ($FeCl_3$) on an adsorbent, involve the use of sulfuric acid and urea adduction. Although to be truthful, the urea adduction method is in reality a physical method but insofar as it provides a separation that is specific to certain constituents of crude oil, it is included here. There are many other methods that have been used for the chemical separation of crude oil that are now included as refinery processes (Parkash, 2003; Gary et al., 2007; Speight, 2014; Hsu and Robinson, 2017; Speight, 2017). These acid treatment (particularly the sulfuric acid method) and the urea addition method are still widely used in the laboratory and hence the inclusion of these methods here.

4.5.1 Acid Treatment

The method of chemical separation commonly applied is treatment with sulfuric acid. Marcusson and Eickmann made an early reference to the use of sulfuric acid in 1908 and used a procedure to precipitate asphaltene constituents from asphaltic materials by treatment of the sample with low-boiling naphtha, followed by fractionation of the naphtha-soluble material with concentrated sulfuric acid. The precipitate produced by the sulfuric acid treatment was actually material that had been converted to an asphaltene type of product by interaction of the asphaltic constituents with the sulfuric acid. It is nevertheless possible that some of the acid-precipitated material originated as asphaltenes that were incompletely precipitated by the naphtha. The constitution of the naphtha was unknown; most likely it was not pure *n*-pentane and it may even have contained hexane(s) or higher paraffins. The addition of only 20 volumes of solvent to heavy feedstocks is not a sufficient amount to completely precipitate asphaltene material.

The chemical precipitation method is claimed to provide the needed subdivision of the resins and the oils to yield chemically related fractions. The names given to the individual fractions are descriptive of the steps used in the procedure. For example, the name paraffins is used for the saturated nonreactive fraction, whereas the next two groups in ascending reactivity are second acidaffins and first acidaffins. The term second acidaffins denotes the group of hydrocarbons having affinity for strong acid as represented by fuming sulfuric acid. The term first acidaffins denotes that the constituents have an affinity for ordinary concentrated sulfuric acid. The next fraction, nitrogen bases, separated by sulfuric acid of 85% concentration, includes the most reactive components and may contain, among other components, substantially all the nitrogen-containing compounds.

4.5.2 Molecular Complex Formation

The formation of crystalline molecular complexes between urea ($H_2NC{=}ONH_2$) or thiourea ($H_2NC{=}SNH_2$) and hydrocarbons has been known since the 1940s, and it is not surprising that the technique has received considerable attention with respect to its use in crude oil chemistry.

The previously described techniques of adsorption and distillation differentiate molecules by class and volatility (or size within any one homologous series), respectively, whereas adduct formation separates on the basis of molecular shape and, to a lesser extent, by size and class. When combined with the older fractionation methods, adduct formation can often be useful for solving separation problems, provided the limitations of the method are realized. Urea and thiourea adduction are not completely selective, as was first supposed, and there is an overlapping of structural types that adduct, especially among the higher molecular weight hydrocarbons.

Adduct formation may be achieved merely by bringing together the adduct-former and reagent under a wide range of reaction conditions. Preferably the reactant is an inert hydrocarbon solvent, a reagent solvent, referred to as the activator, present in varying amounts to increase the rate of reaction, and the crystalline product that precipitates can be conveniently separated by filtration. Furthermore, the adduct may be decomposed easily and the adducting material recovered by one of several procedures.

4.5.2.1 Urea Adduction

For a hydrocarbon to form an adduct with urea, it is mandatory that the compound contain a long, unbranched chain, usually a chain of at least six carbon atoms, under the conditions most often employed, such as ambient temperature and atmospheric pressure. As side chains or ring structures are added to the molecule, the chain must be lengthened if adduction is to occur. Generally, the larger the size or number of the substituents or the farther the substituent is from the terminal carbon atom, the longer must be the unbranched portion of the chain. Olefinic unsaturation usually has little effect on adduct formation, but the stability of the adduct of *n*-olefins are somewhat less than those of the corresponding *n*-paraffins. Thus, *n*-pentane does not form an adduct under normal conditions. However, an adduct can be obtained at low temperature with some pressure, and *n*-hexane is the first member of the *n*-alkane series that forms an adduct at room temperature and atmospheric pressure; under similar conditions, 1-octene is the lowest olefin reported to adduct. There is usually no upper limit to chain length other than that imposed by the solubility of the hydrocarbon; however, as the reaction temperature is increased to meet the requirements of hydrocarbon solubility and preferred reaction rate, the stability of the urea lattice in the adduct decreases.

In any given homologous series, the ease of adduct formation and adduct stability increases with increasing straight-chain length. Single methyl side chains, such as those present in compounds such as 3-methyl heptane [$CH_3CH_2CH(CH_3)CH_2CH_2CH_2CH_3$] interfere with adduct formation, presumably because the dimensions of the molecules at the methyl group are too great to allow the hydrocarbons to fit properly into the urea adduct channel. However, in the presence of an adducting *n*-paraffin, such as *n*-decane, the methyl-substituted paraffins are assisted into the adduct (induction). 3-Methyl eicosane [$CH_3CH_2CH(CH_3)CH_2(CH_2)_{15}CH_3$] presumably because of its sufficiently long unbranched chain, forms adducts readily.

When the carbon chain is sufficiently long, ring structures do not prevent adduct formation. For example, 1-phenyl octadecane [$C_6H_5CH_2(CH_2)_{16}CH_3$] forms an adduct with urea but 1-phenyl octane [$C_6H_5CH_2(CH_2)_6CH_3$] does not participate in adduct formation. The stability of the adduct formed from the phenyl-paraffins or cyclohexyl-paraffins is usually considerably less than those of the corresponding straight chain hydrocarbons having the same number of carbon atoms.

Many classes of organic compounds form adducts, and some examples of these are ketones, acids, esters, halides, mercaptans, and ethers; both saturated and unsaturated structures adduct, provided the chain is sufficiently long. The carbonyl oxygen in ketones, acids, and esters does not appear to interfere with adduct formation and, to some extent, even aids the process since shorter carbon chains are adducted in oxygenated compounds. Thus, acetone, with a straight chain of three carbon atoms, *n*-butyric acid with four carbon atoms, and their higher homologs form complexes with urea. Among the alkyl halides, 1-bromohexane, but not 2-bromohexane, forms an adduct; 2-bromodecane can be adducted, indicating that an unsubstituted chain of eight carbon atoms is sufficient to overcome substituent effects.

4.5.2.2 Thiourea Adduction

The relationship among structure, adductability, and adduct stability is not as well defined for the thiourea-adductable compounds as for the urea-reactive hydrocarbons. The stability of any thiourea adduct is quite low, even at 0 °C (32 °F), and corresponds approximately to the stability of the urea completes of the lower *n*-paraffins. The high stability of the urea completes of the higher *n*-paraffins has no parallel in thiourea completes. Among the higher hydrocarbons containing a ring or branching and a long chain, there is a lesser tendency for the thiourea adducts to form than in the lower molecular weight homologs, as the alkyl chain apparently reduces the stability. The higher *n*-paraffins adduct fairly readily and become more stable with increasing molecular weight, but the stability of these adducts is still of the same low order of magnitude as the lower molecular weight *iso*-paraffins and naphthenes.

4.5.2.3 Adduct Composition

Analysis of the complexes for the ratio of reagent to reactant may be made by several methods. The urea (or thiourea) content of the crystalline complex may be established by nitrogen determination, or the amount of organic component may be determined from carbon content or by the weight loss upon dissociation of the adduct. Also, measurement of concentration changes in the reaction liquid is applicable to certain mixtures.

For urea adducts of normal hydrocarbons, the mol ratio of urea to hydrocarbon in the adduct can be represented by the equation:

$$m = 0.65n + 1.5$$

where m is the mol ratio of urea to hydrocarbon and n is the number of carbon atoms in the hydrocarbon. For all practical purposes, the use of the ratio of 3.3 g of urea per gram of normal hydrocarbon is more convenient. The greater variety of hydrocarbon structures adducting with thiourea than with urea adversely affects development of an equation relating the mol ratio of thiourea to reactant. For a given number of carbon atoms, the more compact molecules are generally associated with less thiourea in the adduct than are the less compact molecules. The weight ratio decreases from 2.7 to 2.8 for aliphatic compounds and monocyclic naphthenes, to 2.5 for dicyclic naphthenes, to 2.2–2.3 for condensed ring systems; the weight ratio for unsaturated compounds is generally very similar to that of the corresponding saturated structure.

4.5.2.4 Adduct Structure

X-ray diffraction patterns have been obtained for many different urea complexes, and it appears that the hydrocarbon in the adduct does not make any contribution to the structure. In the adduct, the urea molecules are held together by spirals of hydrogen bonds between an oxygen atom and a nitrogen atom of adjacent molecules; these adjacent molecules are turned 120 °C (248 °F) with respect to one another. A distance of 3.7 D separates adjacent non-hydrogen-bonded molecules along the axis of the spiral, and the edge length of the unit cell is 4.8 D. In this molecular arrangement, there exists a hexagonal channel, or canal, into which the reactant molecule must fit, which has been calculated to 4.9 D in diameter. The diameter of an *n*-paraffin chain is of the order of 3.8 × 4.2 D, thus allowing the *n*-paraffins to fit into the channel.

The crystal structure of urea in the adduct is entirely different from that of the pure reagent. There are six urea molecules in the hexagonal unit cell of the adduct, whereas pure urea is a tetragonal, close-packed crystal with no canal or available free space in which other molecules could be enclosed. Thus, the change in crystal structure from a tetragonal to a hexagonal system occurs at some time during the adduction process, and it is considered likely that the urea molecules grow in a spiral around the hydrocarbon. X-ray studies also indicate that many of the thiourea adducts have crystalline structures similar to one another and quite analogous to those of the urea adducts. Although pure thiourea crystals have an orthorhombic structure, the unit cell of the thiourea complexes is usually rhombohedral. A few cases are known in which the unit cell of the complex is orthorhombic, and with the thiourea adducts different crystalline forms can apparently be precipitated under different reaction conditions. The channel concept also applies to the thiourea adducts as it does to those of urea. However, the larger size of the sulfur atom in thiourea in relation to the oxygen in urea results in a channel with a larger cross section. The unit cell constants of the thiourea adducts apparently vary with the nature of the adducted molecule, and hence the channel cross-sectional dimensions also vary and are presumed to be of the order of 5.8–6.8 Å.

It is generally accepted that, in the urea adducts, the urea molecules are connected into spirals by hydrogen bonds between the oxygen and the amino groups of adjacent urea molecules, resulting in a channel into which the adducting compound can fit. The size of the channel limits the molecules that may adduct to those having cross-sectional dimensions equal to, or less than, those of the channel in the urea adduct, but in certain cases, distortion of the urea lattice may occur and slightly larger molecules can adduct. The structure of the thiourea adducts is similar to that of the urea adducts. The larger size of the sulfur atom results in a channel of somewhat larger cross-sectional dimensions, hence allowing larger molecules to adduct with thiourea than are able to do so with urea. In general, urea forms adducts with organic compounds containing a long unbranched chain, such as the *n*-paraffins, whereas thiourea forms complexes with compounds that contain a moderate amount of branching of cyclization. However, especially among the higher molecular weight hydrocarbons, urea adducts can be formed from *n*-paraffins.

4.5.2.5 Adduct Properties

The amount of dissociation that occurs when an adduct is contacted with solvents is a function of the solvent type. These types may be classified as: (1) hydrocarbon solvents such as benzene, (2) urea or thiourea using solvents such as water, and (3) hydrocarbon and urea (or thiourea) solvents (methanol or benzene–methanol mixtures).

As expected, the last two types cause the greatest dissociation. Solvents in which urea or thiourea is relatively insoluble, but in which the hydrocarbon is readily dissolved, exert relatively little influence. With any solvent, increasing temperature increases the amount of dissociation.

The most practical method of decomposition of the adduct is by solution, generally with hot water. The adducted hydrocarbons form an immiscible layer on top of the aqueous urea solution and may be readily separated. Volatile reactants may be recovered by heating the adduct, either dry or with steam, and collecting the liberated hydrocarbon as it is released. Actually, a crude fractional dissociation may be accomplished in this manner or by partial solvent extraction. The least stable adduct-formers are released first and may be collected. Separations of compounds from mixtures are carried out, in effect, on the basis of their stability. The most desirable results are generally obtained if all the possible material is first precipitated as the adduct; the recovered adducted material is then fractionally re-adducted. For such fractionation, the reprecipitation may be performed by: (1) using a quantity of reagent insufficient to adduct all the material first adducted, (2) at a different temperature using a more optimum reagent to reactant ratio, and (3) by applying both techniques at the same time.

In a system containing essentially only one homologous series, such as the *n*-paraffins in paraffin waxes, the use of a fractional adduction technique furnishes a separation on the basis of molecular weight.

In a narrow molecular weight distillation fraction containing several adductable hydrocarbon types, separation can be accomplished as a result of the difference in stability of the adducts of the different hydrocarbon homologous series. With urea the *n*-paraffins react first, followed by the slightly branched *iso*-paraffins and by the more highly branched *iso*-paraffins and the cyclic structures last. However, working with a wide molecular weight fraction, such as crude oil waxes, the stability of the urea adducts of lower *n*-paraffins present are of the same order of magnitude as those of the adducts of higher *iso*-paraffins and monocyclic compounds; a mixture of types is of different molecular weights. Hence, fractional distillation of the hydrocarbon samples either before or after adduction allows better analysis of the various fractions.

Fractionation may also be effected by selective decomposition of the adduct or by selective replacement. In the former method, the adduct is extracted using a solvent of relatively poor dissociating power whereby the least stable adducts tend to dissociate first. In selective replacement, the hydrocarbons of the less stable adducts are gradually displaced by slightly more stable adductors, and these in turn are displaced by more stable adduct-formers until the most stable *n*-paraffins are employed.

The relative instabilities of the thiourea adducts render the method less useful than the urea method. In addition, the less selective nature of thiourea adduction and the fact that the large differences in stability within a homologous series with the urea adducts do not apply to the thiourea adducts.

4.6 The Asphaltene Fraction

As a refresher comment to the information presented above, the SARA method of analysis is a method used for the fractionation of crude oils and the viscous feedstocks in which a feedstock sample is separated into four fractions, with each fraction having a different composition and each fraction consists of a "class" containing a range of different molecular-weight species, thus: saturates, aromatics, resins, and asphaltene. The "saturates" fraction typically contains nonaromatic hydrocarbon derivatives such as alkane derivatives, *iso*-paraffin derivatives, and cycloparaffin derivatives with minor or trace amounts of heteroatom derivatives. The aromatics fractions contain a higher proportion of aromatic derivatives and a lesser proportion of nonaromatic hydrocarbon derivatives. The resins fraction contains even more aromatic hydrocarbon derivatives along with heteroatom derivatives which have higher molecular weights that the constituents of the saturate fraction and the aromatics fraction—the resins fraction is sometimes (incorrectly) referred to as containing low molecular weight asphaltene derivative. The asphaltene fraction is the most complex (and highest molecular weight) of the SARA fractions and in addition to a higher aromaticity also contain higher amounts of heteroatom constituents as well as metal-containing derivatives, such as nickel-containing and vanadium-containing constituents. These four fractions form a continuum of molecular species with increasing molecular weight, aromaticity, and heteroatom content.

The asphaltene fraction of refinery feedstocks has drawn considerable attention from investigators because of the role of the constituents of this fraction in coke formation as well as catalyst poisoning and catalyst destruction.

From the early days of a study of the science of the asphaltene fraction, there has been a major focus on the derivation of the structure of the constituents but, unfortunately, the focus has been on the development of a so-called "average structure" for the fraction. Isolation of the asphaltene fraction is achieved from a feedstock by the addition of a hydrocarbon liquid (such as *n*-pentane or *n*-heptane) (Speight, 1994, 2014, 2015, 2021). The fraction is isolated as dark brown to black

Table 4.2 Separation of Asphaltene Constituents Based on Functionality (Polarity)

Source	Process	Products
Asphaltene fraction	Ion-exchange resin	Acids
		Neutral polars
		Bases
Bases	Basic resin	Bases
		Amphoterics
Or		
Acids	Acidic resin	Acids
		Amphoterics

friable solids that have no definite melting point and usually foam and swell on heating to leave a carbonaceous residue.

On the basis of the different types of constituents in the asphaltene fraction, it is not only misleading but erroneous to deduce (by no matter what method) that the asphaltene fraction (of any feedstock) which is composed of a variety of constituents that have the highest molecular weight and the most polar constituents in the feedstock (Speight, 1994, 2014, 2015, 2021). In fact, the asphaltene fraction can be defined in terms of functional group composition (Table 4.2) (Francisco and Speight, 1984). This method can be applied to the asphaltene fractions from a variety of feedstocks with a relationship between the yield of the various subfractions and the source and type of the feedstock.

But in conclusion, there is no one parameter that is operational in the separation of asphaltene constituents; the relevant parameters for asphaltene separation are physical and chemical in nature (Table 4.3) (Girdler, 1965; Mitchell and Speight, 1973; Long, 1981; Speight et al., 1984; Speight, 1994, 2014, 2021).

In summary, the asphaltene fraction can be defined by the hydrocarbon liquids used to separate the fraction—that is, pentane asphaltenes, heptane asphaltenes—or by the relative amounts of the content of the functional groups (Table 4.2).

4.7 Carbenes and Carboids

While the asphaltene fraction can be defined by the hydrocarbon liquids used to separate the fraction—that is, pentane asphaltenes, heptane asphaltenes—or by the relative amounts of the content of the functional groups (Table 4.2), defining the carbene fraction and the carboid fraction is somewhat more difficult.

Table 4.3 Operative Parameters in the Separation of the Asphaltene Fraction

Parameter	Description
Polarity	The presence of functional groups derived from the presence of heteroatoms in the asphaltene constituents
Aromaticity	The presence of polynuclear aromatic systems in the asphaltene constituents
Molecular weight	Molecular size
3-D structure	The asphaltene constituents as they exist in relationship with the other constituents of the feedstock
Solvent power	The solvent strength of the precipitating/extracting liquid used for the separation
Ratio	The ratio of the precipitating/extracting liquid to feedstock that dictates the yield and character of the asphaltene product
Time	The time required to allow the precipitating/extracting liquid to penetrate the micelle which is dependent upon the ability of the hydrocarbon liquid to penetrate the micelle, indicating that the process is diffusion-controlled
Temperature	The temperature may have an effect on the induction period that is a requirement of diffusion-controlled processes
Pressure	As employed in several refinery processes, pressure is a means of maintaining the low-boiling liquid hydrocarbon in the liquid phase

By definition in the context of this book, the carbene constituents and the carboid constituents of a feedstock are not naturally occurring constituents of crude oil or the viscous feedstocks. They are, in fact, the products of the thermal treatment of feedstocks and are defined by solubility/insolubility in pyridine or carbon disulfide (Table 4.2). The formation of these two fractions can be speculatively assigned to thermal reactions that occur during thermal recovery of the crude oil or thermal processing as can occur during refining.

For example, the initial reactions in the thermal decomposition of the asphaltene constituents involve the thermolysis of aromatic-alkyl systems to produce volatile species (paraffin derivatives and olefin derivatives) and nonvolatile species (aromatic derivatives) (Speight, 1987; Roberts, 1989; Schabron and Speight, 1997). In fact, thermal cracking of model compounds confirm that volatility of the fragments is a major influence in carbon residue formation. It has also been reported that the formation of a coke-like substance during thermal decomposition of aromatic compounds and related heterocyclic compounds are dependent upon the structure, and hence the volatility, of the polynuclear aromatic hydrocarbon but more specifically on (1) the degree of polynuclear condensation

Table 4.4 Representation of the Multi-Reaction Sequence for the Thermal Decomposition of Asphaltene Constituents

Source	Primary products	Secondary products	Tertiary products
Asphaltene	Gas	Gas	
	Liquids	Gas	
		Liquids	
	Carbenes	Gas	
		Liquids	
		Carboids	Gas
			Coke
	Carboids	Gas	
		Coke	

in petroleum, (2) the average number of alkyl groups on the polynuclear aromatic systems, (3) the occurrence of heteroatoms in the ring, and (4) the type of heteroatom (Madison and Roberts, 1958; Lewis and Edstrom, 1963; Lewis and Singer, 1964, 1967; Lewis, 1980, 1982). Nitrogen species also contribute to the pattern of the thermal cracking. For example, the hydrogen or carbon–carbon bonds adjacent to a ring nitrogen undergo thermolysis quite readily, as if promoted by the presence of the nitrogen atom (Fitzer et al., 1971).

In addition, analytical data are indicative of the thermal expulsion of volatile product created by the thermal scission of reaction of alkyl carbon–carbon bonds in a position beta to the aromatic ring system (Table 4.4). This would give rise to the carbene constituents while further reaction of the carbene constituents would lead to carbon constituents which may also be formed directly by multiple dealkylation of the polynuclear aromatic systems or by interaction of the various systems to produce even more complex polynuclear aromatic systems (the carboids) that are well on the way to forming coke.

4.8 Use of the Data

In the simplest sense, crude oil can be considered composites of four major operational fractions. However, it must never be forgotten that the nomenclature of these fractions lies within the historical development of crude oil science and that the fraction names are operational and are related more to the general characteristics than to the identification of specific compound types. Nevertheless, once a convenient fractionation technique has been established, it is possible to compare a variety of different feedstocks varying from a conventional crude oil to a propane asphalt (Corbett and Petrossi, 1978).

It is noteworthy here that, throughout the history of studies related to crude oil composition, there has been considerable attention paid to the asphaltene fraction and the resin fraction. This is due in no small part to the tendency of the asphaltene constituents to be responsible for high yields of thermal coke and also for shortened catalyst lifetimes in refinery operations (Speight, 2000; Parkash, 2003; Gary et al., 2007; Speight, 2014; Hsu and Robinson, 2017; Speight, 2017). In fact, it is the unknown character of the asphaltenes that has also been responsible for drawing the attention of investigators for the last five decades (Speight, 1984, 1994). Residua contain the majority of all of the potential coke-forming constituents and catalyst poisons that were originally in the crude oil because the distillation process is a concentration process (Parkash, 2003; Gary et al., 2007; Speight, 2014; Hsu and Robinson, 2017; Speight, 2017). Most of the feedstock constituents that form (during processing) and catalyst poisons are nonvolatile and, hence, the asphaltene fraction contains most of the coke-forming constituents and catalyst poisons that were originally present in the feedstock or in the residuum.

One of the early findings of composition studies was that the behavior and properties of any feedstock are dictated by composition (Speight, 2000; Parkash, 2003; Gary et al., 2007; Speight, 2014; Hsu and Robinson, 2017; Speight, 2017). Although the early studies were primarily focused on the composition and behavior of asphalt, the techniques developed for those investigations have provided an excellent means of studying heavy feedstocks. Later studies have focused not only on the composition of crude oil and its major operational fractions but on further fractionation, which allows different feedstocks to be compared on a relative basis to provide a very simple but convenient feedstock map. However, such a map does not give any indication of the complex interrelationships of the various fractions, such as the interactions between the asphaltene and resin constituents (Koots and Speight, 1975; Speight, 1994, 2014), although predictions of feedstock behavior are possible using such data. It is necessary to take the composition studies one step further using subfractionation of the major fractions to obtain a more representative indication of crude oil composition.

Thus, by careful selection of an appropriate technique, it is possible to obtain an overview of crude oil composition that can be used for behavioral predictions. By taking the approach one step further and by assiduous collection of various subfractions, it becomes possible to develop the crude oil map and add an extra dimension to compositional studies (Long and Speight, 1989). Crude oil and heavy feedstocks then appear more as a continuum than as four specific fractions.

Such a concept has also been applied to the asphaltene fraction of crude oil in which asphaltenes are considered a complex state of matter based on molecular weight and polarity (Long, 1981; Long and Speight, 1989, 1998; Speight, 1994).

The advantage of such a concept is that it can be used to explain differences in asphaltene yield with different hydrocarbon derivatives (such as *n*-pentane and *n*-heptane) and also differences in the character of asphaltenes from different feedstocks.

Furthermore, crude oil can be viewed as consisting of two continuous distributions, one of molecular weight and the other of molecular type. Using data from molecular weight studies and elemental analyses, the number of nitrogen and sulfur atoms in the aromatic and polar aromatic fractions was also exhibited. These data showed that not only can every molecule in the resins and asphaltenes have more than one sulfur or nitrogen but also some molecules probably contain both sulfur and nitrogen. As the molecular weight of the aromatic fraction decreases, the sulfur and nitrogen contents of the fractions also decrease. In contrast to the sulfur containing molecules, which appear in both the naphthene aromatics and the polar aromatic fractions, the oxygen compounds present in the heavy fractions of crude oil are normally found in the polar aromatics fraction.

Other work (Long and Speight, 1989) has presented the development of a different type of compositional map using the molecular weight distribution and the molecular type distribution as coordinates. The separation involved the use of an adsorbent (a clay mineral) and the fractions were characterized by the solubility parameter as a measure of the polarity of the molecular types.

The solubility parameter is a numerical value that indicates the relative solvency behavior of a specific solvent and is derived from the cohesive energy density of the solvent, which in turn is derived from the heat of vaporization. In addition, the Hildebrand solubility parameter provides a numerical estimate of the degree of interaction between materials and can be a good indication of solubility, particularly for nonpolar materials such as many polymers. Materials with similar values of the solubility parameter are likely to be miscible.

The molecular weight distribution can be determined by gel permeation chromatography. Using these two distributions, a map of composition can be prepared using molecular weight and solubility parameter as the coordinates for plotting the two distributions. Such a composition map can provide insights into many separation and conversion processes used in crude oil refining.

The molecular type was characterized by the polarity of the molecules, as measured by the increasing adsorption strength on an adsorbent. At the time of the original concept, it was unclear how to characterize the continuum in molecular type or polarity. For this reason, the molecular type coordinate of the first maps was the yield of the molecular types ranked in order of increasing polarity. However, this type of map can be somewhat misleading because there is not a direct relationship to the amounts of material in a given type. The horizontal distance on the plot is a measure of the yield and there is not a continuous variation in polarity for the horizontal coordinate. It was suggested that the solubility parameter of the different fractions could be used to characterize both polarity and adsorption strength.

In order to attempt to remove some of these potential ambiguities, more recent developments of this concept have focused on the solubility parameter. The simplest map that can be derived using the solubility parameter is produced with the solubility parameters of the solvents used in solvent separation procedures, and equating these parameters to the various fractions. However, the solubility parameter boundaries determined by the values for the eluting solvents that remove the fractions from the adsorbent offer a further step in the evolution of crude oil maps (Long and Speight, 1989).

Measuring the overall solubility parameter of a crude oil fraction is a time-consuming chore. Therefore, it is desirable to have a simpler, less time-consuming measurement that can be made on crude oil fractions that will correlate with the solubility parameter and thus give an alternative continuum in polarity. In fact, the hydrogen-to-carbon atomic ratio and other properties of crude oil fractions that can be correlated with the solubility parameter (Speight, 1994) also provide correlation for the behavior of crude oils.

References

ASTM 2021. *Annual Book of Standards*. ASTM International, West Conshohocken, Pennsylvania.

ASTM D482. 2021. Standard Test Method for Ash from Petroleum Products. Annual Book of Standards. ASTM International, West Conshohocken, Pennsylvania.

ASTM D893. 2021. Standard Test Method for Insolubles in Used Lubricating Oils. Annual Book of Standards. ASTM International, West Conshohocken, Pennsylvania.

ASTM D2007. 2021. Standard Test Method for Characteristic Groups in Rubber Extender and Processing Oils and Other Petroleum-Derived Oils by the Clay-Gel Absorption Chromatographic Method. Annual Book of Standards. ASTM International, West Conshohocken, Pennsylvania.

ASTM D2415. 2021. Standard Test Method for Ash in Coal Tar and Pitch. Annual Book of Standards. ASTM International, West Conshohocken, Pennsylvania.

ASTM D3279. 2021. Standard Test Method for n-Heptane Insolubles. Annual Book of Standards. ASTM International, West Conshohocken, Pennsylvania.

ASTM D3593. 2021. Standard Test Method for Molecular Weight Averages and Molecular Weight Distribution of Certain Polymers by Liquid Size-Exclusion Chromatography (Gel Permeation Chromatography-GPC) Using Universal Calibration. Annual Book of Standards. ASTM International, West Conshohocken, Pennsylvania.

ASTM D4055. 2021. Standard Test Method for Pentane Insolubles by Membrane Filtration. Annual Book of Standards. ASTM International, West Conshohocken, Pennsylvania.

ASTM D4072. 2021. Standard Test Method for Toluene-Insoluble (TI) Content of Tar and Pitch. Annual Book of Standards. ASTM International, West Conshohocken, Pennsylvania.

ASTM D4124. 2021. Standard Test Method for Separation of Asphalt into Four Fractions. Annual Book of Standards. ASTM International, West Conshohocken, Pennsylvania.

ASTM D4312. 2021. Standard Test Method for Toluene-Insoluble (TI) Content of Tar and Pitch (Short Method). Annual Book of Standards. ASTM International, West Conshohocken, Pennsylvania.

ASTM D4628. 2021. Standard Test Method for Analysis of Barium, Calcium, Magnesium, and Zinc in Unused Lubricating Oils by Atomic Absorption Spectrometry. Annual Book of Standards. ASTM International, West Conshohocken, Pennsylvania.

ASTM D4927. 2021. Standard Test Methods for Elemental Analysis of Lubricant and Additive Components—Barium, Calcium, Phosphorus, Sulfur, and Zinc by Wavelength-Dispersive X-Ray Fluorescence Spectroscopy. Annual Book of Standards. ASTM International, West Conshohocken, Pennsylvania.

ASTM D5185. 2021. Standard Test Method for Multielement Determination of Used and Unused Lubricating Oils and Base Oils by Inductively Coupled Plasma Atomic Emission Spectrometry (ICP-AES). Annual Book of Standards. ASTM International, West Conshohocken, Pennsylvania.

ASTM D6443. 2021. Standard Test Method for Determination of Calcium, Chlorine, Copper, Magnesium, Phosphorus, Sulfur, and Zinc in Unused Lubricating Oils and Additives by Wavelength Dispersive X-ray Fluorescence Spectrometry (Mathematical Correction Procedure). Annual Book of Standards. ASTM International, West Conshohocken, Pennsylvania.

ASTM D6560. 2021. Standard Test Method for Determination of Asphaltenes (Heptane Insolubles) in Crude Petroleum and Petroleum Products. Annual Book of Standards. ASTM International, West Conshohocken, Pennsylvania.

Corbett, L.W., and Petrossi, U. 1978. Differences in Distillation and Solvent Separated Asphalt Residua. Industrial and Engineering Chemistry Product Research and Development, 17: 342–346.

Fitzer, E., Mueller, K., and Schaefer, W. 1971. The Chemistry of the Pyrolytic Conversion of Organic Compounds to Carbon. In: *The Chemistry and Physics of Carbon*. Volume 7. p. 237.

Francisco, M.A., and Speight, J.G. 1984. Asphaltene Characterization by a Non-Spectroscopic Method. *American Chemical Society, Division of Fuel Chemistry, Preprints*, 29(1): 36.

Gary, J.G., Handwerk, G.E., and Kaiser, M.J. 2007. *Petroleum Refining: Technology and Economics*. 5th Edition. CRC Press, Taylor & Francis Group, Boca Raton, Florida.

Girdler, R.B. 1965. Constitution of Asphaltenes and Related Studies. *Proceeding of the Association of Asphalt Paving Technologists*, 34: 45.

Halvorsen, I.J., and Skogestad, S. 2000. Distillation Theory. In: *Encyclopedia of Separation Science*. Ian D. Wilson (Editor-in-chief). Academic Press Inc., New York. pp. 1117–1134.

Hildebrand, J.H., Prausnitz, J.M., and Scott, R.L. 1970. *Regular Solutions*. Van Nostrand-Reinhold, New York.

Hirsch, D.E., Hopkins, R.L., Coleman, H.J., Cotton, F.O., and Thompson, D.J. 1972. *Analytical Chemistry*, 44: 915–919.

Hoiberg, A.J. 1964. *Bituminous Materials: Asphalts, Tars and Pitches*. Interscience, New York.

Hsu, C.S., and Robinson, P.R. (Editors). 2017. *Handbook of Petroleum Technology*. Springer International Publishing AG, Cham, Switzerland.

Jewell, D.M., Weber, J.H., Bunger, J.W., Plancher, H., and Latham, D.R. 1972. Ion-Exchange, Coordination, and Adsorption Chromatographic Separation of Heavy-End Petroleum Distillates. *Analytical Chemistry*, 44: 1391.

Koots, J.A., and Speight, J.G. 1975. The Relation of Petroleum Resins to Asphaltenes. *Fuel*, 54: 179.

Lei, Z., Chen, B., and Ding, Z. 2005. *Special Distillation Processes*. Elsevier, Amsterdam, Netherlands.

Lewis, I.C. 1980. Thermal Polymerization of Aromatic Hydrocarbons. *Carbon*, 18: 191–196.

Lewis, I.C. 1982. The Chemistry of Carbonization. *Carbon*, 20: 519–529.

Lewis, I.C., and Edstrom, T. 1963. Thermal Reactivity of Polynuclear Aromatic Hydrocarbons. *The Journal of Organic Chemistry*, 28: 2050–2057.

Lewis, I.C., and Singer, L.S. 1964. Electron Spin Resonance Studies on the Pyrolysis of Aromatic Hydrocarbons. *Carbon*, 2: 115–122.

Lewis, I.C., and Singer, L.S. 1967. Further Electron Spin Resonance Studies on the Pyrolysis of Aromatic Hydrocarbons. *Carbon*, 5: 373–381.

Long, R.B. 1981. The Concept of Asphaltenes. In: *The Chemistry of Asphaltenes*. J.W. Bunger and N. Li (Editors), Advances in Chemistry Series No. 195. American Chemical Society, Washington, DC. p. 17.

Long, R.B., and Speight, J.G. 1989. Studies in Petroleum Composition. I: Development of a Compositional Map for Various Feedstocks. *Revue Institut Français du Pétrole*, 44: 205.

Long, R.B., and Speight, J.G. 1998. The Composition of Petroleum. In: *Petroleum Chemistry and Refining*. J.G. Speight (Editor). Taylor & Francis Publishers, Washington, DC. Chapter 1

Madison, J.J., and Roberts, R.M. 1958. Pyrolysis of Aromatics and Related Heterocycles. *Industrial and Engineering Chemistry*, 50: 237–250.

McKay, J.F., Cogswell, T.E., Weber, J.H., and Latham, D.R. 1975. Analysis of Acids in High-Boiling Petroleum Distillates. *Fuel*, 54: 50–61.

McKay, J.F., Weber, J.H., and Latham, D.R. 1976. Characterization of Nitrogen Bases in High-Boiling Petroleum Distillates. *Analytical Chemistry*, 48: 891–898.

McKay, J.F., Amend, P.J., Cogswell, T.E., Harnsberger, P.M., Erickson, R.B., and Latham, D.R. 1977. *American Chemical Society, Division of Petroleum Chemistry, Preprints*, 22(2): 708.

Mitchell, D.L., and Speight, J.G. 1973. The Solubility of Asphaltenes in Hydrocarbon Solvents. *Fuel*, 52: 149.

Parkash, S. 2003. *Refining Processes Handbook*. Gulf Professional Publishing, Elsevier, Amsterdam, Netherlands.

Petyluk, F.B. 2004. *Distillation Theory and Its Application to Optimal Design of Separation Units*. Cambridge University Press, Cambridge, United Kingdom.

Pfeiffer, J.P. (Editor). 1950. *The Properties of Asphaltic Bitumen*. Elsevier BV, Amsterdam, The Netherlands.

Roberts, I. 1989. The Chemical Significance of Carbon Residue data. *American Chemical Society, Division of Petroleum Chemistry, Preprints*, 34(2): 251.

Rostler, F.S. 1965. Analytical and Functional Significance Bituminous Materials. In: *Bituminous Materials: Asphalts, Tars, and Pitches*. A.J. Hoiberg (Editor). Volume II. Part I. Interscience, New York. pp. 151–223.

Schabron, J.F., and Speight, J.G. 1997. An Evaluation of the Delayed Coking Product Yield of Heavy Feedstocks Using Asphaltene Content and Carbon Residue. *Revue Institut Français du Pétrole*, 52: 73.

Speight, J.G. 1979. *Studies on Bitume: Fractionation – (a) Fractionation by a Cryoscopic Method (b) Effect of Solvent Type on Asphaltene Solubility*, Information Series No. 84. Alberta Research Council, Edmonton, Alberta, Canada.

Speight, J.G. 1984. The Chemical Nature of Petroleum Asphaltenes. In: *Characterization of Heavy Crude Oils and Petroleum Residues*. B.P. Tissot (Editor). Editions Technip, Paris. p. 32.

Speight, J.G. 1987. Initial Reactions in the Coking of Residua. *American Chemical Society, Division of Petroleum Chemistry, Preprints*, 32(2): 413.

Speight, J.G. 1989. A Solvent Method for Feedstock Fractionation: An Alternate to Adsorption Methods of Fractionation. *Neftekhimiya*, 29: 610.

Speight, J.G. 1994. Chemical and Physical Studies of Petroleum Asphaltenes. In: *Asphaltenes and Asphalts. Developments in Petroleum Science, 40*. T.F. Yen and G.V. Chilingarian (Editors). Elsevier, Amsterdam, Netherlands. Chapter 2

Speight, J.G. 2000. *The Desulfurization of Heavy Oils and Residua*. 2nd Edition. Marcel Dekker, New York.

Speight, J.G. 2001. *Handbook of Petroleum Analysis*. John Wiley & Sons Inc., Hoboken, New Jersey.

Speight, J.G. 2014. *The Chemistry and Technology of Petorleum*. 5th Edition. CRC Press, Taylor & Francis Group, Boca Raton, Florida.

Speight, J.G. 2015. *Handbook of Petroleum Product Analysis*. 2nd Edition. John Wiley & Sons Inc., Hoboken, New Jersey.

Speight, J.G. 2017. *Handbook of Petroleum Refining*. CRC Press, Taylor & Francis Group, Boca Raton, Florida.

Speight, J.G. 2021. *Petroleum Refining Technology Series Volume 1: Refinery Feedstocks*. CRC Press, Taylor & Francis Group, Boca Raton, Florida.

Speight, J.G. 2023. *Petroleum Refining Technology Series Volume 2: Dewatering, Desalting, and Distillation*. CRC Press, Taylor & Francis Group, Boca Raton, Florida.

Speight, J.G., Long, R.B., and Trowbridge, T.D. 1984. Factors Influencing the Separation of Asphaltenes from Heavy Petroleum Feedstocks. *Fuel*, 63: 616.

5

Chemical Properties

5.1 Introduction

During the early decades of the refining industry (particularly up to the onset of World War II), there was not the need to understand the composition, properties, and behavior of crude oil in detail that is currently required to apply the necessary processing options. Refining involved distillation of the valuable kerosene fraction that was then sold as an illuminant. After the commercialization of the internal combustion engine, the desired product became gasoline and it was also obtained by distillation. Even when crude oil that contained little natural gasoline was used, cracking (i.e., thermal decomposition with simultaneous removal of distillate) became the modus operandi.

However, with the startling demands on the crude oil industry during and after World War II, and the emergence of the age of petrochemicals and plastics, the crude oil industry needed to produce materials not even considered as products in the decade before the war. Thus, refining took on the role of technological innovator as new and better processes were invented and advances in the use of materials for reactors were developed. In addition, there became a necessity to find out more about crude oil so that refiners might be able to enjoy the luxury of predictability and plan a product slate that was based on market demand. A difficult task when the character of the crude oil was unknown! The idea that crude oil refining should be a hit and miss affair was not acceptable.

Moreover, in addition to tests for physical and chemical properties of refinery feedstocks, it became apparent that refining required not only knowledge of its chemical and physical properties but also knowledge of its chemical and physical reactivity. The former is dealt with in this chapter and the latter, because of the chemical and physical structure (including the constituents' interactions), is dealt with elsewhere in this book (Chapter 8). Because refinery feedstocks vary markedly in its properties and composition according to the source, it also varies in its

Handbook of Heavy Oil Properties and Analysis, First Edition. James G. Speight.

chemical and physical reactivity. Thus, knowledge of feedstock reactivity is required for optimization of existing processes as well as for the development and design of new processes.

For example, valuable information can be obtained from the true boiling point (TBP) curve, which is a function of percent weight distilled and temperature, that is, a boiling point distribution and may be more specific to the gas oil fraction rather than the extra heavy crude oil and tar sand bitumen materials. However, the boiling range does not convey much detail about chemical reactivity of refinery feedstocks. In addition to the boiling point distribution, it is possible to measure bulk physical properties, such as specific gravity and viscosity that have assisted in the establishment of certain empirical relationships for feedstock processing from the curve. Many of these relationships include assumptions that are based on experience with a range of feedstocks. However, the chemical aspects of refining feedstocks that contain different proportions of chemical species emphasizes the need for more definitive data that would enable more realistic predictions to be made of feedstock behavior in refinery operations.

The history of analysis of the chemical constituents in refinery feedstocks started more than 150 years ago, when in 1865 several aromatic hydrocarbons were identified as constituents of crude oil. Identification of individual constituents of crude oil continued and the rapid advances in analytic techniques have allowed the identification of large numbers of crude oil constituents. At this time, the major chemical types of compounds that exist in crude oil have been identified and many members of the various homologous series have been separated or conclusively identified by various techniques (Speight, 2014a). This indicates a wealth of organic compounds that are either resistant to mild hydrogenation (such as paraffins) or are reactive under destructive hydrogenation conditions.

As the boiling points of feedstock fractions increase, the complexity of the constituents in these fractions also increases, and the differences in chemical reactivity between the main classes of constituents may become less well pronounced. Nevertheless, the properties of any material are dictated by its chemical composition, and once the chemical composition of a material has been established, it becomes possible to correlate composition and properties.

It is generally recognized that crude oils are composed of four major fractions (saturate constituents, aromatic constituents, resin constituents, and asphaltene constituents) that differ from one another sufficiently in chemical character that identification can be achieved by application of relevant methods (Chapter 3). Indeed, although these four fractions are chemically complex, the methods of separation have undergone several modifications to such an extent that the evolution of the separation techniques is a study in itself (Speight, 2014a). And because the fractions are in a balanced relationship in crude oil, the chemical character of these main fractions needs reference in this chapter.

Questions may also be raised regarding the usefulness of attempting to determine the exact structure of each complex material, such as asphaltene constituents, resin constituents, and nonvolatile oils. Perhaps the answers lie not only in determining the locations of the heteroatoms (nitrogen, oxygen, and sulfur) but also in understanding the chemical reactions of the various constituents of crude oil. In fact, valuable data can be derived from chemical reactions and it is the hope that this allows a complementary evaluation of the structural types present in the hydrocarbon portions of these molecules in the hope of deriving as much benefit as possible during processing of these complex materials.

It is the purpose of this chapter to review a selection of methods that can be used for the identification of the chemical reactivity of crude oil constituents. Particular emphasis has been placed on the techniques that have offered ready information and will continue to provide valuable information about the chemical constituents of crude oil. In this respect, the reactions that cause the removal of the heteroatoms have been of particular interest in the last four decades.

Thus, this chapter presents methods that can be applied to study the makeup of the feedstock in terms of chemical structures as well as methods that might be preferred for refining. The chapter deals with the older physical property methods that were used in the early days of structural group analysis, and some are used even now. The evolution of the available spectroscopic techniques that could be applied to structural group analysis commenced in the 1960s and these are included in the section following the physical property methods.

5.2 Acid Number

In the refining industry, the acid value is a number used to quantify the acidity of (in this context) a refinery feedstock [i.e., the number of acid carboxylic acid groups(——C(=O)OH) in the feedstock]. Depending upon the location and the method used, the acid number may also include the number of phenolic groups (Ar–OH), where Ar is an aromatic moiety in the compound. The acid number is the quantity of base, expressed as milligrams of base (typically potassium hydroxide, KOH) that is required to neutralize the acidic constituents in 1 g of a sample and which is usually referred to as the total acid number (TAN) (ASTM D664, 2021). In fact, the determination of the acid number of a feedstock is a form of functional group analysis.

Thus, the acid number is a measure of the acidity of a product and is used as a guide in the quality control of the properties of viscous feedstocks as well as a guide to the extent of the oxidation of the material. Since a variety of oxidation products contribute to the acid number and the organic acids vary widely in service properties, the test is not sufficiently accurate to predict the precise behavior of asphalt in service. In addition, the acid number is used as a guide in the quality control of lubricating oil formulations that are typically prepared from viscous feedstocks such as vacuum gas oil.

Viscous feedstocks often contain organic acids and saponifiable material that is largely determined by the percentage of naphthenic acids (cycloparaffin acids) of higher molecular weight that are originally present in the crude oil. With increased hardness, the residuum from a particular crude oil or viscous feedstock normally decreases in acid number as more of the naphthenic acids are removed during the distillation process and other thermal processes (such as visbreaking) which can cause decarboxylation of any carboxylic acid functions:

$$[R]{-}CO_2H \rightarrow [R]{-}H + CO_2$$

Acidic constituents may also be present as additives or as degradation products formed during service, such as oxidation products (ASTM D5770, 2021). The relative amount of these materials can be determined by titrating with bases.

The "saponification number" (SN, also referred to as the "saponification value", SV) is the amount of base that will react with (neutralize) 1 g of the sample when the test method is performed in a specific manner (ASTM D94, 2021). More specifically, it (SV or SN) represents the number of milligrams of potassium hydroxide (KOH) or sodium hydroxide (NaOH) required to saponify 1 g of the sample under the conditions specified in the test method. Since certain compounds may be added to, for example, a viscous material (such as a resid or the asphalt produced therefrom) to consume alkali and acids, the results obtained indicate the effect of these extraneous materials in addition to the saponifiable material present.

By way of further explanation, the difference between acid value and saponification number is that the acid value gives the mass of potassium hydroxide that is required to neutralize 1 g of a chemical substance whereas the saponification number gives the mass of potassium hydroxide required to saponify 1 g of sample.

In the test, a known weight of the sample is dissolved in methyl ethyl ketone or a mixture of suitable solvents and the mixture is heated with a known amount of standard alcoholic potassium hydroxide between 30 and 90 min at 80 °C (176 °F). The excess alkali is titrated with standard hydrochloric acid and the saponification number calculated.

The determination of the saponification number by the standard test method (ASTM D1962, 2021), which is often applied to natural products such as fats and natural oils, may need to be modified if a crude oil fraction of a viscous oil is to be tested. In the test method (ASTM D1962, 2021), the sample is completely saponified with an excess of alkali, which excess is then determined by titration (in mg KOH/g). The saponification number depends on the molecular weight and the percentage concentration of acid components present in the sample and is used to determine the average relative molecular mass (molecular weight) of the sample. Typically, a sample with a high saponified number may have a high tendency to foam.

In a more general sense, the acidity of a feedstock increases with viscosity for heavy crude oil, extra heavy crude oil, and tar sand bitumen but for residua, the acidity depends upon (1) the source and type of the parent crude oil and (2) the temperature since, under conditions in the distillation column in which the residuum is produced, there may have been thermal decarboxylation which can cause a reduction of the acid number:

$$RCO_2H \rightarrow RH + CO_2$$

In addition, severely weathered samples such as those from any of the viscous feedstocks that occur in an outcrop may be more acidic than unoxidized bitumen from deeper within the deposit due to oxidation and the incorporation of carboxylic functions or phenol functions:

$$RCH + O_2 \rightarrow RCO_2H$$
$$RH + [O] \rightarrow ROH$$

In addition, an increase in the acid number of crude oil and the heavy feedstocks can lead to increasing emulsion stability and therefore emulsion-treating difficulties before introduction of the feedstock to refinery operations.

Briefly, high acid crude oils are considered to be those with an acid content >1.0 mg KOH/g sample (many refiners consider the TAN greater than 0.5 mg KOH/g to be high) and refiners looking for discounted supplies of feedstocks for a refinery will import and use greater volumes of high TAN crude oils (Speight, 2014b).

Thus, the TAN (ASTM D664, 2021) represents a composite of acids present in the feedstock and is often also expressed as the neutralization number. Crude oils having a high acid number account for an increasing percentage of refinery feedstocks and, in fact, the increase in world production of heavy, sour, and feedstocks with a high TAN impact many world refineries, especially refineries in the United States (Shafizadeh et al., 2003; Sheridan, 2006).

However, some acidic constituents are relatively inert and the TAN is not always a true reflection of the corrosive properties of the crude oil. Furthermore, different naphthenic acids (a broad group of organic acids in crude oil) will react at different temperatures—making it difficult to pinpoint the processing units within the refinery that will be affected by a particular high TAN crude oil.

However, the impact of corrosive, high TAN, crude oils can be overcome by blending higher and lower TAN crude oils, installing or retrofitting equipment with anticorrosive materials, or by developing low-temperature catalytic decarboxylation processes using metal catalysts such as copper.

In the test method, the sample normally dissolved in toluene/*iso*-propyl alcohol/water is titrated with potassium hydroxide and the results are expressed as milligrams of potassium hydroxide per gram of sample (mg KOH/g). Crude oils having high acid numbers have a high potential to cause corrosion problems in the refineries, especially in the atmospheric and vacuum distillation units where the hot crude oil first comes into contact with hot metal surfaces. Crude oil typically has a TAN value on the order of 0.05–6.0 mg KOH/g of sample.

One method (ASTM D664, 2021) involves dissolution of the sample in a toluene/water/isopropanol (50.0/0.5/49.5) solution titrated potentiometrically with 0.1 N alcoholic potassium (KOH) using a glass indicating electrode and a calomel reference electrode. The endpoint is determined either graphically, or if no inflections in the curve are apparent, by titration to a meter reading for a standard buffer solution.

Two other methods (ASTM D974, 2021; ASTM D3339, 2021) also involve colorimetric titration to determine the acid number of crude oil products and lubricants by titration with 0.1 N potassium hydroxide in a toluene-isopropyl alcohol mixture to a *p*-naptholbenzein indicator end point. The application for the titration of heavy crude oil, extra heavy crude oil, tar sand bitumen, and residua has been hampered because of the difficulty in detecting the color change from orange to green-brown. However, the alternate method (ASTM D974, 2021) is an older method and used for distillates while the ASTM D664 (2021) test method is more accurate but measures acid gases and hydrolyzable salts in addition to organic acids. These differences are important for crude oil residua and the viscous feedstocks but less significant for distillates. A potentiometric perchloric acid titration method (ASTM D2896, 2021) is also available for the determination of acid number.

In addition, application of any of the methods to heavy oil and bitumen (or even to whole crude oil) can be influenced by precipitation of asphaltene constituents on the electrodes during a potentiometric titration. This effect can delay the response of the electrode. When the precipitation problem is severe, inflections in the titration curve cannot be identified and titration to a buffer endpoint becomes slow and imprecise.

Indeed, determination of the acid number of heavy oil and bitumen is subject to uncertainties due to the problem of precipitation of components during titration. Reduction of test sample size, use of alternate solvent systems, and minimizing titration time, possibly by using a more concentrated titrant, may improve precision. These factors should be examined to establish a method suitable for these heavy feedstocks.

There is also a method for the determination of the basic constituents in crude oil products (ASTM D4739, 2021). The data are presented as a base number that is defined as the quantity of acid, expressed in milligrams of potassium hydroxide per gram of sample, which is required to titrate a sample to a specified end point. The base number can be used to indicate the relative changes that occur in crude oil or a crude oil product during use (or storage) under service (or oxidizing) conditions regardless of color or other properties of the resulting product.

The neutralization value can be obtained either by potentiometric titration (ASTM D664, 2021) or by color-indicator titration (ASTM D974, 2021) and provides base numbers as well as acid numbers. Further functionality can be determined by use of saponification numbers using the color-indicator titration method (ASTM D94, 2021).

5.3 Elemental Analysis and Metals

The elemental analysis (sometimes referred to as the ultimate analysis, especially when considering coal technology) of a feedstock is the percentage of carbon, hydrogen, nitrogen, oxygen, and sulfur (as well as metals) and is perhaps the first method used to examine the general nature, and perform an evaluation, of a feedstock.

As already indicated (Chapter 2), the atomic ratios of the elements in various feedstocks have been used as a measure of the general amounts of hydrogen used to convert the feedstock to a given slate of products. In fact, the feedstocks for the percentages of carbon, hydrogen, nitrogen, oxygen, and sulfur is perhaps the first method used to examine the general nature, and perform an evaluation of a feedstock, often referred to as "elemental analysis" or "ultimate analysis." The atomic ratios of the various elements to carbon (i.e., H/C, N/C, O/C, and S/C) are frequently used for indications of the overall character of the feedstock as well as the hydrogen requirements to convert the heteroatoms to their hydrogen analogs (NH_3, H_2O, and H_2S). The H/C ratio has been of particular use in the past as a means of estimating the aromaticity of feedstocks. This ratio finds lesser use now because of the influx of heavier feedstocks into refineries as well as the need for more accurate estimation of feedstock character.

However, elemental analysis is still of considerable value to determine the amounts of elements in feedstocks, especially the trace elements such as vanadium and nickel since these materials can have serious deleterious effects on catalyst performance during refining by catalytic processes. Indeed, one may wonder amount the use or futility of measuring elements that are present in trace amounts. However, considering the nature of many continuous (as opposed to batch) processes in which a feedstock might be recycled to extinction and blended on the way with fresh feedstock (until no more of the desired product is made), the concentration of the trace elements (metals) on the catalyst can become sufficient to modify, if not completely change, the catalyst behavior, and overall process efficiency.

It has also become apparent, with the introduction of the viscous feedstocks into refinery operations, that these ratios are not the only requirement for predicting feedstock character before refining (Speight, 2014a, 2017). The use of more complex feedstocks (in terms of chemical composition) has added a new dimension to refining operations. Thus, although atomic ratios, as determined by elemental analyses, may be used on a comparative basis between feedstocks, there is now no guarantee that a particular feedstock will behave as predicted from these data (Speight, 2017). In fact, the product slate cannot be predicted accurately, if at all, from the various elemental ratios.

The ultimate analysis (elemental composition) of crude oil is not reported to the same extent as for coal because of the difficulty of determining the type of crude oil or feedstock from the elemental rations alone (Speight, 2014a). In fact, in contrast to coal which had a wide carbon range, data that are available for crude oils show that the proportions of the elements vary only slightly over narrow limits:

Carbon	83.0–87.0% w/w
Hydrogen	10.0–14.0% w/w
Nitrogen	0.1–2.0% w/w
Oxygen	0.1–1.5% w/w
Sulfur	0.1–6.0% w/w

Nevertheless, there is a wide variation in physical properties from the lighter more mobile crude oils at one extreme to the heavier viscous materials at the other extreme. The majority of the more aromatic constituents and the heteroatom-containing constituents occur in the higher boiling fractions of the viscous feedstocks (Speight, 2014a, 2017).

There are several standard test methods that can be used for the ultimate analysis of crude oil and crude oil products but many such methods may have been designed for other materials. Indeed, some of these methods are used for the elemental analysis of coal. For example, carbon content and hydrogen content can be determined simultaneously by the method designated for coal and coke (ASTM D5373, 2021). Indeed, this method is also suitable for the determination of carbon and hydrogen in heavy oil and bitumen.

The method chosen for the analysis may be subject to the peculiarities or character of the feedstock under investigation and should be assessed in terms of accuracy and reproducibility. The methods that are commonly used for the elemental analysis of refinery feedstocks are:

1) carbon and hydrogen content (ASTM D5373, 2021; ASTM D3343, 2021; ASTM D3701, 2021; ASTM D5291, 2021),
2) nitrogen content (ASTM D3228, 2021; ASTM D5291, 2021),
3) oxygen content (ASTM E385, 2021), and
4) sulfur content (ASTM D129, 2021; ASTM D139, 2021; ASTM D1266, 2021; ASTM D1552, 2021; ASTM D2622, 2021; ASTM D3120, 2021; ASTM D4045, 2021; ASTM D4294, 2021).

The hydrogen content of crude oil products including gas oils and vacuum residua can also be measured by low-resolution magnetic resonance spectroscopy (ASTM D3701, 2021; ASTM D4808, 2021). The method is claimed to provide a simple and more precise alternative to existing test methods, specifically combustion techniques (ASTM D5291, 2021) for determining the hydrogen content of a variety of crude oil-related materials.

Many crude oil products do not specify a particular oxygen content but if the oxygen compounds are present as acidic compounds such a phenols (Ar–OH) and naphthenic acids (cycloalkyl–COOH), they are controlled in different specifications by a variety of tests. In fact, the total acidity of a sample or whole feedstock (ASTM D974, 2021) is determined for many products, especially fuels, but the method may lack sufficient sensitivity for other products and may need a modification if (in the current context) the sample is one of the viscous oils.

Managing the metals in refinery feedstocks is an important aspect of refinery practice, especially as the amount of high-metal viscous feedstocks submitted to refineries is increasing. If not removed prior to the onset of refinery operations, the metals in the feedstock can ultimately (1) induce coke or sediment formation in the distillation section of the refinery, (2) lower feedstock conversion in the fluid catalytic cracking unit due to the adverse impact on catalyst activity, and (3) reduce the efficiency of hydrotreater and hydrocracking operations by adversely affecting the activity of the catalysts.

Thus, metals cause particular problems because they poison catalysts used for sulfur and nitrogen removal as well as other processes such as catalytic cracking. In addition, the levels of nickel and vanadium in crude oil, heavy oil, and bitumen need to be considered by the geochemist when the origins of crude oil, heavy oil, and bitumen are considered. Thus, the trace components that are also present in crude oils can produce adverse effects in refining either by causing corrosion, by affecting the quality of refined products, or by exerting a deleterious influence on the efficiency of various processing catalysts. This has become of increasing importance due to modern developments in refinery processing.

A number of heavy metals such as nickel, vanadium, copper, and iron can also be effectively bound in large organic molecules characteristic of those found in the asphaltene fraction (pentane- or heptane-insoluble portion of the feedstock) and resins.

Nickel and vanadium porphyrins are commonly found and show high thermal stability allowing them to pass through the extraction process into the upgrading process. Porphyrins are the major, but certainly not the only, organo-metallic complexes present. Metals may simply be entrapped or loosely bound in the very large molecules present in the asphaltenes and resins. Although iron is present as organo-metallic compounds, it occurs mostly in the form of process-accumulated rust or is scrounged from pipelines by the crude oil during shipping and pipelining. Thus, one should not be caught in worrying about the geochemical significance of metal such as iron in crude oil.

Vanadium compounds can cause refractory damage in furnaces and adverse effects in glass manufacture, steel failure in turbines, as well as catalyst poisoning when present in distillate feedstocks. Arsenic and lead are also active catalyst poisons in reforming processes, while the presence of sodium in fuel oils causes failures in furnace brickwork. It is necessary, therefore, for crude oils and distillation unit feedstocks to be examined for the presence of these harmful contaminants, and some form of treatment devised for reducing their effect during or before processing.

Thus, it is important to monitor process streams for metals content for several purposes in order to (1) track the degree of potential catalyst poisoning that may occur within the reactor due to feed metals content, (2) monitor the degree of catalyst physical or chemical breakdown into the product streams, and (3) provide one of many indicators of change in the process operation.

Thus, a variety of tests (ASTM D482, 2021; ASTM D1318, 2021; ASTM D3341, 2021; ASTM D3605, 2021) either directly or as the constituents of combustion ash have been designated to determine metals in crude oil products based on a variety of techniques. At the time of writing, the specific test for the determination of metals in whole feeds has not been designated. However, this task can be accomplished by combustion of the sample so that only inorganic ash remains (ASTM D482, 2021). The ash can then be digested with an acid and the solution examined for metal species by atomic absorption (AA) spectroscopy or by inductively coupled argon plasma (ICP) spectrometry (ASTM C1109, 2021; ASTM C1111, 2021).

The analytical method should be selected depending on the sensitivity required, the compatibility of the sample matrix with the specific analysis technique, and the availability of facilities. Sample preparation, if it is required, can present problems. Significant losses can occur, especially in the case of organo-metallic complexes, and contamination of environmental sample is of serious concern.

5.4 Emulsion Formation

Emulsions involving viscous feedstocks—such as water in oil (W/O), oil in water (O/W), and water in oil in water (W/O/W)—can be formed during the development of viscous feedstock reservoirs/deposit or at any time when the feedstock has contact with water and the parameters are at the right levels for such interaction. Moreover, the naturally occurring viscous feedstocks (prior to entering the refinery) contain natural emulsifiers such as resin constituents and asphaltene constituents, which when adsorbed at the oil–water interface film form W/O emulsion and interconnect to form a three-dimensional (3D) network structure that accumulates on the surface of water droplets which enhances the mechanical properties of the interface film intensity.

A W/O emulsion is a stable dispersion of small droplets of water in oil. When formed from crude oils spilled at sea, these emulsions can have very different characteristics from their parent crude oils. This has important implications for the fate and behavior of the oil and its subsequent cleanup. It is desirable, therefore, to determine if oil is likely to form an emulsion, and if so, whether that emulsion is stable, and the physical characteristics of the emulsion.

In an older test method, the tendency for a crude oil to form a W/O emulsion was measured using a method based on the rotating flask apparatus (Mackay and Zagorski, 1982). All numerical values (mostly ones or zeroes) based on this method have subsequently been reduced to *yes* or *no*, respectively, and indicate the formation (or not) of an emulsion that remained stable 24 hr after settling. In a newer variation, the reproducibility is considerably improved and several parameters (1) the water-to-oil ratio, (2) the fill volume, and (3) the orientation of the vessels were found to be important parameters affecting emulsion formation.

However, such effects are not lasting. Emulsion formation and behavior is influenced by the oxidation of crude oil constituents (Speight, 2014a). The inclusion of polar functions such as hydroxyl groups (–OH) or carbonyl groups (>C=O) (a result of the oxidation process) causes an increase in the density of the emulsion (relative to the original unoxidized crude oil) and with an increased propensity to form emulsions. As a result, the emulsion and sinks to various depths or even to the seabed, depending on the extent of the oxidation and the resulting density. This may give the erroneous appearance (leading to erroneous deductions with catastrophic consequences) that the crude oil spill (as evidenced from the crude oil remaining on the surface of the water) is less than it actually was. The so-called *missing* oil will undergo further chemical changes and eventually reappear on the water surface or on a distant beach.

5.5 Evaporation

Evaporation is the removal of the lower-boiling constituents from crude oil or a crude oil product usually under ambient conditions or, in the current context, under the conditions prevalent at the spill site. Also, rate of the evaporation process can increase as the spilled material spreads due to the increased surface area of the material. High wind speed and a high temperature also tend to increase the rate of evaporation and the proportion of an oil lost by this process.

Evaporation rate and loss are of importance for all volatile constituents of crude oil and crude oil products. While standard test methods such as those designated for distillation and vapor pressure determination are often used to determine evaporation properties, test methods for determining evaporation loss are available for higher boiling crude oil products (ASTM D972, 2021). Although not necessarily applicable to crude oil and crude oil products in general, evaporation loss data can be obtained at any temperature in the range from 93 to 316 °C (200 to 600 °F). Viscous samples can be analyzed using a water vaporizer accessory that heats the sample in the evaporation chamber, and the vaporized water is carried into the Karl Fischer titration cell by a dry inert carrier gas.

Crude oil and crude oil products evaporate at a logarithmic rate which is attributed to the overall logarithmic appearance of many components evaporating at different linear rates. Crude oil products with fewer constituents (such as diesel fuel) evaporate at a rate which is square root with respect to time, which is a result of the number of components evaporating. The evaporation process, as evidenced by crude oil and crude oil products is not strictly boundary-layer regulated, which is largely a result of the high saturation concentrations of volatile feedstock constituents in air and is associated with a rate that is regulated by the boundary-layer effect, which is the thin layer of fluid in the immediate vicinity of a bounding surface formed by the fluid flowing along the surface.

Some volatile constituents show some effect of boundary-layer regulation at the start of the evaporation process, but after a short time (often several minutes), the evaporation decreases because of the loss of the more volatile components, at which point evaporation ceases to be boundary-layer regulated.

It must also be recognized that as evaporation occurs, the density and viscosity of the crude or the viscous feedstocks increase thereby causing behavioral changes in the feedstock which can, for example, be reflected in an increase in the viscosity (and hence flow) of the feedstock.

Finally, as it relates to the viscous feedstocks, evaporation is an important process for spills into the environment. In a matter of days, a spilled conventional crude oil can be reduced by up to 75% of the spilled volume and a medium crude oil can be reduced by up to 40% of the spilled volume. On the other hand, a viscous feedstock, such as heavy crude oil or a resid, may only lose as little as 5% of the spilled material volume which leaves a nonvolatile highly immobile residue in the environment which is extremely difficult to remove. Also, the thicker the layer of the spilled material (which is particularly applied to the low mobility viscous materials), the slower the water diffusion through it to the surface, so the slower the evaporation.

5.6 Flash Point and Fire Point

The flash point of crude oil or a crude oil product is the temperature to which the sample must be heated to produce a vapor/air mixture above the liquid fuel that is ignitable when exposed to an open flame under specified test conditions. In North America, flash point is used as an index of fire hazard.

The "flash point" is an extremely important factor in relation to the safety of spill cleanup operations insofar as naphtha and gasoline as well as low density crude oils and other low-boiling crude oil-derived liquids can be ignited under most ambient conditions and therefore pose a serious hazard when spilled. Many freshly spilled crude oils also have low flash points until the lighter components have evaporated or dispersed. This does not typically apply to the viscous feedstocks unless there is an amount of low boiling constituents in the feedstocks.

There are several ASTM methods for measuring flash points (ASTM D56, 2021; ASTM D93, 2021) are among the most commonly used. The minimum flash point that can be determined (ASTM D93, 2021) is 10 °C (50 °F). One method (ASTM D56, 2021) is intended for liquids with a viscosity less than 9.5 cSt at 25 °C (77 °F). The flash point and fire point of lubricating oil is determined by a separate method (ASTM D92, 2021). However, the flash point of a viscous feedstock is typically (if not, always) above 50 °C (122 °F) and the density is typically (if not, always) always higher than 0.900 25.7°API.

On the other hand, the fire point is the lowest temperature, corrected to one atmosphere pressure (14.7 psi), at which the application of a test flame to the crude oil or crude oil product sample surface causes the vapor of the oil to ignite and burn for at least 5 s.

At any time after a spill of crude oil or a crude oil product, fire should always be considered an imminent hazard. Related to fire point, the flash point is a measure of the tendency of the crude oil or a crude oil product to form a flammable mixture with air under controlled laboratory conditions (ASTM D92, 2021) and is only one of a number of properties that should be considered in assessing the overall flammability hazard of any feedstock whether in the refinery or spilled into an ecosystem (Arjoon and Speight, 2023).

The ignition temperature (sometimes called the autoignition temperature) is the minimum temperature at which the material will ignite without a spark or flame being present (ASTM E659, 2021). Also related to fire point, the flammability limits of vapor in air is an expression of the percent concentration in air (by volume) given for the lower and upper limit. These values give an indication of relative flammability. The limits are sometimes referred to as lower explosive limit (LEL) and upper explosive limit (UEL).

5.7 Functional Group Analysis

The presence of a functional group in an organic compound influences the physical and chemical properties of the compound. Hence, the identification of functional group is essential to determine the complete structure of the organic compound. Apart from spectroscopic techniques, there are the so-called classical methods of detecting functional group but, in the context of the viscous feedstocks, may suffer from the relative presence of functional groups in such a large mass/volume.

In the functional group analysis of organic compounds, it is assumed that an organic molecule can be considered as a sum of virtually independent functional groups and it is the properties of the functional group that determine the physical and chemical properties of the compound. The viscous feedstocks generally follow such a conclusion and, where possible, the functional group analysis of the viscous feedstocks is becoming an integral part of the examination of the viscous feedstocks because the analysis forms the basis of identification of unknown organic functional groups and hence the behavior of the feedstock during refining.

Traditional methods of analyzing functional groups employ reagents and spectroscopic methods for the determination of structural groups (or entities) within the feedstock (Chapter 9). More specifically, in terms of functional groups rather than sectoral groups, the investigations typically focus on the alternative methods for the detection of some of the functional groups such as carboxylic acid, alcohol, phenol, carbonyl, ester, and carbohydrates (Figure 5.1). The procedures adopted are in accordance with the properties of the feedstock and must take into account (when environmental issues are under consideration) the influence of aerial oxygen on the distribution and types of functional group which can influence the properties of the spilled material and, hence, the ability of the environment to tolerate such a spill.

However, the functional group analysis of the higher boiling fractions of crude oil and the viscous feedstocks continues to be a challenge because of the wide variety of molecular types and structures present. Molecular types in the residua, heavy oil, and bitumen range from nonpolar, nonaromatic species to highly aromatic hydrocarbons, the molecular structures of which contain varying amounts of heteroatoms (nitrogen, oxygen, and sulfur) together with parts-per-million amounts of metals such as vanadium and nickel. The heteroatoms are often associated with polar strongly interacting chemical functionality or functional groups that have a disproportionately large effect on the properties of each fraction and, therefore, on the whole feedstock (Speight, 2000).

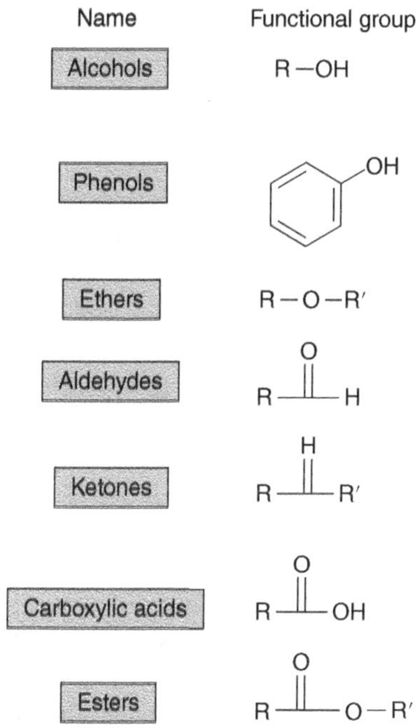

Figure 5.1 Examples of oxygen-containing functional groups.

Because the number of molecules in residua, heavy oil, and bitumen with different chemical structures and reactivity is extremely large, determination of composition by separation of the feedstock into its molecular components is generally considered impractical if not impossible by many of the available techniques. However, if the different chemical functionalities (ASTM F1186, 2021) that comprise the heavy feedstocks and therefore dominate its properties are considered, the number of types of functionalities that need to be considered decreases and become identifiable. Indeed, the heavy feedstocks can be safely assumed to contain these same functionalities, albeit in different proportions depending upon the origin and maturation conditions and the recovery and treatment of the feedstock. In turn, it is the interactions of these functions that can play a role in the structure of the feedstocks and its behavior under different conditions (Speight, 2014a).

One of the principal agents in the behavior of the viscous feedstocks is the various oxygen functions. These functions also play a major role in the chemical and physical structure of the feedstocks. Thus, the need to identify the presence (or the formation) of highly polar and strongly interacting chemical functional groups containing oxygen are necessary steps for assessing the effects of composition on properties and, thus, the performance of the feedstock during refining.

During the past several decades, there have been many studies that relate to the identification and characterization of the polar, heteroatom-containing chemical functionality in various feedstocks. Infrared spectrometry has been a very prominent technique in this research because it can be applied to complex mixtures without alteration or destruction of the sample (Speight, 2014a).

Earlier researchers who explored infrared spectrometry for the characterization of high molecular weight constituents of feedstocks found the technique useful for determining the general chemical structural types present. However, the strongly associating polar functionalities present often required extra and more conscientious effort. The inability to adequately characterize these polar functionalities resulted from inherent problems such as overlapping and ill-defined absorption bands, and the shifting of absorption bands from hydrogen bonding. These inherent problems have been overcome by the combined use of selective chemical reactions and differential spectrometry. Also, there is a variety of solvent systems that can be used to break up hydrogen bonding, eliminating the complicating effects of hydrogen bonding on the spectra (Moschopedis and Speight, 1976).

As previously mentioned, the most polar and strongly interacting functional groups in the viscous feedstocks and residua occur in relatively small amounts and their infrared absorption bands are often complicated by hydrogen bonding and overlap with other strong absorption bands (Moschopedis and Speight, 1976) making their detection difficult and their quantitative assessment virtually impossible without the use of special techniques.

Phenol derivatives and carboxylic acid derivatives in feedstocks are hydrogen-bonding functionalities. Thus, in solutions of these feedstocks in typical infrared spectral solvents, an equilibrium exists between the free absorption bands of these functional groups and their hydrogen-bonding bands (the latter bands overlap) that is dependent on the concentration, the solvent characteristics, and the basic character of the constituents in the feedstock with which the acidic hydrogen of the functionalities interacts (Moschopedis and Speight, 1976).

Two other important, naturally occurring functionalities are carboxylic acids and 2-quinolone-type oxygen. These functionalities are extremely strong hydrogen bonders. Not only do they form dimers readily but also they interact strongly with each other to form a mixed dimer, yielding six absorption bands, of which five overlap and become virtually indistinguishable. However, it must be remembered that sample history can play a major role in the occurrence of such functions in the feedstock. Oxidation during heavy oil recovery can, and does, cause oxygen incorporation into the feedstock and the occurrence of carboxylic acid function in residua and heavy oil may be diminished by the thermal treatment (even distillation) used to produce the heavy feedstock. The incorporation of oxygen functions can have a noticeable influence on the inter- and intra-molecular bonding arrangements in the feedstock.

To overcome the problems just described and to obtain spectra suitable for quantitative analysis, several specialized techniques and procedures are employed. Tetrahydrofuran (THF) solvent has been used to eliminate interference from hydrogen bonding on the carbonyl absorption region in the determination of ketones, carboxylic acids, anhydrides, and 2-quinolones. However, caution is advised because this solvent is not always capable of dissolving the asphaltene constituents and may lead to faulty conclusions because of the occurrence of suspended organic matter.

Whatever solvent is used, advantage can be taken of the hydrogen bonding of phenolic and pyrrolic functionality in the analysis of these functional groups. However, because the solvent employed may have absorption bands in the frequency ranges used in the determinations, a solvent compensation technique must be used.

Thus, when the problem of hydrogen bonding has been eliminated, the problem of overlapping bands in the carbonyl absorption region (about 1,800 cm^{-1} to about 1,600 cm^{-1}) is addressed by using selective chemical reactions and differential infrared spectrometry. The techniques applied are as follows. To reveal the absorption band of interest and eliminate from the spectra other bands with absorption at the same frequency, the sample is treated with a selective reagent that eliminates or shifts the absorption band of interest to another frequency.

A differential spectrum is then taken with the treated sample in one beam and the untreated sample in the other beam of a double-beam infrared spectrophotometer. This procedure reveals the absorption band of the functionality of interest in the differential spectrum and nulls or cancels out all other absorption bands in the same region. Quantitative analysis can also be applied to the absorption band of interest using calculations of the areas under the absorption bands.

The apparent integrated absorption intensity (B) in units of L mol^{-1} cm^{-2} of an infrared absorption band obtained on a spectrophotometer having a monochromatic energy source and finite slit width is defined as follows:

$$B = -l / cl \int \ln(T_0 / T)_\nu \, d\nu$$

In this equation, c is the concentration of functional group type (mol L^{-1}), l is the cell path length (cm), ν is the absorption frequency (cm^{-1}), T_0 is the incident radiation, and T is the transmitted radiation.

The area under the absorbance vs. absorption–frequency curve for the absorption bands of interest is represented by the term $\ln(T_0/T)\nu d\nu$. The term $\ln(T_0/T)$ is equivalent to the commonly used term absorbance, designated by the notation A.

The band area, $\int A\nu d\nu$, can be estimated by counting squares on the recording chart paper or by weighting the paper that represents the respective peak areas. The concentrations of functional group types in the sample are then estimated by using the equation:

$$c = \int \Delta A_{\nu} d\nu / Bl \times 1.05 / 0 / 0.05$$

The potential applications of the functional group analysis in composition-related technology are numerous. Thus, the combined use of differential infrared spectrometry and selective chemical reactions provided the basis for the development of an analytical method for the quantitative analysis of important chemical functionality in feedstocks. The method can be applied to the solution of a variety of composition-related problems in crude oil technology.

The bromine number is the number of grams of bromine consumed by a weighed amount (100 g) of sample when reacted under specified conditions (ASTM D1159, 2021). In theory, the method gives an indication of the amount of unsaturation in the sample by assuming that addition of bromine across multiple carbon–carbon bonds are the only reactions. However, with the heavier feedstocks, caution is needed in the interpretation of the data because of the ready reactivity of bitumen constituents with bromine (Moschopedis and Speight, 1971).

The method uses a polarized electrode pair. As soon as there is a small amount of excess bromine, the solution conductivity increases sharply, producing a large clearly defined deflection at the end point. In the method, the sample is initially dissolved in a solvent consisting of glacial acetic acid, 1,1,1-trichloroethane, methanol, and sulfuric acid. The solution is titrated with bromide-bromate titrant using dual platinum electrodes and a 10 (microamp) 10 μA polarizing current for end point detection.

Alternatively, the bromine index, which is the number of milligrams of bromine that will react with 100 g of the sample, is used mostly by the chemical industry for stocks that have an unusually low olefin content.

5.8 Halogenation

Halogenation is a chemical reaction that involves the introduction of one or more halogen atoms into a compound to produce halide derivatives (also referred to as halogeno-derivatives). Several pathways exist for the halogenation of organic compounds, including (1) free radical halogenation, (2) electrophilic halogenation, and (3) halogen addition and the nature of the substrate determines the pathway. The facility of the halogenation reaction is influenced by the halogen insofar as fluorine and chlorine are the more aggressive halogenating agents while bromine is a weaker halogenating agent and iodine is the least reactive of the halogens. On the other hand, dehydrohalogenation follows the reverse trend and iodine is most easily removed from organic compounds while the organo-fluorine compounds are highly stable.

Thus, the halogenation of refinery feedstocks and feedstock fractions occurs readily to afford the corresponding halo-derivatives; the physical properties of the halogenated materials are markedly different from those of the parent materials. Analysis for the various halogens is possible (ASTM E422, 2021).

In the care of the asphaltene constituents, which have again been the recent focal point of the investigations, for example, the unreacted asphaltene constituents are dark brown, amorphous, and readily soluble in toluene ($C_6H_5CH_3$), nitrobenzene ($C_6H_5NO_2$), and carbon tetrachloride (CCl_4), but the products are black, shiny, and only sparingly soluble, if at all, in these solvents.

There are also several features that distinguish the individual halogen reactions from one another. For example, during chlorination of asphaltene constituents, there is a cessation of chlorine uptake by the asphaltene constituents after 4 hr. Analytical data indicate that more than 37% of the total chlorine in the final product is introduced during the first 0.5 hr, reaching the maximum after 4 hr. Furthermore, the H/C ratio of 1.22 in the parent asphaltene constituents [(H + Cl)/C ratio in the chlorinated materials] remains constant during the first 2 hr of chlorination, by which time chlorination is 88% complete. This is interpreted as substitution of hydrogen atoms by chlorine in the alkyl moieties of the asphaltene constituents; the condensed aromatic sheets remain unaltered since substitution of aryl hydrogen appears to occur readily only in the presence of a suitable catalyst, such as ferric chloride ($FeCl_3$), or at elevated temperatures. It is only after more or less complete reaction of the alkyl chains that addition to the aromatic rings occurs, as evidenced by the increased atomic (H + Cl)/C ratios in the final stages of chlorination.

Bromine uptake by the asphaltene constituents is also complete in a comparatively short time (<8 hr). However, in contrast to the chlorinated products, the atomic (H + halogen)/C ratio remains fairly constant (1.23 and 1.21 in the bromo-asphaltene constituents or 1.22 in the unreacted asphaltene constituents) over the prolonged periods (up to 24 hr) of the bromination.

Iodination of asphaltene constituents is different insofar as a considerable portion of the iodine, recorded initially as iodine uptake, can be removed by extraction with ether or with ethanol, whereas very little weight loss is recorded after prolonged exposure of the material to a high vacuum. The net result is the formation of a product with an atomic (H + I)/C ratio of 1.24 after an 8 hr reaction; a more prolonged reaction period affords a product with a (H + I)/C ratio of 1.17. This latter may be the result of iodination of the alkyl or naphthenic moieties of the asphaltene constituents with subsequent elimination of hydrogen iodide.

Alternatively, dehydrogenation of naphthene rings to aromatic systems or coupling of aromatic nuclei would also account for lower (H + I)/C ratios. In fact, this latter phenomenon could account, in part, for the insolubility of the products in solvents that are normally excellent for dissolving the unchanged asphaltene constituents. However, it will be appreciated that these aforementioned reactions are only a few of many possible reactions that can occur, and undoubtedly halogenation of the asphaltene constituents is much more complex than would appear from the product data.

The iodine value of various oils and their derivatives can be measured thereby giving data that will complement the bromine number. Halogenation of the asphaltene constituents can also be achieved by use of sulfonyl chloride, iodine monochloride, and N-bromosuccinimide, or, indirectly, via the Gomberg reaction. As an addition to this segment, it is worthy of note that iodine monochloride is used as the chemical means for detecting lead in gasoline (ASTM D3341, 2021).

Halogens react with the oils relatively slowly, especially if there is no olefinic (>C=C<) or aromatic unsaturation; essentially substitution reactions occur. At elevated temperatures, dehydrohalogenation occurs, giving rise to resin and asphaltene products. It has been suggested (Prado and de Klerk, 2014) that halogenation of tar sand bitumen is a potential method for the metalloporphyrin structures can be disrupted by halogenation thereby indicating halogenation as a possible method for demetallation.

5.9 Hydrogenation

The purpose of hydrogenating feedstock constituents is to (1) improve existing crude oil products or develop new products or even new uses, (2) convert inferior or low-grade materials into valuable products, and (3) transform higher molecular weight constituents into liquid fuels (Speight, 2014a, 2017). Indeed, it is generally recognized that the higher the hydrogen content of a refinery feedstock, the better the quality of the resulting refinery products. This knowledge has stimulated the use of hydrogen-adding processes in the refinery. The hydrogenation process involves sparging the oil at high temperature and pressure with hydrogen in the presence of a catalyst, typically a nickel-based inorganic compound.

In the context of this chapter, that is, in the chemical laboratory, the hydrogenation of crude oil and the viscous feedstocks (heavy crude oil, extra heavy crude oil, tar sand bitumen, and residua) is somewhat limited by the nature or requirements of the reaction conditions. For example, high temperatures and pressures are usually required that cause changes to the hydrocarbon skeleton. However, there are mild hydrogenating processes that can be applied to crude oil thereby providing information about the constituents. In fact, the focus of hydrogenation in the laboratory tends to be oriented toward the category often referred to as nondestructive or mild hydrogenation.

The basic chemical concept of the hydrogenation process as applied to crude oil is to convert the organic constituents to liquid fuels and, at the same time, to remove the organic heteroatoms in the feedstock by conversion to their respective hydrogen derivatives:

$$\text{feedstock}_{\text{sulfur}} + H_2 \rightarrow H_2S + \text{products}_{\text{sulfur-deficient}}$$
$$\text{feedstock}_{\text{oxygen}} + H_2 \rightarrow H_2O + \text{products}_{\text{oxygen-deficient}}$$
$$\text{feedstock}_{\text{nitrogen}} + H_2 \rightarrow NH_3 + \text{products}_{\text{nitrogen-deficient}}$$

It is the chemical aspects of this reaction that can be most important as a means of building the basis for refining viscous feedstocks. As an example of the hydrogenation reaction is a mild degradation method as is the case in the hydrovisbreaking process in which hydrogen is added to the feedstocks and the product is more amenable to other refinery processes. However, there may be drawbacks to the use of such a method insofar as the asphaltene constituents form aggregates (Speight, 1994, 2014a) that may cause interference with the method and provide data that is, at best, questionable.

The asphaltene constituents may also be hydrogenated to produce resin constituents and oil constituents at elevated temperatures (>250 °C, >480 °F) together with a reduction in the heteroatom content (Speight, 2000, 2014a, 2017). Chemical hydrogenation under much milder conditions, for example, with lithium-ethylenediamine or sodium-liquid ammonia, also produces lower molecular weight species.

In a more general chemical sense, the effect of hydrogen on naphthenic hydrocarbons is mainly that of ring scission followed by immediate saturation of each end of the fragment produced. The ring is preferentially broken at favored positions, although generally all the carbon–carbon bond positions are attacked to some extent. Aromatic hydrocarbon derivatives are resistant to hydrogenation under mild conditions, but under more severe conditions, the main reactions are conversion of the aromatic to naphthenic rings and scissions within the alkyl side chains. The naphthene derivatives may also be converted to paraffin derivatives. Polynuclear aromatic hydrocarbon derivatives are more readily attacked than the single-ring compounds, the reaction proceeding by a stepwise process in which one ring at a time is saturated and then opened.

Starting in the 1950s, there was considerable focus on hydrogenation as a means of sulfur removal from crude oil and this will be the focus of the present discussion as it illustrates many of the aspects of the hydrogenation of crude oil, heavy oil, and bitumen.

Thus, the thermodynamics of the hydrogenation reaction can be evaluated from the equilibrium constants of typical desulfurization or partial desulfurization reactions such as: (1) hydrogenation of model compounds to yield saturated hydrocarbons (R—H) and hydrogen sulfide, H_2S; (2) decomposition of model compounds to yield unsaturated hydrocarbons, R—CH=CH–R^1, and hydrogen sulfide, H_2S; (3) decomposition of alkyl sulfides to yield thiol derivatives, RSH, and olefin derivatives, R—CH=CH–R^1; (4) condensation of thiol derivatives, R—SH, to yield alkyl sulfide derivatives, RSR^1, and hydrogen sulfide, H_2S; and (5) hydrogenation of disulfide derivatives, $RSSR^1$, to yield thiol derivatives, RSH and R^1SH.

The logarithms of the equilibrium constants for the reduction of sulfur compounds to saturated hydrocarbons over a wide temperature range (Speight, 2000) are almost all positive, indicating that the reaction can virtually proceed to completion if hydrogen is present in the stoichiometric quantity. The equilibrium constant does, however, decrease with increasing temperature for each particular reaction but still does retain a substantially positive value at 425 °C (795 °F) which is approaching the maximum temperature at which many of the hydrodesulfurization reactions (especially the nondestructive reactions) would be attempted. The data also indicate that the decomposition of sulfur compounds to yield unsaturated hydrocarbons and hydrogen sulfide is not thermodynamically favored at temperatures below 325 °C (615 °F) and such a reaction has no guarantee of completion until temperatures of about 625 °C (1,155 °F) are reached. However, substantial decomposition of thiols can occur at temperatures below 300 °C (570 °F); in fact (with only few exceptions), the decomposition of all saturated sulfur compounds is thermodynamically favored at temperatures below 425 °C (795 °F). In addition, kinetic studies using individual compounds, especially for the hydrodesulfurization reaction, have usually indicated that first-order kinetics with respect to sulfur is the predominant mechanism by which sulfur is removed as hydrogen sulfide.

The structural differences between the various sulfur-containing molecules make it difficult (if impractical) to derive a single rate expression applicable to all of the hydrodesulfurization reactions. Each sulfur-containing molecule has reaction-specific hydrogenolysis (hydrodesulfurization) kinetics that is usually complex because several successive equilibrium stages are involved and these are often controlled by internal diffusion limitations during refining.

Thiophene derivatives are the most refractory of the sulfur compounds. Consequently, thiophene is frequently chosen as representative of the sulfur compounds in feedstocks. The hydrogenolysis of thiophene takes place according to two distinct paths. The first path leads through thiophane to butyl mercaptan in equilibrium with butene and dibutyl thioether, and finally to butene and hydrogen sulfide. It is considered unlikely that the thiophene and the dibutyl sulfide can undergo direct hydrogenolysis with production of hydrogen sulfide. However, it is possible that the butyl mercaptan can be decomposed according to the two parallel paths namely: (1) desulfurization of the mercaptan on the active metal sulfides and acid sites of alumina followed by hydrogenation of the intermediate butene and (2) direct hydrogenolysis of the C—SH bond on the active metal sulfides. As complex as the desulfurization of thiophene might appear, projection of the kinetic picture to benzothiophene and dibenzothiophene, and to their derivatives, is even more complex. However, kinetic data derived from model compounds cannot be expected to include contributions from the various steric effects than are a consequence of complex molecules containing three-dimensional structures.

Furthermore, the complexity of the individual reactions occurring in an extremely complex mixture and the interference of the products with those from other components of the mixture is unpredictable. Alternatively, the interference of the secondary products and the tertiary products with the course of a reaction and, hence, with the formation of the primary products, may also be cause for concern. Hence, caution is advised when applying the data from model compound studies to the behavior of crude oil, especially the molecularly complex heavy oils. These have few, if any, parallels in organic chemistry. And all such contributions may be missing from the kinetic data that must be treated with some degree of caution.

However, there are several generalizations that come from the available thermodynamic data and investigations of pure compounds. Thus, at room temperature, hydrogenation of sulfur compounds to hydrogen sulfide is thermodynamically favorable and the reaction will essentially proceed to completion in the presence of a stoichiometric amount of hydrogen. Sulfides, simple thiophenes, and benzothiophenes are generally easier to desulfurize than the dibenzothiophenes and the higher molecular weight condensed thiophenes.

Nevertheless, the development of general kinetic data for the hydrogenation of different feedstocks is complicated by the presence of a large number of compounds, each of which may react at a different rate because of structural differences as well as differences in molecular weight and may be reflected in the appearance of a complicated kinetic picture for in which the kinetics is not, apparently, first order. The overall reaction may be satisfied by a second-order kinetic expression when it can, in fact, also be considered as two competing first-order reactions.

Thus, it has become possible to define certain general trends that occur in the hydrogenation of refinery feedstocks. One of the more noticeable facets of the process is that the rate of reaction declines markedly with the molecular weight of the feedstock. It should be noted here that, because of the nature of the reaction, steric influences would be anticipated to play a lesser role in the hydrocracking process.

The hydrogenation of the viscous feedstocks is considerably more complex than the hydrodesulfurization of model organic sulfur compounds or, for that matter, narrow-boiling crude oil fractions. In published studies of the kinetics of residua hydrogenation, one of three approaches has generally been taken which are (1) the reactions can be described in terms of simple first-order expressions; (2) the reactions can be described by use of two simultaneous first-order expressions, one expression for easy-to-hydrogenate systems and a separate expression for difficult-to-hydrogenate systems, with special reference to the remove sulfur; and (3) the reactions can be described using a pseudo-second-order treatment. Each of the three approaches has been used to describe hydrogenation of residua under a variety of conditions with varying degrees of success, but it does appear that pseudo-second-order kinetics are favored. In this particular treatment, the rate of hydrogenation (hydrodesulfurization) is expressed by a simple second-order equation:

$$C/1-C=k(1/\mathrm{LHSV})$$

In this equation, C is the (% w/w sulfur in product)/(% w/w sulfur in the feedstock), k is the reaction rate constant, and LHSV is the liquid hourly space velocity (volume of liquid feed per hour per volume of catalyst).

On this basis, the use of two simultaneous first-order equations may be more appropriate. The complexity of the constituents tends to increase with an increase in boiling point and the reactivity tends to decrease with complexity. It is anticipated that such an approach is consistent with the relative reactivity of various compound types observed for model compounds and for the various crude oil fractions that have been investigated.

Other kinetic work has shown that, for a fixed level of sulfur removal, the order of a reaction at constant temperature can be defined with respect to pressure:

$$k=1/\mathrm{LHSV}(P_\mathrm{h})^n$$

P_h is the hydrogen partial pressure, LHSV is the liquid volume hourly space velocity, k is a constant, and n is the order of the reaction. It has been concluded, on the basis of this equation, that the hydrodesulfurization of residuum is first-order with respect to pressure over the range 800–2,300 psi, although it does appear that the response to pressure diminishes markedly (and may even be minimal) above 1,000 psi.

During hydrogenation at higher temperatures, essentially all the initial reactions of catalytic cracking occur, but some of the secondary reactions are inhibited or stopped by the presence of hydrogen. For example, the yield of olefin derivatives and the secondary reactions that result from the presence of these materials are substantially diminished and any branched-chain paraffin derivatives tend to undergo demethanation. The methyl groups attached to secondary carbons are more easily removed than those attached to tertiary carbon atoms, whereas methyl groups attached to quaternary carbons are the most resistant to hydrocracking.

On an individual hydrocarbon type basis, and without the effects of hydrodesulfurization, the effect of hydrogen on naphthenic hydrocarbons is mainly that of ring scission followed by immediate saturation of each end of the fragment produced. The ring is preferentially broken at favored positions, although generally all the carbon–carbon bond positions are attacked to some extent. For example, methyl-cyclopentane is converted (over a platinum-carbon catalyst) to 2-methylpentane, 3-methylpentane, and *n*-hexane.

Aromatic hydrocarbon derivatives are resistant to hydrogenation under mild conditions, but under more severe conditions the main reactions are conversion of the aromatic to naphthenic rings and scissions within the alkyl side chains. The naphthene derivatives may also be converted to paraffin derivatives. Polynuclear aromatic derivatives are more readily attacked than the single-ring compounds, the reaction proceeding by a stepwise process in which one ring at a time is saturated and then opened.

In terms of the analytical chemistry, once the hydrogenation reaction has been carried out, it is then the focus of the chemist or analytical chemist to determine the reaction pathway using the various analytical techniques that can be brought to bear on the issue.

First, there is the analysis of gaseous products such as hydrogen sulfide (ASTM D4084, 2021; ASTM D4810, 2021) or the composition of the gas as might be applied to the analysis of natural gas (ASTM D1945, 2021) and the composition of the liquid in terms of hydrocarbon types by gas chromatographic analysis (Chapter 7) and high performance liquid chromatographic analysis (Chapter 7) of the organic liquid products might be the first techniques to be applied. Simulated distillation (Chapter 7) as well as the determination of compound class types by adsorption chromatography (Chapter 7) would also be beneficial. This might be followed by elemental analysis of the individual fractions as well as by infrared spectroscopy (Chapter 8) and mass spectrometry (Chapter 8) with the potential for application of nuclear magnetic resonance spectroscopy being very real.

Throughout this section, the focus has been on the kinetic behavior of various organic molecules during refinery operations, and specifically during the hydrodesulfurization process. However, it must be remembered that the kinetic properties of the catalyst also deteriorate because of deposits on its surface. Such deposits typically consist of coke and metals that are products of the various chemical reactions.

5.10 Oxidation

Oxidation is defined as a process in which an electron is removed from a molecule during a chemical reaction. During the oxidation reaction, there is loss of electrons or an increase in the oxidation state of a substrate or an increase of the atoms within the substrate. Also, by definition, a redox reaction is a type of chemical reaction in which the oxidation states of substrate change. Oxidation is the loss of electrons and reduction is the gain of electrons or a decrease in the oxidation state of a substrate or the atoms within the substrate.

Oxidation is an important reaction of crude oil-related materials, even the viscous feedstocks. When the feedstocks are stored (or spilled during transportation), the materials not only pollute the ecosystem by collecting in hazardous concentrations in the soil or wastewater coming out of cities and other populated areas. These materials will, more than likely oxidize on the surface after which the form oil-in-water emulsions. The inclusion of polar functions such as hydroxyl groups (—OH) or carbonyl groups (>C=O) (a result of the oxidation process) causes an increase in the density of the emulsion (relative to the original unoxidized crude oil) and with an increased propensity to form emulsions.

Also, predicting the fate of (1) the polynuclear aromatic systems; (2) the heteroatom systems, which are principally constituents containing nitrogen and sulfur; and (3) the metal-containing systems (principally compounds of vanadium, nickel, and iron in the feedstocks) is the subject of many studies and migration models. These constituents generally cause processing problems and knowledge of the behavior of these elements is essential for process improvements, process flexibility, and environmental compliance.

Thus, initial inspection of the nature of the spilled feedstock will provide deductions about the most logical means of clean up and any subsequent environmental effects. Indeed, careful evaluation of crude oil from physical property data is a major part of the initial study of any crude oil that has been released to the environment. Proper interpretation of the data resulting from the inspection of crude oil requires an understanding of their significance.

The data derived from any one or more of the analytical methods will present an indication of the nature of the spilled material and can (or should) be employed to give the environmental scientist or engineer an indication of the means by which the spilled material can be, or should be, recovered. Thus, it then possible to consider the nature of the spill followed by development of preferred cleanup methods from one (but preferably more) of the physical properties as determined by the evaluation test methods.

Furthermore, since the viscous feedstocks have high proportions of asphaltene constituents, it is not surprising that the oxidation of asphaltene constituents has been studied widely not only from the perspective of asphalt production but also from the perspective of structural studies (Chapter 9).

Oxidation of asphaltene constituents with common oxidizing agents, such as acid and alkaline peroxide, acid dichromate, and alkaline permanganate, is a slow process. The occurrence of a broad band centered at 3,420 cm^{-1} and a band at 1,710 cm^{-1} in the infrared spectra of the products indicates the formation of phenolic and carboxyl groups during the oxidation. Elemental analyses of the products indicate that there are two predominant oxidation routes, notably (1) the oxidation of naphthene moieties to aromatics as well as the oxidation of active methylene groups to ketones and (2) severe oxidation of naphthene and aromatic functions resulting in degradation of these systems to carboxylic acid functions.

Oxidation of asphaltene constituents in solution, by air, and in either the presence or absence of a metal salt, is also possible (Moschopedis and Speight, 1978). There is some oxygen uptake, as can be seen from the increased O/C atomic ratios, but the most obvious effect is the increase in the amount of *n*-heptane-insoluble material. And analysis of the data show that it is the higher heteroatom (more polar constituents) of the asphaltene constituents that are more susceptible to oxidation leaving the suggestion that the polarity of the constituents may be determined by the incorporation of the heteroatoms into ring systems.

Air blowing of asphaltene constituents at various temperatures brings about significant oxygen uptake. This is accompanied by a marked decrease in the molecular weight (vapor pressure osmometry) of the product. This indicates that intermolecular hydrogen bonding of oxygen functionality may play a part in the observed high molecular weights and physical structure of crude oil (Moschopedis and Speight, 1978).

Aromatic derivatives undergo condensation with formaldehyde to afford a variety of products. This process can be extended to the introduction of various functions into the asphaltene molecules, such as sulfomethylation, that is, introduction of the $-CH_2SO_3H$ group. This latter process, however, usually proceeds more readily if functional groups are present within the asphaltene molecule.

Thus, oxidation of asphaltene constituents produces the necessary functional groups, and subsequently sulfomethylation can be conveniently achieved. Sulfomethylation of the oxidized asphaltene constituents can be confirmed from three sources: (1) an overall increase in the sulfur content of the products relative to those of the starting material, (2) the appearance of a new infrared absorption band at 1,030 cm^{-1} attributable to the presence of sulfonic acid group(s) in the molecule(s), and (3) the water solubility of the products, a characteristic of this type of material. These sulfomethylated oxidized asphaltene constituents even remain in solution after parent oxidized asphaltene constituents can be precipitated from alkaline solution by acidification to pH 6.5.

The facile sulfomethylation reaction indicates the presence in the starting materials of reactive sites ortho or para to a phenolic hydroxyl group. The related reaction, sulfonation, is also a feasible process for oxidized asphaltene constituents. The ease with which this reaction proceeds suggests the presence of quinoid structures in the oxidized materials. Alternatively, active methylene groups in the starting materials facilitate sulfonation since such groups have been known to remain intact after prolonged oxidation.

The chemical reactions of the resin constituents of feedstocks have received much less attention than the reactions of the asphaltene constituents. However, one area of resin chemistry that has received attention is the interaction with oxygen. Thus, the oxidation of resin constituents in benzene solution with air in the presence or absence of various metal salts proceeds readily to yield asphaltene products (Moschopedis and Speight, 1978). Substantial uptake of oxygen occurs, and from the atomic heteroatom-carbon ratios in the starting material and products, it appears that preferential reaction of the more polar entities occurs. Resin constituents also undergo condensation with formaldehyde, which is especially rapid after introduction of oxygen functions by oxidation. Resin constituents can also be sulfonated to yield water-soluble or oil-soluble sulfonates. Resin constituents react with nitric acid to yield complex mixtures of oxidation and nitration products. Reactions with sulfur cause dehydrogenation as well as the formation of complex sulfides. The overall result is the production of higher molecular weight material with low hydrogen content.

Just as the asphaltene constituents and resin constituents react with oxygen at low temperatures (<100 °C, <212 °F), so do the oils; oxidation in benzene solution by air, in either the presence or absence of metal salts, proceeds smoothly to yield asphaltene and resin products (Moschopedis and Speight, 1978). In general, the majority of the oil recovered after oxidation has lower atomic N/C and S/C ratios than the untreated oil; this indicates that it is the more polar components of the oil fraction that are susceptible to oxidation. It is also evident that the presence of catalytic amounts of metal salts causes an acceleration in the oxidation rate of the oil fraction.

Selective oxidation with sodium periodate ($NaIO_4$) catalyzed by ruthenium (Ru^{8+}) ions can give more specific information on the nature of asphaltene constituents. This method selectively oxidizes alkyl-substituted aromatic carbons to carbon dioxide leaving behind (1) alkyl substituents as monocarboxylic acids and (2) polymethylene bridges between aromatic rings as α,ω-dicarboxylic acids. Some of the condensed aromatic rings survive the treatment as the polycarboxylic acid derivatives of benzene. The various acids are made volatile by esterification with diazomethane, and the esters are analyzed by gas chromatography-mass spectrometry.

This technique provides the distribution of alkyl substituents attached to aromatic rings, that of paraffin chains connecting two aromatic rings or ring systems, and information on the nature of the aromatic and naphthenic ring systems in the original sample. The amounts of alkyl chains found this way seem to be reasonably quantitative. Those of aromatic and naphthenic rings appear more tentative because the fate of the aromatic hetero-compounds is not known except that some of the sulfur compounds are recovered in the form of sulfones.

Oxidation is also used for the separation of neutral sulfur compounds from hydrocarbons. The sulfur compounds are converted to sulfoxides and sulfones that are strongly retained on chromatographic columns and can, thus, be separated from the hydrocarbons. After recovery, they can be reconverted to the sulfides by reduction with lithium aluminum hydride ($LiAlH_4$) for further analysis. By combining such oxidation and reduction procedure, it was possible to distinguish between thiophene derivatives and three different types of sulfides.

Once the oxidation reaction has been carried to completion or to the desired point, it is then the focus of the chemist or analytical chemist to determine the reaction pathway using the various analytical techniques that can be brought to bear on the issue. For example, gas chromatographic analysis (Chapter 7) and high performance liquid chromatographic analysis (Chapter 7) of the organic liquid products might be the first techniques to be applied. Simulated distillation (Chapter 7) as well as the determination of compound class types by adsorption chromatography (Chapter 7) would also be beneficial. This might be followed by elemental analysis of the individual fractions as well as by infrared spectroscopy (Chapter 7) and mass spectrometry (Chapter 7) with the potential for application of nuclear magnetic resonance spectroscopy being very real.

5.11 Thermal Methods

There are different methods of thermal analysis that can be applied to crude oil and the viscous feedstock: (1) thermogravimetric analysis (TGA), (2) thermogravimetric analysis with evolved gas analysis (TG-EGA), (3) differential scanning calorimetry (DSC), (4) thermomechanical analysis (TMA), and (5) dynamic mechanical analysis (DMA). However, within the context of the thermal analysis of crude oil and the viscous feedstocks, this section focuses on the production of a carbon residue.

The simplest thermal methods of analysis of crude oil, heavy crude oil, extra heavy crude oil, tar sand bitumen, residua, and their respective fractions are, of course, the methods for the determination of the carbon residue of the sample (Chapter 7).

Briefly, and by way of introduction to this chapter, the carbon residue of crude oil and crude oil products is a measure of the propensity of the sample to form coke when combusted under specific conditions that require a limited supply of oxygen.

The carbon residue of a crude oil sample might be considered to be analogous to the determination of fixed carbon for coal (ASTM D3172, 2021; ASTM D3173, 2021; ASTM D3174, 2021; ASTM D3175, 2021). There is a range of other thermal tests that can also be applied but the major focus of this chapter is on the actual chemistry of the thermal decomposition process. Since the standard tests for such work are extremely limited, even nonexistent, it is necessary to rely on data from published works other than that of standards organizations.

Asphaltene constituents can be thermally decomposed under conditions similar to those employed for visbreaking (viscosity breaking; ca. 470 °C, 880 °F) to afford, on the one hand, light oils that contain higher (to C_{36}) paraffins and, on the other hand, coke. In fact, the thermal decomposition of asphaltene constituents provides an excellent example of inconsistencies in the derivation of structural types from spectroscopic materials (i.e., magnetic resonance) in which alkyl side chains are deduced to contain approximately four carbon atoms (Speight, 1970, 1971). Asphaltene pyrolysis (350–800 °C, 660–1,470 °F) produces substantial amounts of alkane derivatives (having up to 40 carbon atoms in the molecule) in the distillate that can only be presumed to reflect the presence of such chains in the original asphaltene. Transalkylation studies also provide evidence for longer alkyl chains. Obviously, recognition of the inconsistencies of the spectroscopic method with respect to the paraffinic moieties must lead to the recognition of similar inconsistencies when considering the aromatic nucleus.

The application of thermal techniques to study the nature of the volatile thermal fragments from crude oil asphaltene constituents has produced some interesting data relating to the polynuclear aromatic systems (Speight, 1971). These thermal techniques have produced strong evidence for the presence of small (one- to four-ring) polynuclear aromatic systems (Speight and Pancirov, 1984) and now, application of the technique to the various functional fractions confirmed the general but unequal distribution of these systems throughout asphaltene constituents.

Each asphaltene fraction produced the same type of polynuclear aromatic systems (Moschopedis and Speight, 1978) in the volatile matter but the distribution was not constant. It was also possible to compute the hydrocarbon distribution; a noteworthy point here is the overall preponderance of single-ring derivatives (cycloparaffin derivatives and alkylbenzene derivatives) species as well as the domination of saturated material over aromatic material. The preponderance of the low-molecular-weight material in the volatile products is anticipated on the basis that more complex systems remain as nonvolatile material and, in fact, are converted to coke. One other point worthy of note is that the py/gc/ms program does not accommodate nitrogen and oxygen species. This matter is resolved, in part, by the concentration of nitrogen and oxygen in the nonvolatile material (coke) and the overall low proportions of these heteroatoms originally present in the asphaltene constituents.

The major drawback to the application of a pyrolysis-gas chromatography-mass spectrometry (py/gc/ms) technique to the study of the polynuclear aromatic systems in crude oil asphaltene constituents is the amount of material that remains as a nonvolatile residue. Aside from speculation about the polynuclear aromatic systems in the residue, it should be noted that the majority of the nitrogen (>90%), oxygen (>50%), and sulfur (>60%) in the natural asphaltene remains in the coke (Speight, 1971; Speight and Pancirov, 1984).

Paraffin derivatives are not the only hydrocarbon products of the thermal reactions of asphaltene constituents. The reaction paths are extremely complex and difficult to predict although spectroscopic investigations indicate an overall dealkylation of the aromatic derivatives to (predominantly) methyl groups with lesser yields of ethyl groups. This is in keeping with a mass spectroscopic examination of asphaltene fractions (by direct introduction into the ionization chamber), which indicates a progressive increase with increasing temperature (50–350 °C, 120–660 °F) of ions attributable to low-molecular-weight hydrocarbons. Higher temperatures (500 °C, 930 °F) favor the formation of benzene and naphthalene nuclei as the predominant aromatics in the light oil, but unfortunately an increase in coke production is noted.

In conclusion, thermal decomposition of asphaltene constituents affords a light oil having a similar composition to that from the heavy oil and a hydrocarbon gas composed of the lower paraffins, which, after the removal of the by-products (water, ammonia, and hydrogen sulfide) has good burning properties. The formation of these paraffins can be ascribed to the generation of hydrogen within the system that occurs during the pyrolysis of condensed aromatic structures.

Data from pyrolysis investigations and from HPLC-UV investigations show that the condensed aromatic hydrocarbon fragments in the resin derivatives and asphaltene derivatives match those in the gas oil fraction. It now appears that the asphaltene constituents are not different types (i.e., polynuclear aromatic sheets containing 12 or more condensed rings) to the remainder of the oil. The asphaltene constituents are compatible with the structural types with the lower molecular weight fractions and also compatible with natural product origins of crude oil (Speight, 1986, 1994). It is however, in addition to the increasing molecular weight of the resin and asphaltene constituents, the frequency of occurrence of the heteroatom functions that increase. This is the main difference between the constituents of the resin and asphaltene constituents fractions and the remainder of the constituents in crude oil.

The nonvolatile oil fraction is thermally stable at ordinary temperatures and is quite resistant to attack by many chemicals. At elevated temperatures, fission of side chains takes place. In the presence of oxygen (or air), carbon dioxide, water, and products containing carbonyl >C=O and hydroxyl —O—H groups are found. If the temperature is sufficiently high, the products containing oxygen functions are unstable and dehydrogenation is the overall result.

In pyrolysis, the simplest degradation method, the sample is rapidly heated under exclusion of air to a temperature high enough to break some of the chemical bonds. Typical reactions are dealkylation and the breakage of (aliphatic) sulfur–sulfur and sulfur–carbon bonds. The problem here is to find a compromise between specificity and yield, which means that the temperature and other reaction conditions must be carefully chosen.

A common procedure is to heat the sample in a tube to the desired temperature for several hours. The tube may be evacuated or swept with an inert gas. Typical temperatures reported for these procedures are 300–400 °C (570–750 °F), and sample sizes are in the order of 100 mg or less. The reaction products are collected in various ways and usually analyzed afterward. It is only with gas-swept arrangements that the products be analyzed on-line by gas chromatography.

Flash pyrolysis on a thin, electrically heated wire with properly chosen Curie temperature, for example, 610 °C (1,130 °F), ensures rapid warm-up and brief exposure time for suppression of side reactions. In this procedure, a very small amount (about 100 μg) of the sample is pressed onto the wire. Warm-up time is 0.1 s, and the gaseous products are directly transferred to a gas chromatograph, preferably with MS as detector.

In a variation of flash pyrolysis, the sample (in toluene solution) is dripped from a dropping funnel at a slow rate (<1 ml/min) to the bottom of a flask which is heated by a temperature-controlled sand bath, typically at 430 °C (800 °F). A flow of nitrogen is used to carry the reaction products and the solvent to a cooled collection flask for subsequent concentration and analysis and there was no evidence for degradation of the toluene. The pyrolysis oil can be separated into a number of compound groups by a combination of liquid chromatography with another associated physical identification or by a chemical method.

More recent work (Mothé et al., 2013) has reported an investigation of the thermal various samples (such as a medium crude oil, a heavy crude oil, and extra heavy crude oil) not only to study the thermal behavior of the feedstocks but also to study the data as a contribution to the characterization of feedstocks that have different API values.

5.12 Miscellaneous Methods

Reactions of asphaltene derivatives with sulfur have also received some attention and have yielded interesting results. For example, treatment of the asphaltene constituents with oxygen or with sulfur at 150–250 °C (300–480 °F) yields a condensed aromatic product (H/C = 0.97; H/C [asphaltene constituents] = 1.20) containing very little additional sulfur. The predominant reaction here appears to be condensation between the aromatic and aliphatic moieties of the asphaltene constituents effected by elemental sulfur, which are in turn converted to hydrogen sulfide. Condensation appears to proceed in preference to molecular degradation, and treatment of the condensed products at 200–300 °C (390–570 °F) for 1–5 hr again affords good grade cokes (H/C = 0.54–0.56). In all instances, the final products contain only very low amounts of elements other than carbon and hydrogen (N + O + S = <5% w/w), a desirable property of good grade coke.

Reductive alkylation is ordinarily used to make asphaltene constituents more soluble and less prone to aggregation in hydrocarbon solvents. The sample is dissolved in tetrahydrofuran (although not all asphaltene constituents are soluble in this solvent) an excess of metallic potassium under exclusion of air. Assuming no side reactions, this method produces the final alkylated product.

References

Arjoon, K.K., and Speight, J.G. 2023. *Petroleum Biodegradation and Oil Spill Bioremediation*. CRC Press, Taylor & Francis Group, Boca Raton, Florida.

ASTM C1109. 2021. Standard Practice for Analysis of Aqueous Leachates from Nuclear Waste Materials Using Inductively Coupled Plasma-Atomic Emission Spectrometry. Annual Book of Standards. ASTM International, West Conshohocken, Pennsylvania.

ASTM C1111. 2021. Standard Test Method for Determining Elements in Waste Streams by Inductively Coupled Plasma-Atomic Emission Spectroscopy. Annual Book of Standards. ASTM International, West Conshohocken, Pennsylvania.

ASTM D56. 2021. Standard Test Method for Flash Point by Tag Closed Cup Tester. Annual Book of Standards. ASTM International, West Conshohocken, Pennsylvania.

ASTM D92. 2021. Standard Test Method for Flash and Fire Points by Cleveland Open Cup Tester. Annual Book of Standards. ASTM International, West Conshohocken, Pennsylvania.

ASTM D93. 2021. Standard Test Methods for Flash Point by Pensky-Martens Closed Cup Tester. Annual Book of Standards. ASTM International, West Conshohocken, Pennsylvania.

ASTM D94. 2021. Standard Test Methods for Saponification Number of Petroleum Products. Annual Book of Standards. ASTM International, West Conshohocken, Pennsylvania.

ASTM D129. 2021. Standard Test Method for Sulfur in Petroleum Products (General High Pressure Decomposition Device Method). Annual Book of Standards. ASTM International, West Conshohocken, Pennsylvania.

ASTM D139. 2021. Standard Test Method for Float Test for Bituminous Materials. Annual Book of Standards. ASTM International, West Conshohocken, Pennsylvania.

ASTM D482. 2021. Standard Test Method for Ash from Petroleum Products. Annual Book of Standards. ASTM International, West Conshohocken, Pennsylvania.
ASTM D664. 2021. Standard Test Method for Acid Number of Petroleum Products by Potentiometric Titration. Annual Book of Standards. ASTM International, West Conshohocken, Pennsylvania.
ASTM D972. 2021. Standard Test Method for Evaporation Loss of Lubricating Greases and Oils. Annual Book of Standards. ASTM International, West Conshohocken, Pennsylvania.
ASTM D974. 2021. Standard Test Method for Acid and Base Number by Color-Indicator Titration. Annual Book of Standards. ASTM International, West Conshohocken, Pennsylvania.
ASTM D1159. 2021. Standard Test Method for Bromine Numbers of Petroleum Distillates and Commercial Aliphatic Olefins by Electrometric Titration. Annual Book of Standards. ASTM International, West Conshohocken, Pennsylvania.
ASTM D1318. 2021. Standard Test Method for Sodium in Residual Fuel Oil (Flame Photometric Method). Annual Book of Standards. ASTM International, West Conshohocken, Pennsylvania.
ASTM D1552. 2021. Standard Test Method for Sulfur in Petroleum Products by High Temperature Combustion and Infrared (IR) Detection or Thermal Conductivity Detection (TCD). Annual Book of Standards. ASTM International, West Conshohocken, Pennsylvania.
ASTM D1945. 2021. Standard Test Method for Analysis of Natural Gas by Gas Chromatography. Annual Book of Standards. ASTM International, West Conshohocken, Pennsylvania.
ASTM D1962. 2021. Standard Test Method for Saponification Value of Drying Oils, Fatty Acids, and Polymerized Fatty Acids. Annual Book of Standards. ASTM International, West Conshohocken, Pennsylvania.
ASTM D2622. 2021. Standard Test Method for Sulfur in Petroleum Products by Wavelength Dispersive X-ray Fluorescence Spectrometry. Annual Book of Standards. ASTM International, West Conshohocken, Pennsylvania.
ASTM D2896. 2021. Standard Test Method for Base Number of Petroleum Products by Potentiometric Perchloric Acid Titration. Annual Book of Standards. ASTM International, West Conshohocken, Pennsylvania.
ASTM D3120. 2021. Standard Test Method for Trace Quantities of Sulfur in Light Liquid Petroleum Hydrocarbons by Oxidative Microcoulometry. Annual Book of Standards. ASTM International, West Conshohocken, Pennsylvania.
ASTM D3172. 2021. Standard Practice for Proximate Analysis of Coal and Coke. Annual Book of Standards. ASTM International, West Conshohocken, Pennsylvania.
ASTM D3173. 2021. Standard Test Method for Moisture in the Analysis Sample of Coal and Coke. Annual Book of Standards. ASTM International, West Conshohocken, Pennsylvania.

ASTM D3174. 2021. Standard Test Method for Ash in the Analysis Sample of Coal and Coke from Coal. Annual Book of Standards. ASTM International, West Conshohocken, Pennsylvania.

ASTM D3175. 2021. Standard Test Method for Volatile Matter in the Analysis Sample of Coal and Coke. Annual Book of Standards. ASTM International, West Conshohocken, Pennsylvania.

ASTM D3228. 2021. Standard Test Method for Total Nitrogen in Lubricating Oils and Fuel Oils by Modified Kjeldahl Method. Annual Book of Standards. ASTM International, West Conshohocken, Pennsylvania.

ASTM D3339. 2021. Standard Test Method for Acid Number of Petroleum Products by Semi-Micro Color Indicator Titration. Annual Book of Standards. ASTM International, West Conshohocken, Pennsylvania.

ASTM D3341. 2021. Standard Test Method for Lead in Gasoline-Iodine Monochloride Method. Annual Book of Standards. ASTM International, West Conshohocken, Pennsylvania.

ASTM D3343. 2021. Standard Test Method for Estimation of Hydrogen Content of Aviation Fuels. Annual Book of Standards. ASTM International, West Conshohocken, Pennsylvania.

ASTM D3605. 2021. Standard Test Method for Trace Metals in Gas Turbine Fuels by Atomic Absorption and Flame Emission Spectroscopy. Annual Book of Standards. ASTM International, West Conshohocken, Pennsylvania.

ASTM D3701. 2021. Standard Test Method for Hydrogen Content of Aviation Turbine Fuels by Low Resolution Nuclear Magnetic Resonance Spectrometry. Annual Book of Standards. ASTM International, West Conshohocken, Pennsylvania.

ASTM D4045. 2021. Standard Test Method for Sulfur in Petroleum Products by Hydrogenolysis and Rateometric Colorimetry. Annual Book of Standards. ASTM International, West Conshohocken, Pennsylvania.

ASTM D4084. 2021. Standard Test Method for Analysis of Hydrogen Sulfide in Gaseous Fuels (Lead Acetate Reaction Rate Method). Annual Book of Standards. ASTM International, West Conshohocken, Pennsylvania.

ASTM D4294. 2021. Standard Test Method for Sulfur in Petroleum and Petroleum Products by Energy Dispersive X-ray Fluorescence Spectrometry. Annual Book of Standards. ASTM International, West Conshohocken, Pennsylvania.

ASTM D4739. 2021. Standard Test Method for Base Number Determination by Potentiometric Hydrochloric Acid Titration. Annual Book of Standards. ASTM International, West Conshohocken, Pennsylvania.

ASTM D4808. 2021. Standard Test Methods for Hydrogen Content of Light Distillates, Middle Distillates, Gas Oils, and Residua by Low-Resolution Nuclear Magnetic Resonance Spectroscopy. Annual Book of Standards. ASTM International, West Conshohocken, Pennsylvania.

ASTM D4810. 2021. Standard Test Method for Hydrogen Sulfide in Natural Gas Using Length-of-Stain Detector Tubes. Annual Book of Standards. ASTM International, West Conshohocken, Pennsylvania.

ASTM D5291. 2021. Standard Test Methods for Instrumental Determination of Carbon, Hydrogen, and Nitrogen in Petroleum Products and Lubricants. Annual Book of Standards. ASTM International, West Conshohocken, Pennsylvania.

ASTM D5373. 2021. Standard Test Methods for Determination of Carbon, Hydrogen and Nitrogen in Analysis Samples of Coal and Carbon in Analysis Samples of Coal and Coke. Annual Book of Standards. ASTM International, West Conshohocken, Pennsylvania.

ASTM D5770. 2021. Standard Test Method for Semiquantitative Micro Determination of Acid Number of Lubricating Oils During Oxidation Testing. Annual Book of Standards. ASTM International, West Conshohocken, Pennsylvania.

ASTM E385. 2021. Standard Test Method for Oxygen Content Using a 14-MeV Neutron Activation and Direct-Counting Technique. Annual Book of Standards. ASTM International, West Conshohocken, Pennsylvania.

ASTM E422. 2021. Standard Test Method for Measuring Net Heat Flux Using a Water-Cooled Calorimeter. Annual Book of Standards. ASTM International, West Conshohocken, Pennsylvania.

ASTM E659. 2021. Standard Test Method for Autoignition Temperature of Chemicals. Annual Book of Standards. ASTM International, West Conshohocken, Pennsylvania.

ASTM F1186. 2021. Standard Classification System for Chemicals According to Functional Groups. Annual Book of Standards. ASTM International, West Conshohocken, Pennsylvania.

Mackay, D., and Zagorski, W. 1982. Studies of Water-in-Oil Emulsions. Report No. EE-34. Environment Canada, Ottawa, Ontario, Canada.

Moschopedis, S.E., and Speight, J.G. 1971. Halogenation of Athabasca Asphaltenes with Elemental Halogen. *Fuel*, 50: 58.

Moschopedis, S.E., and Speight, J.G. 1976. An Investigation of Hydrogen Bonding by Oxygen Functions in Athabasca Bitumen. *Fuel*, 55: 187.

Moschopedis, S.E., and Speight, J.G. 1978. The Influence of Metal Salts on Bitumen Oxidation. *Fuel*, 57: 235.

Mothé, M.G., Carvalho, C.H.M., Sérvulo, E.F.C., and Mothé, C.G. 2013. Kinetic Study of Heavy Crude Oils by Thermal Analysis. *Journal of Thermal Analysis and Calorimetry*, 111: 663–668.

Prado, G.H.C., and de Klerk, A. 2014. Halogenation of Oil Sands Bitumen, Maltenes, and Asphaltenes. *Energy & Fuels*, 28(7): 4458–4468.

Shafizadeh, A., McAteer, G., and Sigmon, J. 2003. High-Acid Crudes. Proceedings. Crude Oil Quality Group Meeting, New Orleans, Louisiana. January 30.

Sheridan, M. 2006. California Crude Oil Production and Imports. Staff Paper, Report No. CERC-600-2006-006. Fossil Fuels Office, Fuels and Transportation Division, California Energy Commission. Sacramento, California. April.

Speight, J.G. 1970. A Structural Investigation of the Constituents of Athabasca Bitumen by Proton Magnetic Resonance Spectroscopy. *Fuel*, 49: 76.

Speight, J.G. 1971. Structural Analysis of Athabasca Asphaltenes by Proton Magnetic Resonance Spectroscopy. *Fuel*, 50: 102.

Speight, J.G. 1986. Polynuclear Aromatic Systems in Petroleum. *American Chemical Society, Division of Petroleum Chemistry. Preprints*, 31(4): 818.

Speight, J.G. 1994. Chemical and Physical Studies of Petroleum Asphaltenes. In: *Asphaltenes and Asphalts. I. Developments in Petroleum Science, 40*. T.F. Yen and G.V. Chilingarian (Editors). Elsevier, Amsterdam, The Netherlands. Chapter 2

Speight, J.G. 2000. *The Desulfurization of Heavy Oils and Residua*. 2nd Edition. Marcel Dekker Inc., New York.

Speight, J.G. 2014a. *The Chemistry and Technology of Petroleum*. 5th Edition. CRC Press, Taylor & Francis Group, Boca Raton, Florida.

Speight, J.G. 2014b. *High Acid Crudes*. Gulf Professional Publishing Company, Elsevier, Oxford, United Kingdom.

Speight, J.G. 2017. *Handbook of Petroleum Refining*. CRC Press, Taylor & Francis Group, Boca Raton, Florida.

Speight, J.G., and Pancirov, R.J. 1984. Some Aspects of the Structure of Petroleum Asphaltenes. *Liquid Fuels Technology*, 2: 287.

6

Physical Properties, Electrical Properties, and Optical Properties

6.1 Introduction

Crude oil, heavy crude oil, extra heavy crude oil, and tar sand bitumen exhibit a wide range of physical properties and whereas the properties such as viscosity, density, boiling point, and color of crude oil, heavy crude oil, extra heavy crude oil, and tar sand bitumen vary widely, the elemental (ultimate) analysis varies over a narrow range for a large number of samples. In fact, relative to another fossil energy resource (coal; Speight, 2013), the carbon content is fairly constant.

The hydrogen content and the heteroatom content in these feedstocks vary to a greater extent than the carbon content and are often considered as being responsible for the major differences between crude oil and the viscous feedstocks. However, this does not take into consideration the relative volatility of the various constituents of these three feedstocks. The nitrogen, oxygen, and sulfur can be present in only trace amounts in some crude oil, which as a result consists primarily of hydrocarbons. In fact, bitumen containing 9.5% by weight heteroatoms may contain very few, if any, true hydrocarbon constituents insofar as the constituents contain *at least one or more* nitrogen, oxygen, and/or sulfur atoms within the various molecular structures. Coupled with the changes brought about to the feedstock constituents by refinery operations, it is not surprising that crude oil characterization is a monumental task.

Refinery processes can be conveniently divided into three fundamentally different types which are (1) separation processes, which involve the division of the feedstock into various streams or fractions depending upon the nature of the feedstock; (2) conversion processes, which involve the production of salable materials from the feedstock by skeletal alteration, or even by alteration of the chemical type of the feedstock constituents; and (3) finishing processes, which involve the purification of the various product streams by a variety of processes that remove impurities from the product. In some case, a fourth category can be added and

Handbook of Heavy Oil Properties and Analysis, First Edition. James G. Speight.

includes processes such as the reforming (molecular rearrangement) processes (Parkash, 2003; Speight, 2014; Hsu and Robinson, 2017; Speight, 2017).

The analytical chemist attached to a refinery or process research laboratory must be able to respond to the requirements of each of the above process categories through use of a variety of analytical methods that produce data for process predictability and monitoring. The data must be acquired by methods that are quantitatively reproducible and offer a high degree of precision. As a side comment, there is reason to question the value of any method that might be reproducible but only offers accuracy to within 10% above or below the true value. That is, a mean data point that offers 10% variability is of very questionable value.

Although it is possible to classify refinery operations using the general categories outlined above, the behavior of the various feedstocks in refinery operations is not simple because of the elemental and compositional differences between the feedstocks. The elemental atomic ratios that are obtained by elemental analysis (sometimes referred to as the ultimate analysis) can give an "indication" of the nature of a feedstock and the generic hydrogen requirements to satisfy the refining chemistry (Chapter 3) (Parkash, 2003; Speight, 2014; Hsu and Robinson, 2017; Speight, 2017). In spite of this apparent benefit of knowing the atomic ratios, it is difficult, if not impossible, to predict with any degree of certainty how the feedstock will behave during refining. Any deductions made from data such as the atomic ratios are speculative and are open to justifiable doubt.

The chemical composition of a feedstock is a much truer indicator of refining behavior (Chapter 2). Whether the composition is represented in terms of compounds types or in terms of generic compound classes, it can offer the refiner a measure of predictability through an understanding of the nature of refinery chemistry. Hence, chemical composition not only plays a large part in determining the nature of the products that arise from the refining operations but also plays a role in determining the means by which a particular feedstock should be processed (Parkash, 2003; Speight, 2014; Hsu and Robinson, 2017; Speight, 2017).

To satisfy specific needs with regard to the type of a refinery feedstock to be processed, as well as to the nature of the product, most refiners have, through time, developed their own methods of feedstock analysis and evaluation (Speight, 2014, 2015, 2017, 2021). However, such methods are considered proprietary and are not normally available. A complete discussion of the large number of routine tests available for crude oil fills an entire book (ASTM, 2021). However, it seems appropriate that in any discussion of the physical properties of crude oil and crude oil products, reference be made to the corresponding test method and, accordingly, the various test numbers have been included in the text.

Crude oil and the viscous feedstocks are exceedingly complex and structured mixtures consisting of hydrocarbon derivatives as well as sulfur-containing constituents, nitrogen-containing constituents, oxygen-containing constituents, and metal-containing constituents. As a result of this complexity, the importance of the correct sampling of the feedstock cannot be overstressed. Properties such as elemental analysis, metals content, density (specific gravity), and viscosity (to mention only a few properties) are affected by the homogeneity (or heterogeneity) of the sample. In addition, adequate records of the circumstances and conditions during sampling have to be made, for example, in sampling from oilfield separators, the temperatures and pressures of the separation plant, and the atmospheric temperature would be noted. An accurate sample handling and storage log should be maintained and should include information such as (1) the precise source of the sample, that is, the exact geographic location or refinery locale from which the sample was obtained; (2) a description of the means by which the sample was obtained; (3) the protocols that have been used to store the sample; (4) chemical analyses, such as elemental composition; (5) physical property analyses, such as API gravity, pour point, and distillation profile; and (6) the methods used to determine the properties in items 4 and 5, the number of times that the samples have been retrieved from storage to extract a portion, that is, indications of exposure to the air or oxygen. Attention to factors such as these enables standardized comparisons to be made when subsequent samples are taken.

However, before this occurs, there are several protocols involved in initial isolation and cleanup of the sample. In fact, considerable importance attaches to the presence of water or sediment in crude oil (ASTM D1796, 2021; ASTM D4007, 2021) for they lead to difficulties in the refinery, for example, corrosion of equipment, uneven running on the distillation unit, blockages in heat exchangers, and adverse effects on product quality.

The sediment consists of finely divided solids that may be dispersed in the oil or carried in water droplets. The solids may be drilling mud or sand or scale picked up during the transport of the oil, or may consist of chlorides derived from evaporation of brine droplets in the oil. In any event, the sediment can lead to (1) serious plugging of the equipment, (2) corrosion due to chloride decomposition, and (3) a lowering of feedstock and product quality.

Water may be found in the crude either in an emulsified form or in large droplets and can cause flooding of distillation units and excessive accumulation of sludge in tanks. The quantity is generally limited by pipeline companies and by refiners, and steps are normally taken at the wellhead to reduce the water content as low as possible. However, water can be introduced during shipment and, in any form, water and sediment are highly undesirable in a refinery feedstock and the relevant tests (ASTM D96, 2021; ASTM D1796, 2021) are regarded as important in crude oil quality examinations.

The viscous feedstocks often form persistent emulsions which are difficult to separate, while in testing wax-bearing crude oils for sediment and water, care has to be taken to ensure that wax suspended in the sample is brought into solution prior to the test, otherwise it will be recorded as sediment. Careful sampling of a crude oil is necessary when examination has to be made for suspended impurities of the type discussed.

The physical and chemical characteristics of refinery feedstocks and the yields and properties of products or factions prepared from them vary considerably and are dependent on the concentration of the various types of hydrocarbons and minor constituents present. Some types of crude oil have economic advantages as sources of fuels and lubricants with highly restrictive characteristics because they require less specialized processing than that needed for production of the same products from many types of crude oil. Others may contain unusually low concentrations of components that are desirable fuel or lubricant constituents, and the production of these products from such crude oils may not be economically feasible.

In the current context, the evaluation of refinery feedstocks typically involves an examination of one or more of the physical properties of the material. By this means, a set of basic characteristics can be obtained that can be correlated with utility. The physical properties of crude oil and crude oil products are often equated with those of the hydrocarbons for although crude oil is indeed a very complex mixture, there is gasoline, produced by nondestructive distillation (straight-run gasoline), in which fewer than a dozen hydrocarbons make up at least 50% of the material. However, for the purposes of this chapter, data relating to various physical properties have been included as illustrative examples (listed alphabetically below), but theoretical discussions of the physical properties of hydrocarbons were deemed irrelevant and are omitted.

In addition to the properties examined during a feedstock assay (Chapter 2), there are other properties that (because of the acceptance of viscous feedstocks by refineries) also need attention and include (1) physical properties, which include adhesion, density, and specific gravity, surface tension and interfacial tension, and viscosity; (2) electrical properties, which include conductivity, dielectric constant, dielectric strength, dielectric loss and power factor, and static electrification; and (3) optical properties, which include optical activity and refractive index.

These additional properties, which are presented (alphabetically) in the following subsections, are not only relevant to understanding the nature of the existence of the crude oil in the reservoir as well as the existence of viscous feedstocks in the deposit but also the various aspects of the recovery and transportation of any of these materials as well as the processability of crude oil and the viscous feedstocks as they are moved around the refinery from one processing option to another. In all of these respects, flow assurance is a much challenging issue in producing and transporting viscous feedstocks.

Moreover, an understanding of these properties (i.e., the physical properties, the electrical properties, and the optical properties) is very relevant to understanding the behavior of the feedstocks after an inadvertent spill of crude oil or the viscous feedstocks into the environment.

6.2 Physical Properties

Heavy crude oil, extra heavy crude oil, tar sand bitumen, and residua are characterized by (in addition to a high sulfur and a high content of metal-containing components) (1) high density, which converts to a low API gravity; (2) high viscosity; and (3) high amounts of high molecular weight components such as the asphaltene fraction and the associated resin fraction. Due to these unfavorable properties, the viscous feedstocks are characterized as low-quality feedstocks because of the technological challenges that occur during production, transportation, and processing. All of these aforementioned characteristics relate to difficulties in the selection of (1) the manner of production from the reservoir/deposit, (2) the manner of transportation, and (3) the selection of the most appropriate processing options. As a result, it is necessary to understand these properties before the most appropriate processing options are chosen (Huc, 2011; Speight, 2014, 2017).

Thus, any feedstock entering a refinery must, in addition to defining and understanding the chemical properties (Chapter 5), be analyzed for suitability as refinery feedstocks. This is determined by means of a feedstock assay (Chapter 2) which examines a series of mostly a series of physical properties and (to a lesser extent) a series of chemical properties. There are two types of assays that are used to determine feedstock quality and suitability: (1) a detailed assay, often referred to as a comprehensive assay and (2) a preliminary assay, also often referred to as an inspection assay (Chapter 2). Testing can include characterization of whole feedstocks crude oils and the various boiling range fractions that are produced from physical distillation or simulated distillation by various procedures. Information obtained from the petroleum assay is used for detailed refinery engineering and are an important aspect of refinery operations.

The detailed assay consists of the application of a wide range of test methods that assist in the refiner gaining a detailed understanding of the properties of the incoming (new) feedstock (Chapter 2). A refinery may choose to apply the detailed assay on an annual basis or only to feedstocks that are relatively new to the refinery.

The preliminary assay is applied to feedstocks that are already within the refinery operations and are specifically for recent shipment to that feedstock to the refinery to ensure that are no major property changes. The most common properties examined in the preliminary assay are (1) the density, typically the API gravity; (2) the viscosity; and (3) the boiling ranges of the various fractions of the feedstock. Other properties may be examined as part of the preliminary assay based on the type of feedstock and the potential refinery options for that feedstock. In other words, the preliminary assay is a spot-check of the properties of the incoming feedstock.

Thus, an assay of an incoming feedstock is an evaluation of a refinery feedstock since each incoming feedstock type has a series of unique chemical and physical characteristics. No two crude oil types are identical and there are crucial differences in feedstock quality (Chapter 1) and the results of the assay provide detailed analysis data for refinery operators. Assay data are used to determine if an incoming feedstock is compatible for a particular refinery or if the feedstock could cause yield, quality, production, environmental, and other problems. These data are especially true when the refinery uses a blend of several feedstocks, one of which may not be compatible with the other feedstocks in the blend thereby resulting in incompatibility problems leading to blend instability and shut down of refinery equipment (Chapter 11).

6.2.1 Adhesion

Adhesion is the tendency of particles and/or surfaces that are dissimilar in properties to cling to one another—on the other hand, cohesion is the tendency of similar or identical particles/surfaces to cling to one another. By way of further elucidation of the definition, adhesion and cohesion are important properties of refinery feedstocks because of the tendency for adhesion to occur not only in the refinery but also when a feedstock is spilled into the environment. On the other hand, while adhesion occurs when the feedstock constituents are attracted to other substances, cohesion occurs when the feedstock constituents are attracted to each other. The forces that cause adhesion and cohesion can be divided into several types: (1) chemical adhesion, (2) dispersive adhesion, and (3) diffusive adhesion.

One of the properties of the viscous feedstocks is that they tend to adhere to surfaces to a greater or lesser degree which is true not only for refinery operations but also when the material is spilled into the environment.

Accordingly, a test has been developed, which uses a standard surface that gives a semiquantitative measure of the adhesive property of the various crude oil-related materials.

For the purposes of this test, the adhesion of the viscous crude oil-related feedstocks is the mass of feedstock per unit area that will remain on a standard test surface, after "dunking and draining" for 30 min, under prescribed conditions (Jokuty et al., 1996). Test parameters that can be evaluated include temperature, oil viscosity, time, and test-surface area. A recent study has also determined that the relative adhesiveness of different oils is independent of the type of surface material used.

The usefulness of the test lies in the produced data which allow the investigator to make reasonable deductions on the behavior of crude oil and crude oil products under standard conditions thereby removing much of the guesswork from the behavior of the spilled constituents. Such data, with a knowledge of the composition of the material, will assist the investigator to identify the principal constituents that are responsible for such behavior.

The adhesion phenomenon is particularly important when crude oil or a viscous feedstock is spilled into the marine environment and undergoes natural weathering processes that result in oil components being dissolved into the water column or present in particulate form as dispersed droplets. Feedstock components dissolved in seawater are typically considered as more bioavailable to pelagic marine organisms and the main driver of oil toxicity.

Briefly, the pelagic zone is divided into two areas (sometimes referred to as provinces) which are (1) the neritic province, which corresponds to all of the water from the low tide line to the shelf break; and (2) the oceanic province, also referred to as the oceanic zone which represents all of the other water in the open ocean regions and which is subdivided into depth zones.

Adhesion of feedstock constituents to oceanic rock and to oceanic life can cause serious effects and influence the well-being of the ocean.

6.2.2 Density, Specific Gravity, and API Gravity

The density of a substance is the mass of a unit volume of the substance at a specified temperature and has the dimensions of grams per cubic centimeter (a close approximation to grams per milliliter). Density is measured at a standard temperature, mostly 15.6 or 20 °C (60 or 68 °F) and is often written in the form of d_4^{20} to indicate that it was measured at 20 °C (68 °F) and in pycnometer calibrated with water at 4 °C (49 °F), that is, with a medium of density equal to 1.0000. The pycnometer method (ASTM D70, 2021) is also applicable to the broader range of samples covered by that procedure. However, only high viscosity samples, with density equal to or greater than water, are measured using this method. A densimeter is usually used to measure the density of lighter-than-water samples.

On the other hand, the specific gravity is the ratio of the mass of a volume of the substance to the mass of the same volume of water and is dependent on two temperatures, those at which the masses of the sample and the water are measured. The standard temperatures for a specific gravity in the crude oil industry in North America are 60/60 °F (15.6/15.6 °C). The specific gravity values of crude oil samples are close to those of density and range from about 0.80 for light paraffinic crude oils, about 0.98 for heavy oil, and about 1.00–1.03 for bitumen.

In the early days of the crude oil refining industry, the measurement of density and specific gravity data were used to (1) indicate feedstock quality and (2) correlate with aromatic character, naphthenic character, and paraffinic character; the specific gravity is highest for aromatics and lowest for paraffins—the API gravity reverses this relationship. Thus, the density (or API gravity) of a refinery feedstock is not only a major criterion in setting the price of crude oil (and the viscous feedstocks) but also for determining the means by which the feedstock should be processed and the potential slate of products. For the convenience of the refiner, conversion tables (ASTM D1250, 2021) are frequently used for converting feedstock density to the 15 °C reference temperature.

The density of refinery feedstocks is generally expressed in terms of the API gravity (°API) (Parkash, 2003; Speight, 2014; Hsu and Robinson, 2017; Speight, 2017). When detailed physical and chemical analyses are available, °API can be used as a rigorous quality index of the petroleum against a similar composition in its natural state without the effects of mixing and dilution. As the API gravity decreases, the value of the distillation product decreases due to the higher production of constituents that have a lower added value, such as in the current context the gas oil and the residuum. When different types of feedstocks are mixed (blended), the API gravity cannot be used for anything other than as a measure of fluid density.

As a brief note, the viscous feedstocks typically have densities in excess of 1,000 kg/m^3 (<10°API).

The density and specific gravity of a feedstock are two properties that have found wide use in the industry for preliminary assessment of the character of the crude oil and can be measured by selection of an appropriate test method crude oil (such as ASTM D70, 2021; ASTM D71, 2021; ASTM D287, 2021; ASTM D1217, 2021; ASTM D1298, 2021; ASTM D1480, 2021; ASTM D1481, 2021; ASTM D1555, 2021; ASTM D1657, 2021; ASTM D4052, 2021).

Briefly, the use of a digital density meter is a quick method for application to the viscous feedstocks (ASTM D4052, 2021) and the alternate methods such as the pycnometer method (ASTM D70, 2021) is reliable but may be time-consuming while the hydrometer method (ASTM D1298, 2021) is also reliable and precise for viscous feedstocks but typically requires a larger sample and is also time-consuming.

The use of density or specific gravity has largely been used replaced by the API (American Crude oil Institute) gravity is the preferred property. It is one criterion that is used in setting prices for crude oil. In the United States, the API gravity has been used for the classification of a reservoir and the accompanying tax and royalty consequences. This property was derived from the Baumé scale:

$$\text{Degrees Baumé} = 140/\text{sp gr}@60/60°\text{F} - 130$$

However, a considerable number of hydrometers calibrated according to the Baumé scale were found at an early period to be in error by a consistent amount, and this led to the adoption of the equation:

$$\text{Degrees API} = 141.5/\text{sp gr}@60/60°\text{F} - 131.5$$

There is also a scale that has been termed the "incremental density" (Altgelt and Boduszynski, 1994) and is defined by the relationship:

$$D_\text{i} = (d - 0.5)(200)$$

Using this scale, typical refinery feedstocks fall into a range on the order of 60–100, which is a similar span (40 units) as the span of the API gravity scale. Whether or not this scale will be accepted for consideration to replace or supplement the API gravity scale, in spite of the claims of simplicity, remains to be seen. Currently it seems to offer little advantage over the API gravity scale.

The variation of density with temperature, effectively the coefficient of expansion, is a property of great technical importance, since most crude oil products are sold by volume and specific gravity is usually determined at the prevailing temperature (21 °C, 70 °F) rather than at the standard temperature (60 °F, 15.6 °C).

The specific gravity of heavy crude oil, extra heavy crude oil, and tar sand bitumen shows a fairly wide range of variation. The largest degree of variation is usually due to local conditions that affect material close to the faces, or exposures, occurring in surface oil sand beds. There are also variations in the specific gravity of the bitumen found in beds that have not been exposed to weathering or other external factors. The range of specific gravity variation is usually on the order of 0.9–1.04.

An important feature of the inverse of density, the specific volume, is the cumulative effect (additivity) of atomic contributions. The specific volume is, in reality, the molar volume, MV, which is equal to the density divided by the molecular weight, which is the sum of the atomic volumes. The additivity feature has played an important part in the use of density for average structure determinations. Other features in addition to atomic volumes are needed for the calculation of molar volumes. In alkyl benzenes, and even more so in condensed saturated as well as aromatic-ring systems, the junction carbon atoms differ from the regular carbons in their volume contribution.

Finally, density, alone or in combination with other physical properties such as boiling point, has been used for the attempted classification of crude oil and its fractions (Speight, 2014, 2015, 2017). Examples of such composite parameters are the Watson K factor, also called the UOP K factor, the viscosity gravity constant (VGC), and the refractivity intercept, R_i. For example, paraffin hydrocarbon derivatives might have a VGC on the order of 0.74–0.75, a Watson factor in the range 13.1–13.5, and a refractivity intercept of 1.048–1.05. Naphthenic hydrocarbons might have a VGC on the order of 0.89–0.95, a Watson factor in the range 10.5–13.2, and a refractivity intercept of 1.03–1.046. Aromatic hydrocarbons might have a VGC on the order of 0.95–1.13, a Watson factor in the range 9.5–12.53, and a refractivity intercept of 1.07–1.1.105. The precise meaning of such numbers and attempted classification systems remain unanswered, especially when the system utilizes (predominantly) a single property (i.e., paraffinic, naphthenic, and aromatic) and ranges for complex mixtures such as crude oil heavy oil and bitumen are the norm with overlap between the ranges. In such cases, not only are the accuracy and precision of the analytical method called into question but also the deductive reasoning behind any conclusions.

Environmentally, density is also important as it indicates whether a particular oil will float or sink in water (ASTM D139, 2021) (Speight, 2005; Speight and Arjoon, 2012). As the density of water is 1.0 g/cm^3 at 15 °C and the density of most viscous feedstocks ranges from 0.1.01 to 1.04 g/cm^3, the viscous feedstocks will tend to sink when spilled on water. However, since the density of seawater is on the order of 1.03 g/cm^3, some viscous feedstocks may float on seawater. However, the density of an oil will increase with time of exposure because of (1) the incorporation of oxygen-containing functional groups into the constituents as well as (2) the evaporation of any lower boiling constituents, although in the case of the viscous feedstocks this latter event is unlikely because of the high boiling nature of the feedstocks. This effect may cause the oil to (apparently) disappear over time when it is, in fact, below the level of the water surface and migrating to the shoreline with the tidal flow of the water.

6.2.3 Surface and Interfacial Tension

The surface tension of a feedstock is a measure of the force acting at a boundary between two phases (i.e., the feedstock and another phase). If the boundary is between a liquid and a solid or between a liquid and a gas (air), the attractive forces are referred to as surface tension, but the attractive force between two immiscible liquids is referred to as interfacial tension. The surface tension and the interfacial tension of feedstocks are important properties because they are indicative of the ease of formation and stability of emulsions and foams.

The surface and interfacial tensions are determined by measuring the force necessary to detach a planar ring of platinum-iridium wire from the surface of the liquid. For interfacial tension, the ring is detached from the liquid of higher surface tension. The force is measured by a direct reading of torsion-type balance. This reading is corrected by a factor which depends upon the force applied, the density of the solutions, and the dimensions of the ring.

The surface tension of crude oil or (in this context) a viscous feedstock, together with the respective viscosity data, is illustrative of the rate at which an oil spill (or a product spill) spreads and can be estimated by the spreading coefficient. Thus:

$$\text{Spreading coefficient} = S_{\text{WA}} - S_{\text{OA}} - S_{\text{WO}}$$

S_{WA} is the water/air interfacial tension, S_{OA} is the oil/air interfacial tension, and S_{WO} is the water/oil interfacial tension. A single test method (ASTM D971, 2021) is applicable to the measurement of the oil/water interfacial tension.

A limited number of data are available for bitumen from Venezuela, Mexico, and Borneo (Pfeiffer, 1950). Thus:

	Surface tension, dyn/cm (mN/m)		
Source	**100 °C**	**120 °C**	**150 °C**
Venezuela	28.8	27.7	26.0
Mexico	29.4	28.1	26.2
	28.7	27.4	25.5
Borneo	30.0	28.8	26.9

The data, like the data for viscosity, tend to show a linear variation with temperature.

On the other hand, the interfacial tension of a material is the force of attraction between the molecules at the interface of two fluids.

Generally, the interfacial tension for crude oil falls in the range 13.6–34.3 dyn/cm at 21 °C (70 °F) and the interfacial tension is generally higher for paraffinic crude oil. Heavy oil and bitumen containing appreciable quantities of aromatic constituents, naphthenes, and polar species is at the lower end of the range.

6.2.4 Viscosity

Viscosity is a measure of the internal resistance to motion of a fluid by reason of the forces of cohesion between molecules or molecular groupings. More specifically, the viscosity is the force in dynes that (in the SI system of units) is required to move a plane of 1 cm^2 area at a distance of 1 cm from another plane of 1 cm^2 area through a distance of 1 cm in 1 s. In the centimeter-gram-second (cgs) system, the unit of viscosity is the poise (P) or centipoise (cP = 0.01 P). Two other terms in common use are kinematic viscosity and fluidity. The kinematic viscosity is the viscosity in centipoises divided by the specific gravity, and the unit is the stoke (cm^2/s), although centistokes (0.01 cs) is in more common usage; fluidity is simply the reciprocal of viscosity.

Thus, one of the most important properties of refinery feedstocks (especially the viscous feedstocks) is the viscosity since the high viscosity of heavy oils directly not only impacts the recovery and productivity of the reservoir but also impacts behavior during refinery processes. Although there is no direct relationship between density and viscosity, a reduction in degrees API (°API) is generally accompanied by an increase in viscosity. Heavy oils display viscosities ranging from several hundred centipoises to numbers on the order of tens of millions of centipoises under reservoir (low temperature) conditions (Speight, 2014, 2017). Thus, the transport of viscous feedstocks through pipelines and porous media almost always requires additional energy, and often requires the addition of heat or diluents to guarantee an acceptable flow rate.

In more general terms, the lower the viscosity, the more readily the oil will flow. The viscosity of a feedstock is largely determined by the amount of lower molecular weight constituents and higher molecular weight constituents that are contained in the feedstock. The greater the percentage of lower molecular weight constituents, the lesser the amount of higher molecular weight constituents (i.e., asphaltene constituents) and the lower the viscosity.

As with other physical properties, viscosity is affected by temperature, the feedstock having a higher viscosity at a lower temperature. For most feedstocks (or products), the viscosity varies as the logarithm of the temperature and feedstocks (or products) that flow readily at high temperatures can (as might or should be anticipated) become a slow-moving, viscous mass at a low temperature.

In terms of oil spill cleanup, viscous feedstocks (1) do not spread rapidly, (2) do not disperse readily in water, (3) do not penetrate soil as readily as the less viscous materials, and (4) affect the ability of pumps and skimmers to handle the oil.

Thus, the viscosity of a viscous feedstock crude affects not only the viability of movement of the feedstock in the refinery (as well as pipelining) but is an indicator of the need for dilution of the feedstock. In fact, one of the most important aspects of the viscous feedstocks is the viscosity, since the high viscosity of heavy oils directly impacts the recovery and productivity of the reservoir (Parkash, 2003; Speight, 2014; Hsu and Robinson, 2017; Speight, 2017).

Thus, the measurement of the viscosity of crude oil has been regarded by the crude oil industry to have particular importance, because of the role it plays in the assessment of crude oil properties and, hence, the producibility of a reservoir. In addition, the viscosity of a feedstock must also be considered when determining the amount of diluent that will permit pipeline transportation of the oil.

The changes in viscosity with temperature, pressure, and rate of shear are pertinent not only in lubrication but also for such engineering concepts as heat transfer. The viscosity and relative viscosity of different phases, such as gas, liquid oil, and water, are determining influences in producing the flow of reservoir fluids through porous oil-bearing formations. The rate and amount of oil production from a reservoir are often governed by these properties.

The viscosity (ASTM D88, 2021; ASTM D341, 2021; ASTM D445, 2021; ASTM D1665, 2021; ASTM D2161, 2021; ASTM D2170, 2021; ASTM D2171, 2021; ASTM D2270, 2021; ASTM E102, 2021) of crude oils varies markedly over a very wide range. Values vary from less than 10 centipoises at room temperature to many thousands of centipoises at the same temperature. In the present context, oil sands bitumen occurs at the higher end of this scale, where a relationship between viscosity and density between various crude oils has been noted (Speight, 2014, 2015).

The kinematic viscosity is a measure of the resistive force of a fluid under gravity. Many crude oil products are used as lubricants and the operation of equipment depends on the proper viscosity of the liquid. Thus, the accurate measurement of kinematic viscosity is essential for many product specifications. Also, the kinematic viscosity of many crude oil products is an important parameter for calculating the flow of the liquids through nozzles, orifices, and pipelines.

The time for a volume of liquid to flow through a marked zone in a calibrated glass capillary at a defined temperature is measured, and is directly related to the kinematic viscosity. Absolute kinematic viscosity values are obtained through comparison with standards of known viscosity. The Cannon-Fenske method (ASTM D445, 2021) determines the kinematic viscosity of transparent and opaque crude oil products exhibiting the properties of a Newtonian liquid. The dynamic viscosity can also be calculated if the sample density is known. The kinematic viscosity is reported in terms of centistokes (cs) to nearest zero and the dynamic viscosity is reported in terms of centipoises to nearest 0.01.

As noted, the simplest and most widely used viscometers are capillary types (ASTM D445, 2021), and the viscosity (μ) is derived from the equation:

$$\nu = \mathrm{B}r^4 P/8nl$$

In this equation, r is the tube radius, l the tube length, P the pressure difference between the ends of a capillary, n the coefficient of viscosity, and: the quantity discharged in unit time.

The Saybolt universal viscosity (SUS) (ASTM D88, 2021) is the time in seconds required for the flow of 60 ml of crude oil from a container, at constant temperature, through a calibrated orifice. The Saybolt Furol viscosity (SFS) (ASTM D88, 2021) is determined in a similar manner except that a larger orifice is employed.

As a result of the various methods for viscosity determination, it is not surprising that much effort has been spent on interconversion of the several scales, especially converting Saybolt to kinematic viscosity (ASTM D2161, 2021),

$$\text{Kinematic viscosity} = a \times \text{Saybolt s} + b/\text{Saybolt s}$$

In this equation, a and b are constants.

There are also tables for converting kinematic viscosity to Saybolt Universal viscosity (ASTM D446, 2021) and to Saybolt Furol viscosity (ASTM D2161, 2021). Standard charts are also available (ASTM D341, 2021) for estimating the viscosity of crude oil or a crude oil product at any temperature, provided the viscosity is known at two temperatures.

The Saybolt universal viscosity that is equivalent to a specific kinematic viscosity varies with the temperature at which the determination is made. However, conversion factors are used to convert kinematic viscosity from 2 to 70 cs at 38 °C (100 °F) and 99 °C (210 °F) to equivalent Saybolt universal viscosity in seconds (Speight, 2014, 2015, 2017). Appropriate multipliers are listed to convert kinematic viscosity over 70 cs. For a kinematic viscosity determined at any other temperature, the equivalent Saybolt universal value is calculated by use of the Saybolt equivalent at 38 °C (100 °F) and a multiplier that varies with the temperature. Thus:

$$\text{Saybolt seconds at } 100°\text{F}(38°\text{C}) = \text{cs} \times 4.635$$
$$\text{Saybolt seconds at } 210°\text{F}(99°\text{C}) = \text{cs} \times 4.667$$

The kinematic viscosity of a fluid of viscosity greater than 10 cs is given by the expression:

$$v = Ct$$

In this equation, C is the viscometer constant and t is the time of flow in seconds.

When the fluid has a viscosity of 10 cs, or less, a second coefficient is used to correct for changes in kinetic energy at the exit to the capillary. The expression then becomes:

$$v = Ct - B/t$$

In this equation, B is the coefficient of kinetic energy, which may be determined experimentally, or eliminated by choosing long flow times.

Viscosity decreases with increasing temperature for crude oil or crude oil products there being a decrease in viscosity with an increase in temperature. Compared to temperature, viscosity has only a minor influence on natural samples. Pressure does play a more important role where dissolved gases are involved because of the dependence of gas solubility on pressure. For lubricants and other crude oil fractions, the ASTM viscosity–temperature chart is the standard method of plotting the viscosity–temperature data for interpolation and extrapolation.

The rate of change appears to depend primarily on the nature or composition of the crude oil, but other factors, such as volatility, may also have a minor effect. The effect of temperature on viscosity is generally represented by the equation

$$\log\log(v + c) = A + B\log T$$

In this equation, ν is absolute viscosity, T is temperature, and A and B are constants. This equation has been sufficient for most purposes and has come into very general use. The constants A and B vary widely with different oils, but c remains fixed at 0.6 for all oils having a viscosity over 1.5 cs; it increases only slightly at lower viscosity (0.75 at 0.5 cs). The viscosity–temperature characteristics of any oil, so plotted, thus create a straight line, and the parameters A and B are equivalent to the intercept and slope of the line. To express the viscosity and viscosity–temperature characteristics of an oil, the slope and the viscosity at one temperature must be known; the usual practice is to select 38 °C (100 °F) and 99 °C (210 °F) as the observation temperatures.

As with the density, the viscosity of feedstocks increases on the application of pressure. The pressure coefficient of viscosity correlates with the temperature coefficient, even when oil samples of widely different types are compared. A plot of the logarithm of the kinematic viscosity against pressure for several oils has given reasonably linear results up to about 20,000 psi, and the slopes of the isotherms are such that extrapolated values for a given oil intersect. At higher pressures, the viscosity decreases with increasing temperature, as at atmospheric pressure; in fact, viscosity changes of small magnitude are usually proportional to density changes, whether these are caused by pressure or by temperature. In summary, an increase in total pressure results in an increase in viscosity of heavy oil and bitumen at any given temperature. This increase is, however, small compared to viscosity at atmospheric condition.

6.3 Electrical Properties

Understanding of how a feedstock behaves from physical and chemical property data and the focus on the relationship between the properties of a material and the electrical properties allows uses to be designed and also provides a knowledge base for a variety of chemical and engineering applications.

6.3.1 Conductivity

The electrical conductivity of crude oil and related water-in-oil emulsion play a vital role to design internal electric field and external large transformers in the electrostatic dehydrator/desalter system (Mohammadi et al., 2019; Speight, 2021).

From the fragmentary evidence that is available, the electrical conductivity of feedstock fractions is small but measurable (Penzes and Speight, 1974; Fotland et al., 1993; Fotland and Anfindsen, 1996). For example, the normal hydrocarbon derivatives (from hexane up) have an electrical conductivity smaller than 10^{-16} Ω/cm; benzene itself has an electrical conductivity of 4.4×10^{-17} Ω/cm, and cyclohexane has an electrical conductivity of 7×10^{-18} Ω/cm. It is generally recognized that hydrocarbon derivatives do not usually have an electrical conductivity larger than 10^{-18} Ω/cm. Thus, it is not surprising that the electrical conductivity of hydrocarbon oils is also exceedingly small—on the order of 10^{-19} to 10^{-12} Ω/cm.

Available data indicate that the observed conductivity is frequently more dependent on the method of measurement and the presence of trace impurities than on the chemical type of the oil. Conduction through refinery feedstocks is not ohmic insofar as the current is not proportional to field strength: in some regions, it is observed to increase exponentially with the latter. Time effects are also observed, the current being at first relatively large and decreasing to a smaller steady value because of electrode polarization and partly because of ions removed from the solution. Most oils increase in conductivity with rising temperatures.

6.3.2 Dielectric Constant

The dielectric constant (ε) of a substance may be defined as the ratio of the capacity of a condenser with the material between the condenser plates C to that with the condenser empty and under vacuum C_0:

$$\varepsilon = C/C_0$$

The dielectric constant of a feedstock product may be used to indicate the presence of various constituents, such as asphaltene constituents, resin constituents, or oxidized materials.

The dielectric constant of hydrocarbon derivatives and hence most feedstocks and their products is usually low and decreases with an increase in temperature. It is also noteworthy that for hydrocarbon derivatives, hydrocarbon derivative fractions, and products, the dielectric constant is approximately equal to the square of the refractive index. Polar materials have dielectric constants greater than the square of the refractive index.

6.3.3 Dielectric Strength

The dielectric strength, or breakdown voltage (ASTM D877, 2019), is the greatest potential gradient or potential that an insulator can withstand without permitting an electric discharge. The property is, in the case of oils as well as other dielectric materials, somewhat dependent on the method of measurement, that is, on the length of path through which the breakdown occurs, the composition, shape, and condition of the electrode surfaces, and the duration of the applied potential difference.

The standard test used in North America is applied to oils of crude oil origin for use in cables, transformers, oil circuit breakers, and similar apparatus. Oils of high purity and cleanliness show nearly the same value under standard conditions, generally ranging from 30 to 35 kV. For alkane derivatives, the dielectric strength has been shown to increase linearly with liquid density, and the value for a mineral oil fits the data well. For *n*-heptane, a correlation was found between the dielectric strength and the density changes with temperature. There are many reasons that the dielectric strength of an insulator may fail. The most important appears to be the presence of some type of impurity, produced by corrosion, oxidation, thermal or electrical cracking, or gaseous discharge; invasion by water is a common trouble.

6.3.4 Dielectric Loss and Power Factor

A condenser insulated with an ideal dielectric shows no dissipation of energy when an alternating potential is applied. The charging current, technically termed the *circulating current*, lags exactly 90° in phase angle behind the applied potential, and the energy stored in the condenser during each half-cycle is completely recovered in the next. No real dielectric material exhibits this ideal behavior; that is, some energy is dissipated under alternating stress and appears as heat. Such a lack of efficiency is broadly termed "dielectric loss."

Ordinary conduction comprises one component of dielectric loss. Here the capacitance-held charge is partly lost by short circuit through the medium. Other effects in the presence of an alternating field occur, and a dielectric of zero conductivity may still exhibit losses. Suspended droplets of another phase undergo spheroidal oscillation by electrostatic induction effects and dissipate energy as heat as a consequence of the viscosity of the medium. Polar molecules oscillate as electrets and dissipate energy on collision with others. All such losses are of practical importance when insulation is used in connection with alternating-current equipment.

The measure of the dielectric loss is the power factor. This is defined as the factor k in the relation:

$$k = W/EI$$

W is the power in watts dissipated by a circuit portion under voltage E and passing current, I.

From alternating current theory, the power factor is recognized as the cosine of the phase angle between the voltage and current where a pure sine wave form exists for both; it increases with a use in temperature. When an insulating material serves as the dielectric of a condenser, the power factor is an intrinsic property of the dielectric. For practical electrical equipment, low-power factors for the insulation are of course always desirable; feedstock oils are generally excellent in this respect, having values on the order of 0.0005, comparable with fused quartz and polystyrene resin constituents. The power factor of pure hydrocarbon derivatives is extremely small. Traces of polar impurities, however, cause a striking increase. All electrical oils, therefore, are drastically refined and handled with care to avoid contamination; insoluble oxidation products are particularly undesirable.

6.3.5 Static Electrification

Dielectric liquids, particularly light naphtha, may acquire high static charges on flowing through or being sprayed from metal pipes. The effect seems to be associated with colloidally dispersed contaminants, such as oxidation products, which can be removed by drastic filtration or adsorption. Since a considerable fire hazard is involved, a variety of methods have been studied for minimizing the danger.

For large-scale storage, avoidance of surface agitation and the use of floating metal roofs on tanks are beneficial. High humidity in the surrounding atmosphere is helpful in lowering the static charge, and radioactive materials have been used to try to induce discharge to ground. A variety of additives have been found that increase the conductivity of feedstock liquids, thus lowering the degree of electrification; chromium salts of alkylated salicylic acids and other salts of alkylated sulfo-succinic acids are employed in low concentrations, say 0.005%.

6.4 Optical Properties

The optical properties of a material define how the material interacts with light. The optical properties of matter are studied in optical physics, a subfield of optics. Major application fields based on optical properties include light emission and detection, solar cells, photocatalysis, photoelectronic and imaging, and biosensing. The basic understanding of the fundamental optical properties and related spectroscopic techniques can help distinguish the nanomaterial.

In the context of this section of the text, the term "optical property" refers to the response of a feedstock to exposure to electromagnetic radiation and, in particular, to visible light. Such properties, while not often used in the past are now finding use in environmental cases when crude oil or a viscous feedstock has been spilled into an ecosystem.

The optical properties of crude oil, heavy crude oil, extra heavy crude oil, tar sand bitumen and their respective products are important in determining the perceived quality of emulsion-based products (Park et al., 2013). The overall appearance of an emulsion is determined by the way that it interacts with electromagnetic radiation in the visible region of the spectrum, for example, reflection, transmission, adsorption, and scattering. These interactions are principally determined by the characteristics of emulsion droplets (size, concentration, and refractive index). The measurement of optical properties can be used to characterize the optical properties and, as a result, rheological behavior of water-in-oil emulsions, especially the behavior of macroemulsions.

As an example of the measurement of the optical properties of feedstocks, the spectrophotometer is the instrument of choice. This instrument can be used to measure the optical properties of materials across "spectrum" of light energy wavelengths (250–1,700 nm). Originally the range of colors of visible light, but now applied to the entire range of electromagnetic transmission.

Among their many impacts, oil pollutants modify light fields above and below the water surface. These modifications are manifested by the attenuation of the light passing through an oiled water surface by changes in light absorption in the seawater column due to the formation of an emulsion, and by the scattering of light by particles of such an emulsion. The optical properties of an emulsion can be described by the attenuation specific cross-section and the absorption specific cross-section, which depend on the optical characteristics of the oil and the seawater, on the size distribution of oil droplets, and on their concentration.

Weathering is a process causing changes in crude oil, which is reflected in the modification of its optical properties. The increase in the refractive index and the light absorption coefficient of crude oil during its presence in the natural environment is the reason for the increase in light scattering, especially in the case of large angles. Depolarization of linearly polarized light is greater for fresh crude oil in the case of most scattering angles. The weathering of petroleum has only weak polarization effects: an emulsion of weathered oil absorbs more light and attenuates the light in the water column to a greater extent than an emulsion of fresh oil. The considerable variability in these modifications at the onset of the weathering process indicates the appropriateness of taking the optical parameters characterizing aged petroleum into consideration.

Crude oil and the viscous feedstocks contain diverse hydrocarbons and organic molecules that could contribute to UV absorbance and fluorescence signatures in seawater and could be readily determined and characterized by UV–visible spectroscopy and fluorescence spectroscopy techniques (Zhou and Guo, 2012). Similarly, the composition and sources of dissolved organic matter in aquatic environments can be effectively characterized by its optical properties, including UV–visible absorbance and fluorescence spectra. Indeed, fluorescence techniques have been used in many previous studies to characterize, fingerprint, and monitor oil in coastal and marine environments.

In fact, the transmittance and the optical properties play an important role in detection of crude oil in the environment. The transmittance spectra of crude oil can be measured by a Fourier transform infrared spectrometer. Thus, the optical properties of crude oil can be calculated by a double thickness method, and the effects of different temperature on transmittance spectra and the optical properties (the refractive index, the absorption index, the absorption coefficient α, and the reflectivity) can be analyzed. Also, temperature has a significant influence on the transmission spectrum and the optical properties of crude oil (Zhou et al., 2019).

In the case of pollution with fresh crude, the oil undergoes weathering, which alters its properties and, hence, changes in the optical properties of a feedstock-seawater emulsion due to weathering and oxidation. The properties depend on the spillage concentration, the optical characteristics of the seawater and oil in question, and on the size distribution of the oil droplets. They may be described by the attenuation specific cross-sections and absorption specific cross-sections. Specific cross-sections and other optical parameters for droplets of a crude oil are available (Król et al., 2006; Rudz et al., 2013).

6.4.1 Optical Activity

The occurrence of optical activity in crude oil is universal and is a general phenomenon not restricted to a particular type of feedstock, such as the paraffinic or naphthenic feedstocks. Crude oil is usually *dextrorotatory*, that is, the plane of polarized light is rotated to the right, but there are known *laevorotatory* crude oils, that is, the plane of polarized light is rotated to the left, and some crude oils have been reported to be optically inactive.

Examination of the individual fractions of optically active crude oils shows that the rotatory power increases with molecular weight (or boiling point) to pronounced maxima and then decreases again. The rotatory power appears to be concentrated in certain fractions, the maximum lying at a molecular weight of approximately 350–400; this maximum is approximately the same for all crude oils. The occurrence of optically active compounds in unaltered natural crude oil has been a strong argument in favor of a rather low temperature origin of crude oil from organic raw materials.

A magnetic field causes all liquids to exhibit optical rotation, usually in the same direction as that of the magnetizing current; this phenomenon is known as the Faraday effect (θ) and it may be expressed by the relation:

$$\theta = pth$$

θ is the total angle of rotation, t is the thickness of substance through which the light passes, and h is the magnetic field; the constant p is an intrinsic property of the substance, usually termed the Verdet constant (minutes of arc/cm per G); there have been some attempts to use the Verdet constant in studying the constitution of hydrocarbon derivatives by physical property correlation.

6.4.2 Refractive Index

The refractive index is the ratio of the velocity of light in a vacuum to the velocity of light in the substance. The measurement of the refractive index is very simple (ASTM D1218, 2021), requires small quantities of material, and, consequently, has found wide use in the characterization of hydrocarbon derivatives and feedstock samples.

For closely separated fractions of similar molecular weight, the values increase in the order paraffin, naphthene, and aromatic. For polycyclic naphthenes and polycyclic aromatics, the refractive index is usually higher than that of the corresponding monocyclic compounds. For a series of hydrocarbon derivatives of essentially the same type, the refractive index increases with molecular weight, especially in the paraffin series. Thus, the refractive index can be used to provide valuable information about the composition of hydrocarbon (feedstock) mixtures; as with density, low values indicate paraffinic materials and higher values indicate the presence of aromatic compounds. However, the combination of refractive index and density may be used to provide even more definite information about the nature of a hydrocarbon mixture and, hence, the use of the refractivity intercept ($n - \mathrm{d}/2$).

The refractive and specific dispersion as well as the molecular and specific refraction have all been advocated for use in the characterization of a refinery feedstock.

The refractive dispersion of a substance is defined as the difference between its refractive indices at two specified wavelengths of light. Two lines commonly used to calculate dispersions are the C (6563 D, red) and F (4861 D, blue) lines of the hydrogen spectrum. The *specific dispersion* is the refractive dispersion divided by the density at the same temperature:

$$\text{Specific dispersion} = n_{\mathrm{F}} - n_{\mathrm{C}} / d$$

This equation is of particular significance in feedstock chemistry because all the saturated hydrocarbon derivatives, naphthene and paraffin, have nearly the same value irrespective of molecular weight, whereas aromatics are much higher and unsaturated aliphatic hydrocarbon derivatives are intermediate.

Specific refraction is the term applied to the quantity defined by the expression:

$$n - 1 / (n^2 + 2)d = C$$

where n is the refractive index, d is the density, and C is a constant independent of temperature.

Molecular refraction is the specific refraction multiplied by molecular weight; its particular usefulness lies in the fact that it is very nearly additive for the components of a molecule; that is, numerical values can be assigned to atoms and structural features, such as double bonds and rings. The value for any pure compound is then approximately the sum of such component constants for the molecule.

References

Altgelt, K.H., and Boduszynski, M.M. 1994. *Composition and Analysis of Heavy Petroleum Fractions*. Marcel Dekker Inc., New York.

ASTM 2021. *Annual Book of Standards*. ASTM International, West Conshohocken, Pennsylvania.

ASTM D70. 2021. Standard Test Method for Density of Semi-Solid Asphalt Binder (Pycnometer Method). Annual Book of Standards. ASTM International, West Conshohocken, Pennsylvania.

ASTM D71. 2021. Standard Test Method for Relative Density of Solid Pitch and Asphalt (Displacement Method). Annual Book of Standards. ASTM International, West Conshohocken, Pennsylvania.

ASTM D88. 2021. Standard Test Method for Saybolt Viscosity. Annual Book of Standards. ASTM International, West Conshohocken, Pennsylvania.

ASTM D96. 2021. Standard Test Method for Water and Sediment in Crude Oil by Centrifuge Method (Field Procedure). Annual Book of Standards. ASTM International, West Conshohocken, Pennsylvania.

ASTM D139. 2021. Standard Test Method for Float Test for Bituminous Materials. Annual Book of Standards. ASTM International, West Conshohocken, Pennsylvania.

ASTM D287. 2021. Standard Test Method for API Gravity of Crude Petroleum and Petroleum Products (Hydrometer Method). Annual Book of Standards. ASTM International, West Conshohocken, Pennsylvania.

ASTM D341. 2021. Standard Practice for Viscosity-Temperature Equations and Charts for Liquid Petroleum or Hydrocarbon Products. Annual Book of Standards. ASTM International, West Conshohocken, Pennsylvania.

ASTM D445. 2021. Standard Test Method for Kinematic Viscosity of Transparent and Opaque Liquids (and Calculation of Dynamic Viscosity). Annual Book of Standards. ASTM International, West Conshohocken, Pennsylvania.

ASTM D446. 2021. Standard Specifications and Operating Instructions for Glass Capillary Kinematic Viscometers. Annual Book of Standards. ASTM International, West Conshohocken, Pennsylvania.

ASTM D877. 2019. Standard Test Method for Dielectric Breakdown Voltage of Insulating Liquids Using Disk Electrodes. Annual Book of Standards. ASTM International, West Conshohocken, Pennsylvania.

ASTM D971. 2021. Standard Test Method for Interfacial Tension of Insulating Liquids against Water by the Ring Method. Annual Book of Standards. ASTM International, West Conshohocken, Pennsylvania.

ASTM D1217. 2021. Standard Test Method for Density and Relative Density (Specific Gravity) of Liquids by Bingham Pycnometer. Annual Book of Standards. ASTM International, West Conshohocken, Pennsylvania.

ASTM D1218. 2021. Standard Test Method for Refractive Index and Refractive Dispersion of Hydrocarbon Liquids. Annual Book of Standards. ASTM International, West Conshohocken, Pennsylvania.

ASTM D1250. 2021. Standard Guide for Use of the Petroleum Measurement Tables. Annual Book of Standards. ASTM International, West Conshohocken, Pennsylvania.

ASTM D1298. 2021. Standard Test Method for Density, Relative Density, or API Gravity of Crude Petroleum and Liquid Petroleum Products by Hydrometer Method. Annual Book of Standards. ASTM International, West Conshohocken, Pennsylvania.

ASTM D1480. 2021. Standard Test Method for Density and Relative Density (Specific Gravity) of Viscous Materials by Bingham Pycnometer. Annual Book of Standards. ASTM International, West Conshohocken, Pennsylvania.

ASTM D1481. 2021. Standard Test Method for Density and Relative Density (Specific Gravity) of Viscous Materials by Lipkin Bicapillary Pycnometer. Annual Book of Standards. ASTM International, West Conshohocken, Pennsylvania.

ASTM D1555. 2021. Standard Test Method for Calculation of Volume and Weight of Industrial Aromatic Hydrocarbons and Cyclohexane. Annual Book of Standards. ASTM International, West Conshohocken, Pennsylvania.

ASTM D1657. 2021. Standard Test Method for Density or Relative Density of Light Hydrocarbons by Pressure Hydrometer. Annual Book of Standards. ASTM International, West Conshohocken, Pennsylvania.

ASTM D1665. 2021. Standard Test Method for Engler Specific Viscosity of Tar Products. Annual Book of Standards. ASTM International, West Conshohocken, Pennsylvania.

ASTM D1796. 2021. Standard Test Method for Water and Sediment in Fuel Oils by the Centrifuge Method (Laboratory Procedure). Annual Book of Standards. ASTM International, West Conshohocken, Pennsylvania.

ASTM D2161. 2021. Standard Practice for Conversion of Kinematic Viscosity to Saybolt Universal Viscosity or to Saybolt Furol Viscosity. Annual Book of Standards. ASTM International, West Conshohocken, Pennsylvania.

ASTM D2170. 2021. Standard Test Method for Kinematic Viscosity of Asphalts (Bitumens). Annual Book of Standards. ASTM International, West Conshohocken, Pennsylvania.

ASTM D2171. 2021. Standard Test Method for Viscosity of Asphalts by Vacuum Capillary Viscometer. Annual Book of Standards. ASTM International, West Conshohocken, Pennsylvania.

ASTM D2270. 2021. Standard Practice for Calculating Viscosity Index from Kinematic Viscosity at 40 °C and 100 °C. Annual Book of Standards. ASTM International, West Conshohocken, Pennsylvania.

ASTM D4007. 2021. Standard Test Method for Water and Sediment in Crude Oil by the Centrifuge Method (Laboratory Procedure). Annual Book of Standards. ASTM International, West Conshohocken, Pennsylvania.

ASTM D4052. 2021. Standard Test Method for Density, Relative Density, and API Gravity of Liquids by Digital Density Meter. Annual Book of Standards. ASTM International, West Conshohocken, Pennsylvania.

ASTM E102. 2021. Standard Test Method for Saybolt Furol Viscosity of Bituminous Materials at High Temperatures. Annual Book of Standards. ASTM International, West Conshohocken, Pennsylvania.

Fotland, P., and Anfindsen, H. 1996. Electrical Conductivity of Asphaltenes in Organic Solvents. *Fuel Science and Technology International*, 14: 101–115.

Fotland, P., Anfindsen, H., and Fadnes, F.H. 1993. Detection of Asphaltene Precipitation and Amounts Precipitated by Measurement of Electrical Conductivity. *Fluid Phase Equilibria*, 82: 157–164.

Hsu, C.S., and Robinson, P.R. (Editors). 2017. *Handbook of Petroleum Technology*. Springer International Publishing AG, Cham, Switzerland.

Huc, A.Y. 2011. *Heavy Crude Oils: From Geology to Upgrading: An Overview*. Editions Technip, Paris, France.

Jokuty, P., Whiticar, S., McRoberts, K., and Mullin, J. 1996. Oil Adhesion Testing—Recent Results. Proceedings. 19th Arctic and Marine Oil Spill Program Technical Seminar, Environment Canada, Ottawa, Ontario, Canada. pp. 9-27.

Król, T., Stelmaszewski, A., and Włodzimierz, F. 2006. Variability in the Optical Properties of a Crude Oil-Seawater Emulsion. *Oceanologia*, 48: 203–211. http://www.iopan.gda.pl/oceanologia/48Skrol.pdf

Mohammadi, F., Mohammadi, M., and Behrouz, M. 2019. A Comprehensive Electrical Model for the Electrostatic Desalting Process of Crude Oil. *Petroleum and Coal*, 61(4): 738–748. https://www.researchgate.net/publication/334192732_Article_Open_Access_A_COMPREHENSIVE_ELECTRICAL_MODEL_FOR_THE_ELECTROSTATIC_DESALTING_PROCESS_OF_CRUDE_OIL

Park, B.-G., Park, I.-J., Han, J.-S., Lee, S.-M., Lee, C.-G., and Ha, C.-S. 2013. Characterization of Optical Properties in Water-in-Oil Emulsion. *Journal of Dispersion Science and Technology*, 34(4): 560–565. https://www.tandfonline.com/doi/abs/10.1080/01932691.2012.680833; https://doi.org/10.1080/01932691.2012.680833

Parkash, S. 2003. *Refining Processes Handbook*. Gulf Professional Publishing, Elsevier, Amsterdam, Netherlands.

Penzes, S., and Speight, J.G. 1974. Electrical Conductivities of Bitumen Fractions in Non-Aqueous Solvents. *Fuel*, 53: 192.

Pfeiffer, J.P. 1950. *The Properties of Asphaltic Bitumen*. Elsevier Publishing Co., New York. p. 80.

Rudz, K., Darecki, M., and Toczek, H. 2013. Modelling the Influence of Oil Content on Optical Properties of Seawater in the Baltic Sea. *Journal of the European Optical Society—Rapid Publications*, 8: 13063. http://www.jeos.org/index.php/jeos_rp/article/view/13063

Speight, J.G. 2005. *Environmental Analysis and Technology for the Refining Industry*. John Wiley & Sons Inc., Hoboken, New Jersey.

Speight, J.G. 2013. *The Chemistry and Technology of Coal*. 3rd Edition. Marcel Dekker Inc., New York.

Speight, J.G. 2014. *The Chemistry and Technology of Petroleum*. 5th Edition. CRC Press, Taylor & Francis Group, Boca Raton, Florida.

Speight, J.G. 2015. *Handbook of Petroleum Product Analysis*. 2nd Edition. John Wiley & Sons Inc., Hoboken, New Jersey.

Speight, J.G. 2017. *Handbook of Petroleum Refining*. CRC Press, Taylor & Francis Group, Boca Raton, Florida.

Speight, J.G. 2021. *Petroleum Refining Technology Series Volume 1: Refinery Feedstocks*. CRC Press, Taylor & Francis Group, Boca Raton, Florida.

Speight, J.G., and Arjoon, K.K. 2012. *Bioremediation of Petroleum and Petroleum Products*. Scrivener Publishing, Salem, Massachusetts.

Zhou, Z., and Guo, L. 2012. Evolution of the Optical Properties of Seawater Influenced by the Deepwater Horizon Oil Spill in the Gulf of Mexico. *Environmental Research Letters*, 7: 025301. https://iopscience.iop.org/article/10.1088/1748-9326/7/2/025301

Zhou, W., Zhang, Y., Qi, H., Wang, Q., Yao, B., and Li, D. 2019. Optical Properties of Crude Oil with Different Temperatures. *Optik*, 196: 162946. https://doi.org/10.1016/j.ijleo.2019.162946; https://www.sciencedirect.com/science/article/pii/S003040261930823X

7

Thermal Properties

7.1 Introduction

The viscous feedstocks are very complex and structured mixtures that consist predominantly of non-hydrocarbon derivatives (often referred to as hydrocarbonaceous derivatives that contain one or more atoms other than carbon and hydrogen) with very few pure hydrocarbon derivatives that contain constituents having hydrocarbon skeletons as the molecular backbone (Chapter 2). The hydrocarbonaceous derivatives in the viscous feedstocks have a predominantly hydrocarbon skeleton but also contain varying amounts of heteroatoms (such as nitrogen, oxygen, sulfur, and metals) (Parkash, 2003; Speight, 2014a; Hsu and Robinson, 2017; Speight, 2017).

The physical and chemical characteristics of the viscous feedstocks and the yields and properties of products or factions prepared from them vary considerably and are dependent on the concentration of the various types of hydrocarbons and minor constituents present. Thus, evaluation of the viscous feedstocks for use in the refinery involves an examination of several of the physical properties of each material. It is no longer sufficient to use one property only for evaluation purposes as has been the practice in the past and, indeed, the slate of properties required to produce evaluation data may vary from the more convenient physical inspections (Parkash, 2003; Speight, 2014a; Hsu and Robinson, 2017; Speight, 2017) to the more complex physical properties, of which some are described in this chapter. By this means, a set of basic characteristics can be obtained that can be correlated with the potential of the feedstock for producing specific products.

This chapter presents descriptions for the standard (ASTM) test methods that will produce data relating to the various thermal properties of heavy crude oil, extra heavy crude oil, tar sand bitumen, and residua. It is these properties (listed alphabetically below rather than in any order of preference) that provide valuable information relating to the movement of feedstocks around a refinery as well as indications of the yields of various products.

Handbook of Heavy Oil Properties and Analysis, First Edition. James G. Speight.

7.2 Ash Production

Contrary to many general (and incorrect) comments, the analysis for crude oil for "ash" is a poor choice of words. The correct term would be "ash production" which refers to the combustion procedure which procures ash (metal oxides) from the metal-containing constituents. More specifically, the ash produced is defined as the inorganic residue that remains after combustion of the oil in air at specific high temperature.

Typically, the ash produced from crude oil and crude oil products is low and is on the order of (or less than) 0.5–0.7% by weight, although in certain crude oils a higher yield of ash may be produced. A feedstock that produces more than 0.05% w/w ash in the combustion test is considered (arbitrarily) to be a high ash feedstock while a feedstock containing less than 0.05% w/w ash in the combustion test is considered (again, arbitrarily) to be a low ash feedstock. The ash yields from the viscous feedstocks are typically much higher and, hence, determination of the ash yield is a necessary part of the evaluation of the viscous feedstocks.

The ash is produced by the combustion of the feedstock and consists of extraneous solids, residues of organometallic compounds in solution, and salts dissolved in water present in the fuel. The ash yield, which is related to the mineral matter content, is one of the properties used to evaluate coke and indicates the amount of undesirable residue present. Some sample of coke may be declared to have an acceptable ash content but this varies with the intended use of the coke. Moreover, the ash produced from any feedstock should not be ignored as it is a valuable source material for the identification of trace metals in the feedstock.

The ash yield of a feedstock is a measure of the amount of inorganic noncombustible material that is present in the feedstock which occurs either (1) naturally in the feedstock or (2) is present as a result of refining or contamination during storage or transportation. For instance, the ash could be due to the formation of oxides (formed during the combustion process) of the following metals (alphabetically and not by preference or occurrence): calcium, iron, lead, magnesium, nickel, vanadium, sodium, and zinc. A content of metals that is in excess of 200 ppm is considered to be an adverse property but the variations are very large. Also, the higher the mineral matter content (i.e., the higher the ash yield) the higher is the tendency of the feedstock to form sludge or sediment.

For the test method, the preparation and sampling of the analytical sample must neither remove nor add mineral matter (ASTM D346, 2021). Improper dividing, sieving, and crushing equipment, and some muffle furnace lining material can contaminate the coke and lead to erroneous results. In addition, a high sulfur content of the furnace gases, regardless of the source of the sulfur, can react with an alkaline ash to produce erratic results. To counteract such an effect, the furnace should be swept with air.

In the test method, a sample of petroleum coke is dried, ground, and heated in a muffle furnace at 700–775 °C (1,292–1,427 °F). The non-carbonaceous residue is weighed and reported as the percent by weight ash. As already noted, the ash must not be understood to be the same as the mineral content of the petroleum coke.

In addition, ashing procedures can be used as a preliminary step for determination of the trace elements in coke, and by inference in the higher boiling fractions of the crude oil. Among the techniques used for trace element determinations are flameless and flame atomic absorption (AA) spectrophotometry (ASTM D5863, 2021) and inductively coupled argon plasma (ICP) spectrophotometry (ASTM D5708, 2021).

The inductively coupled argon plasma emission spectrophotometry method (ASTM D5708, 2021) has an advantage over the AA spectrophotometry method (ASTM D5863, 2021) because it can provide more complete elemental composition data than the AA method. Flame emission spectroscopy is often used successfully in conjunction with AA spectrophotometry. X-ray fluorescence spectrophotometry is also sometimes used but matrix effects can be a problem. The method to be used for the determination of metallic constituents is often a matter of individual preference.

In the ashing procedure, the sample is ignited with air in order to completely remove any organic material which leaves behind inorganic ash. The procedure involved the use of an elevated temperature, with subsequent dissolution of inorganic ash in mineral acids. Typically, the system consists of an open inert vessel (pure quartz beaker or platinum crucible) containing the sample that undergoes thermal decomposition at a temperature on the order of 450–600 °C (840–1,110 °F).

More specifically, the two test methods that are presented (ASTM D5863, 2021)—which proposes ashing of 1–20 g of sample using sulfuric acid as an ashing agent (0.5 ml of acid per each gram of oil)—and cover the determination of nickel, vanadium, iron, and sodium in crude oils and residual fuels by flame atomic absorption spectrometry (AAS). A similar method (ASTM D5708, 2021) is also recommended for the determination of nickel, vanadium, and iron in residual fuels, determined by flame atomic absorption spectroscopy (FAAS) which is a very sensitive technique that can be used to detect the concentration of different metals of low concentration up to 1 ppm.

Another option is the dry ashing method (Gab-Allah et al., 2020) which offers the possibility of reconstitution of the obtained ash residue in small volume of dilute mineral acid for analyte enrichment. Also, sulfated (wet) ashing is an alternative mean that involves treatment of the sample with a moderate amount of sulfuric acid, as ashing aid, before charring. In such cases, the losses by volatilization are avoided because metal sulfate derivatives are relatively nonvolatile at the high furnace temperatures and, as a result, are retained in the sulfated ash.

7.3 Carbon Residue

The carbon residue of a refinery feedstock (or a refinery product such as a gas oil fraction, an atmospheric residuum, or a vacuum residuum) gives an indication of the propensity for that product to form a carbonaceous residue (a coke-like product) under thermal conditions. The carbonaceous residue is correctly referred to as the carbon residue but is also often referred to as coke or thermal coke.

Thus, the term "carbon residue" as used here denotes the carbonaceous residue formed after evaporation of any volatile constituents and pyrolysis of the feedstock. The residue is not entirely composed of carbon, but is a coke that can be changed further by heating. The carbon residue of a viscous feedstock is a measure of the propensity of the feedstock to form coke when heated under specific (as defined by the test method) conditions that require a limited supply of oxygen (ASTM D3172, 2021; ASTM D3173, 2021; ASTM D3174, 2021; ASTM D3175, 2021).

There are two alternate (and older) methods for determining the carbon residue of a crude oil, a crude oil product, or a viscous feedstock which are (1) the Conradson carbon method (ASTM D189, 2021) and (2) the Ramsbottom carbon method (ASTM D524, 2021). Both methods are readily applicable to the relatively nonvolatile portion of crude oil and the viscous feedstocks which partially decompose when distilled at a pressure of 1 atm. However, feedstocks that contain ash-forming constituents have erroneously high carbon residues by either method unless the ash is first removed from the oil; the degree of error is proportional to the amount of ash.

The Conradson carbon method is the more usual test for determining the amount of carbon residue left after evaporation and pyrolysis (destructive distillation) of residua, heavy oil, and bitumen and is intended to provide some indication of relative coke-forming propensity. The amount of Conradson Carbon Residue (CCR) is also corrected for the presence of mineral matter in the sample.

In the method (ASTM D189, 2021), a weighed quantity of sample is placed in a crucible and subjected to destructive distillation. The equipment consists of a large porcelain crucible, placed inside two iron crucibles and heated by a gas burner until all petroleum vapors are driven off. A known weight of the feedstock (e.g., 10 g) is placed inside the porcelain crucible and the apparatus is assembled and heated at such a rate that the vapors start to burn within about 10 min. The flame is then adjusted so that they burn at the top of the chimney for a further 21–23 min. The crucibles are then heated to a cherry-red heat for a further 7–8 min giving a total time of approximately 30 min. The residue undergoes a cracking and coking reaction during severe heating for a fixed period of time. The crucible containing the carbonaceous residue is cooled in a desiccator and weighed and the residue is subsequently reduced to an ash by heating in a muffle furnace at 750 °C (1,380 °F) after which it is allowed to cool and weighed. The amount of residue is corrected for the ash content and reported as CCR (% w/w CCR).

In the Ramsbottom carbon method (ASTM D524, 2021), a known weight of the petroleum sample (proportional to the anticipated weight of residue) is heated in a hemispherical glass bulb of standard dimensions by a bath of molten solder at 550 °C (1,020 °F) for 20 min. The carbon residue is weighed and recorded as % wt., Ramsbottom carbon.

The mechanical design and operating conditions of such equipment have such a profound influence on carbon deposition during service that comparison of carbon residues between oils should be considered as giving only a rough approximation of relative deposit-forming tendencies. Recent work has focused on the carbon residue of the different fractions of crude oils, especially the asphaltenes (Speight, 2014a, 2015a). These data can provide more precise information about the anticipated behavior of a variety of feedstocks in thermal processes. Although there is no exact correlation between the two methods, it is possible to interconnect the data (ASTM D189, 2021; ASTM D524, 2021), but caution is advised when using that portion of the curve below 0.1% w/w CCR.

Another method (ASTM D4530, 2021, often referred to as the "microcarbon method") that is appropriate to those situations where only small amounts of the sample are available. The method only requires small amounts of sample and was originally developed as a thermogravimetric method.

The agreement between the data from the three methods is acceptable and within the specified limits making it possible to interrelate all of the data from these carbon residue tests (Long and Speight, 1989; Speight, 2014a, 2015a).

Like the CCR test and the Ramsbottom carbon residue test, the Microcarbon residue method also provides a measure of the relative coke-forming propensity of the sample under degradation conditions that simulate commercial coking operations. Ash-forming constituents and/or nonvolatile constituents present in the sample will add to the residue value determined. In the method, a weighed quantity of sample is placed into a glass vial and heated to 500 °C (930 °F) under a nitrogen atmosphere for a specific time during which the sample undergoes cracking and coking. Nitrogen is used to remove the volatile materials. The carbonaceous residue is reported as % microcarbon residue of the original sample.

There is a relationship between the carbon residue and the atomic H/C ratio and this can be refined to a fairly universal linear correlation between the carbon residue (Conradson) and the H/C ratio:

$$H/C = 171 - 0.0115CR\ (\text{Conradson})$$

This equation holds within two limits; at H/C values = 1.71, where the carbon residue is zero (no coke formation) and H/C = 0.5, where the carbon residue is 100 (all the material converts to coke under test conditions). There is a relationship between the carbon residue (Conradson) and the nitrogen content.

The carbon residue (ASTM D189, 2021; ASTM D524, 2021; ASTM D4530, 2021) is a property that can be correlated with several other properties of crude oil and the viscous feedstocks (Speight, 2014a, 2017) and determination of the carbon residue of a feedstock is an good indication of the coke-forming propensity or (by inference) the distillate-producing propensity of the feedstock.

The fact that these correlations exist can best be explained in terms of the distribution of chemical species in these materials. In each, there are some molecules with a higher proportion of aromatic carbon and more heteroatoms that, because of these characteristics, have low H/C ratios and a high propensity for producing coke. The proportion of these coke precursors, of course, is higher in the higher-boiling fractions than in the lower-boiling fractions.

The application of the CCR determination to heavy feedstocks, such as Athabasca bitumen, requires that a correction be made for inorganic matter by determining the ash content of the carbon residue. In addition, the high CCR of such feedstocks allows use of a smaller sample size together with correspondingly shorter burning and flame times.

7.4 Critical Properties

A study of the pressure, volume, and temperature relationships of a pure component reveals a particular unique state where the properties of a liquid and vapor become indistinguishable from each other (Section 7.10). At that state, the latent heat of vaporization becomes zero and no volume change occurs when the liquid is vaporized. This state is called the critical state and the appropriate parameters of state are termed the critical pressure (P_C), critical volume (V_C), and critical temperature (T_C). It is an important characteristic of the critical state for a pure component that with values of P or T greater than either P_C or T_C, the vapor and liquid states cannot coexist at equilibrium, and thus P_C and T_C represent the maximum values of P and T at which phase separation can occur.

Since the critical state of a component is unique, it is perhaps not surprising that knowledge of P_C, T_C, and V_C allows many predictions to be made concerning the physical properties of substances. These predictions are based on the Law of Corresponding States that states that substances behave in the same way when they are in the same state with reference to the critical state. The particular corresponding state is characterized by its reduced properties, thus:

$$T_r = T/T_C, P_r = P/P_C, V_r = V/V_C$$

The use of this concept permits generalized plots in terms of reduced properties to be drawn that are then applicable to all substances (which obey the law) and can be of great value in determining thermodynamic relationships. It is rare in petroleum engineering to have to deal with pure substances, and unfortunately the application of the Law of Corresponding States to mixtures is complicated by the fact that use of the true critical point for a mixture does not yield correct values of reduced properties for accurate prediction from generalized charts. For a mixture, the critical state no longer represents the maximum temperature and pressure at which a liquid and vapor phase can coexist, and phase separation can occur under retrograde conditions.

For engineering purposes, this difficulty is resolved by the use of pseudo-critical conditions, which are based on the molal average critical temperatures and pressures of the compounds of the mixture. Although use of pseudo-reduced conditions for mixtures of hydrocarbons is generally satisfactory, this is not true for states near the true critical, nor, in general, for mixtures of vapor and liquid.

The temperature, pressure, and volume at the critical state are of considerable interest in petroleum physics, particularly in connection with modern high-pressure, high-temperature refinery operations and in correlating pressure–temperature–volume relationships for other states. Critical data are known for most of the lower molecular weight pure hydrocarbons, and standard methods are generally used for such determinations.

The critical point of a pure compound is the equilibrium state in which its gaseous and liquid phases are indistinguishable and coexistent; they have the same intensive properties but site-specific (localized) variations in the phase properties may be determined experimentally. The definition of the critical point of a mixture is the same although a mixture may have a maximum temperature or maximum pressure at other than the true critical point—the word "maximum" as used here denotes the greatest value at which two phases can coexist in equilibrium.

Thus, when a pure compound is heated at atmospheric pressure, it eventually reaches its boiling point and is completely vaporized at a constant temperature unless the pressure is increased. If the pressure is increased, the compound is completely condensed and cannot be vaporized again unless the temperature is also increased. This mechanism, alternately increasing the pressure and temperature, functions until at some high temperature and pressure and it is found that the material cannot be condensed regardless of the amount of pressure applied. This point is the critical point and the temperature and pressure at the critical point are, respectively, referred to as the "critical temperature" and the "critical pressure."

The liquid phase and vapor phase merge at the critical point so that one phase cannot be distinguished from the other. No volume change occurs when a liquid is vaporized at the critical point, and no heat is required for vaporization but the coefficient of expansion has become large.

Limited information concerning the behavior of complex mixtures has required that the pseudo-critical temperature and pseudo-critical pressure be used for many petroleum fractions and products. The pseudo-critical point is defined as the molal average critical temperature and pressure of the several constituents that make up a mixture. It may be used as the critical point of a mixture in computing reduced temperatures and pressures. However, in computing the pressure–volume–temperature (PVT) relations of mixtures by use of the pseudo-critical point, it must he recognized that the values are not accurate in the region of the critical point and that it cannot be applied to mixtures of gas and liquid.

In the correlation of many properties, reduced properties are useful. Reduced properties are defined as the ratio of the actual value of the property to its critical value. Thus, for volume, temperature, or pressure, the relationships are:

$$\text{Reduced volume, } V_R = V/V_c$$

V is the volume at specified conditions and V_c is the volume at the critical point. Similarly,

$$\text{Reduced temperature, } T_R = T/T_c$$

and

$$\text{Reduced volume, } P_R = P/P_c$$

T and P are the temperature and volume, respectively, at specified conditions and T_c and V_c are the temperature and volume, respectively, at the critical point.

7.5 Enthalpy

The enthalpy (the heat content) of a compound is a thermodynamic quantity equivalent to the total heat content of a system and is equal to the internal energy of the system plus the product of pressure and volume. It is an important property since a change in enthalpy is associated with a particular chemical process.

Thus, the enthalpy is the heat energy necessary to bring a system from a reference state to a given state and is a measure of the amount of heat in a quantity of material relative to a suitable base temperature. Enthalpy is typically measured at constant pressure and may thus be referred to as enthalpy. It is the sum of the total latent heat changes and the product of specific heats and temperatures for the system. The unit used is the same as that for latent heat and the uses are in the preparation of heat balances.

Enthalpy is a function only of the end states and is the integral of the specific heats with respect to temperature between the limit states, plus any latent heats of transition that occur within the interval. The usual reference temperature is

0 °C (32 °F). Enthalpy data can be derived from the specific heat data by graphic integration or, if the empirical equation given for specific heat is sufficiently accurate, from the following equation:

$$H = 1/d\,(0.388 + 0.000225\,t^2 - 12.65)$$

Generally, only differences in enthalpy are required in engineering design, that is, the quantity of heat necessary to heat (or cool) a unit amount of material from one temperature to another. Such calculations are very simple since the quantities are arithmetically additive, and the enthalpy for such a change of state is merely the difference between the enthalpies of the end states.

7.6 Heat of Combustion

The heat of combustion (ASTM D240, 2021) is a direct measure of fuel energy content and is determined as the quantity of heat liberated by the combustion of a quantity of fuel with oxygen in a standard bomb calorimeter.

Chemically, it is the energy (heat) released when an organic compound is burned to produce water (H_2O_{liquid}), carbon dioxide (CO_{2gas}), sulfuric acid ($H_2SO_{41iquid}$), and nitric acid ($HNO_{31iquid}$). The value can be calculated using a theoretical equation based upon the elemental composition of the feedstock:

$$H_g / 4.187 = 8{,}400\text{C} + 2{,}7765\text{H} + 1{,}500\text{N} + 2{,}500\text{S} - 2{,}650\text{O}$$

H_g is given in kilo joules per kilogram (1.0 kJ/kg = 0.43 Btu/lb), C, H, N, S, and O are the normalized weight fractions for these elements in the sample.

The gross heats of combustion of crude oil and its products are also given with reasonable accuracy by the equation:

$$Q = 12{,}400 - 2{,}100d^2$$

In this equation, d is the 60/60 °F specific gravity. Deviation is generally less than 1% although many highly aromatic crude oils show considerably higher values; the range for crude oil is 10,000–11,600 cal/g (Speight, 2014a). For gasoline, the heat of combustion is 11,000–11,500 cal/g (1.0 cal/g = 1.80 Btu/lb) and for kerosene (and diesel fuel) it falls in the range 10,500–11,200 cal/g. Finally, the heat of combustion for fuel oil is on the order of 9,500–11,200 cal/g. Heavy oil and bitumen have a heat of combustion 9,800–10,200 cal/g. Heats of combustion of petroleum gases may be calculated from the analysis and data for the pure compounds. Experimental values for gaseous fuels may be obtained by measurement in a water flow calorimeter, and heats of combustion of liquids are usually measured in a bomb calorimeter.

For thermodynamic calculation of equilibria useful in hydrocarbon research, combustion data of extreme accuracy are required because the heats of formation of water and carbon dioxide are large in comparison with those in the hydrocarbons. Great accuracy is also required of the specific heat data for the calculation of free energy or entropy. Much care must be exercised in selecting values from the literature for these purposes, since many of those available were determined before the development of modern calorimetric techniques.

An alternative criterion of energy content is the aniline gravity product (AGP) that is in reasonable agreement with the calorific value. It is the product of the API gravity and the aniline point (ASTM D611, 2021) of the sample.

As a control measure, the smoke point test (ASTM D1322, 2021) gives the maximum smokeless flame height in millimeters at which a material (typically, a fuel) will burn in a wick-fed lamp under prescribed conditions. Also, the combustion performance under combustion conditions correlates well with the smoke point when a volatility factor is included, since carbon formation tends to increase with boiling point.

$$\text{SVI} = \text{Smoke Point} + 0.42(\text{per cent distilled below } 204°\text{C} / 400°\text{F})$$

Various alternative laboratory test methods have previously been specified such as the lamp burning test (ASTM D187, 2021) and a limit on the polynuclear aromatic content (ASTM D1840, 2021).

7.7 Latent Heat

The latent heat of a material is the heat required to convert a solid into a liquid or vapor, or a liquid into a vapor, without a change of temperature. Another way of defining the latent heat of a substance is that is the heat evolved or absorbed in a process (such as fusion or vaporization) other than a change of temperature. Thus, the latent heat is the energy released or absorbed by a body or a thermodynamic system during a constant-temperature process and is typically a first-order phase transition. Latent heat can be understood as energy in hidden form which is supplied or extracted to change the state of a substance without changing its temperature.

There are two properties that represent phase transformations: the latent heat of fusion and the latent heat of vaporization. The latent heat of fusion, defined as the quantity of heat necessary to change a unit weight of solid to a liquid without any temperature change, has received only intermittent attention but, nevertheless, some general rules have been formulated.

Phase transformations for viscous feedstocks (such as melting and solidification) are not accompanied by a sensible heat change (rise in temperature), although they are associated with heat absorption or evolution. Such changes are typically normally measured at constant pressure and are thus enthalpy changes, although values for the constant volume (internal energy) changes are obtainable in thermodynamic tables. The unit is Btu/lb (1 Btu/lb = 2.326 kJ/kg).

The latent heat of vaporization is the amount of heat required to vaporize a unit weight of a liquid at its atmospheric boiling point and is perhaps the most important property of the two and has received considerably more attention because of its connection with equipment design. The latent heat of vaporization at the atmospheric boiling point generally increases with increasing molecular weight and, for the normal paraffins, generally decreases with increasing temperature and pressure.

The heat of vaporization is perhaps the most important latent heat in petroleum engineering. It decreases as the temperature rises and becomes zero at the critical temperature (this phenomenon forming, in fact, a useful indication of the critical point). The importance of latent heat has led to many methods of obtaining such data from temperature and the more readily obtainable physical properties such as density. These correlations, although they contain a degree of empiricism, are generally based on the Clausius–Clapeyron equation.

As with specific heat, values of latent heat are needed in order to calculate heat effects, particularly for distillation equipment.

The latent heat of vaporization at any temperature can be described in terms of the latent heat at the normal boiling point:

$$L = \gamma L_{\mathrm{B}} / (T / T_{\mathrm{B}})$$

In this equation, L is the latent heat of vaporization at the absolute temperature T, γ is a temperature correction factor obtained from graphical data, and L_{B} is the heat of vaporization at absolute normal boiling point T_{B}.

The latent heat decreases with temperature until it has a value of zero at the critical temperature. Furthermore, the critical temperature of a high-API gravity crude oil is lower than the critical temperature of a low-API gravity crude oil. Hence, the latent heat curves of various crude oil will exhibit cross-over points.

7.8 Liquefaction and Solidification

Liquefaction is a process that generates a liquid from a solid or a gas or that generates a nonliquid phase which behaves in accordance with fluid dynamics. On the other hand, solidification is a process in which atoms are converted into an ordered solid state from a liquid disordered state. The conversion rate for the process of solidification can be achieved by following the kinetic laws and the movement of atoms for the conversion of liquid can be observed by these laws.

Freezing (a common form of solidification) is a phase transition where a liquid turns into a solid when the temperature (of the liquid) is lowered to below the freezing point of the liquid. Thus, by definition, freezing is the solidification phase change of a liquid or the liquid content of a substance, usually due to cooling.

Data on the liquefaction and solidification of viscous feedstocks (including residua) are important in refinery management. The melting point is a test (ASTM D87, 2021; ASTM D127, 2021) that is applied to the highly paraffinic or crystalline waxes and can be applied to viscous feedstocks.

The softening point (ASTM D36, 2021) is the temperature at which a disk of the material softens and sags downward a distance of 25 mm under the weight of a steel ball under strictly specified conditions, and finds wider use for the viscous feedstocks and residua. The dropping point (ASTM D566, 2021) is the near-equivalent test that is used for lubricating greases and may, with suitable modification, be applicable to the viscous feedstocks.

Although the melting points of the viscous feedstocks are of some (but often limited) usefulness, except to estimate the purity or perhaps the composition of waxes, the reverse process, solidification, has received some, but again limited, attention in petroleum chemistry. For example, the solidification point of paraffin wax is an important monitor for the behavior and performance of wax. However, the test is somewhat limited in its application to heavy oil or bitumen because the complexity of these materials prevents careful and conscientious application of the test. Hence, more emphasis is placed on the reverse phenomena such a test such as melting point, dropping point, softening point, and pour point.

The pour point has also found use as an indicator of the temperature at which crude oil or heavy oil will flow during in situ recovery operations (Wallace, 1988, p. 183). Under similar conditions, tar sand bitumen is immobile. For example, for asphaltic crude oils where paraffin precipitation will not occur, if 21 °C (70 °F) is the pour point of a crude oil in a reservoir where the temperature is 38 °C (100 °F), the oil is liquid under reservoir conditions and will be mobile and will flow under those conditions. On the other hand, tar sand bitumen (pour point: 60 °C, 140 °F) in a deposit (temperature: 10 °C, 50 °F) will be solid and immobile. This state of the oil in the reservoir can also have consequences on the ability of gases and liquids (e.g., steam and hot water) used for recovery operations to penetrate the reservoir/deposit. Although pressure can have some influence on the pour point, the effect is not large and unlikely to affect any general conclusions. Indeed, for specific types of petroleum (usually asphaltic and non-waxy), there is a relationship between API gravity and pour point. Thus, any increase in pour point due to an increase in pressure (surface compared to reservoir or deposit pressure) will be most likely be negated as the API gravity decreases with increase in temperature (60 °F compared to reservoir temperature).

7.9 Pour Point

The pour point of a liquid is the temperature below which the liquid loses its flow characteristics and the minimum temperature in which the oil has the ability to pour down from a container and which is the lowest temperature at which the oil will flow under gravity. At a temperature above the pour point, the liquid will flow without stirring, under standard conditions.

Thus, the pour point of a crude oil or a crude oil product or a crude oil-like material (such as a viscous feedstock) is the lowest temperature at which the oil will just flow, under standard test conditions (ASTM D97, 2021). The failure to flow at the pour point is usually attributed to the separation of waxes from the oil, but can also be due to the effect of viscosity in the case of the viscous feedstocks. Also, particularly in the case of residual fuel oil, the pour point may be influenced by the thermal history of the sample, that is, the degree and duration of heating and cooling to which the sample has been exposed.

Because refinery feedstocks are composed of many different types of constituents—some of which may still be liquid at the pour point—the pour point is not the temperature at which the feedstock is a solid mass. The nature of some constituents within the feedstock will separate and form a solid surface layer, or crust, can create a situation in which the bulk of the feedstock will flow at temperatures below the measured pour point, yet will not flow under the conditions of the test.

In the case of a spill into the environment, it must be emphasized that the tendency of the oil to flow will be influenced by the size and shape of the container, the head of the oil, and the physical structure of the solidified oil. The pour point of an oil is, therefore, an indication, and not an exact measure, of the temperature at which the flow of the oil ceases (Dyroff, 1993).

7.10 Pressure–Volume–Temperature Relationships

PVT properties of a refinery feedstock are not only necessary for reservoir engineering calculations and pipe line flow calculations but are also valuable for understanding the behavior of the feedstock within the refinery system. Thus, an accurate understanding of the prediction of the PVT properties is one of the most important tasks for studying feedstock character and behavior. While a large number of PVT correlations have been established for several hydrocarbon systems, in most cases, overall accuracy of these correlations are often limited due to properties such as compositional variation and/or impurities in the feedstock, the PVT properties of crude oil need to be determined through experimental analyses (Rahman et al., 2016).

Knowledge of the pressure, volume, and temperature of a system forms the basis of the calculation of the thermodynamic relationships of the system. These properties are easily measured, and since they are sufficient to define the thermodynamic state, they are termed the Parameters of State. However, they are not independent, and in fact any two are sufficient to define the state of a system of known mass. A simple treatment of a mathematical model of a gas based on kinetic theory yields a relationship:

$$PV = nRT$$

In this equation, n is the number of moles of gas and R a universal constant equal to 8.314 kJ/kmol °K. This mathematical model is called the Perfect Gas, and real gases that behave in this way are called Ideal Gases. Most light petroleum gases, air, flue gases, and so on, may be treated as ideal up to pressures on the order of 250–300, provided the temperature is well above the critical. For heavier vapors and for lower temperatures, a more accurate equation of state is needed.

Sufficient accuracy is normally obtained for engineering purposes by the use of the compressibility coefficient (Z), where $PV = ZnRT$. The compressibility coefficient is a function of temperature, pressure, and the nature of the gas, but values can be obtained from generalized charts by use of reduced temperatures and pressures (pseudo-values in the case of mixtures).

In the calculation of *PVT* relationships, it is often useful to use the concept of standard conditions at 0 °C (32 °F) and 1 atm pressure (approximately $10^5\,N/m^2$) 1 kmol of the perfect gas occupies 22.4 m^3 (the corresponding value in British units is 359 ft^3 and 1 atm pressure or 379 ft^3 at 60 °F and 1 atm pressure [both values for 1 lb mole]). When using these relationships, it is important that both pressure and temperature be expressed in absolute scales. Gauge pressures and or °C or °F must not be used.

Hydrocarbon vapors, like other gases, follow the ideal gas law (i.e., $PV = RT$) only at relatively low pressures and high temperatures, that is, far from the critical state. Several more empirical equations have been proposed to represent the gas laws more accurately, such as the well-known van der Waals equation, but they are either inconvenient for calculation or require the experimental determination of several constants. A more useful device is to use the simple gas law and to induce a correction, termed the compressibility factor, so that the equation takes the form:

$$PV = \mu RT$$

For hydrocarbons, the compressibility factor is very nearly a function only of the reduced variables of state, that is, a function of the pressure and temperature divided by their respective critical values.

7.11 Softening Point

The softening point is the temperature at which a material softens beyond some arbitrary softness and can be determined, for example, by the ring and ball apparatus which is also suitable for the determination of softening point of bituminous materials. As a general guide, the softening point of the viscous feedstocks lies between 35 and 70 °C (95 and 158 °F).

Thus, the softening point of a viscous material or a solid material is the temperature at which the material exhibits a particular degree of softness under the conditions that are specified by the test method (e.g., ASTM D36, 2021). Typically, viscous feedstocks (especially residua) do not go through a solid–liquid phase change when heated, and therefore does not have true melting point. As the temperature is raised, the material gradually softens or becomes less viscous.

There are several tests available to determine the softening point (ASTM D36, 2021; ASTM D2319, 2021; ASTM D3104, 2012; ASTM D3461, 2021). In the test method (ASTM D36, 2021), a steel ball of specified weight is laid on a layer of sample contained in a ring of specified dimensions. In this test method, the softening point of the sample is the temperature at which the material surrounding the ball deforms and contacts a base plate, during heating under specified conditions.

On the other hand, the melting point is the temperature where all the crystallinity in the bags and film is destroyed. When this happens, the bag will completely disperse into the rubber. The softening point describes a temperature where the bag will begin to experience noticeable changes in physical properties. The dropping point is another characteristic property of, in this contest, the viscous feedstocks. In the test method (ASTM D566, 2021; ASTM D2265, 2021), which has been designed for lubricating greases but can be modified for the viscous feedstocks, samples are heated until they transform from a solid into a liquid state and the dropping point is the temperature at which the first drop of a molten feedstock precipitates from a standardized cup with a defined orifice under controlled testing conditions in a furnace.

The softening point and the dropping point are properties that are available to thermally characterize the viscous feedstocks. Analysis of the softening point and the dropping point are mainly conducted during the quality control process and is valuable for the determination of the operating temperatures and process parameters of the viscous feedstocks.

7.12 Specific Heat

The specific heat of a material is the quantity of heat required to raise a unit mass of material through one degree of temperature which, in fact, describes specific heat capacity, and is not related to the heat capacity of water. The more common units of Btu/lb °F or cal/g °C (which are numerically equal, being effectively the ratio of the heat required to heat the substance to that required to heat water) are related to the SI unit by 1 Btu/lb °F = 1 cal/g °C = 4.1868 kJ/kg °K. In addition, to the standard method for the determination of specific heat, differential scanning calorimetry has also been used (ASTM E1269, 2021).

The property is often required in estimating the amount of heat necessary to preheat a viscous feedstock in the reservoir (or deposit) before recovery. Knowledge of the specific heat of a fraction is of importance in determining net heat fluxes in all heating and cooling problems. Thus, specific heat data are extremely important engineering quantities in refinery practice because they are used in all calculations on heating and cooling petroleum products.

Many measurements have been made on various samples and the data for most purposes may be summarized by the general equation:

$$C = 1/d(0.388 + 0.00045t)$$

In this equation, C is the specific heat at t °F of an oil whose specific gravity 60/60 °F is d; thus, specific heat increases with temperature and decreases with specific gravity. For bitumen and heavy oil, the equation has been modified. Thus:

$$C_p = [0.758 + 0.00339(t + 273)] / d^{0.5}$$

In this equation, C_p is the heat capacity (J/g °K, 1.0 = 0.24 Btu/lb °F), t is the temperature (°C), and d is the density (g/cm^3). Another relationship (Bland and Davidson, 1967) is:

$$C_p = (0.6811 - 0.308d) + t(0.000815 - 0.000306d) \times (0.055K + 0.35)$$

Again, C_p is the heat capacity (J/g °K, 1.0 = 0.24 Btu/lb °F), t is the temperature (°F), K is the Watson characterization factor (Speight, 2014a), and d is the specific gravity (60 °F/60 °F).

On the other hand, and more specifically, after extensive testing of bitumen samples from the Athabasca deposit, the following temperature-dependent equation was derived for the specific heat of bitumen covering the range 50–300 °C (122–572 °F):

$$C_p = 1.5574 \; 5.219 \times 10^3 \; t - 8.686 \times 10^{-6} \; t^2$$

where C_p is the specific heat (heat capacity) (J/g °K, 1.0 = 0.24 Btu/lb °F) and t is the temperature (°C). For most hydrocarbons, specific heat is almost a linear function of temperature (increasing as temperature increases). It decreases, however, as the density increases.

For liquids, specific heat normally implies measurements at constant (atmospheric) pressure. The value is not very dependent upon pressure, and the value given is almost identical with the value obtained at constant volume. For gases, however, the values at constant pressure and constant volume are quite different, the value at constant pressure being the greater by an amount equal to the specific gas constant.

7.13 Thermal Conductivity

The thermal conductivity of a material is a measure of the ability of the material to transfer heat to another material. For example, if there are two surfaces on either side of the material and a temperature difference exists between the surfaces, the heat energy transferred per unit time and per unit surface area, divided by the temperature difference between the two substances is the thermal conductivity.

The thermal conductivity and the viscosity are transport properties that are required for the design of a variety of processes and process units in a refinery, also including in pipelines, and heat exchangers (Magomadov, 2004). Momentum and heat are transferred by means of either molecular movement (conduction) or bulk fluid movement (convection); or a combination of both.

In a refinery, it is often necessary to establish the optimum conditions for the transfer of heat as well as the cooling function of these materials. Thus, a knowledge of the specific heat is of considerable value in estimating the likely rise in temperature for a given set of heat transfer conditions. In many heat transfer problems, the viscosity may have a decided influence over the establishment of convection currents or volume flow if forced circulation is involved. Thus, the thermal conductivity of the oil becomes a significant factor.

The thermal conductivity K of feedstocks (in cgs units) is given by the equation:

$$K = 0.28 / d(1 - 0.00054) \times 10^{-3}$$

In this equation, d is the specific gravity. The units of thermal conductivity have been expressed as W/mK (1.0 W/mK = 0.86 kcal/h m °C = 0.58 Btu/h ft °F). The thermal conductivity of a series of feedstocks can be estimated from the API gravity that shows a variation with temperature. The API gravity of bitumen and heavy oil varies from 6 to about 20. Thus, at reservoir and deposit temperatures 0–70 °C (32–158 °F), the value of bitumen thermal conductivity would thus be approximately 0.11 to about 0.12.

7.14 Volatility

Volatility is a property that indicates the tendency of a material (specifically in this context the viscous feedstocks, including atmospheric residua and vacuum residua) to vaporize or more specifically for the material to change from the liquid to the vapor state (often referred to as the gaseous state). If at all possible, acquiring a distillation profile is a good indicator of the relative amounts of the material that will be in the gaseous state at a specific temperature.

The volatility of the viscous feedstocks do not always qualify as volatile (or even potentially volatile) materials but the volatility (or potential volatility) of the material is still an important aspect of safety and quality. It would be unsafe to attempt to store highly volatile materials in the open sunlight or in an enclosed space where temperature can rise to be in excess of 37.8 °C (100 °F). And yet without any indications of when the material might vaporize and spontaneously ignite, there is no way of even considering the correct storage and handling conditions. Similarly, there must also be some estimate of the ability of the constituents of the feedstock to distill from the mixture if refining is to be achieved in the most effective manner. Thus, the volatility of any refinery feedstock in terms of the amounts of the constituents' boiling fractions became of very prominent test for crude oil quality.

The vaporizing tendency of any feedstock is the basis for the general characterization of a viscous feedstock (ASTM D6, 2021). Another test (ASTM D20, 2021) that is a designed for testing the volatility of "road tar" can also be applied to estimating the volatility of high molecular weight unknown viscous feedstocks.

By definition, the volatility of a feedstock is the tendency of constituents of the feedstock to vaporize, that is, to change from the liquid to the vapor or gaseous state. In the context of the viscous feedstocks, the potential for the volatility of any of the constituents of the feedstock should not be ignored. In addition, before any volatility tests (or, for that matter, any physical tests) are carried out on a feedstock, there is the need to recognize that the presence of more than 0.5% w/w (or 0.5% v/v) water in a feedstock can cause several problems not only during the various test procedures and produce erroneous results but also contribute to, for example, fouling and the corrosion of the equipment, during the refining operations (Speight, 2014b, 2015b). For example, during various thermal tests, water (which has a high heat of vaporization) requires the application of additional thermal energy to the distillation flask. In addition, water is relatively easily superheated and therefore excessive *bumping* can occur, leading to erroneous readings and the potential for destruction of the glass equipment is real. Steam formed during distillation can act as a carrier gas and high boiling point components may end up in the distillate (often referred to as steam distillation).

Removal of water (and sediment) can be achieved by centrifugation if the viscous sample is dissolved in an appropriate solvent that can be easily removed when required. Other methods that are used to remove water include the following: (1) heating in a pressure vessel to control loss of light ends, (2) addition of calcium chloride as recommended in ASTM D1160 (2021), (3) addition of an azeotroping agent such as *iso*-propanol or *n*-butanol, (4) removal of water in a preliminary low-efficiency or flash distillation followed by reblending the hydrocarbon which co-distills with the water into the sample, and (5) separation of the water from the hydrocarbon distillate by freezing. However, the method chosen must be appropriate to use with the viscous feedstock under investigation. Other options that involve heating the feedstock to remove water (which must be applied with caution) include a test method (ASTM D6, 2021) that also exists for determining the loss of material when a feedstock is heated. Another test (ASTM D20, 2021) is a method for the distillation of viscous feedstocks and residua might also be applied to estimating the volatility of high molecular weight residues.

For some purposes, it is necessary to have information on the initial stage of vaporization and the potential hazards that such a property can cause. To supply this need, flash and fire, vapor pressure, and evaporation methods are available. The data from the early stages of the several distillation methods are also useful. For other uses, it is important to know the tendency of a product to partially vaporize or to completely vaporize, and in some cases to know if small quantities of high-boiling components are present. For such purposes, chief reliance is placed on the distillation methods.

The flash point of a feedstock is the temperature to which the product must be heated under specified conditions to give sufficient vapor to form a mixture with air that can be ignited momentarily by a specified flame (ASTM D56, 2021; ASTM D92, 2021; ASTM D93, 2021; ASTM D1310, 2021; ASTM D3828, 2021). The Pensky–Martens apparatus that uses a closed system or an open system (ASTM D93, 2021) is the standard instrument for determining a flash point that is above 50 °C (122 °F). In addition, the Cleveland Open-Cup method (ASTM D92, 2021) is used for the determination of the fire point (the temperature at which the sample will ignite and burn for at least 5 s).

The Pensky–Martens apparatus consists of a brass cup, mounted in an air bath and heated by a gas flame. A propeller-type stirrer, operated by a flexible drive, extends from the center of the cover into the cup. The cover has four openings: one for a thermometer, and the others fitted with sliding shutters for the introduction of a pilot flame and for ventilation. The temperature of the material in the cup is raised at 5–6 °C/min (9–11 °F/min) and the stirrer is rotated at approximately 60 rpm. When the temperature has risen to approximately 15 °C (59 °F) from the anticipated flash point, the pilot flame is dipped into the oil vapor for 2 s for every 1 °C (1.8 °F) rise in temperature up to 105 °C (221 °F). Above 105 °C

(221 °F), the flame is introduced every 2 °C (3.6 °F) rise in temperature. The flash point is the temperature at which a distinct flash is observed when the pilot flame meets the vapor in the cup.

Another criterion that deserves attention is the fire point which is the temperature to which the material must be heated (under the prescribed conditions outlined in the method) to burn continuously when the mixture of vapor and air is ignited by a specified flame (ASTM D92, 2021).

The flash point of a feedstock is also used to detect contamination since a substantially lower flash point than expected for a viscous feedstock is a reliable indicator that a material has become contaminated with a more volatile product. In fact, as it pertains to the safety and behavior of a feedstock within the refinery system, the flash point is often considered to be of the most significance at or slightly above the maximum temperatures (30–60 °C, 86–140 °F) that may be encountered in storage and/or transportation. In this temperature range, the relative fire and explosion hazard can be estimated from the flash point. If the feedstocks contain constituents that have a flash point below 40 °C (104 °F), special precautions are necessary for safe handling. However, as is the case with most viscous feedstocks, flash points above 60 °C (140 °F) gradually lose their safety significance until they become indirect measures of some other quality.

Another aspect that depends upon the volatility of the sample is the vapor pressure which is the force exerted on the walls of a closed container by the vaporized portion of a feedstock. Conversely, it is the force that must be exerted on the liquid to prevent it from vaporizing further (ASTM D323, 2021). The vapor pressure increases with temperature and the temperature at which the vapor pressure of the sample is equal to 1 atm (14.7 psi) is designated as the boiling point of the liquid.

In fact, distillation was at one time the predominant method by which refinery feedstocks were evaluated as suitable for various refinery options. It is now one of several methods used for evaluation purposes.

Distillation involves the general procedure of vaporizing the feedstock in a suitable flask either at atmospheric pressure (ASTM D86, 2021; ASTM D216, 2021; ASTM D2892, 2021) or at reduced pressure (ASTM D1160, 2021). There are also test methods for the distillation of cutback asphalt, which is usually a nonvolatile distillation residue, which can be applied to heavy crude oil, extra heavy crude oil, and tar sand bitumen. However, most of the methods specify an upper atmospheric equivalent temperature (AET) limit of 360 °C (680 °F) and therefore are too limited to be of value in the analysis of viscous feedstocks as defined herein.

In the simplest case, the distillation method involves using a standard round-bottom distillation flask of 250-ml capacity attached to a water-cooled condenser. The thermometer bulb is placed at the opening to the side arm of the flask. One hundred milliliters of sample are placed in the flask and heated by a small gas flame so as to produce 10 ml of distillate every 4 or 5 min. The temperature of initial distillation is recorded; the temperature at which each further 10 ml distils and the final boiling point are also recorded.

The distillation procedures that are commonly used by laboratories either singly or in combination, to determine the distillation curve of, or produce fractions from heavy oil and bitumen are: (1) ASTM D1160 (2021) and (2) ASTM D5236 (2021).

The ASTM D1160 (2021) method is used to determine the boiling ranges of petroleum products to a maximum liquid temperature of 400 °C (750 °F) at pressures as low as 1 mmHg. In the method, a 200-ml sample is weighed to the nearest 0.1 g in a distillation flask. After the distillation assembly has been evacuated to the desired pressure, heat is applied to the flask and when any refluxing liquid appears, the rate of heating is adjusted so that the distillate is recovered at 4–8 ml/min until the distillation is complete. However, because of the thermal sensitivity of the viscous feedstocks, cracking will most likely occur before the liquid temperature reaches 400 °C (750 °F). An increase in distillation rate accompanied by a drop in head temperature, loss of vacuum in the system which is restored when heat to the still is reduced, and production of vapor clouds in the system are all evidence of cracking.

The vacuum pot still method (ASTM D5236, 2021) is a procedure for the distillation of high boiling samples having an initial boiling point above 150 °C (300 °F) which includes heavy crude oil, extra heavy crude oil, tar sand bitumen, and residua. The method employs a pot still with a low-pressure drop entrainment separator. Another method (ASTM D2892, 2021) describes the procedure for distilling petroleum-type materials up to 400 °C (750 °F) AET and is often referred to as the true boiling point (TBP) distillation method.

There is also another method that is increasing in popularity for application to a variety of feedstocks and that is the method commonly known as simulated distillation (ASTM D2887, 2021). This is a low-resolution, temperature-programmed gas chromatographic analysis intended to simulate the time-consuming true boiling point distillation method. It is based on the observation that hydrocarbons generally are eluted from a nonpolar column in the order of their boiling point, with the retention time being converted to equivalent distillation temperatures. The accumulated detector response is related to the amount of sample distilled. In practice, the nonpolar stationary phases do discriminate to some extent among different classes of hydrocarbon. The retention time is a function of the adsorptive property of the stationary phase, the vapor pressure, and the heat of vaporization. However, the empirical correlation between the retention time and boiling point, usually established by running a series of *n*-alkane derivatives, is not strictly valid for other hydrocarbon classes such as aromatic derivatives and cycloalkane derivatives which show a tendency to elute earlier than *n*-alkanes having the same boiling points.

The benefits of the technique include good comparisons with other ASTM distillation data as well as the application to higher boiling fractions of petroleum. In fact, data output includes the provision of the corresponding Engler profile (ASTM D86, 2021) as well as the prediction of other properties such as vapor pressure and flash point.

In the spinning band method, fractions of feedstocks such as viscous feedstocks (i.e., heavy crude oil, extra heavy crude oil, tar sand bitumen, and residua) with an initial boiling point above room temperature at atmospheric pressure can be prepared. The distillation is terminated at an AET of 525 °C (975 °F) while some viscous crude oil may necessitate termination of the distillation at a pot temperature on the order of 360 °C (680 °F).

Samples are distilled under atmospheric and reduced pressures of 20.0 and 0.60 mmHg in a distillation unit equipped with a spinning band column. The spinning band, which effectively provides a large contact area between the liquid phase and the vapor phase, increases the number of theoretical plates in the column and thus its fractionating efficiency. Readings of vapor temperature (that is convertible to AET) and distillate volume (that is convertible to per cent by volume) are used to plot a distillation curve. Distillate yields for naphtha, light gas oil, heavy gas oil, and residue fractions are determined on a gravimetric basis.

When fractionating a viscous feedstock, the pot temperature cannot exceed 360 °C (680 °F). In the method, samples are distilled under atmospheric and reduced pressures in a still equipped with a spinning band column. Vapor temperatures are converted to AETs and can be plotted as a function of volume or weight percent distilled to yield a distillation profile.

Another method that is often used when the sample under investigation is a viscous (essentially nonvolatile) feedstock is the short-path distillation method (sometimes referred to as the molecular distillation method or, on occasion, fractional distillation) which involves the distillate traveling a short distance, often only a short distance (often described as a "few centimeters") and is normally done at reduced pressure. The advantage of the method is that a lower temperature can be used because of the reduced pressure thereby producing a relatively lower boiling point of the sample that at standard pressure and, moreover, the distillate only has to travel a short distance before condensation occurs. In order to increase the evaporation rate without increasing temperature, there are several modern techniques that increase the surface area of the liquid such as (1) the thin film method; (2) the wiped film method, also referred to as the "wiper" film method; and (3) the rolled film method all of which involve mechanically spreading a film of the liquid over a large surface.

In the short path distillation as applied to viscous feedstocks, a single distillate and a single residue fraction are produced as defined by the operating temperature and pressure of the still. The procedure is used to generate high boiling point fractions with endpoints up to 700 °C (1,290 °F) for further analysis.

Since only one cut temperature is used in each run, generation of a distillation curve using this equipment would be time-consuming. In the method, the material to be fractionated is introduced at a constant rate onto the hot inner wall of the evaporator under high vacuum. Rotating (Teflon) rollers ensure that the film on the wall is kept thin. The feedstock is progressively distilled at the fixed conditions of temperature and pressure. The distillate vapors condense on a concentric cold surface (60 °C, 140 °F) placed a short distance from the hot wall inside the still. The condensate then drains by gravity to the base of the cold finger where it is collected. The residue drains down the hot wall and is collected through a separate port.

Short path distillation permits production and collection of a higher boiling distillate (often referred to as a "deeper cut") and a higher boiling residue (i.e., a residue with a higher initial boiling point, referred to as the IBP) than the conventional vacuum distillation method. AET endpoints of up to 660 °C (1,220 °F) can be attained without thermal cracking of the material. The distillation, however, is conducted under nonequilibrium conditions so the cut point is not as specific as in other vacuum distillation methods.

Owing to the wide variation in the types of viscous feedstocks and to the fact that most companies have evolved their own assay methods, it has proved difficult to devise a rigid procedure to be followed in all cases, although there are considerable advantages to be gained if this were possible. The procedure is, in essence, that followed by many companies.

For a more detailed distillation analysis of feedstocks, a low-resolution, temperature-programmed gas chromatographic analysis has been developed to simulate the time-consuming true boiling point distillation. The method relies on the general observation that hydrocarbons are eluted from a nonpolar adsorbent in the order of their boiling points. The regularity of the elution order of the hydrocarbon components allows the retention times to be equated to distillation temperatures (Butler, 1979).

In comparison to an actual distillation, simulated distillation by gas chromatography requires a very small sample (on the order of several microliters) to produce a full boiling range distribution profile for prediction of distillate yield and residue. In addition, simulated distillation by gas chromatography might take as little as 2 hr whereas a full distillation may take 2 days.

Simulated distillation for distillate fractions (ASTM D2887, 2021) is designed predominantly for samples with a final boiling point below 538 °C (1,000 °F) and is not suitable for the analysis of higher boiling residua, heavy crude oil, extra heavy crude oil, and tar sand bitumen.

The test method for the boiling range distribution of petroleum fractions by gas chromatography (ASTM D2887, 2021) cannot be used to map the distillation profile of heavy oil and bitumen because it does not provide for the determination of the proportion of sample having a boiling point in excess of 538 °C (>1,000 °F) residue in the sample. The use of internal standards and combinations of actual distillations and gas chromatographic distillation which provide a distillate fraction and can then be analyzed by gas chromatography have already gained wide acceptance in the industry.

Finally, the test method (ASTM D2887, 2021) and another method (ASTM D7096, 2021) use external standards composed of *n*-alkane derivatives. Another test method (ASTM D7169, 2021) requires two determinations to be made with each sample, one of which uses an internal standard. The amount of material boiling above 538 °C (1,000 °F) (reported as residue) is calculated from the differences between the two determinations.

References

ASTM D6. 2021. Standard Test Method for Loss on Heating of Oil and Asphaltic Compounds. Annual Book of Standards. ASTM International, West Conshohocken, Pennsylvania.

ASTM D20. 2021. Standard Test Method for Distillation of Road Tars. Annual Book of Standards. ASTM International, West Conshohocken, Pennsylvania.

ASTM D36. 2021. Standard Test Method for Softening Point of Bitumen (Ring-and-Ball Apparatus). Annual Book of Standards. ASTM International, West Conshohocken, Pennsylvania.

ASTM D56. 2021. Standard Test Method for Flash Point by Tag Closed Cup Tester. Annual Book of Standards. ASTM International, West Conshohocken, Pennsylvania.

ASTM D86. 2021. Standard Test Method for Distillation of Petroleum Products and Liquid Fuels at Atmospheric Pressure. Annual Book of Standards. ASTM International, West Conshohocken, Pennsylvania.

ASTM D87. 2021. Standard Test Method for Melting Point of Petroleum Wax (Cooling Curve). Annual Book of Standards. ASTM International, West Conshohocken, Pennsylvania.

ASTM D92. 2021. Standard Test Method for Flash and Fire Points by Cleveland Open Cup Tester. Annual Book of Standards. ASTM International, West Conshohocken, Pennsylvania.

ASTM D93. 2021. Standard Test Methods for Flash Point by Pensky-Martens Closed Cup Tester. Annual Book of Standards. ASTM International, West Conshohocken, Pennsylvania.

ASTM D97. 2021. Standard Test Method for Pour Point of Petroleum Products. Annual Book of Standards. ASTM International, West Conshohocken, Pennsylvania.

ASTM D127. 2021. Standard Test Method for Drop Melting Point of Petroleum Wax, Including Petrolatum. Annual Book of Standards. ASTM International, West Conshohocken, Pennsylvania.

ASTM D187. 2021. Standard Test Method for Burning Quality of Kerosene. Annual Book of Standards. ASTM International, West Conshohocken, Pennsylvania.

ASTM D189. 2021. Standard Test Method for Conradson Carbon Residue of Petroleum Products. Annual Book of Standards. ASTM International, West Conshohocken, Pennsylvania.

ASTM D216. 2021. Method for Distillation of Natural Gasoline. Annual Book of Standards. ASTM International, West Conshohocken, Pennsylvania.

ASTM D240. 2021. Standard Test Method for Heat of Combustion of Liquid Hydrocarbon Fuels by Bomb Calorimeter. Annual Book of Standards. ASTM International, West Conshohocken, Pennsylvania.

ASTM D323. 2021. Standard Test Method for Vapor Pressure of Petroleum Products (Reid Method). Annual Book of Standards. ASTM International, West Conshohocken, Pennsylvania.

ASTM D524. 2021. Standard Test Method for Ramsbottom Carbon Residue of Petroleum Products. Annual Book of Standards. ASTM International, West Conshohocken, Pennsylvania.

ASTM D566. 2021. Standard Test Method for Dropping Point of Lubricating Grease. Annual Book of Standards. ASTM International, West Conshohocken, Pennsylvania.

ASTM D611. 2021. Standard Test Method for Dropping Point of Lubricating Grease. Annual Book of Standards. ASTM International, West Conshohocken, Pennsylvania.

ASTM D1160. 2021. Standard Test Method for Distillation of Petroleum Products at Reduced Pressure. Annual Book of Standards. ASTM International, West Conshohocken, Pennsylvania.

ASTM D1310. 2021. Standard Test Method for Flash Point and Fire Point of Liquids by Tag Open-Cup Apparatus. Annual Book of Standards. ASTM International, West Conshohocken, Pennsylvania.

ASTM D1322. 2021. Standard Test Method for Smoke Point of Kerosene and Aviation Turbine Fuel. Annual Book of Standards. ASTM International, West Conshohocken, Pennsylvania.

ASTM D1840. 2021. Standard Test Method for Naphthalene Hydrocarbons in Aviation Turbine Fuels by Ultraviolet Spectrophotometry. Annual Book of Standards. ASTM International, West Conshohocken, Pennsylvania.

ASTM D2265. 2021. Standard Test Method for Dropping Point of Lubricating Grease Over Wide Temperature Range. Annual Book of Standards. ASTM International, West Conshohocken, Pennsylvania.

ASTM D2319. 2021. Standard Test Method for Softening Point of Pitch (Cube-In-Air Method). Annual Book of Standards. ASTM International, West Conshohocken, Pennsylvania.

ASTM D2887. 2021. Standard Test Method for Boiling Range Distribution of Petroleum Fractions by Gas Chromatography. Annual Book of Standards. ASTM International, West Conshohocken, Pennsylvania.

ASTM D2892. 2021. Standard Test Method for Distillation of Crude Petroleum (15-Theoretical Plate Column). Annual Book of Standards. ASTM International, West Conshohocken, Pennsylvania.

ASTM D3104. 2012. Standard Test Method for Softening Point of Pitches (Mettler Softening Point Method). Annual Book of Standards. ASTM International, West Conshohocken, Pennsylvania.

ASTM D3172. 2021. Standard Practice for Proximate Analysis of Coal and Coke. Annual Book of Standards. ASTM International, West Conshohocken, Pennsylvania.

ASTM D3173. 2021. Standard Test Method for Moisture in the Analysis Sample of Coal and Coke. Annual Book of Standards. ASTM International, West Conshohocken, Pennsylvania.

ASTM D3174. 2021. Standard Test Method for Ash in the Analysis Sample of Coal and Coke from Coal. Annual Book of Standards. ASTM International, West Conshohocken, Pennsylvania.

ASTM D3175. 2021. Standard Test Method for Volatile Matter in the Analysis Sample of Coal and Coke. Annual Book of Standards. ASTM International, West Conshohocken, Pennsylvania.

ASTM D3461. 2021. Standard Test Method for Softening Point of Asphalt and Pitch (Mettler Cup-and-Ball Method). Annual Book of Standards. ASTM International, West Conshohocken, Pennsylvania.

ASTM D3828. 2021. Standard Test Methods for Flash Point by Small Scale Closed Cup Tester. Annual Book of Standards. ASTM International, West Conshohocken, Pennsylvania.

ASTM D4530. 2021. Standard Test Method for Determination of Carbon Residue (Micro Method). Annual Book of Standards. ASTM International, West Conshohocken, Pennsylvania.

ASTM D5236. 2021. Standard Test Method for Distillation of Heavy Hydrocarbon Mixtures (Vacuum Potstill Method). Annual Book of Standards. ASTM International, West Conshohocken, Pennsylvania.

ASTM D5708. 2021. Standard Test Methods for Determination of Nickel, Vanadium, and Iron in Crude Oils and Residual Fuels by Inductively Coupled Plasma (ICP) Atomic Emission Spectrometry. Annual Book of Standards. ASTM International, West Conshohocken, Pennsylvania.

ASTM D5863. 2021. Standard Test Methods for Determination of Nickel, Vanadium, Iron, and Sodium in Crude Oils and Residual Fuels by Flame Atomic Absorption Spectrometry. Annual Book of Standards. ASTM International, West Conshohocken, Pennsylvania.

ASTM D7096. 2021. Standard Test Method for Determination of the Boiling Range Distribution of Gasoline by Wide-Bore Capillary Gas Chromatography. Annual Book of Standards. ASTM International, West Conshohocken, Pennsylvania.

ASTM D7169. 2021. Standard Test Method for Boiling Point Distribution of Samples with Residues Such as Crude Oils and Atmospheric and Vacuum Residues by High Temperature Gas Chromatography. Annual Book of Standards. ASTM International, West Conshohocken, Pennsylvania.

ASTM E1269. 2021. Standard Test Method for Determining Specific Heat Capacity by Differential Scanning Calorimetry. Annual Book of Standards. ASTM International, West Conshohocken, Pennsylvania.

Bland, W.F., and Davidson, R.L. 1967. *Petroleum Processing Handbook*. McGraw-Hill Book Company, New York. pp. 12–43.

Butler, R.D. 1979. In: *In Chromatography in Petroleum Analysis*. K.H. Altgelt and T.H. Gouw (Editors). Marcel Dekker Inc., New York.

Dyroff, G.V. (Editor). 1993. *Manual on Significance of Tests for Petroleum Products*. 6th Edition. American Society for Testing and Materials, West Conshohocken, Pennsylvania.

Gab-Allah, M.A., Goda, E.S., Shehata, A.B., and Gamal, H. 2020. Critical Review on the Analytical Methods for the Determination of Sulfur and Trace Elements in Crude Oil. *Critical Reviews in Analytical Chemistry*, 50(2): 161–178.

Hsu, C.S., and Robinson, P.R. (Editors). 2017. *Handbook of Petroleum Technology*. Springer International Publishing AG, Cham, Switzerland.

Long, R.B., and Speight, J.G. 1989. Studies in Petroleum Composition. I: Development of a Compositional Map for Various Feedstocks. *Revue Institut Français du Pétrole*, 44: 205–217.

Magomadov, A.S. 2004. Viscosity and Thermal Conductivity of Heavy Oils in the Region of High Temperatures and Pressures. *High Temperature*, 42: 238–241. https://journals.scholarsportal.info/details/0018151x/v42i0002/238_vatcohrohtap.xml

Parkash, S. 2003. *Refining Processes Handbook*. Gulf Professional Publishing, Elsevier, Amsterdam, Netherlands.

Rahman, Md S., Miah, M., Ahmed, S., Rahman, M., and Hossain, M.E. 2016. PVT Properties of Heavy Oil—A Critical Review. Proceedings. International Conference on Petroleum Engineering 2016 (ICPE-2016) Bangladesh University of Engineering & Technology (BUET) Dhaka-1000, Bangladesh. https://www.researchgate.net/publication/313560785_PVT_Properties_of_Heavy_Oil_-_A_Critical_Review

Speight, J.G. 2014a. *The Chemistry and Technology of Petroleum*. 5th Edition. CRC Press, Taylor & Francis Group, Boca Raton, Florida.

Speight, J.G. 2014b. *Oil and Gas Corrosion Prevention*. Gulf Professional Publishing Company, Elsevier, Oxford, United Kingdom.

Speight, J.G. 2015a. *Handbook of Petroleum Product Analysis*. 2nd Edition. John Wiley & Sons Inc., Hoboken, New Jersey.

Speight, J.G. 2015b. *Fouling in Refineries*. Gulf Professional Publishing Company, Elsevier, Oxford, United Kingdom.

Speight, J.G. 2017. *Handbook of Petroleum Refining*. CRC Press, Taylor & Francis Group, Boca Raton, Florida.

Wallace, D. (Editor). 1988. *A Review of Analytical Methods for Bitumens and Heavy Oils*. Alberta Oil Sands Technology and Research Authority, Edmonton, Alberta, Canada.

8

Chromatographic Properties

8.1 Introduction

By definition, the term "chromatography" refers to an analytical method that is a technique commonly used to separate a mixture of chemical substances into the individual components, so that the individual components can be analyzed, identified, and whenever necessary be purified. There are many types of chromatographic methods such as liquid chromatography (LC), gas chromatography (GC), and ion-exchange chromatography all of which employ the same basic principles. Chromatography has found wide use in the area of crude oil analysis and has become an important biophysical technique that enables the separation, identification, and purification of the components of a mixture for qualitative and quantitative analysis.

By way of introduction, the common terminology used in this field of analysis is as follows: (1) the mobile phase or carrier is solvent moving through the column, (2) the stationary phase or adsorbent is the substance that stays fixed inside the column, (3) the eluent is the fluid entering the column, and (4) the eluate is the fluid exiting the column that is collected in flasks for further examination. Elution is the process of washing out a compound through a column using a suitable solvent and the analyte is the mixture whose individual components have to be separated and analyzed.

The history of the use of chromatography goes back to the 16th century when, in the year 1512, a Strasbourg surgeon (by the name of Brunschwig) purified ethyl alcohol by a chromatographic technique, and Day, an American geochemist, separated crude oils using a clay mineral, commonly referred to as Fuller's earth (1898–1903).

Historically, the work carried out by a technique that is now commonly referred to as column chromatography is also often referred to as adsorption chromatography. Many variations have been developed since the original discovery. The more common chromatographic procedures are GC, adsorption chromatography, gel

Handbook of Heavy Oil Properties and Analysis, First Edition. James G. Speight.

permeation chromatography (GPC), ion-exchange chromatography, and high performance chromatography that will be discussed in more detail here. Paper chromatography and thin layer chromatography (TLC) are used less frequently for feedstock analysis and are omitted from this work.

Due to the startling demands on the industry during and after World War II, and the emergence of the age of petrochemicals and plastics, the industry needed to produce materials not even considered as products in the decade before the war. Thus, crude oil refining took on the role of technological innovator as new and better processes were invented and advances in the use of materials for reactors were developed. In addition, there became a necessity to find out more about refinery feedstocks so that refiners might be able to enjoy the luxury of predictability and plan a product slate that was based on market demand. A difficult task was when the character of the feedstock was unknown or was (in the present context) a highly viscous material. The idea that refining should be a hit and miss affair was not acceptable. The demand arose for a more detailed inspection of the character of the feedstock which can vary markedly in composition and properties according to the source of the feedstock and it is these variations in feedstock character that caused changes to optimization of existing processes as well as for the development and design of new processes.

For example, valuable information can be obtained from the true boiling point (TBP) curve, which is a function of percent weight distilled and temperature, that is, a boiling point distribution. However, there are boiling point limitations on this function that are well below the final boiling point (FBP) of a crude oil. In addition to the boiling point distribution, bulk physical properties, such as specific gravity and viscosity, have assisted in the establishment of certain empirical relationships for refining from the TBP curve. Many of these relationships include assumptions that are based on experience with a range of feedstocks. However, movement of the refining to feedstocks that contain higher proportions of coke-forming materials emphasizes the need for more definitive data that would enable more realistic predictions to be made of crude oil behavior in refinery operations.

However, regardless of the type of chromatography employed for separation and identification of the feedstock constituents, the mechanisms involved in the separation are basically the same. The two major mechanisms at work during a chromatographic separation are displacement and partition. Few instances involve only one mechanism, but certain types of separations emphasize one mechanism over the other.

Chromatographic procedures are often referred to as displacement procedures. In the simplest procedure (that relates more to the historical work), a glass column is packed tightly with powder-size dry particles, the inert phase. A thin layer of a sample mixture consisting of weakly adsorbed compounds and strongly adsorbed compounds is added to the top of the inert phase. A liquid, which is more strongly adsorbed to the inert phase than any of the compounds in the sample, is now poured into the top of the column. This is called the mobile phase and serves to wash the sample down the column, a process called elution.

The inert particles insofar as they are mainly silicate derivatives and aluminum derivatives contain active sites due to charge distribution, and molecules of the solute attach themselves to the particles based on their relative adsorption tendencies derived from size or polarity and pass down the column as far as necessary until all of the molecules are adsorbed. An equilibrium is established with the strongly adsorbed molecules remaining on the column and the weakly adsorbed molecules passing out at the bottom of the column.

The analysis of the constituents in feedstocks started over 100 years ago, when in 1865 several aromatic hydrocarbons were identified as constituents of feedstocks. Identification of individual constituents of feedstocks continued and the rapid advances in analytic techniques have allowed the identification of large numbers of feedstocks constituents. At this time, the major chemical types of compounds that exist in crude oil have been identified and many members of the various homologous series have been separated or conclusively identified by various techniques.

Thus, chromatographic analysis of complex sample such as the crude oil-related feedstocks has become an important method for the characterization of crude oils and viscous feedstocks. Both gas chromatographic analysis (often referred to as GC analysis) and liquid chromatographic analysis (often referred to as LC analysis) are widely used in many laboratories for the analysis of crude oil, the viscous feedstocks, and their respective products. The major area of growth in chromatographic analysis in the past four decades has been in the development of high-performance liquid chromatography (HPLC) and high-performance anion-exchange chromatography (HPAEC), which is a form of IO chromatography.

It is the purpose of this chapter to review the methods that can be used for the identification of the constituents of refinery feedstocks. Particular emphasis has been placed on the chromatographic techniques that have offered valuable information in the past (Altgelt and Gouw, 1975) and will continue to offer in the future valuable information about the constituents of the feedstocks. Spectroscopic and spectrometric methods are not included here (unless there is some mention of use as a detector system) and, to emphasize the importance of spectroscopy and spectrometry are included as a separate chapter as well as their use in structural determination. Statistical methods, as applied to analytical chemistry, are beyond the scope of this chapter and have been omitted; the readers are referred to other more comprehensive works on this subject (e.g., Meier and Zünd, 2000).

As with other chapters, the nature of the subject matter of this chapter dictates that there is overlap with the subject matter of other chapters. This is inevitable, and the readers are referred to other relevant chapters to place the subject matter of all chapters into the proper perspective.

8.2 Adsorption Chromatography

The adsorption process involves the adhesion of molecules, ions, or atoms from a gas, a liquid, or a dissolved solid (the adsorbate) on to a surface (the adsorbent) which creates a film of the adsorbate on the surface of the adsorbent. It is worthy of note here, that adsorption differs from absorption which is a process in which a fluid is dissolved by or permeates a liquid or solid. Thus, adsorption chromatography is a method used for the separation of components from a mixture using their differential adsorption on a stationary component applied over a stationary solid surface.

Adsorption chromatography is the oldest type of chromatographic method and employs the use of a mobile phase which is, for the samples used in the current context of the analysis of high boiling fractions of crude oil and the viscous feedstocks, in the liquid form. In this method, the mobile phase is adsorbed onto the surface of a stationary solid phase.

The stationary phase is the adsorbent which is instrumental in removing solutes from the adsorbent so that they can move with the mobile phase. The mobile phase is either a liquid or a gas is used to remove solutes from the adsorbent so that they can move with the mobile phase. When a liquid is used as a mobile phase, the method is liquid–solid chromatography (LSC) but when a gas is used as a mobile phase, the method is gas–solid chromatography (GSC).

Typically, in the laboratory method (referred to as adsorption chromatography or column chromatography), a glass column of 1–4cm in diameter and 15–60cm in length with a mobile phase passes through the inert phase by the force of gravity, mild suction, or at most, the application of mild pressure. The column (which should be a wide-bore column to prevent plugging) may or may not be equipped with an attached solvent reservoir for the mobile phase but does not require an attached stopcock to control the flow rate with viscous mobile phases. Removable bottoms are used for those separations that require long times. The bottom can be removed and the column packing pushed out with a plunger. The column packing then can be sectioned, and the desired compounds extracted or washed from the inert phase for further use.

Adsorption chromatography has helped to characterize the group composition of crude oils and hydrocarbon products since the beginning of this century (Altgelt and Gouw, 1979; Altgelt and Boduszynski, 1994; Masel, 1995). In fact, it is probably the most used and abused of all of the chromatographic methods that have been applied to feedstock characterization.

The analytical chemist needs to choose and rationalize the number of fractions that are to be the outcome of the separation procedure. And this rationalization must be based on the manner in which the data will be used. The techniques used to produce fractions that will be used for process predictability should be approached differently to the techniques used to search for geochemical markers. The advisability of separating viscous feedstocks into a multitude of fractions must be given serious pre-application consideration. The analytical chemist must ask themselves, and be prepared to advise the process chemist or process engineer, about the number of fractions necessary to accomplish the goal. Four fractions might suffice, thus the tendency to proceed and separate the feedstock into 20 or more discreet fractions is always a dream! Such temptation must be avoided and fractionation schemes that are meaningful to the task are to be preferred.

The apparatus for the various chromatographic separation schemes usually consists of a glass column 1–5 cm in diameter and 20–100 cm long, packed with a solid material (called the inert phase, although it seldom is) such as silica gel or diatomaceous earth (ASTM D2007, 2021) (Chapter 3). The sample is placed as a thin band on top of the packing and washed down the column with a liquid (mobile phase). As the compounds move down the column, small differences in their molecular shape and polarity allow them to become separated.

The method can be applied to conventional crude oil and also to the viscous feedstocks but subtle differences may be applied to the procedure depending on the nature of the material to be separated. For example, when conventional petroleum is to be separated, the feedstock may first be topped (the volatile constituents removed) before separation. The procedure separates the feedstock into saturate, aromatic, resin, and asphaltene fractions or for specific chemical types such as sulfonate derivatives (ASTM D3712, 2021).

In general, the method involves deasphaltening the feedstock and the resulting deasphaltened oil passed through or refluxed on a recycle adsorption apparatus packed with Attapulgus clay. Resins adsorb on the Attapulgus clay while the saturates and aromatics are carried to the kettle. The resins are eluted from the Attapulgus clay with methyl ethyl ketone and tetrahydrofuran containing 50% water by volume water. Separation of the saturates and aromatics is performed on a chromatographic column packed with alumina and silica gel. The saturates and aromatics are eluted with *n*-pentane and benzene, respectively. Any resin constituents that may have remained are eluted with methyl ethyl ketone. All fractions are determined gravimetrically after evaporation of the solvent.

The fractions denoted as saturates, aromatics, resins, and asphaltenes in this procedure are defined operationally by the specific test conditions described in the method and separations according to molecular type. Instead, the fractions represent a separation based on solubility and strength of adsorption and must be identified with the procedure by which they are generated. This operational definition of the fractions is emphasized since similar terminology for designating fractions is employed by methods using different separation conditions.

In practice, the separation is reproducible and hence useful for evaluating similarities and differences among different feedstock samples.

One of the common problems of such a fractionation scheme is the nature of the adsorbent. In the early reports on feedstock fractionation, clay often appeared as an adsorbent to effect separation of the feedstock into various constituent fractions. However, clay (fuller's earth, Attapulgus clay attapulgite, and the like) is often difficult to define with any degree of precision from one batch to another. Variations in the nature and properties of the clay can, and will, cause differences not only in the yields of composite fractions but also in the distribution of the compound types in those fractions. In addition, irreversible adsorption of the more polar constituents on the clay can be a serious problem when further investigations of the constituent fractions are planned.

One option for resolving this problem has been to use more standard adsorbents such as alumina and silica. These materials are easier to define and are often accompanied by guarantees of composition and type by various manufacturers. They also tend to adsorb less of the feedstock than a clay. Once the nature of the adsorbent is guaranteed, reproducibility becomes a reality. Without reproducibility the analytical method does not have credibility.

The type and relative amount of certain hydrocarbon classes in the matrix can have a profound effect on the quality and performance of the hydrocarbon product. The fluorescent indicator adsorption (FIA) method (ASTM D1319, 2021) has served for over 30 years as the official method of the industry for measuring the paraffinic, olefinic, and aromatic content of gasoline and jet fuel. The technique consists of displacing a sample under *iso*-propanol pressure through a column packed with silica gel in the presence of fluorescent indicators specific to each hydrocarbon family.

The segregation of individual components from a mixture can be achieved by application of adsorption chromatography in which the adsorbent is either packed in an open tube (column chromatography) (Altgelt and Gouw, 1979) or shaped in the form of a sheet (TLC) and a solvent is chosen that will elute the various constituents for the adsorbent. Chromatographic separations are usually performed for the purpose of determining the composition of a sample (Snyder, 1968). Even with such complex samples as the viscous feedstocks, much information about the chemical structure of a fraction can be gained from the separation data.

In the present context, the challenge is the nature of the heteroatomic species in the heavier feedstocks. It is these constituents that are largely responsible for coke formation and catalyst deactivation during refining operations. Therefore, it is these constituents that are the focus of much of the study. An ideal integrated separation scheme for the analysis of the heteroatomic constituents should, therefore, meet several criteria:

1) The various compound types should be concentrated into a reasonable number of discrete fractions, and each fraction should contain specific types of the heteroatomic compounds. It is also necessary that most of the heteroatom constituents are separated from the hydrocarbons and sulfur compounds that may constitute the bulk of the sample.

2) Perhaps most important, the separation should be reproducible insofar as the yields of the various fractions and the distribution of the compound types among the fractions should be constant within the limits of experimental error.
3) The separation scheme should be applicable to high-boiling distillates and heavy feedstocks such as residua since heteroatomic compounds often predominate in these feedstocks.
4) The separation procedures should be relatively simple to perform and free of complexity.
5) Finally, the overall separation procedure should yield quantitative or, at worst, near-quantitative recovery of the various heteroatomic species present in the feedstock. There should be no significant loss of these species to the adsorbent or, perhaps more important, any chemical alteration of these compounds. Should chemical alteration occur, it will give misleading data that could have serious effects on refining predictions or on geochemical observations.

Group type analysis by means of chromatography has been applied to a wide variety of feedstock types, such as residua and asphalt (Chapter 7). These types of analysis are often abbreviated by the names PONA (paraffins, olefins, naphthenes, and aromatics), PIONA (paraffins, *iso*-paraffins, olefins, naphthenes, and aromatics), PNA (paraffins, naphthenes, and aromatics), PINA (paraffins, *iso*-paraffins, naphthenes, and aromatics), or SARA (saturates, aromatics, resins, and asphaltenes).

The USBM-API (US Bureau of Mines-American Petroleum Institute) method allows fractionation of samples into acids, bases, neutral nitrogen compounds, saturates, and mononuclear aromatic compounds, di-nuclear aromatic compounds, and polynuclear aromatic compounds. Multidimensional techniques, that is, the combination of two or more chromatographic techniques, can be very useful to gain further information about the individual components of chemical groups. Compounds can be isolated and identified from complex matrices and detailed fingerprinting of feedstock constituents is feasible (Altgelt and Gouw, 1975).

In normal phase chromatography (above), the mobile phase is nonpolar and the column packing is polar in nature. Separation occurs because sample molecules dissolved in the mobile phase interact with the packing. The molecules are temporarily adsorbed and, after some time, returned to the mobile phase. This process of adsorption–desorption occurs many times for a molecule on its way through the column. The greater the compound's polarity, the stronger it interacts with the sorbent surface, the more it is retained, and the later it elutes. Various types of interactions occur, from a weak induced dipole-induced dipole interaction (London forces) to hydrogen bonding to strong acid–base attraction.

Normal phase partitioning chromatography is a variant of the normal phase chromatographic method. In the method, the column packing is coated with a layer of a polar liquid that is immiscible with the nonpolar mobile phase and which will remain on (adhere to) on the packing as the stationary phase. Sample molecules move back and forth between the two phases and are partitioned between these according to their respective solubility in the phases. Polar sample molecules ordinarily have a greater solubility ratio in favor of the stationary phase than less polar ones and spend more time there. Consequently, they elute later than nonpolar molecules. During the last two decades, so-called bonded phases have replaced the liquid stationary phases in partitioning chromatography.

Bonded phase packings are usually silica particles with chemically modified surfaces. The hydroxyl (OH) groups on the silica are either silanized or esterified and, in effect, replaced by organic groups, mostly paraffinic hydrocarbons of eight or more carbon atoms. These hydrocarbons are nonpolar and offer no advantage in normal phase chromatography. However, they are very useful in reversed phase chromatography where the sample is eluted with a very polar solvent such as acetonitrile and even mixtures of acetonitrile or similar solvents with methanol or water. Here, the sample molecules can partition again between a polar phase and a nonpolar one, except that the phases are reversed from those in normal phase chromatography. Here, the nonpolar molecules are more soluble in the stationary phase than in the polar mobile phase and are, therefore, retained more strongly than polar molecules.

The advantage of reversed phase chromatography with its combination of an inert packing surface and a polar mobile phase is that, theoretically at least, it is a very mild method and prevents sample loss due to irreversible adsorption of highly polar molecules, which can be a serious problem in normal phase chromatography of high-boiling materials. Unfortunately, it is difficult to completely cover the silica surface with the organic phase. Often a small percentage of the silica hydroxyl groups remains unreacted and is available for interaction with polar sample molecules that may then be strongly adsorbed.

Fractionation of aromatics by ring number is best done on column packings with bonded phases, for example, amino-derivatized silica, cyano-derivatized silica, phenyl-derivatized silica, or nitro-derivatized silica (Altgelt and Boduszynski, 1994). Separation on these packings proceeds only nominally by ring number but the method really discriminates by the number of pi electrons or double bonds in the ring system. For cata-condensed aromatics, this is equivalent to ring number separation. However, peri-condensed molecules have fewer double bonds than cata-condensed ones with the same ring number. For instance, pyrene (four rings, eight double bonds) elutes after phenanthrene (three rings, seven double bonds) and before chrysene (four rings, nine double bonds).

Molecules with nonplanar aromatic ring systems, usually sterically hindered, elute before planar ones and long, stretched-out polynuclear aromatic systems elute later than more compact ones. Generally, molecules with a larger (aromatic) surface area can interact with more polar groups on the packing surface and will, therefore, be more strongly retained.

An extension of adsorption chromatography, ligand exchange chromatography (LEC), has been used as a way to separate sulfur compounds from hydrocarbons. The method exploits the specific interactions of sulfur compounds or other heteroatoms with specific metal ions. Silver, mercury, copper, and zinc ions, for example, form coordination complexes with (aliphatic) sulfides, whereas palladium ions interact strongly with (aromatic) thiophenic compounds. These metals, deposited as chlorides or other salts on silica, can, therefore, selectively slow down the elution of sulfur compounds relative to hydrocarbons on a chromatographic column.

8.3 Gas Chromatography

GC (also referred to as gas–liquid chromatography, GLC) is a chromatographic method for separating the volatile components of various mixtures in which the mobile phase is a gas (sometimes referred to as the "carrier gas") (Fowlis, 1995; Grob, 1995).

GC, which is based on the principle of a stationary phase and a mobile phase, remains a primary technique for determining hydrocarbon distribution in petroleum products through the identification of individual hydrocarbons. Although a measure of purity by GC is often sufficient for many purposes, it is not always sufficient for measuring absolute purity—not all possible impurities will pass through the chromatographic column, and not all those that do will be measured by the detector. Absolute purity is best measured by distillation range or freeze or solidification points. Despite this disadvantage, the technique is still widely used and is the basis of many current standard test methods for the determination and measurement of hydrocarbons in petroleum products. When classes of hydrocarbons such as olefins need to be measured, techniques such as bromine index are used (ASTM D1492, 2021; ASTM D2710, 2021; ASTM D5776, 2021).

Briefly, *GLC* is a method for separating the volatile components of various mixtures. It is, in fact, a highly efficient fractionating technique, and it is ideally suited to the quantitative analysis of mixtures when the possible components are known and the interest lies only in determining the amounts of each present. In this type of application, GC has taken over much of the work previously done by the other techniques; it is now the preferred technique for the analysis of hydrocarbon gases, and gas chromatographic in-line monitors are having increasing application in refinery plant control. GLC is also used extensively for individual component identification, as well as percentage composition, in the gasoline boiling range.

It is, in fact, a highly efficient fractionating technique, and it is ideally suited to the quantitative analysis of mixtures when the possible components are known and the interest lies only in determining the amounts of each present. In this type of application, GC has taken over much of the work previously done by the other techniques; it is now the preferred technique for the analysis of hydrocarbon gases, and GC in-line monitors are having increasing application in refinery plant control.

The longer compounds remain in a column, the more time they have to diffuse and remix and the best separation occurs if the stationary phase was very thin and the mobile phase was changed from a liquid to a gas. This would provide fewer foreign molecules for the sample molecules to hit, thus reducing the mixing by diffusion. The detectors available at this time (1952) had poor response, and in order to separate sufficient material to be detected, a relatively thick layer of stationary phase was required. However, the liquid mobile phase could be changed to a gas. Rapid development followed.

The importance of GLC lies in its powerful separating ability. Columns can have the equivalent of up to 1,000,000 theoretical plates compared to conventional distillation that normally operates in the range of 5–10 theoretical plates. In the case of gas–liquid chromatography (GLC or GC), the two phases are a flowing gas phase and a liquid phase held stationary on an inert solid support.

The mobile phase is the carrier gas, and the gas selected has a bearing on the resolution. Nitrogen has very poor resolution ability, while helium or hydrogen is the better choice with hydrogen being the best carrier gas for resolution. However, hydrogen is reactive and may not be compatible with all sets of target analytes. There is an optimum flow rate for each carrier gas to achieve maximum resolution. As the temperature of the oven increases, the flow rate of the gas changes due to thermal expansion of the gas. Most modern gas chromatographs are equipped with constant flow devices that change the gas valve settings as the temperature in the oven changes, so changing flow rates are no longer a concern. Once the flow is optimized at one temperature, it is optimized for all temperatures.

In practice, for the older style using packed columns, a small sample (1–5 μl) is injected by means of a syringe or sample loop into a heated injection port, where the liquid is vaporized into the flowing gas stream (usually nitrogen, helium, or argon). The carrier gas transports the sample through the column (glass, copper, or stainless steel, 3–6 mm diameter, 1–2 m long), which is packed with a solid of large surface area. This solid support, which is commonly a diatomaceous earth or crushed firebrick, is impregnated with a nonvolatile liquid phase. Separation of the injected mixture occurs by partitioning between the gas stream and the liquid phase. A large number of partitions occur and even very small differences in physical and chemical properties allow the components to become separated.

The gas stream, containing the separated compounds, then is passed through a detector; the simplest type is one that measures the difference in thermal conductivity between the sample plus helium (He) or nitrogen (N_2) gas and a reference helium or nitrogen gas. The detector signal is monitored continuously by a recorder that plots the components as a function of time, resulting in Gaussian-shaped peaks. More recently, capillary columns, 0.05–0.53 mm in diameter and 10–100 in long, having as many as 4,000 plates, are becoming common and finding use as analytical methods (ASTM D2268, 2021).

The elapsed time between the injection and the center of a peak is called the retention time of that compound. Although a gas chromatograph separates compounds, the method does not identify them but the retention time may be used as corroborating evidence in the identification of a compound by injecting a known sample under identical conditions and observing identical retention times. Another outcome of the chromatographic method is the identification of individual components of a mixture as well as the calculation of GC response factors for individual compounds (ASTM D4626, 2021). The unknown–known comparison then should be repeated on another column of different polarity to substantiate the previous agreement of retention times. Failure to use columns of different polarity to corroborate the identification of compounds can lead to embarrassing situations.

The evolution of GC has been a major factor in the successful identification of feedstock constituents. It is, however, almost impossible to apply this technique to the higher boiling feedstock constituents because of the comparatively low volatility of these materials. It is this comparative lack of volatility in the higher molecular weight, asphaltic constituents of feedstocks that brought about another type of identification procedure, namely, carbon-type analysis.

Thus, it is not surprising that GC has been used extensively for individual component identification, as well as percentage composition, in the gaseous boiling ranges (ASTM D2163, 2021; ASTM D2426, 2021; ASTM D2504, 2021; ASTM D2505, 2021; ASTM D2593, 2021; ASTM D2712, 2021; ASTM D4424, 2021; ASTM D5303, 2021; ASTM D6159, 2021), in the gasoline boiling range (e.g., ASTM D2426, 2021; ASTM D2427, 2021; ASTM D3525, 2021; ASTM D3606, 2021; ASTM D4815, 2021; ASTM D5134, 2021; ASTM D5441, 2021; ASTM D5443, 2021; ASTM D5501, 2021; ASTM D5580, 2021; ASTM D5599, 2021; ASTM D5623, 2021; ASTM D5845, 2021; ASTM D5986, 2021; ASTM D7096, 2021), in higher boiling ranges such as diesel fuel (ASTM D3524, 2021), aviation gasoline (ASTM D3606, 2021), and wax (ASTM D5442, 2021), as well as for the boiling range distribution of feedstock fractions (ASTM D2887, 2021) (Speight, 2014, 2015).

GC analysis is the method of choice for light and middle distillates because of its high resolution. However, for high-boiling viscous feedstocks, the method is somewhat limited in use because of the propensity of the method to separate the constituents primarily by carbon number and because of the immense number of compound types and homologs in these materials, let alone their low volatility

(Altgelt and Boduszynski, 1994). The two main applications here are simulated distillation and the separation and identification of compound types and their homologs in well-defined LC fractions.

The technique has proved to be an exceptional and versatile instrumental tool for analyzing compounds that are of low molecular weight and that can be volatilized without decomposition. However, these constraints limit the principal applicability in crude oil science to feedstock identification when the composition is known to be in the low to medium boiling range. The use of this technique for direct component analysis in the heavy fractions of the feedstock is not only difficult but also subject to many limitations.

For example, the number of possible components of a certain molecular weight range increases geometrically with increasing molecular weight. Furthermore, there is a corresponding sharp decrease in physical property differences between isometric structures as the molecular weight increases. Thus, it is very difficult, and on occasion almost impossible, to separate and identify single components in the heavier fractions of the feedstock by GC. Indeed, the molecular weights of the constituents dictate that long residence times are inevitable. This is inevitably accompanied by the requirement of increased column temperature, which decreases the residence time on the column but, at the same time, increases the possibility of thermal decomposition.

In all senses, the instrumentation for GC is fairly straightforward. A carrier gas passes through a controller to the column (packed with an adsorbent) at the opening of which is a sample injector. The carrier gas then elutes the components of the mixture through the column to the detector at the end of which may be another gas flow monitor.

Any gas, such as helium, argon, nitrogen, or hydrogen that is easily distinguishable from the components in the mixture, may be used as the carrier gas. Column dimensions vary, but for analytic purposes a packed column may be 6 ft (2 m) long by 3 in. (6 mm) in diameter. It is also necessary to use a dissolving liquid as part of the column substance. This remains stationary on the adsorbent and effects partition of the components of the mixture. The solid support is usually a porous material that allows passage of the gas. For example, kieselguhr (diatomaceous earth), which can absorb up to 40% by weight of a liquid without appearing to be overly moist, is commonly used. The supporting material should not adsorb any of the components of the mixture and must, therefore, be inert.

Individual components of mixtures are usually identified by their respective retention times, that is, the time required for the component to traverse through the column under the specified conditions. Although tables for retention time data are available, it is more common in practice to determine the retention times of the pure compounds. The retention time of any component is itself a function of the many variables of column operation, such as the flow rate of the carrier gas and column temperature, and exact duplication of other operator's conditions may be difficult, if not impossible.

The sample size used in GC may vary upward from a microliter, and there is no theoretical upper limit to the size of the sample that may be handled if the equipment is built to accommodate it. The technique can be used for the analysis of mixtures of volatile vaporizable compounds boiling at any temperature between absolute zero (−273 °C, −459 °F) and 450 °C (840 °F). Identification of any substance that can be heated sufficiently without decomposing to give a vapor pressure of a few millimeters mercury is also possible.

The components of the feedstock have been studied extensively by GC. This includes the determination of saturated compounds and olefins in cracked naphtha, as well as the identification of alkylbenzenes and the higher molecular weight normal paraffins (to at least n-C_{40}) in higher boiling feedstocks. Previous methods usually involved the somewhat slow and careful fractional distillation followed by various analyses (physical and spectroscopic) of the fractions.

GC can also be used to determine the three main hydrocarbon types in gasoline: saturates, olefins, and aromatics. The procedure has been used for simple characterization of gasoline and naphtha. This application demonstrates the use of two techniques, a highly selective column and a chemical reaction to remove one group of compounds. Sometimes it is advantageous to follow such a type of separation with further separations. For example, the saturate fraction after separation from aromatics and olefins can be sent to a high-efficiency column for determination of individual isomers. Determination of individual saturate isomers is difficult or impossible when olefins and aromatics are present.

Distillation analyses by GC depend on the fact that certain nonpolar silicone liquid phases elute hydrocarbons nearly in order of boiling point. The analyses are made with temperature programming, and boiling point can be correlated with elution time or with column temperature at the time of emergence. Thus, a distillation curve can be constructed from gas-chromatographic data. Analyses by GC usually can be as accurate as they are by precision distillation.

The use of GC for direct component analysis in the heavier fractions, such as the viscous feedstocks (including residua), is beset by many problems, not the least of which is the low volatility and high adsorption tendencies of many of the high molecular weight constituents of the feedstock. The number of possible components in any given molecular weight range increases markedly with the molecular weight, and there is a significant drop in the differences in physical properties among similar structural entities. This limits the ability of GC, and unless the sample has been fractionated by other techniques to reduce the complexity, complete component analysis is difficult, if not impossible.

In view of the molecular characterizing nature of spectrometric techniques, it is not surprising that considerable attention has been given to the combined use of

GC and these techniques. In recent years, the use of the mass spectrometer to monitor continuously the effluent of a chromatographic column has been reported, and considerable progress has been made in the development of rapid scan infrared spectrometers for this purpose. The mass spectrometer, however, has the advantage that the quantity of material required for the production of a spectrum is considerably less than that necessary to produce an infrared spectrum.

A gas chromatography–mass spectrometry system is used to measure concentrations of target volatile and semi-volatile petroleum constituents. It is not typically used to measure the amount of total petroleum hydrocarbons. The advantage of the technique is the high selectivity or ability to confirm compound identity through retention time and unique spectral pattern. The method is used for identification and quantification of the constituents of petroleum fractions and petroleum products.

To reduce the possibility of false positives, the intensities of one-to-three selected ions are compared to the intensity of a unique target ion of the same spectrum. The sample ratios are compared to the ratios of a standard. If the sample ratios fall within a certain range of the standard, and the retention time matches the standard within specifications, the analyte is considered present. Quantification is performed by integrating the response of the target ion only.

Mass spectrometers are among the most selective detectors, but they are still susceptible to interferences. Isomers have identical spectra, while many other compounds have similar mass spectra. Heavy petroleum products can contain thousands of major components that are not resolved by the gas chromatograph. As a result, multiple compounds are simultaneously entering the mass spectrometer. Different compounds may share many of the same ions, confusing the identification process. The probability of misidentification is high in complex mixtures such as petroleum products.

The mass spectrometer identifies chemical compounds principally in terms of molecular type and molecular weight, and for many problems, therefore, it becomes necessary to use additional means of identification; the integrated GC infrared system is a very valuable complement to the mass spectrometer route.

Considerable attention has also been given to trapping devices to collect GC fractions for examination by one or more of the spectroscopic techniques. At the same time, developments in preparative GC have contributed even more to the compositional studies of the feedstock and its products. With column size of 4–6 in. diameter capable of dealing with sample sizes of 200 ml or more, there is every possibility that GC will replace distillation in such areas as standard crude oil assay work.

GC also provides a simple and convenient method for determining *n*-paraffin distribution throughout the feedstock distillate range. In this method, the *n*-paraffins are first separated by activated chemical destruction of the sieve with hydrofluoric acid, and the identity of the individual paraffins is determined chromatographically. This allows *n*-paraffin distribution throughout the boiling range 170–500 °C (340–930 °F) to be determined.

GC has also been found useful for the preparation of simulated distillation curves. By integrating increments of the total area of the chromatogram and relating these to the boiling points of the components within each increment, which are calculated from the known boiling points of the easily recognizable *n*-paraffins, simulated boiling point data are produced.

GC process analyzers have become very important in refineries. In some refineries, more samples are analyzed automatically by process chromatographs than are analyzed with laboratory instruments. These chromatographs are usually fully automatic. In some cases, after an analysis the instrument even makes automatic adjustments to the refinery unit. The chromatographs usually determine from 1 to 10 components, and the analyses are repeated at short intervals (15–20 min) over 24 hr.

A more recent, very important development in GC is its combination with a mass spectrometer as the detector. The technique in which GC is combined with spectrometry (GC/MS) has proved to be a powerful tool for identifying many compounds at very low levels in a wide range of boiling matrix (ASTM D5769, 2021). By the combination of the two techniques in one instrument, the onerous trapping of fractions from the GC column is avoided and higher sensitivities can be attained. In passing through the GC column, the sample is separated more or less according to its boiling point.

Although insufficient component resolution is observed in most cases, the eluting compounds at any time are usually closely related to each other in boiling point and molecular weight or both and are free from interfering lower and higher molecular weight species. Because of the reduced complexity of the GC fractions, mass spectrometric scans carried out at regular intervals yield simpler spectra from which compound classes can more easily be determined.

Pyrolysis GC can be used for information on the gross composition of viscous feedstocks. In this technique, the sample under investigation is pyrolyzed and the products are introduced into a GC system for analysis. There has also been extensive use of pyrolysis GC by geochemists to correlate crude oil with source rock and to derive geochemical characterization parameters from oil-bearing strata.

In the technique of inverse GC, the sample under study is used as the stationary phase and a number of volatile test compounds are chromatographed on this column. The interaction coefficient determined for these compounds is a measure of certain qualities of the liquid phase. The coefficient is, therefore, indicative of the chemical interaction of the solute with the stationary phase.

Despite successful applications to some viscous feedstocks, GC itself is generally inadequate for the primary separation of such high molecular weight materials. Although its temperature range may well be high enough, the resolution is not sufficient for the large number of compound types and homologs in these complex materials. Resolution is further impaired by the peak broadening caused by high molecular weight species. For these heavy materials, LC and supercritical fluid chromatography (SFC) are superior and are often the method of choice. The wide choice of mobile phases for these methods permits adaptation of the methods to the compositional subtleties of the feedstocks in a manner that is not possible by GC. In addition, SFC has the additional advantage of performance with the application of pressure.

The prime use of GC in the analysis of high-boiling fractions is in simulated distillation (simdis) with flame ionization detection (FID), in GC with atomic emission detection and other element specific detection (Lee et al., 1984), and in combination with mass spectrometry as gas chromatography–mass spectrometry.

8.4 Gel Permeation Chromatography

GPC (also referred to as “size exclusion chromatography” and on occasion as “gel filtration chromatography”) is a chromatographic technique in which the separation is based on differences in the size of the sample molecules. The method is used to separate analytes on the basis of size, typically in organic solvents. The column packing is made from beads of a porous gel shown which separations according to molecular size.

Large molecules cannot fit into the gel at all and are washed out of the column. The lowest molecular weight molecule that does not enter the gel is called the exclusion limit, and the volume required to elute the large molecules is called the void volume. The smaller molecules can permeate the gel—the smaller the molecule, the farther the penetration and the greater the retardation on or within the gel. This results in a separation because it requires more solvent to wash the smaller molecules out of the gel.

The size of the pores determines the molecular weight range of the compounds that can be separated. Size exclusion is often used to clean up an extracted solution such as removing waxes from the residua that are produced in a variety of methods. Also, there is a general rule is that compounds that differ by 10% in size can be separated in the same column. A series of columns, each fitted with a different exclusion limit gel, can be used to separate a multicomponent mixture.

Gel filtration chromatography (GFC) has been successfully employed for application to aqueous systems by biochemists for more than three decades. This technique was developed using soft, cross-linked dextran beads. The other technique, GPC (also called size exclusion chromatography, SEC), employs semirigid, cross-linked polystyrene beads. In either technique, the packing particles swell in the chromatographic solvent and form a porous gel structure.

The distinction between the methods is based on the degree of swelling of the packing insofar as, for example, the dextran swells to a much greater extent than the polystyrene. Subsequent developments of rigid porous packings of glass, silica, and silica gel have led to their use and classification as packings for GPC.

GPC, also called SEC, involves the use of a column that is packed with a gel material of suitable pore size. Under conditions of constant flow, the solutes are injected onto the top of the column, whereupon they appear at the detector in order of decreasing molecular weight. The separation is based on the fact that the larger solute molecules cannot be accommodated within the pore systems of the gel beads and thus are eluted first. On the other hand, the smaller solute molecules have increasing volume within the beads, depending upon their relative size, and may require more time to elute from the adsorbent.

Thus, it is possible, with careful flow control, calibration, injection, and detection (usually by refractive index or UV absorption), to obtain an accurate chromatographic representation of the molecular weight distribution of the solute. This premise involves the assumption that there is no chemical or physical interaction between the solute and the gel that negates the concept of solute size and pore size. For example, highly polar, small molecules that could associate in solution and are difficult to dissociate could conceivably appear in the incorrect molecular weight range.

In theory, GPC is an attractive technique for the determination of the number of average molecular weight (M_n) distribution of fractions. However, it is imperative to recognize that the viscous feedstocks contain constituents of widely differing polarity, including nonpolar paraffins and naphthenes (alicyclic compounds), moderately polar aromatics (mononuclear and condensed), and polar nitrogen, oxygen, and sulfur species. Each particular compound type interacts with the gel surface to a different degree. The strength of the interaction increases with increasing polarity of the constituents and with decreasing polarity of the solvent. It must, therefore, be anticipated that the ideal linear relationship of log M_n against elution volume V_e that may be operative for nonpolar hydrocarbon species cannot be expected to remain in operation. It must also be recognized that the lack of realistic standards of known number average molecular weight distribution and of chemical nature similar to that of the constituents used for calibration purposes may also be an issue. However, GPC has been employed in the study of feedstock constituents, especially the heavier constituents, and has yielded valuable data (Albaugh et al., 1971).

The adoption of GPC represents a novel approach to the identification of the constituents since the method is not limited by the vapor pressure of the constituents. However, the viscous feedstocks are not homologous mixtures and differ only in molecular weight. In any feedstock, a large variety of molecular species, varying from paraffinic molecules to the polynuclear aromatic ring systems, may not follow the assumed physical relationships that the method dictates from use with polymers.

SEC is the separation method that comes closest to differentiating by molecular weight only, and is almost unaffected by chemical composition. The method actually separates by molecular size and has been used to measure molecular weights (Altgelt and Gouw, 1979; Hagel and Johnson, 1992) although there is some question about the value of the data when the method is applied to asphaltenes (Speight et al., 1985).

SEC is usually practiced with refractive index detection and yields a mass profile (concentration vs. time or elution volume) that can be converted to a mass vs. molecular weight plot by means of a calibration curve. The combination of SEC with element-specific detection has widened this concept to provide the distribution of hetero-compounds in the sample as a function of elution volume and molecular weight.

The use of SEC with reverse phase HPLC with a graphite furnace atomic absorption (GFAA) detector has been described for measuring the distribution of vanadium and nickel in high molecular weight feedstock fractions, including the asphaltene fraction. Using variants of this technique, inductively coupled and direct current plasma atomic emission spectroscopy (ICP and DCP), the method was extended and improved the former size exclusion chromatography-graphite furnace atomic absorption (SEC-GFAA) which allows the separation to be continuously monitored.

The combination of GPC with another separation technique also allows the fractionation of a sample separately by molecular weight and by chemical structure. This is particularly advantageous for the characterization of the heavier fractions of feedstock materials because there are limitations to the use of other methods. Thus, it is possible to obtain a matrix of fractions differing in molecular weight and in chemical structure. It is also considered advisable to first fractionate a feedstock by GPC to avoid overlap of the functionality that might occur in different molecular weight species in the separation by other chromatographic methods.

In short, the GPC technique concentrates all of a specific functional type into one fraction, recognizing that there will be a wide range of molecular weight species in that fraction. This is especially true when the chromatographic feedstock is a whole feed rather than a distillate fraction.

8.5 High-Performance Liquid Chromatography

High-performance liquid chromatography (often referred to as HPLC and formerly referred to as high-pressure liquid chromatography) is a method that is widely used technique in analytical chemistry to separate, identify, and quantify each component in a mixture. The technique, particularly in the normal phase mode, has found great utility in separating different hydrocarbon group types and for identifying specific types of constituents.

Normal-phase high-performance liquid chromatography (often referred to as NP-HPLC) utilizes a polar stationary phase (usually silica, SiO_2) and less polar (nonaqueous) eluting solvents (such as *n*-hexane or ethyl acetate as the mobile phase). The separation is based on the ability of the analyte to participate in polar interactions (such as hydrogen bonding interactions or dipole–dipole-type interactions) with the surface of the sorbent. Typically, the polar compounds in the mixture being analyzed are adsorbed more strongly to the polar silica than the nonpolar compounds, which pass more quickly through the column and are eluted quicker than the polar compounds.

A HPLC system can be used to measure concentrations of target semi-volatile and nonvolatile petroleum constituents. The system only requires that the sample be dissolved in a solvent compatible with those used in the separation. The detector most often used in petroleum environmental analysis is the fluorescence detector. These detectors are particularly sensitive to aromatic molecules, especially polynuclear aromatic hydrocarbons. An ultraviolet detector may be used to measure compounds that do not fluoresce.

In the method, polynuclear aromatic hydrocarbons are extracted from the sample matrix with a suitable solvent, which is then injected into the chromatographic system. Usually, the extract must be filtered because fine particulate matter can collect on the inlet frit of the column, resulting in high back-pressures and eventual plugging of the column. For most hydrocarbon analyses, reverse phase HPLC (i.e., using a nonpolar column packing with a more polar mobile phase) is used. The most common bonded phase is the octadecyl (C18) phase. The mobile phase is commonly aqueous mixtures of either acetonitrile or methanol.

After the chromatographic separation, the analytes flow through the cell of the detector. A fluorescence detector shines light of a particular wavelength (the excitation wavelength) into the cell. Fluorescent compounds absorb light and reemit light of other, higher wavelengths (emission wavelengths). The emission wavelengths of a molecule are mainly determined by its structure. For polynuclear aromatic hydrocarbons, the emission wavelengths are mainly determined by the arrangement of the rings and vary greatly between isomers.

Some of the polynuclear aromatic hydrocarbons, such as phenanthrene, pyrene, and benzo(g,h,i)perylene, are commonly seen in products boiling in the middle to heavy distillate range. In a method for their detection and analysis (EPA 8310), an octadecyl column and an aqueous acetonitrile mobile phase are used. Analytes are excited at 280 nm and detected at emission wavelengths of >389 nm. Naphthalene, acenaphthene, and fluorene must be detected by a less-sensitive UV detector because these condensed compounds emit light at wavelengths below 389 nm.

The methods using fluorescence detection will measure any compounds that elute in the appropriate retention time range and which fluoresce at the targeted emission wavelength(s). In the case of one method (EPA 8310), the excitation wavelength excites most aromatic compounds. These include the target compounds and also many aromatics derivatives, such as alkyl aromatics, phenols, anilines, and heterocyclic aromatic compounds containing the pyrrole (indole, carbazole, etc.), pyridine (quinoline, acridine, etc.), furan (benzofuran, naphthofuran, etc.), and thiophene (benzothiophene, naphthothiophene, etc.) structures. In petroleum samples, alkyl polynuclear aromatic hydrocarbons are strong interfering compounds. For example, there are five methyl phenanthrene derivatives and more than 20 dimethyl phenanthrene derivatives. The alkyl substitution does not significantly affect either the wavelengths or intensity of the phenanthrene fluorescence. For a very long time after the retention time of phenanthrene, the alkyl phenanthrene derivatives will interfere, affecting the measurements of all later-eluting target polynuclear aromatic hydrocarbons.

Interfering compounds will vary considerably from source to source and samples may require a variety of cleanup steps to reach required method detection limits. The emission wavelengths used (EPA 8310) are not optimal for sensitivity of the small ring compounds. With modern electronically controlled monochromator, wavelength programs can be used which tune excitation and emission wavelengths to maximize sensitivity and/or selectivity for a specific analyte in its retention time window.

Of particular interest is the application of the HPLC technique to the identification of the molecular types in nonvolatile feedstocks, such as residua and the molecular species in the asphaltene fraction, has been of considerable interest (Speight, 1986, 2014, 2015).

Several recent HPLC separation schemes are particularly interesting since they incorporate detectors that are not usually associated with conventional hydrocarbon group types of analyses. The merits of the use of a dielectric constant detector as an integral part of a hydrocarbon group analyzer system have also been demonstrated.

In general, the amount of information that can be derived from any chromatographic separation, however effective the separation, depends on the detectors. As the field of application for HPLC has increased, the limitations of commercially available conventional detectors, such as ultraviolet/visible absorption (UV/VIS) and refractive index (RI) have become increasingly restrictive to the growth of the technique. This has led a search for detectors capable of producing even more information. The so-called hyphenated techniques are the outcome of this search.

The general advantages of HPLC method are (1) each sample is analyzed as received, (2) the boiling range of the sample is generally immaterial, (3) the total time per analysis is usually of the order of minutes, and (4) the method can be adapted for on-stream analysis.

8.6 Ion Exchange Chromatography

Ion exchange is a process in which one type of ion in a compound is exchanged for a different type, cation for cation and anion for anion. For example, calcium, iron, and magnesium ions can be removed from water by being exchanged for sodium ion. Ion exchange chromatography employs the exchange process in which the desired ions are exchanged in sequence and are eluted from a column much the same as compounds are eluted from a column in column chromatography (Snyder, 1968). The technique is often used to separate charged molecules, especially the zwitterion compounds that contain both positively and negatively charged chemical groups.

Ion chromatography is the same as ion exchange chromatography, except that it includes a provision for removing the ions in the eluting agent, so sensitive electrochemical detection is possible. The equipment used in ion exchange chromatography is similar to that used in column chromatography and can be identical for most separations.

An interesting historical aside to the use of ion exchange chromatography occurs in the biblical account of Moses at Marah (Christian Bible, Exodus, Chapter 15, Verses 23–25; see also Christian Bible, Numbers, Chapter 33, Verses 8–9), one of the stations of the Exodus where the Israelites found bitter water. Moses reputedly:

> was shown a tree which, when he had cast it into the water the waters were made sweet.

Marah actually means bitter and the explanation is that the oxidized cellulose exchanged with the salt ions that gave the water its peculiar taste. Whether or not this is true can be questioned but what does matter is that the people of those times were aware of such effects!

Other records as far back as Aristotle (384–322 BC) indicate that soils and sand filters were used to treat impure drinking water. Sir Roger Bacon (1561–1626), an English philosopher, observed in the early 1600s that:

> salt water passed through earth through 10 vessels, one within another, and yet it hath not lost its saltiness as to become potable, but when drayned [sic] through 20 vessels hath become fresh.

The first systematic studies involved the determination of the failure of ammonium sulfate [$(NH_4)_2SO_4$] and potassium chloride (KCl) to wash out of soil. It was observed that if these compounds were passed through a column of soil, calcium sulfate ($CaSO_4$) and calcium chloride ($CaCl_2$) would be obtained in the filtrate.

Ion exchange chromatography is widely used in the analyses of feedstock fractions for the isolation and preliminary separation of acid and basic components. This technique has the advantage of greatly improving the quality of a complex operation, but it can be a very time-consuming separation.

Ion exchange resins are prepared from aluminum silicates, synthetic resins, and polysaccharides. The most widely used resins have a skeletal structure of polystyrene cross-linked with varying amounts of divinylbenzene. They have a loose gel structure of cross-linked polymer chains through which the sample ions must diffuse to reach most of the exchange sites. Since ion exchange resins are usually prepared as beads that are several hundred micrometers in diameter, most of the exchange sites are located at points quite distant from the surface. Because of the polyelectrolytic nature of these organic resins, they can absorb large amounts of water or solvents and swell to volumes considerably larger than the dried gel. The size of the species that can be used through the particle is determined by the intermolecular spacing between the polymeric chains of the three-dimensional polyelectrolyte resin.

Although ion exchange resins have a wide use, the major issue in the industry is with nonaqueous chromatography on ion exchange resins. The technique has found widespread use in fractionating feedstocks into fractions that are comprised of various functional types. Thus, acidic and basic components can be separated from the viscous feedstocks.

Before a separation procedure is initiated, a column needs to be flushed with four void volumes of mobile phase to mitigate bleeding. The sample, dissolved at about 20 wt% concentration in the appropriate solvent, is introduced into a pair of columns connected in series, usually with the anion column first and the cation column second. The neutral fraction is eluted at 40 °C (104 °F) with 2.5 times the void volume of solvent.

After elution, the resins are transferred from the columns to a Soxhlet apparatus. The acids are extracted from the anion exchange resins with a benzene-formic acid azeotrope, and the bases from the cation exchange resins, with benzene plus propylamine. The amine must be added slowly, in contrast to the formic acid. Modifications of resin preparation and elution and extraction conditions allow the separation and isolation of neutral materials, weak acids and strong acids, as well as weak bases and strong bases.

The technique is complex and tedious. The preparation and conditioning of the resins is critical for their stability as well as for their performance. Bleeding (i.e., the release of decomposition products from the ion exchange resins) may be an issue. The amphoteric character of certain compounds causes them to be classified either as acids or as bases; depending on the sequence of the ion exchange resins, the sample has to pass must also be given consideration. With anion exchange resin first, the amphoteric species are collected as part of the acid fraction but with cation exchange resins first, they are collected as part of the base fraction. By performing the analysis both ways, the amount of amphoteric material can be measured. In fact, using the correct sequence of ion exchange columns allows the amphoteric species fraction to be collected as a distinct and separate fraction.

Ion exchange chromatography procedures often advocate use of the whole feedstock rather than use of deasphaltened oil thereby foregoing the precipitation of asphaltenes as a first step in the separation of the feedstock.

The technique can also be employed for the separation and isolation of nitrogen compounds from crude oil fractions. This gives fractions that are, chemically, much better defined as concentrates of acids and one or several concentrates of bases. Neutral hetero-compounds can also be isolated by charge-transfer chromatography on Attapulgus clay coated with ferric chloride ($FeCl_3$) and hydrocarbon species can be subdivided into saturates and aromatics in a separate column.

For example, cation-exchange chromatography has been used successfully to isolate the nitrogen constituents in a feedstock fraction. The relative importance of these compounds has increased because of the deleterious effects in many refining processes in which catalytic activity is reduced (often markedly reduced). The activity of cracking catalysts and hydrocracking catalysts are reduced and contribute to gum formation, color, odor, and poor storage properties of the products. However, not all basic compounds isolated by cation-exchange chromatography are nitrogen compounds. Anion-exchange chromatography is used to isolate the acid components (such as carboxylic acids and phenols) from feedstock fractions.

If no subdivision of the polar species into acid and base fractions is desired, much simpler methods can be used for the separation of feedstocks fractions into saturate constituents, aromatic constituents, and polar constituents.

8.7 Simulated Distillation

Simulated distillation (often referred to as "SimDist") is a GC method that is designed and developed to simulate the actual physical distillation and TBP distribution of crude oils and crude oil products. It is particularly adaptable to the high-boiling fraction and products.

Distillation is the most widely used separation process in the refining (Nelson, 1958; Speight, 2000, 2014, 2023). In fact, knowledge of the boiling range of crude feedstocks and finished products has been an essential part of the determination of feedstock quality since the start of the refining industry and the various test methods have been noted elsewhere (Chapters 3 and 4). The technique has been used for control of plant and refinery processes as well as for predicting product slates. Thus, it is not surprising that routine laboratory-scale distillation tests have been widely used for determining the boiling ranges of crude feedstocks and a wide variety of refinery products (ASTM D323, 2021; ASTM D1160, 2021; ASTM D1298, 2021; ASTM D2887, 2021; ASTM D7169, 2021).

There are some limitations to the routine distillation tests. For example, although heavy crude oils contain volatile constituents, it is not always advisable to use distillation for identification of these volatile constituents. Thermal decomposition of the constituents of the feedstock is known to occur at approximately 350 °C (660 °F). Thermal decomposition of the constituents of the heavier, but immature, crude oils have been known to commence at temperatures as low as 200 °C (390 °F), however. Thus, thermal alteration of the constituents and erroneous identification of the decomposition products as natural constituents is always a possibility.

On the other hand, the limitations to the use of distillation as an identification technique may be economic, and detailed fractionation of the sample may also be of secondary importance. There have been attempts to combat these limitations, but it must be recognized that the general shape of a one-plate distillation curve is often adequate for making engineering calculations, correlating with other physical properties, and predicting the product slate (Nelson, 1958).

A low-resolution, temperature-programmed GC analysis has been developed to simulate the time-consuming TBP distillation (ASTM D2887, 2021). The method relies on the general observation that hydrocarbons are eluted from a nonpolar adsorbent in the order of their boiling points. The regularity of the elution order of the hydrocarbon components allows the retention times to be equated to the distillation temperature, and the term simulated distillation by GC (or simdis) is used throughout the industry to refer to this technique (Altgelt and Boduszynski, 1994).

The method has been well researched in terms of method development and application. The benefits of the technique include good comparisons with other ASTM distillation data as well as application to higher boiling fractions of the feedstock. This offers a valuable aid to process control operations and product testing. In fact, since the introduction of the method (ASTM D2887, 2021), there has been widespread acceptance by process chemists and process engineers for operational control of the plant.

The full development of simulated distillation as a routine procedure has been made possible by the massive expansion in GC instrumentation (such as the introduction of automatic temperature programming) since the 1960s. In fact, a fully automated simulated distillation system, under computer control, can operate continuously to provide finished reports in a choice of formats that agree well with TBP data. For example, data output includes the provision of the corresponding Engler profile (ASTM D86, 2021) as well as the prediction of other properties, such as vapor pressure and flash point.

Simulated distillation by GC is applied in the petrochemical industry to obtain TBP distributions of distillates and crude oil feedstocks. A standardized method (ASTM D2887, 2021) is available for the boiling point determination of feedstock. The method utilizes nonpolar, packed GC columns in conjunction with flame ionization detection. The upper limit of the boiling range covered by this method is to approximately 540 °C (1,000 °F) atmospheric equivalent boiling point (AEBP). Recent efforts in which high-temperature GC was used have focused on extending the scope of the ASTM D2887 (2021) methods for higher boiling materials to 800 °C (1,470 °F) atmospheric equivalent boiling point.

Interestingly, simdis techniques not only have upper temperature limits but also lower limits. In GC-simdis, the lower limit is determined by the resolution of the solvent peak from the first eluting sample peak. For this reason, the ASTM method D 2887 is restricted to samples of initial boiling points of at least 38 °C (100 °F). In the SFC method, the lower boiling limit is dependent on the initial column temperature and pressure. The lowest possible pressure of carbon dioxide (CO_2) is about 800 psi, namely, the vapor pressure in the supply cylinder at room temperature. However, for many high-boiling samples lacking low-boiling components, a higher initial pressure, for example, 850 psi, avoids unnecessarily long analysis times while providing the desirable resolution of the lower members. This restricts the lower boiling temperature of samples to be analyzed by SFC to about 120 °C (250 °F).

This simdis procedure is particularly adaptable to the simulated distillation of heavy oil and bitumen, and samples that contain large amounts of nonvolatile residue. The procedure may also be applied to heavy gas oil samples that have a final boiling point in excess of 525 °C (975 °F).

For the simulated distillation of heavy oil and bitumen, two chromatographic analyses are performed to characterize the sample. The volatile components of the sample are analyzed by a standard simulated distillation procedure and a separate assay is performed to determine the amount of volatile material in the sample by an internal standard technique. The data from the simulated distillation is then corrected for the nonvolatile residue, and results are reported as percentages of the total sample.

8.8 Supercritical Fluid Chromatography

Supercritical fluid chromatography (often referred to as SFC) is a separation technique similar to gas and liquid chromatography, but using a supercritical fluid as mobile phase which behaves like a compressible fluid of very low viscosity, where mass transfer operates very rapidly, allowing the use of high linear velocities and, therefore, very short analysis times without any loss of efficiency.

By way of explanation, a supercritical fluid is a fluid that is above the critical temperature and has properties not usually found at ambient temperatures and pressures. Use of a fluid under supercritical conditions conveys upon the fluid extraction capabilities that allows the opportunity to improve recovery of a solute (Taylor, 1996).

Thus, SFC combines the advantages of gas and liquid chromatography in that its mobile phase has the low viscosity of a gas and the variable solvent strength of liquids. Under the simplest conditions (constant pressure, only one mobile phase), SFC has nearly the resolution of GC, but its performance can be further enhanced by the variation of solvent and by pressure programming (in addition to temperature programming) during a separation.

In SFC, the mobile phase is a substance maintained at a temperature a few degrees above its critical point. The physical properties of this substance are intermediate to those of a liquid and of a gas at ambient conditions. Hence, it is preferable to designate this condition as the supercritical phase.

In a chromatographic column, the supercritical fluid usually has a density about one third to one fourth of that of the corresponding liquid when used as the mobile phase; the diffusivity is about 1/100 that of a gas and about 200 times that of the liquid. The viscosity is of the same order of magnitude as that of the gas. In addition, the high density of the fluid results in a 1,000-fold better solvency than that of a gas. This is especially valuable for analyzing high molecular weight compounds.

A primary advantage of chromatography using supercritical mobile phases results from the mass transfer characteristics of the solute. The increased diffusion coefficients of supercritical fluids compared with liquids can lead to greater speed in separations or greater resolution in complex mixture analyses. Another advantage of supercritical fluids compared with gases is that they can dissolve thermally labile and nonvolatile solutes and, upon expansion (decompression) of this solution, introduce the solute into the vapor phase for detection. Although supercritical fluids are sometimes considered to have superior solvating power, they usually do not provide any advantages in solvating power over liquids given a similar temperature constraint. In fact, many unique capabilities of supercritical fluids can be attributed to the poor solvent properties obtained at lower fluid densities. This dissolution phenomenon is increased by the variability of the solvent power of the fluid with density as the pressure or temperature changes.

The solvent properties that are most relevant for SFC are the critical temperature, polarity, and any specific solute–solvent intermolecular interactions (such as hydrogen bonding) that can enhance solubility and selectivity in a separation. Nonpolar or low-polarity solvents with moderate critical temperatures (e.g., nitrous oxide, carbon dioxide, ethane, propane, pentane, xenon, sulfur hexafluoride, and various Freons) have been well explored for use in SFC. Carbon dioxide has been the fluid of choice in many SFC applications because of its low critical temperature (31 °C, 88 °F), nontoxic nature, and lack of interference with most detection methods.

At the high pressure applied with SFC, *n*-butane and *n*-pentane are good solvents even for polar compounds. Carbon dioxide (CO_2) is only a poor to fair solvent for most high molecular weight fractions separated from the viscous feedstocks but it has the additional virtue of being easily removed from the fractions. It permits the use of a FID for SFC. The FID provides a uniform response over a wide range of sample compounds and comes closest to being a true mass detector. Carbon dioxide is also the solvent of choice for the combination of SFC and mass spectrometry. Commercial equipment for analytical as well as preparative SFC separations is available.

Despite these advantages, SFC has only begun to have its potential realized when applied to viscous feedstocks and the method is starting to be recognized for application to the viscous feedstocks or the high molecular weight fractions. One disadvantage is the tendency of the restrictor to plug. The restrictor is needed to maintain the required high pressure throughout the entire column and consists either of a short, very fine capillary or a plate with a tiny hole positioned at the outlet. Even if complete plugging of the restrictors can be prevented, molecular clusters may still form in it and then cause spikes in the FID output. These annoyances become worse with increasing sample molecular weight and polarity and, therefore, with increasing sample atmospheric equivalent boiling point.

The method (ASTM D2549, 2021) is generally limited to naphtha and light gas oil samples and to samples having a vapor pressure sufficiently low to permit sampling at ambient temperatures. The indications are that SFC will find wide applicability to the problems of identification of the higher molecular weight species in a feedstock.

The sample is injected into a supercritical fluid chromatograph (SFC), equipped with a column of limited efficiency, which separates the hydrocarbons into saturated (nonaromatic) and aromatic compounds. Due to its similar response to both saturated and aromatic compounds, a FID is used to detect the components. At constant pressure and temperature, the area under the chromatogram is measured as a function of retention time.

8.9 Thin Layer Chromatography

TLC is a chromatographic method that is used to separate nonvolatile mixtures. The method involves the use of a sheet of an inert substrate such as glass, plastic, or aluminum foil, which is coated with a thin layer of adsorbent material such as, for example, usually silica gel (SiO_2) or aluminum oxide (Al_2O_3). The method is widely used for both qualitative and quantitative sample analysis.

Column chromatography made separations possible that were otherwise impossible at the time, but complete separation usually required from one day to several weeks. However, biochemists and natural product chemists had realized for some time that a piece of filter paper could be used instead of the column. The glass tube and the tightly packed columns could be eliminated, and the separation times were reduced, usually from a day or two to a few hours. One of the difficulties was that only cellulose could be used as an inert phase. Other difficulties included the fact that the paper was made of fibers that were thicker in some parts than others, which caused irregular flow rates and poorer separations and the paper could be destroyed by strong acids or bases, so the color-developing reagents had to be in mild solutions. These problems severely limited the types of compounds that could be separated insofar as a paper was needed that is made from uniform-sized particles, regularly spaced, and capable of withstanding strong acids and bases.

The principle of what we now know as TLC was first developed in 1938 in a method developed as an alternate to paper chromatography. The method involved dusting alumina (Al_2O_3) on to a glass plate. The particles were uniform and inert to acids, but the dust would blow or wash away unless extreme care was taken, so a photograph of the results had to be obtained. There was no way to store the plates. The method did not receive attention as an analytical technique until the method was described to produce uniform thickness, thin layers of silica gel (SiO_2) held onto glass plates with 7–8% w/w of plaster of Paris ($CaSO_4$). This concept formed the basis of current procedures and reduced separation times to 10–30 min.

The apparatus commonly used is a small glass jar with a screw cap lid and may or may not have filter paper placed around the walls on the inside. The sample, 1–2 drops, is placed on a single spot about 2 cm from one end of the plate, and the spot is indicated with a pencil mark. A scratch mark is made on the coating 12 cm from the bottom end. The developing solvent is added to the developing jar to a height of about 1 cm. The plate is inserted, sample end down, and the jar closed to help maintain a solvent-saturated environment. When the solvent has risen to the 12 cm mark by capillary action (typically requiring 10–30 min), the plate is removed. The solvent is allowed to evaporate or is dried with a hair dryer. The plate then is sprayed with a reagent that will form a color with the desired compounds. After spraying, colored spots become visible, and the distance they have moved compared to the solvent front is measured. The mechanism for the separation is most likely a combination of adsorption and partition. If essentially round spots form that are separated from each other, then the separation is primarily partitioning. If tailing occurs, then adsorption is a major factor.

This technique is perhaps one of the most ignored techniques that are available to the analytical chemist. If offers identification of a wide variety of compounds and compound type. Although still widely used by biochemists and by natural product chemists, its revival in the field of crude oil analysis is perhaps overdue.

In the environmental field, TLC is best used for screening analyses and characterization of semi-volatile and nonvolatile petroleum products. The precision and accuracy of the technique is inferior to other methods, but when speed and simplicity are desired, TLC may be a suitable alternative. For characterizations of petroleum products such as asphalt, the method has the advantage of separating compounds that are too high boiling to pass through a gas chromatograph. While TLC does not have the resolving power of a gas chromatograph, it is able to separate different classes of compounds. TLC analysis is fairly simple and, since the method does not give highly accurate or precise results, there is no need to perform the highest quality extractions.

For the analysis of the hydrocarbon derivatives of crude oil, a moderately polar material stationary phase works well. The plate is placed in a sealed chamber with a solvent (mobile phase). The solvent travels up the plate carrying compounds present in the sample. Compounds with chemical structure and polarity similar to the solvent pass readily into the mobile phase. For example, the saturated hydrocarbons seen in diesel fuel travel readily up a plate in a hexane mobile phase. Polar compounds such as ketone derivatives or alcohol derivatives travel a lesser distance in hexane than saturated hydrocarbon derivatives.

The method is considered to be a qualitative and useful tool for rapid sample screening. Limitations of the method center on its moderate reproducibility, detection limits, and resolving capabilities. Variability between operators can be as high as 30%. Detection limits (without any concentration of the sample extract) are near 50 ppm (mg/kg) for most petroleum products in soils. When the aromatic content of a sample is high, as with bunker C fuel oil, the detection limit can be near 100 ppm. It is often not possible to distinguish between similar products such as diesel and jet fuel. As with all chemical analyses, quality assurance tests should be run to verify the accuracy and precision of the method.

References

Albaugh, E.W., Talarico, P.C., Davis, B.E., and Wirkkal, R. 1971. In: *Gel Permeation Chromatography*. K.H. Altgelt and L. Segal (Editors). Marcel Dekker Inc., New York.

Altgelt, K.H., and Boduszynski, M.M. 1994. *Composition and Analysis of Heavy Petroleum Fractions*. Marcel Dekker Inc., New York.

Altgelt, K.H., and Gouw, T.H. 1975. In: *Advances in Chromatograph*. J.C. Giddings, E. Grushka, R.A. Keller, and J. Cazes (Editors). Marcel Dekker Inc., New York.

Altgelt, K.H., and Gouw, T.H. (Editors). 1979. *Chromatography in Petroleum Analysis*. Marcel Dekker Inc., New York.

ASTM D86. 2021. Standard Test Method for Distillation of Petroleum Products and Liquid Fuels at Atmospheric Pressure. Annual Book of Standards. ASTM International, West Conshohocken, Pennsylvania.

ASTM D323. 2021. Standard Test Method for Vapor Pressure of Petroleum Products (Reid Method). Annual Book of Standards. ASTM International, West Conshohocken, Pennsylvania.

ASTM D1160. 2021. Standard Test Method for Distillation of Petroleum Products at Reduced Pressure. Annual Book of Standards. ASTM International, West Conshohocken, Pennsylvania.

ASTM D1298. 2021. Standard Test Method for Density, Relative Density, or API Gravity of Crude Petroleum and Liquid Petroleum Products by Hydrometer Method. Annual Book of Standards. ASTM International, West Conshohocken, Pennsylvania.

ASTM D1319. 2021. Standard Test Method for Hydrocarbon Types in Liquid Petroleum Products by Fluorescent Indicator Adsorption. Annual Book of Standards. ASTM International, West Conshohocken, Pennsylvania.

ASTM D1492. 2021. Standard Test Method for Bromine Index of Aromatic Hydrocarbons by Coulometric Titration. Annual Book of Standards. ASTM International, West Conshohocken, Pennsylvania.

ASTM D2007. 2021. Standard Test Method for Characteristic Groups in Rubber Extender and Processing Oils and Other Petroleum-Derived Oils by the Clay-Gel Absorption Chromatographic Method. Annual Book of Standards. ASTM International, West Conshohocken, Pennsylvania.

ASTM D2163. 2021. Standard Test Method for Determination of Hydrocarbons in Liquefied Petroleum (LP) Gases and Propane/Propene Mixtures by Gas Chromatography. Annual Book of Standards. ASTM International, West Conshohocken, Pennsylvania.

ASTM D2268. 2021. Standard Test Method for Analysis of High-Purity *n*-Heptane and Isooctane by Capillary Gas Chromatography. Annual Book of Standards. ASTM International, West Conshohocken, Pennsylvania.

ASTM D2426. 2021. Standard Test Method for Butadiene Dimer and Styrene in Butadiene Concentrates by Gas Chromatography. Annual Book of Standards. ASTM International, West Conshohocken, Pennsylvania.

ASTM D2427. 2021. Standard Test Method for Determination of C2 through C5 Hydrocarbons in Gasolines by Gas Chromatography. Annual Book of Standards. ASTM International, West Conshohocken, Pennsylvania.

ASTM D2504. 2021. Standard Test Method for Noncondensable Gases in C2 and Lighter Hydrocarbon Products by Gas Chromatography. Annual Book of Standards. ASTM International, West Conshohocken, Pennsylvania.

ASTM D2505. 2021. Standard Test Method for Ethylene, Other Hydrocarbons, and Carbon Dioxide in High- Purity Ethylene by Gas Chromatography. Annual Book of Standards. ASTM International, West Conshohocken, Pennsylvania.

ASTM D2549. 2021. Standard Test Method for Separation of Representative Aromatics and Nonaromatics Fractions of High-Boiling Oils by Elution Chromatography. Annual Book of Standards. ASTM International, West Conshohocken, Pennsylvania.

ASTM D2593. 2021. Standard Test Method for Butadiene Purity and Hydrocarbon Impurities by Gas Chromatography. Annual Book of Standards. ASTM International, West Conshohocken, Pennsylvania.

ASTM D2710. 2021. Standard Test Method for Bromine Index of Petroleum Hydrocarbons by Electrometric Titration. Annual Book of Standards. ASTM International, West Conshohocken, Pennsylvania.

ASTM D2712. 2021. Standard Test Method for Determination of Hydrocarbon Impurities in High Purity Propylene by Gas Chromatography. Annual Book of Standards. ASTM International, West Conshohocken, Pennsylvania.

ASTM D2887. 2021. Standard Test Method for Boiling Range Distribution of Petroleum Fractions by Gas Chromatography. Annual Book of Standards. ASTM International, West Conshohocken, Pennsylvania.

ASTM D3524. 2021. Standard Test Method for Diesel Fuel Diluent in Used Diesel Engine Oils by Gas Chromatography. Annual Book of Standards. ASTM International, West Conshohocken, Pennsylvania.

ASTM D3525. 2021. Standard Test Method for Gasoline Fuel Dilution in Used Gasoline Engine Oils by Wide-Bore Capillary Gas Chromatography. Annual Book of Standards. ASTM International, West Conshohocken, Pennsylvania.

ASTM D3606. 2021. Standard Test Method for Determination of Benzene and Toluene in Spark Ignition Fuels by Gas Chromatography. Annual Book of Standards. ASTM International, West Conshohocken, Pennsylvania.

ASTM D3712. 2021. Standard Test Method of Analysis of Oil-Soluble Sulfonates by Liquid Chromatography. Annual Book of Standards. ASTM International, West Conshohocken, Pennsylvania.

ASTM D4424. 2021. Standard Test Method for Butylene Analysis by Gas Chromatography. Annual Book of Standards. ASTM International, West Conshohocken, Pennsylvania.

ASTM D4626. 2021. Standard Practice for Calculation of Gas Chromatographic Response Factors. Annual Book of Standards. ASTM International, West Conshohocken, Pennsylvania.

ASTM D4815. 2021. Standard Test Method for Determination of MTBE, ETBE, TAME, DIPE, tertiary-Amyl Alcohol and C1 to C4 Alcohols in Gasoline by Gas Chromatography. Annual Book of Standards. ASTM International, West Conshohocken, Pennsylvania.

ASTM D5134. 2021. Standard Test Method for Detailed Analysis of Petroleum Naphthas through *n*-Nonane by Capillary Gas Chromatography. Annual Book of Standards. ASTM International, West Conshohocken, Pennsylvania.

ASTM D5303. 2021. Standard Test Method for Trace Carbonyl Sulfide in Propylene by Gas Chromatography. Annual Book of Standards. ASTM International, West Conshohocken, Pennsylvania.

ASTM D5441. 2021. Standards Test Method for Analysis of Methyl-Tert-Butyl Ether (MTBE) by Gas Chromatography. Annual Book of Standards. ASTM International, West Conshohocken, Pennsylvania.

ASTM D5442. 2021. Standard Test Method for Analysis of Petroleum Waxes by Gas Chromatography. Annual Book of Standards. ASTM International, West Conshohocken, Pennsylvania.

ASTM D5443. 2021. Standard Test Method for Paraffin, Naphthene, and Aromatic Hydrocarbon Type Analysis in Petroleum Distillates Through 200 °C by Multi-Dimensional Gas Chromatography. Annual Book of Standards. ASTM International, West Conshohocken, Pennsylvania.

ASTM D5501. 2021. Standard Test Method for Determination of Ethanol Content of Denatured Fuel Ethanol by Gas Chromatography. Annual Book of Standards. ASTM International, West Conshohocken, Pennsylvania.

ASTM D5580. 2021. Standard Test Method for Determination of Benzene, Toluene, Ethylbenzene, P/M-Xylene, O-Xylene, C9 and Heavier Aromatics, and Total Aromatics in Finished Gasoline by Gas Chromatography. Annual Book of Standards. ASTM International, West Conshohocken, Pennsylvania.

ASTM D5599. 2021. Standard Test Method for Determination of Oxygenates in Gasoline by Gas Chromatography and Oxygen Selective Flame Ionization Detection. Annual Book of Standards. ASTM International, West Conshohocken, Pennsylvania.

ASTM D5623. 2021. Standard Test Method for Sulfur Compounds in Light Petroleum Liquids by Gas Chromatography and Sulfur Selective Detection. Annual Book of Standards. ASTM International, West Conshohocken, Pennsylvania.

ASTM D5769. 2021. Standard Test Method for Determination of Benzene, Toluene, and Total Aromatics in Finished Gasolines by Gas Chromatography/Mass Spectrometry. Annual Book of Standards. ASTM International, West Conshohocken, Pennsylvania.

ASTM D5776. 2021. Standard Test Method for Bromine Index of Aromatic Hydrocarbons by Electrometric Titration. Annual Book of Standards. ASTM International, West Conshohocken, Pennsylvania.

ASTM D5845. 2021. Standard Test Method for Determination of MTBE, ETBE, TAME, DIPE, Methanol, Ethanol and Tert-Butanol in Gasoline by Infrared Spectroscopy. Annual Book of Standards. ASTM International, West Conshohocken, Pennsylvania.

ASTM D5986. 2021. Standard Test Method for Determination of Oxygenates, Benzene, Toluene, C8–C12 Aromatics and Total Aromatics in Finished Gasoline by Gas Chromatography/Fourier Transform Infrared Spectroscopy. Annual Book of Standards. ASTM International, West Conshohocken, Pennsylvania.

ASTM D6159. 2021. Standard Test Method for Determination of Hydrocarbon Impurities in Ethylene by Gas Chromatography. Annual Book of Standards. ASTM International, West Conshohocken, Pennsylvania.

ASTM D7096. 2021. Standard Test Method for Determination of the Boiling Range Distribution of Gasoline by Wide-Bore Capillary Gas Chromatography. Annual Book of Standards. ASTM International, West Conshohocken, Pennsylvania.

ASTM D7169. 2021. Standard Test Method for Boiling Point Distribution of Samples with Residues Such as Crude Oils and Atmospheric and Vacuum Residues by High Temperature Gas Chromatography. Annual Book of Standards. ASTM International, West Conshohocken, Pennsylvania.

Fowlis, I.A. 1995. *Gas Chromatography*. 2nd Edition. John Wiley & Sons Inc., Hoboken, New Jersey.

Grob, R.L. 1995. *Modern Practice of Gas Chromatography*. 3rd Edition. John Wiley & Sons Inc., Hoboken, New Jersey.

Hagel, L., and Janson, J.-C. 1992. In: *Chromatography: Fundamentals and Applications of Chromatography and Related Migration Methods*. E. Heftmann (Editor). 5th Edition. Elsevier, Amsterdam, Netherlands, Chapter 6

Lee, M.L., Yang, F.J., and Bartle, K.D. 1984. *Open Tubular Column Gas Chromatography. Theory and Practice*. John Wiley & Sons Inc., New York.

Masel, R.I. 1995. *Principles of Adsorption and Reaction on Solid Surfaces*. John Wiley & Sons Inc., New York.

Meier, P.C., and Zünd, R.E. 2000. *Statistical Methods in Analytical Chemistry*. 2nd Edition. John Wiley & Sons Inc., New York.

Nelson, W.L. 1958. *Petroleum Refinery Engineering*. 4th Edition. McGraw-Hill, New York.

Snyder, L.R. 1968. *Principles of Adsorption Chromatography*. Marcel Dekker Inc., New York.

Speight, J.G. 1986. Polynuclear Aromatic Systems in Petroleum. *American Chemical Society, Division of Petroleum Chemistry, Preprints*, 31(4): 818.

Speight, J.G. 2000. *The Desulfurization of Heavy Oils and Residua*. 2nd Edition. Marcel Dekker Inc., New York.

Speight, J.G. 2014. *The Chemistry and Technology of Petroleum*. 5th Edition. CRC Press, Taylor & Francis Group, Boca Raton, Florida.

Speight, J.G. 2015. *Handbook of Petroleum Product Analysis*. 2nd Edition. John Wiley & Sons Inc., Hoboken, New Jersey.

Speight, J.G. 2023. *Petroleum Refining Technology Series Volume 2: Refinery Dewatering. Desalting, and Distillation in Petroleum Refining*. CRC Press, Taylor & Francis Group, Boca Raton, Florida.

Speight, J.G., Wernick, D.L., Gould, K.A., Overfield, R.E., Rao, B.M.L., and Savage, D.W. 1985. Molecular Weights and Association of Asphaltenes: A Critical Review. *Revue de l'Insitut Français du Pétrole*, 40: 51.

Taylor, L.T. 1996. *Supercritical Fluid Extraction*. John Wiley & Sons Inc., Hoboken, New Jersey.

9

Structural Group Analysis

9.1 Introduction

In order to successfully convert any feedstocks into valuable products, knowledge of feedstock properties is required because of the marked variation in properties and composition that are both dependent upon the types of the feedstocks and the source of the feedstock. Knowledge of the properties allows the refiner to optimize use of the existing refinery processes as well as (as may be required) the development and design of new processes.

In the early days of crude oil refining, there was a lesser need to understand the character and behavior of petroleum in detail that is currently required. However, with the startling demands on refineries during, and after, World War II, and the emergence of the age of petrochemicals and plastics, the petroleum industry needed to produce materials not even considered as products in the decade before the war. In addition, the depletion of the reserves of conventional crude oil either through constant use or the advent of the so-called petropolitics (Speight and Islam), refineries started to accept the more viscous materials as refinery feedstocks. Thus, there became the necessity to find out more about these feedstocks so that the refiner might once more enjoy the luxury of predictability of refinery behavior.

Valuable information can be obtained from a series of standard test methods that produce data related to the bulk properties of feedstocks, such as specific gravity and viscosity which have assisted in the establishment of certain empirical relationships. However, movement of the refining industry to feedstocks that contain higher proportions of coke-forming materials emphasizes the need for more definitive data that would enable more realistic predictions to be made of crude oil behavior in refinery operations.

Starting in the 1960s and 1970s, the postulation of the structural aspects of the feedstock constituents became serious area of study with the identification of

Handbook of Heavy Oil Properties and Analysis, First Edition. James G. Speight.

Table 9.1 The Various Compound Classes and Compound Types in Petroleum

Class	Compound types
Saturated hydrocarbons	*n*-Paraffins
	iso-Paraffins and other branched paraffins
	Cycloparaffins (naphthenes)
	Condensed cycloparaffins (including steranes and hopanes)
	Alkyl side chains on ring systems
Unsaturated hydrocarbons	Olefins not indigenous to petroleum
	Present in products of thermal reactions
Aromatic hydrocarbons	Benzene systems
	Condensed aromatic systems
	Condensed aromatic-cycloalkyl systems
	Alkyl side chains on ring systems
Saturated heteroatomic systems	Alkyl sulfides
	Cycloalkyl sulfides
	Alkyl side chains on ring systems
Aromatic heteroatomic systems	Furans (single-ring and multi-ring systems)
	Thiophenes (single-ring and multi-ring systems)
	Pyrroles (single-ring and multi-ring systems)
	Pyridines (single-ring and multi-ring systems)
	Mixed heteroatomic systems
	Amphoteric (acid–base) systems
	Alkyl side chains on ring systems

various chemical types in feedstocks (Table 9.1). As the studies evolved, it became clear that the asphaltene fraction of a feedstock is a complex fraction in which the character of the constituents varies in terms of the range of molecular weight and range of the polarity of the constituents, to mention only two parameters. However, the tendency by researchers to postulate average structures has been born again and raised its head again in the 1990s and has continued to this day (at the time of writing). In addition, the postulation of an average structure based on poorly planned and faulty experimental work has also been applied to the issue.

Finally, before embarking upon such a quest for the search for an average structure (or an average structural parameter), the investigator must determine if the highly variable chemical properties and physical properties of a complex mixture that contains not only hundreds (perhaps thousands) of constituents of different chemical character (Chapters 2 and 3) and molecular weight can be represented by an average structure and to aid the refiner.

It is the purpose of this chapter to present the various historical methods that can be used for the structural group analysis of complex petroleum fractions. Methods are also included that are usually employed to emphasize a particular set of molecular parameters that are derived from physical properties or from particular chemical methods. Each method should be recognized for the particular capabilities that it offers and used in conjunction with one or more of the other methods described in this chapter. In addition, the structural data are typically obtained from a combination of two or more methods (Speight, 2001, 2002). This is preferred insofar as a multidimensional approach yields more valuable information about a particular fraction than the data from one method alone.

However, it is not the intent of the chapter to advocate the use of one method over another. It is, however, the intent to show the methods as they were used historically, with some justification, but it must be remembered that many of the methods (which are listed alphabetically within each category) culminate with the finding of an average structure or an average structural parameter. No such structure or parameter exists and the use of such information must be treated for what it is: a best but inaccurate guess that defies the science of structural analysis that has limited use in the refinery.

The precursors to crude oil and the viscous feedstocks are as diverse as the plant chemicals themselves, with the added recognition that there may have been some evolution of these chemicals over geological time as plants, and their constituents, evolved to modern-day counterparts.

Furthermore, local and regional variations in the nature of the original plant precursors and in the maturation conditions offer a means of variation of any refinery feedstock from field to field and can be used to explain the variations in petroleum types and constituents. Thus, as composition studies advance to the higher molecular weight ranges (especially in relation to the constituents of the viscous feedstocks), it is impractical to consider the determination of the individual compounds and very difficult, if not impossible, to segregate and identify fractions of the required simplicity. For this reason, investigators will be well advised to use information about the molecular types of the feedstocks and it is perhaps in this field that spectroscopy has a valuable application.

Structural group analysis is, in fact, the determination of the statistical distribution of these structural elements in the oil fraction, irrespective of the way in which the elements are combined in molecules. In fact, technique referred to as "structural group analysis" is midway between ultimate analysis, in which atoms are the components, and molecular analysis, in which molecules are the components. A method for structural group analysis seems to be complete only if structural elements are chosen in such a way that the sum of all equals 100% (or unity), for instance by considering the distribution of carbon in aromatic, naphthenic, and paraffinic locations in petroleum, its fractions, and products.

One of the basic tenets of structural group analysis is the conception of a high molecular weight feedstock constituent as complex in structure and hence complex in properties. High molecular weight constituents may contain (and very frequently do contain) aromatic rings as well as naphthenic rings and paraffin side chains as well as heteroatom derivatives, which is reflected in their physical and chemical properties. Moreover, the bulk properties of such feedstocks will, to a large extent, be determined by the relative quantity of each structural element.

Thus, structural group analysis may be considered to be an analytical method giving information somewhere between that obtained by elemental analysis which, unfortunately, gives only limited information about the feedstock constituents because of the remarkable constancy of the elemental composition. Thus, instead of molecules or atoms, certain structural types are considered components of the oil. Thus, the viscous feedstocks can be analyzed in terms of molecular groups.

There are two methods by which the results of a structural group analysis can be reported are (1) to determine the number of rings or other structural groups which can indicate the number of aromatic rings (R_A), naphthene rings (R_N), and the total number of rings ($R_T = R_A + R_N$) in designated ring content; and (2) to determine the number of carbon atoms in aromatic ($\%C_A$), naphthenic ($\%C_N$), and paraffinic structures ($\%C_P$) that are expressed per 100 carbon atoms in the sample. Such figures may be designated carbon distribution because the analysis gives the distribution of carbon over various structures, such as aromatic, naphthenic, and paraffinic structures. If the mean molecular weight of the sample is known and if an assumption is made about the type of rings present, the ring content can be recalculated as carbon distribution; the converse is also applicable. This can lead to the beginnings of an understanding of the refining chemistry and, with the elemental analysis, the amount of hydrogen required to produce the desired product. Even then, there is always the question of the reliability of the data and the next steps that may be necessary that involve (1) physical property methods and (2) spectroscopic methods of feedstock evaluation.

9.2 Physical Property Methods

There have been many proposals for the structural group analysis of refinery feedstocks which are usually based on inspection of the elemental analyses and physical properties of the material. The overall result has been the acceptance of several of these methods on the basis of their convenience and/or relative simplicity, and these methods are described here.

9.2.1 Density Method

This method is based on the use of the density d, specific refraction r_{LL}, and the molecular weight M (Chapter 10). This method is based on correlation of oil composition, determined by the direct method, with the physical properties of a large number of petroleum fractions obtained by various means. The procedure used for this method is analogous to that employed for the Waterman ring analysis, but the main difference is that the density is used instead of the aniline point.

Thus, the density corresponding to the observed molecular weight M and specific refraction r_{LL} is determined; the difference d between this value and the observed value is noted. Thus:

$$\%C_A = 420d/(1+3.2d)$$

C_R is obtained graphically; thence,

$$\%C_p = 100 - \%C_R$$
$$\%C_N = \%C_R - \%C_A$$

Compared with the Waterman ring analysis, the experimental procedure is simplified, the applicability is more general, and there is better agreement with data obtained by the direct method. Disadvantages, which are also inherent in the Waterman ring analysis, are that the molecular weight range is limited to 200–500 and extrapolation of the data to the higher range is uncertain; indeed, a high aromatic content causes the specific refraction to fall outside the correlation. The accuracy of the density method is in general fairly good, although according to the basic data used, the application should be limited to fractions with $\%C_A < 1.5\%\ C_N$ or with $R_A < 0.5\ R_T$.

9.2.2 Density–Temperature Coefficient Method

For the analysis of paraffin–naphthene mixtures with a density below 0.861, it is assumed that in the density–density temperature coefficient diagram, the portion of the intercept (at constant density) between the paraffin and naphthene line is divided by the sample point into parts proportional to paraffin and naphthene content. In this way, the following equation for mixtures of paraffins and naphthenes in the regions below 0.861 density was derived:

$$\text{wt\% rings} = [190.0d - 217.9 - 10^5 \mathrm{d}d/\mathrm{d}t]/(0.593d - 0.249)$$

where d is the density at 20 °C (68 °F) (Chapter 10) and $\mathrm{d}d/\mathrm{d}t$ is the change in density per degree change in temperature.

For mixtures having a density above 0.861, a compromise was reached between the condensed and non-condensed naphthenes, assuming an equal distribution of these two types. The sample point was assumed to divide the line between the limiting paraffin point and a point on the naphthene ring line above 0.861 into parts proportional to paraffin and naphthene content, and for this region the following formula was derived:

$$\text{wt\% rings} = [102.8d - 142.8 - 10^5 \mathrm{d}d/\mathrm{d}t]/0.262$$

Simplification of the experimental procedure could be achieved by deriving the density coefficient from the molecular weight, which in turn can be estimated from other physical properties, such as the density and mid-boiling point, or viscosity at 38 °C (100 °F) and 99 °C (210 °F). However, since:

$$-10^5 \mathrm{d}d/\mathrm{d}t = 53.5 + 3360/M$$

If $d < 0.861$,

$$\text{wt\% rings} = \left[190.0d - 164.4 + 3360/M\right]\left(0.593d - 0.249\right)$$

If $d > 0.861$,

$$\text{wt\% rings} = (102.8d - 89.3 + 3360M)/0.262$$

For the analysis of aromatic mixtures containing no naphthene rings, a procedure is followed that is quite analogous to the ring-chain analysis for naphthenes. The line between the limiting paraffin point and a point of the aromatic ring line in the density–density temperature coefficient diagram is divided by the sample point. It is assumed that the parts are proportional to aromatic ring and paraffinic chain content. The relationship between wt aromatic ring, density coefficient, and density is only given graphically; graphs are used for condensed ring aromatics, for non-condensed ring aromatics, and for mixtures having an equal distribution of these two types.

The applicability of the method for analyzing aromatic fractions is limited to fractions having no naphthene rings along with the aromatic rings and since it is generally accepted that aromatic molecules in the higher boiling straight-run petroleum fractions nearly all contain naphthene rings, the method derived for alkylaromatics is probably not applicable to gas oil and lubricating oil factions.

9.2.3 Direct Method

This method involves direct determination of the required physical properties of the petroleum sample or petroleum product and, unlike other methods, does not usually require prior separation of the oil into aromatic and saturated fractions. Thus, by means of elemental analysis and molecular weight determination before and after hydrogenation of the sample, the percentage carbon in aromatic structure ($\%C_A$) and the average number of rings R_T can be estimated.

If the feedstock is hydrogenated so that only aromatic rings are converted into naphthene rings, each aromatic carbon atom takes up one hydrogen atom. Therefore,

$$\%C_A = 1191(H'M' - HM)/(100 - H)M$$

In this equation, H is the percentage hydrogen of the oil fraction, M is the average molecular weight, and H' and M' are the corresponding data for the hydrogenated product. In several cases, the molecular weights can be omitted since the difference between M and M' may be negligible. The hydrogen content of the hydrogenated product, which is presumed to contain naphthenic and paraffinic carbon only, is an exact measure of the number of rings. As compared with paraffins, each ring closure involves a reduction by two hydrogen atoms. If R_T is the number of rings of the hypothetical molecule,

$$R_T = 1 + (0.08326 - 0.005793H')M'$$

To derive a complete set of carbon distribution and ring content figures from $\%C_A$ and R_T, an assumption must be made about the type of rings present. It has been assumed that all rings are six membered and, in the case of the polycyclic materials, are kata-condensed; this type of condensation only occurs in such a way that the rings have two carbon atoms in common, as, for example, naphthalene, anthracene, or tetracene.

However, it must be remembered that the direct method is based on the assumption that heteroelements (e.g., nitrogen, oxygen, and sulfur) are not present in the petroleum sample. However, if these elements are present, corrections can be applied but this detracts from the basis of the method since rings containing heteroelements may be cleaved during the hydrogenation process.

9.2.4 Dispersion–Refraction Method

This method is analogous to the direct method, but the hydrogenation and the examination of the hydrogenated product are replaced by the determination of specific dispersion, bromine number (ASTM D1159, 2021), and specific refraction (Lorentz–Lorenz) of the original fraction. The analytic results obtained by the dispersion–refraction method are similar to those derived from the direct method, except olefin double bonds are also measured. The method is based on data derived from pure hydrocarbons having fewer than 18 carbon atoms per molecule and requires the following data: (1) carbon and hydrogen content, (2) refractive index, (3) density, (4) dispersion, (5) molecular weight, and (6) bromine number.

Calculation of the structural parameters is fairly involved, and the accuracy of the method does not seem to be very high. However, it should be remembered that the method can be used on materials derived from cracking processes as determination of the olefin content is provided for by inclusion of the bromine number.

9.2.5 Molecular Weight-Refractive Index Method

The samples to be analyzed by the molecular weight-refractive index method must be separated beforehand into aromatic and paraffin–naphthene fractions by, for example, silica gel adsorption chromatography (Chapters 9 and 10). As the name indicates, only the molecular weight M (Chapter 10) and the refractive index $n^{20}{}_{\mathrm{D}}$ (Chapter 10) are the required parameters, and by using data derived from use of pure hydrocarbons, the following relationships appear to hold:

Aromatic oils, non-condensed rings:

$$\%C_{\mathrm{A}} = \left[1060\left[\left(n^{20}{}_{\mathrm{D}} - 1.4750\right)M + 8.79\right]^{0.85}\right] \Big/ M - 2 + 1.01\left[\left(n^{20}{}_{\mathrm{D}} - 1.4750\right)M + 8.79\right]^{0.85}$$

$$R_{\mathrm{A}} = 0.126\left[\left(n^{20}{}_{\mathrm{D}} - 1.4750\right)\mathrm{M} + 8.79\right]^{0.85}$$

Aromatic oils, kata-condensed rings:

$$\%C_{\mathrm{A}} = \left[2150\left[\left(n^{20}{}_{\mathrm{D}} - 1.4750\right)M + 8.79\right]^{0.56}\right] \Big/ \mathrm{M} - 3 + 2.3\left[\left(n^{20}{}_{\mathrm{D}} - 1.4750\right)\mathrm{M} + 8.79\right]^{0.56}$$

$$R_{\mathrm{A}} = 0.165\left[\left(n^{20}{}_{\mathrm{D}} - 1.4750\right)M + 8.79\right]^{0.74}$$

Mixtures of condensed and non-condensed rings:

$$\%C_{\mathrm{A}} = \left[1650\left[\left(n^{20}{}_{\mathrm{D}} - 1.4750\right) + 8.79\right]^{0.67}\right] \Big/ M - 2 + 1.65\left[\left(n^{20}{}_{\mathrm{D}} - 1.4750\right)M + 8.79\right]^{0.67}$$

$$R_{\mathrm{A}} = 0.151\left[\left(n^{20}{}_{\mathrm{D}} - 1.4750\right)M + 8.79\right]^{0.78}$$

Nonaromatic oils, condensed and non-condensed rings:

$$\%C_R = \left[2920\left[\left(n^{20}{}_D - 1.4750\right)M + 8.79\right]^{0.73}\right] \Big/ M - 2 + 0.57\left[\left(n^{20}{}_D - 1.4750\right)M + 8.79\right]^{0.86}$$

$$R_T = 0.284\left[\left(n^{20}{}_D - 1.4750\right)M + 8.79\right]^{0.86}$$

Application of the $M - n^{20}{}_D$ method to saturated portions of oil fractions gives results, which in general are in good agreement with those obtained by other methods.

9.2.6 *n-d-M* Method

Linear relationships between the composition of petroleum fractions (as determined by the direct method) and the refractive index, density, and molecular weight led to the development of a method based on these three physical characteristics of petroleum fractions.

Thus, as the name (*n-d-M* method) implies, the refractive index, *n*, density, *d*, and molecular weight, *M*, of the sample are determined; if the sulfur content of the sample is expected to be greater than 0.206, this value should also be determined. Substitution of the data into the formulae allows an estimation of the carbon distribution and ring content.

The *n-d-M* method is especially intended for petroleum fractions boiling above the gasoline range and for similar products after extraction, hydrogenation, or other treatment. The validity of the method is good for petroleum fractions with up to $\%C_R$ = 75% (aromatic + naphthenic), provided that $\%C_A$ (as found by the *n-d-M* method) is not higher than 1.5 times $\%C_N$. The method is also valid for samples containing up to four rings per molecule if not more than half of these are aromatic. Application to oils with a relatively high naphthene ring content is presumed not to cause serious errors. Although the presence of foreign elements influences its accuracy unfavorably, the method is applicable to samples containing sulfur up to 2% w/w, oxygen up to 0.5% w/w, and nitrogen up to 0.5% w/w.

9.2.7 Waterman Ring Analysis

This method requires the determination of the refractive index, density, molecular weight, and aniline point (Chapter 10). In essence, the elemental analysis used in the direct method is replaced by the specific refraction.

Thus, since molecular refraction r_M is additive for a hydrocarbon C_xH_y,

$$r_M = Mr^{20}{}_D = xr_c = yr_H$$

where r_C and r_H represent the atomic refraction of carbon and hydrogen, and M is the molecular weight. Since

$$y = MH/100.8$$
$$x = M(100 - H)/1201.0$$

H is the percentage of hydrogen, then

$$\%H = (100.8 \times 1201.0r^{20}{}_D - 100.8r_c)/(1201r_h - 100.8r_c)$$

However, correlation of hydrogen content and specific refraction for a large series of petroleum fractions leads to:

$$\%H = 110.48r^{20}{}_D - 22.078$$

The total number of rings R_T is given by

$$R_T = 1 + (0.08326 - 0.005793H')M'$$

Then,

$$R_T = 1 + \left(0.2122 - 0.6401r^{20}{}_D\right)M$$

Since

$$\%C_R = 240.200(2R_T + R_{TS})/M(100 - H)$$

Therefore,

$$\%C_R = 240.200\left(2R_T + 2R_{TS}\right)/M\left(77.922 - 100.48r^{20}{}_D\right)R_{TS} = RT, \text{if } R_T < 1$$

And

$$R_{TS} = 1 \text{if } R_T > 1$$

Thus, a relation between specific refraction, molecular weight, and percentage carbon in naphthene rings for saturated hydrocarbons or atomic- and olefin-free oil fractions can be defined. Similarly, a graphic relation between aniline point, specific refraction, and the molecular weight of saturated oil fractions has also been defined.

The most important sources of error of the Waterman ring analysis are the specific refraction, the molecular weight, and the aniline point, and whereas the accuracy of the Waterman ring analysis is questionable, the method has found wide acceptance.

9.2.8 Miscellaneous Methods

A method has also been proposed for determining the number of aromatic rings R_A and naphthenic rings R_N in aromatic fractions of petroleum and involves measurement of the molecular weight M, density d, and refractive dispersion, s (Chapter 10). From the measured data, the functions are calculated, and R_A and R_N are estimated graphically:

$$F(s,M) = (s - 98)(M + 12)10^{-3}$$
$$F(d,M) = (d - 0.854)(M + 12)$$

The method provides a valuable contribution to the existing methods for structural group analysis. The procedure is simple and rapid, and the method is claimed to hold for aromatic concentrates from both nondestructive distillation and cracking of petroleum fractions; average deviations of 0.1 of an aromatic ring and about 0.2 of a naphthene ring are usual. If there are more than three aromatic rings per molecule, the results are uncertain because of a lack of basic data and knowledge concerning the types of petroleum hydrocarbons with four or more aromatic rings. Saturated hydrocarbons, olefins, some non-condensed polycyclic aromatics, and non-hydrocarbons are reputed to introduce serious errors into the analysis.

Other methods include the derivation of a linear relation between percentage carbon in aromatic structure $\%C_A$, refractive index $n^{20}{}_D$, density, and aniline point (AP):

$$\%C_A = 1039.4n^{20}{}_D - 470.4d^{20} - 0.315AP - 1094.3$$

This formula holds good only if $\%C_A = <30$. When the calculated value for $\%C_A$ exceeds 30, a corrected value must be found by using the formula

$$\%C_{A(corr)} = 0.5\%C_{A(calc)} + 15$$

Alternatively, it has been suggested that the molecular weight determination of the *n-d-M* method be replaced by kinematic viscosity measurements (Chapter 10) leading to the *n-d-V* method.

An equation has been devised that is applicable to lubricating distillates that have not been subjected to thermal cracking, temperatures <350 °C (<660 °F). If the naphthenic carbon is of the order of 25–75% of the total carbon, a relationship exists between the refractivity intercept and the number of carbons in naphthenic locations (C_N):

$$\text{Refractivity intercept} = 1.0502 - 0.00020 \times \%C_N$$

Finally, it has also been proposed that for substances such as the constituents of the asphaltene fraction, which contain condensed ring systems, the following relationships be applied:

$$R = 0.11(9.9\text{C} - 3.1\text{H} - 3.7\text{O} + 1.5\text{N} + 14\text{S} - M/d)$$

where R is the number of rings; C, H, O, N, and S are the numbers of carbon, hydrogen, oxygen, nitrogen, and sulfur atoms; M is the molecular weight; and d is the density.

However, if it is not possible to determine the molecular weight, the relationship has been modified to

$$C/R = 9.2/(9.9 + 3.1\text{H/C} + 3.70/\text{C} + 1.5\text{N/C} + 14\text{S/C} - 1200/\%C, d)$$

C/R is the ring condensation index; H/C, O/C, N/C, and S/C are the various atomic ratios calculated from elemental analyses; $\%C$ is the percentage of carbon obtained by elemental analysis; and d is the density.

The method assumes that atoms other than carbon and hydrogen are not major components, olefins are absent, the number of rings is greater than one, and the rings are the fused naphthalene type rather than biphenyl type.

9.3 Spectroscopic Methods

The advances in spectroscopic techniques after World War II are a better understanding of the chemical nature of refinery feedstocks has been achieved as well as the chemical changes which occur during refining. Thus, although the application of spectroscopic methods to structural analysis is a fairly recent innovation that was developed during the latter half of the 20th century, the methods involved have received considerable attention.

Thus, it is not surprising that spectroscopic methods of structural analysis have come into fairly common usage. This is not to say that distinct divisions should be made and a particular method for structural analysis be assigned to a particular fraction. It is in fact preferred that as many methods of structural analysis as possible be applied to any particular fraction to gain an overall picture of the structural entities present.

It is not intended to convey here that any one of these methods can be used for absolute identification of feedstock constituents but the methods are recognized as methods by which an overall evaluation of the feedstock may be obtained in terms of molecular types.

9.3.1 Electron Spin Resonance

Spectrophotometry in the electron spin resonance range (ESR) is useful to supply supplementary information to that obtained by other methods. Parameters that can be obtained by such methods include I, color intensity; g, Lande g factor; N_g, spin numbers; T_1, spin–lattice relaxation time; P_C, population of aromatic carbon per spin; J, spin-excitation energy; and N_d, doublet-spin concentration.

The free-radical signal can be used as a measure of the asphaltene content of refinery feedstocks. Asphaltene constituents exhibit a signal equivalent to $2–4\times10^{18}$ free radicals per gram, or about one per 100 asphaltene molecules, assuming a molecular weight of about 2,000 for the asphaltene fraction (Wen et al., 1978). The postulate is that the free radicals contribute to the tendency toward micelle formation in the constituents of the asphaltene fraction and their results indicate that the free-radical sites were essentially confined to the asphaltene fraction and the resins contain only 2% of the total free radicals present.

There has also been some focus on the metal species in the constituents of the asphaltene fraction (Tynan and Yen, 1969; Shepherd and Graham, 1986). The data indicate that there are, in addition to other metal species, two different types of vanadium, one bound and the other free, occurring naturally in petroleum, both as tetravalent vanadium. It was suggested that the observed vanadium occurring in an asphaltene fraction is associated with the asphaltene molecule itself, most probably the aromatic portion.

9.3.2 Infrared Spectroscopy

Infrared spectroscopy is a well-established method that was developed for comparative, semiquantitative analysis leading to the present quantitative analysis (ASTM E168, 2021; ASTM E334, 2021; ASTM E1252, 2021). Infrared spectra are displayed either as percent transmittance or as absorptivity versus frequency (cm^{-1}) (Table 9.2). Transmittance, T, defined as the ratio of transmitted light over incident light, or percent transmittance, usually shows more detail over the entire range and is generally the preferred display. Absorbance, A, on the other hand, is proportional to the concentration and is, therefore, used for quantitative measurements.

Table 9.2 Infrared Absorption Frequencies

Bond	Compound type	Frequency range, cm^{-1} (wave number)
O—H	Monomeric alcohols and phenols	3,640–3,160(s, b) stretch
O—H	Hydrogen-bonded alcohols and phenols	3,600–3,200(b) stretch
N—H	Amines, amides	3,560–3,320(m) stretch; non-hydrogen bonded
N—H	Amines	3,400–3,100(m) stretch; hydrogen bonded
C—H	Alkynes	3,333–3,267(s) stretch
C—H	Aromatic rings	3,100–3,000(m) stretch
C—H	Alkenes	3,080–3,020(m) stretch
O—H	Carboxylic acids	3,000–2,500(b) stretch
C—H	Alkanes	2,960–2,850(s) stretch
C≡N	Nitriles	2,260–2,220(v) stretch
C≡C	Alkynes	2,260–2,100(w, sh) stretch
C—H	Phenyl ring substitution	2,000–1,600(w)
C═O	Aldehydes, ketones, carboxylic acids	1,760–1,670(s) stretch
C═O	Esters	
C═C	Alkenes	1,680–1,640(m, w) stretch
NO_2	Nitro-compounds	1,660–1,500(s) asymmetrical stretch
N—H	Amines	1,650–1,580 (m) bend
C═C	Aromatic rings	1,600, 1,500(w) stretch
C—H	Alkanes	1,470–1,350(v) scissoring and bending
NO_2	Nitro-compounds	1,390–1,260(s) symmetrical stretch
C—H	CH3 deformation	1,380(m or w)—isopropyl, t-butyl
C—N	Amines	1,340–1,020(m) stretch
C—O	Alcohols, ethers, carboxylic acids, esters	1,260–1,000(s) stretch
C—H	Alkenes	1,000–675(s) bend
C—H	Phenyl ring substitution	870–675(s) bend
C—H	Alkynes	700–610(b) bend
C—Cl	Chloro-compounds	600–800 (m, w)
C—Br	Bromo-compounds	500–600 (m, w)
C—I	Iodo-compounds	500 (n, w)

v = variable, m = medium, s = strong, br = broad, w = weak.
wave number = 1/wavelength.

Infrared spectroscopy is a simple procedure and is one of several techniques that can provide quick information about the distribution of several structural and functional groups (Drews, 1998; Rand, 2003). In combination with nuclear magnetic resonance (NMR) spectroscopy, it will provide quick yet fairly detail on the distribution of carbon–hydrogen groups. However, in the context of high molecular weight petroleum fractions, conventional infrared spectroscopy yields information about the functional features of various petroleum constituents. For example, infrared spectroscopy will aid in the identification of imino functions (=N—H) and hydroxyl (—O—H) functions, as well as the nature of the various carbonyl (—C=O) functions.

In the older, infrared spectroscopy, light is refracted by a prism or a grating and scanned by a moving slit that takes several minutes for one measurement. In Fourier transform infrared (FIR) spectroscopy, the entire spectrum is obtained by an interferometer in a fraction of a second. Thus, several hundred measurements can be taken in a matter of minutes and averaged by computer. This multiplexing leads to greatly increased (about 100-fold) sensitivity and precision over those achievable with dispersive instruments.

Thus, with the recent progress of Fourier transform infrared spectroscopy, quantitative estimates of the various functional groups can also be made. This is particularly important for application to the higher molecular weight solid constituents of petroleum (i.e., the asphaltene fraction) and for group-type analysis.

Infrared absorption spectroscopy is an excellent method for providing detailed information about the chemical constitution of organic materials. Infrared absorption spectroscopy is one of the most widely used techniques for examining the chemical constitution of organic materials. Whether the material being examined is a single compound or a mixture, infrared absorption spectroscopy offers valuable information about the hydrocarbon skeleton and about the functional groups in feedstocks (Wen et al., 1978).

Thus, it is not surprising that attempts have been made to estimate the structural entities present in petroleum and other carbonaceous materials by means of infrared spectroscopy.

When properly prepared and scanned with suitable instruments, the infrared spectra of feedstock fractions show a number of well-defined bands (Table 9.2) (Rao, 1961). For the purposes of structural group analysis, however, the bands at 1,380, 1,465, 2,880, and 2,920 cm^{-1} are usually employed. This allows an estimation of the number of methyl and methylene groups present in the samples.

Thus, initially, the absorptivity, in grams per liter per centimeter, is calculated as

$$a_b = A_b / cl$$

In this equation, a_b is the absorptivity ($g/L \times cm^{-1}$) at the wavelength b, A_b is the absorbance of the solution at wavelength b, c is the concentration of the solution (g/L), and l is the cell thickness in centimeters.

Following from this, the absorptivity per gram carbon in the standard group absorbing is derived in the following manner:

$$a_c = Ma_b / 12.01n$$

where a_c is the absorptivity (g/L cm^{-1}) per gram carbon of the type absorbing, M is the gram molecular weight, n is the number of carbon atoms of the type absorbing per molecule, 12.01 is the gram atomic weight of carbon, and a_b is the absorptivity derived previously. By use of the a_c term for specific types of carbon (1.50 at 1,465 cm^{-1} for methyl; 3.60–4.00 at 2,880 cm^{-1} for methylene), the number of carbon atoms of that type present in the sample may be calculated.

Another infrared spectroscopic procedure that has been employed to gain insight into the carbon distribution in petroleum residua and related materials involves determination of the molecular extinction coefficient E for the band under investigation.

$$E = (l/cl)\log(I_o/I)$$

where c is the molarity of the solution, l is the path length in centimeters, and I_o/I is determined directly from the infrared spectrum and is actually the ratio of the intensity of the incident light I_o to the intensity of the transmitted light I.

The fraction of hydrogen atoms h_a, attached to aromatic carbons can be derived in the following manner:

$$h_a = H_A / (H_A + H_S)$$

H_A is the number of aromatic hydrogen atoms and H_S is the number of saturated hydrogen atoms.

$$h_a = \left[E_{3030} / (E_{3030} + E_{2920})\right]^k$$

E_{3030} and E_{2920} are the extinction coefficients of the 3,030 and 2,920 cm^{-1} absorption bands, respectively. Studies with standard alkyl aromatic compounds and polymers indicated that there was good agreement between the theoretical and observed hydrogen distribution when k was unity.

The proportion of carbon atoms in methyl groups to the total number of carbon atoms can be determined in a similar manner. Thus,

$$C_{Me}/C_T = \mathrm{H/C} \times C_{1380} / E_{3030} + E_{2920}$$

C_{Me} is methyl carbon, C_T is total carbon, and H/C is the atomic hydrogen–carbon ratio determined from elemental analysis, E_{1380} is the extinction coefficient of the band at 1,380 cm^{-1}, and E_{3030} and E_{2920} are the extinction coefficients of the bands at 3,030 cm^{-1} and 2,920 cm^{-1}, respectively.

The application of applied infrared spectroscopic techniques to the elucidation of the structures of refinery feedstocks has illustrated the presence of many structural groups and their relative positions to one another.

Infrared spectroscopy has also been used to identify various functional groups including hydroxyl (3,450 cm^{-1}) and carbonyl (1,695–1,725 cm^{-1}) (McKay et al., 1978). In combination with nitrogen titration data (Jacobson and Gray, 1987), better representations of the structural groups can be made (Petersen, 1967).

It is also possible to derive structural parameters from infrared spectroscopic data and these are H_S/C_S, saturated hydrogen to saturated carbon ratios; C_P/C, paraffinic character; C_N/C, naphthenic character; %Me (methyl content); and C_{MP}/C_{SU}, chain length.

9.3.3 Mass Spectrometry

Mass spectrometry furnishes the molecular weight and chemical formula of compounds and their relative amounts in a mixture and offers nondestructive examination of the sample (ASTM D2425, 2021; ASTM D2650, 2021; ASTM D2786, 2021; ASTM D2789, 2021; ASTM D3239, 2021; ASTM E1316, 2021). The technique can also provide important information about their molecular structure. The earliest and most common type of mass spectrometry, electron impact mass spectrometry (El-MS), gives a fragmentation pattern displaying both parent ion peaks and fragment ion peaks, characteristic of each molecular type.

Fragmentation is frequently employed to differentiate between isomers of pure compounds and of molecules in fairly simple mixtures. However, it is usually avoided with such complex samples as high molecular weight petroleum fractions, which are composed of such a multitude of closely related compounds that their fragmentation patterns are nondistinctive and cannot be readily interpreted.

Mass spectrometry can play a key role in the identification of the constituents of feedstocks and products either in the laboratory or online (ASTM D2425, 2021; ASTM D2786, 2021; ASTM D2789, 2021; ASTM D3239, 2021). The principal advantages of mass spectrometric methods are (1) high reproducibility of quantitative analyses, (2) the potential for obtaining detailed data on the individual components and/or carbon number homologs in complex mixtures, and (3) a minimal sample size is required for analysis. The ability of mass spectrometry to identify individual components in complex mixtures is unmatched by any modern analytical technique. Perhaps the exception is gas chromatography.

However, there are disadvantages arising from the use of mass spectrometry and these are: (1) the limitation of the method to organic materials that are volatile and stable at temperatures up to 300 °C (570 °F) and (2) the difficulty of separating isomers for absolute identification. The sample is usually destroyed, but this is seldom a disadvantage.

Nevertheless, in spite of these limitations, mass spectrometry does furnish useful information about the composition of feedstocks and products even if this information is not as exhaustive as might be required. There are structural similarities that might hinder identification of individual components. Consequently, identification by type or by homolog will be more meaningful since similar structural types may be presumed to behave similarly in processing situations. Knowledge of the individual isomeric distribution may add only a little to an understanding of the relationships between composition and processing parameters.

Mass spectrometry should be used discriminately where a maximum amount of information can be expected. The higher molecular weight nonvolatile viscous feedstocks are, for practical purposes, beyond the useful range of routine mass spectrometry. At the elevated temperatures necessary to promote volatility of the feedstock constituents, thermal decomposition will occur in the inlet and any subsequent analysis would be biased to the low molecular weight end and to the lower molecular products produced by the thermal decomposition.

High-voltage electron impact mass spectrometry (HVEI-MS or EI-MS) can cause repeated fragmentation of daughter ions. Considering the repeated fragmentation of even the simplest hydrocarbons and the fact that the higher molecular weight fractions of petroleum contains such a broad range of molecular weight species, the patterns obtained from petroleum fractions are so complex as to almost defy interpretation. Thus, non-fragmenting mass spectrometric methods are preferred.

For high molecular weight petroleum fractions, the use of non-fragmenting mass spectrometry (NF-MS) methods is now preferred. These methods, also called *soft ionization* methods, produce predominantly parent ion (molecular ion) peaks and, thus, much simpler spectra than methods producing fragments. By rendering the molecular weight of each compound in a sample, and reasonably well its abundance, non-fragmenting mass spectrometry also can be used to determine the molecular weight distribution, that is, the molar mass profile of a sample.

Indeed, the great advantage of *non-fragmenting mass spectrometry* is the relative simplicity of the spectra. A disadvantage is the relatively low signal intensity of the ions. With *low-voltage electron impact ionization*, the number of parent ions formed increases rapidly as the ionizing voltage increases above the ionizing potential of the molecules in the sample. Thus, higher voltage gives a more intense signal (up to about 20–40 eV, depending on compound type). However, the higher ionizing voltage transfers more energy to the sample molecules and causes fragmentation to increase. The surplus energy transferred to the molecule in excess of the ionizing potential is equilibrated and dissipated in various ways, for example, by increasing the internal energy of the molecule or by breaking atomic bonds.

Thus, the challenge in *non-fragmenting mass spectrometry* is to maximize the number of parent ions and, at the same time, maintain the number of fragment ions at acceptable levels. Paraffins are highly susceptible to fragmentation; an ionizing voltage high enough to generate a good parent ion spectrum also breaks many (paraffinic) carbon–carbon bonds, producing a significant number of fragment ions.

Among the most important non-fragmenting mass spectrometric techniques are field ionization mass spectrometry (FIMS), field desorption mass spectrometry (FDMS), chemical ionization mass spectrometry (CIMS), and low-voltage (10–20 eV) electron impact mass spectrometry (LVEI-MS). A particularly powerful version of electron impact mass spectrometry is high-resolution electron impact mass spectrometry (HR-LEVI-MS or LVRH-MS). All of these are designed to generate *cold* ions of such low excess energy that they do not undergo fragmentation to any great extent, in contrast to the *hot* ions generated by conventional electron impact ionization (70–100 eV). With such ionization methods, a compromise must be found between low fragmentation and sufficiently high sensitivity. Unfortunately, samples rich in aliphatic hydrocarbons are hard to keep completely from fragmenting, but most other compound types, especially aromatic compounds, yield clean parent ion spectra.

FIMS shows extremely low fragmentation because the ionization process imparts little excess energy to the formed ion. This is crucial when ionizing saturates. Both low-voltage electron impact (LVEI) and chemical ionization (CI) transfer more excess energy than FIMS, leading to higher levels of fragmentation with saturates. As long as the analyte is volatile, the FIMS experiment is not sample limited. That is, additional sample can be vaporized as needed to obtain the required signal-to-noise ratio. FDMS can be considered a special form of electron impact mass spectrometry (EIMS) where the sample material is placed on the ionizing surface of the probe before insertion into the sample chamber of the mass spectrometer. It is, however, sample limited and subject to experimental artifacts not normally present in EIMS.

In the EIMS technique, the emitting surface is a cathode consisting of an array of sharp tips on a support such as a wire, a grid, or a razor blade. This surface is located very close to the anode. The combination of an extremely small radius of curvature of the tips, the short distance, and the high electric field creates a very high field gradient. The sample is introduced to this field gradient by evaporation and diffusion from a heated surface nearby.

FDMS is a special form of FIMS. The cathode is the same but here it is coated with the sample before it is placed in the mass spectrometer source. This eliminates the need to evaporate the sample and any losses on transport to the cathode. Instead, the ions are desorbed directly from the solid sample. The emitter is heated to the point where the sample melts and can be drawn by the surface tension to the tip where the molecules are converted to ions. As the ions are ejected from the surface into the mass spectrometer, more sample molecules can replace them at the tip. In this way, labile molecules can be handled without significant fragmentation, and even conventionally nonvolatile samples can be run. Thus, the method is especially suitable for the high-MW fractions of petroleum residua. Because of the low thermal energy to which the sample molecules are exposed, the total excess energy in this process is minimal. Thus, FDMS is an even milder method than FIMS and produces less fragmentation.

Chemical ionization mass spectrometry, another *soft* method, has been used only occasionally for petroleum samples. In the method, a relatively high-pressure (0.1–1 Torr) reagent gas (e.g., methane, *n*-butane, and *iso*-butane) is ionized and allowed to contact the sample. At such pressures, the reagent ions undergo many collisions that help them lose excess energy left after ionization. The ions also exchange charges during collision with one another and with the sample molecules. The latter is the desired reaction and the methods works best when the sample molecules retain the charge better than the reagent molecules.

Fast atom bombardment mass spectrometry (FAB-MS) is a method in which the sample is dissolved in a high-boiling polar liquid matrix (such as glycerol or triethanolamine) and spread on a plate. Rare gas molecules (e.g., xenon or argon) are ionized, accelerated, focused at the sample, and stripped of their charge before they hit their target. The impact of uncharged atoms "sputters" small amounts of the matrix and sample off the sample plate and the liquid matrix evaporates thereby liberating charged sample ions which are propelled by the accelerating voltage into the mass spectrometer.

The method produces substantial amounts of fragments and rarely works for neutral compounds, two disadvantages for our purpose. Other problems and interferences with fast atom bombardment mass spectrometry are that the matrix material is ionized along with the sample, and that both sample and matrix material may form cluster ions that complicate the spectrum.

For high-resolution mass spectrometers, precise standardization of operating conditions is even more important for quantitative measurements than for regular mass spectrometers. Generally, the higher the resolution, the weaker the signal and the longer the data acquisition time. Special mass standards, mostly chlorinated hydrocarbon petroleum fractions, are employed for mass calibration. The parent petroleum fractions guarantee a complete mass sequence that is needed for the large number of hetero-compound peaks. Even with such a standard, there will still be several hetero-compound peaks between each pair of hydrocarbon peaks. The chloro-compounds are preferred because of their higher sensitivities at the low ionization voltage compared to those of the fluorinated standards common with regular EIMS.

Gas chromatography coupled with mass spectrometry (GC–MS) is a powerful method for petroleum distillates. Heavy petroleum fractions (from 345 to 450 °C,

650 to 850 °F) are usually first separated by liquid chromatography into compound class fractions before application of gas chromatography that separates the sample by boiling point. Mass spectrometry, used as a detection method, gives the molecular weight and, with fragmenting ionization, the compound type of each peak.

Petroleum samples boiling higher than about 450 °C (850 °C), other than well-separated narrow fractions, are too complex and high-boiling for gas chromatography. In these cases, non-fragmenting mass spectrometry should be combined with liquid chromatography instead of gas chromatography. However, higher boiling samples (>650 °C, >1,200 °F) may not be seen with the same sensitivity as lower boiling constituents by most mass spectrometric methods. For these cases, mass spectrometric methods not limited by volatility are now available. For example, FDMS, laser desorption (LS), fast atom bombardment (FAB), and electrospray (ES) techniques can be applied to very large (polar) molecules, even of several thousand molecular weight. However, because FAB and ES work only with highly polar, easily-to-ionize compounds, their application to petroleum fractions may be limited.

One of the applications of mass spectrometry is group-type mass spectrometry. The method uses fragment patterns and sets of empirical relations to sort the contributions from different molecules to certain selected peaks and to reconstruct the molecular distribution from these peak areas. One test method (ASTM D2786, 2021) method measures the concentration of eight compound types—paraffins, naphthenes with one to six rings, and mono-aromatics in the saturates fractions. The mass spectrometric group-type method (ASTM D3239, 2021) is used to determine aromatic compound types present in chromatographic fractions of aromatics separated from vacuum gas oils. Other ASTM group-type methods determine various aliphatic and aromatic compound types in middle distillates (ASTM D2425, 2021).

Mass spectrometric group-type methods are fast, relatively cheap, and require only a few milligrams sample. However, they have their limits. They are restricted to samples free of olefins and, ordinarily, with less than 2–5% w/w sulfur, nitrogen, or oxygen compounds and in some cases, with much more stringent limits. Other restrictions, for example, to certain boiling ranges (and molecular weight ranges) also apply, depending on the method. Furthermore, only those compound types are considered in the calculation for which results are given in the output. If any other compound types are present in the sample to any large degree, which may easily be the case, major errors may result.

The mass spectrometric group-type analysis makes use of average fragmentation patterns specific to the different compound types, that is, to certain groups of molecular structures. But the masses, as well as the abundances, of fragment ions are characteristic of the specific parent molecule. The sum of all of the peaks gives the average distribution of the parent molecules. Generally, a limited set of peaks of a molecule is selected for evaluation. Calibration ensures that the system is operating in a reproducible fashion response factor matrix approach is used for quantitative measurement and to compensate for overlaps in fragmentation.

Selected fragment peak intensities within the homologous series of compound types are added together, and the sums are evaluated with a matrix of response factors to give the concentration of each compound type in the sample. Mass spectrometric group-type analyses can be expected to have errors in the order of up to 20%. However, *relative* changes are ordinarily reported with better accuracy.

Thus, mass spectrometry plays a key role in the identification of the constituents of feedstock constituents (Aczel, 1979). Mass spectrometric methods are particularly well suited to the analyses of complex mixtures of relatively simple hydrocarbons that comprise the bulk of petroleum and its products.

The principal advantages of mass spectrometric methods are (1) high reproducibility of quantitative analyses, (2) the potential for obtaining detailed data on the individual components and/or carbon number homologs in complex mixtures, and (3) a minimal sample size is required for analysis. The ability of mass spectrometry to identify individual components in complex mixtures is unmatched by any modern analytic technique, perhaps the exception is gas chromatography.

However, there are disadvantages arising from the use of mass spectrometry: (1) the limitation of the method to organic materials that are volatile and stable at temperatures up to 300 °C (570 °F) and (2) the difficulty of separating isomers for absolute identification. The sample is usually destroyed, but this is seldom a disadvantage.

The availability of high-resolution instrumentation extends the scope of mass spectrometry for both qualitative and quantitative applications. The main advantage of high-resolution techniques is the separation of fragment and molecular ions of the same nominal molecular weight but different elemental composition. In addition, most high-resolution instruments are capable of furnishing precise mass measurements, so that the formulae can be calculated from these measurements.

Further studies using pyrolysis/gas chromatography/mass spectrometry (py/gc/ms) (Speight and Pancirov, 1984) showed that different subfractions of an asphaltene fraction would produce the same type of polycyclic aromatic systems in the volatile matter but the distribution was not constant. It was also possible to compute the hydrocarbon distribution from which a noteworthy point here is preponderance of single ring (cycloparaffin and alkylbenzene) species as well as the domination of saturated material over aromatic material. The emphasis on low molecular weight material in the volatile products is to be anticipated on the basis that more complex systems remain as nonvolatile material and, in fact, are converted to coke.

9.3.4 Nuclear Magnetic Resonance Spectroscopy

NMR has frequently been employed for general studies of hydrogen types in petroleum and petroleum products (ASTM D4808, 2021) as well as for the structural studies of petroleum constituents (ASTM E386, 2021) (Speight, 1994a, 1994b). The technique has recently been adapted to measure the hydrogen content of fuels and other petroleum products (ASTM D3701, 2021; ASTM D4808, 2021). In fact, proton magnetic resonance (PMR) studies (along with infrared spectroscopic studies) were, perhaps, the first studies of the modern era that allowed structural inferences to be made about the polynuclear aromatic systems that occur in the high molecular weight constituents of petroleum.

Thus, NMR methods have gained a prominent place in the compositional and structural analysis of petroleum fractions (Speight, 1994a, 1994b). In its basic applications, NMR is fast and relatively inexpensive. Because of its convenience, speed, and greater wealth of detailed information, particularly from ^{13}C-NMR, it has displaced the *n-d-M* and related methods in most laboratories. NMR presents a direct measure of the aromatic and aliphatic carbon as well as hydrogen distribution. Beyond these results, both carbon and hydrogen in various structural groupings in a molecule can be determined. Thus, proton (^{1}H) and carbon-13 (^{13}C) nuclei are the most common ones used in NMR spectroscopy; nitrogen (^{15}N and ^{14}N) and sulfur (^{33}S) have been employed on occasion with petroleum fractions for special applications.

PMR has been widely employed in the structural analysis of petroleum fractions. It is a relatively inexpensive technique that allows measurement of hydrogen atoms in aromatic and aliphatic groups, even allowing differentiation between hydrogen attached next to an aromatic ring (α-position) and those farther removed from the ring. Atoms in single-ring and multi-ring aromatic compounds as well as those in olefin locations can also be identified.

Only a small amount (<10 mg) of the sample is required, dissolved in a solvent such as deutero-chloroform, contained in a glass tube of 5 mm diameter and placed in a highly homogeneous magnetic field where it is surrounded by one or more coils. The coils serve to subject the sample to a weak radio-frequency (RF) field. The hydrogen nuclei of the sample can be visualized as magnets and when the radio frequency is equal to the processing frequency, resonance occurs between the two, and the spin resonance is detected by a receiver coil. The position of a sample resonance with that of tetramethylsilane (TMS) as a reference difference is reported as chemical shift, δ, which is a dimensional number that is expressed in terms of parts per million (ppm) difference from the reference (TMS).

Quantitative accuracy of PMR for aromatic and aliphatic hydrogen is about 1% for distillates and 2–3% for residua; that for the distinction of aliphatic hydrogen atoms to an aromatic ring from those β and farther away from the ring is somewhat lower. Methyl (CH_3), methylene (CH_2), and methine (CH) hydrogen can ordinarily not be distinguished except for methyl hydrogen γ or farther away from aromatic rings. Even this methyl peak is sometimes difficult to quantify because of interference by naphthenic methine and methylene hydrogen.

Protons attached to single-ring and multi-ring aromatic compounds can usually be distinguished with reasonable accuracy, especially when the sample concentration is 2% or less. At such low concentrations, the dividing line between these protons is at 7.25 ppm. In the spectra of samples with high boiling point, much of the detail is lost and differentiation between group types is difficult.

Thus, in general, the proton (hydrogen) types in petroleum fractions can be subdivided into three types: (1) aromatic ring hydrogen, (2) aliphatic hydrogen adjacent to an aromatic ring, and (3) aliphatic hydrogen remote from an aromatic ring. In other cases, five types of hydrogen locations are identified: (1) aromatic hydrogen, (2) substituted hydrogen next to an aromatic ring, (3) naphthene hydrogen, (4) methylene hydrogen, and (5) terminal methyl hydrogen remote from an aromatic ring. Other ratios are also derived from which a series of structural parameters can be calculated.

However, it must be remembered that the structural details of the carbon backbone obtained from proton spectra are derived by inference, but it must be recognized that protons at peripheral positions can be obscured by intermolecular interactions. This, of course, can cause errors in the ratios that can have a substantial influence on the outcome of the calculations (Ebert et al., 1984, 1987; Ebert, 1990).

It is in this regard that carbon-13 magnetic resonance (CMR) can play a useful role. Since carbon magnetic resonance deals with analyzing the carbon distribution types, the obvious structural parameter to be determined is the degree of the aromaticity of the sample (i.e., f_a, which is the fraction of the carbon atoms that are in aromatic locations). A direct determination from the various carbon type environments is one of the better methods for the determination of aromaticity. Thus, through a combination of proton and carbon magnetic resonance techniques, refinements can be made on the structural parameters and, for the solid-state high-resolution CMR technique, additional structural parameters can be obtained.

The basic instrumentation for CMR is the same as that for PMR except that there are two radio-frequency fields orthogonal (at right angles) to the main magnetic field, one for observing the carbon-13 nuclei and the other for decoupling the proton nuclei. The low abundance of carbon-13 isotopes (1.1%) and the lower gyromagnetic ratio of the carbon-13 nucleus make the signal weaker by more than two orders of magnitude and, moreover, the nuclei have longer relaxation times. The effect is that, even with Fourier transform (FT) data acquisition, CMR measurements can hours to perform.

In contrast to PMR spectroscopy, the peak areas arising from the carbon-13 nuclei in different molecular positions ordinarily are not proportional to their concentration. Quantitative measurements require that two effects must be overcome: (1) the different relaxation times of the carbon-13 nuclei in different chemical groups and (2) the nuclear Overhauser enhancement (NOE). The latter effect refers to the rise in signal intensity when C–H coupled protons are saturated by the decoupling field.

One way to do this is to add a small amount of a paramagnetic relaxation reagent, such as trisacetylacetonatochromium(III), $Cr(AcAc)_3$, which changes the dominant relaxation mechanism into one involving the interaction between unpaired electrons and ^{13}C nuclei. It also reduces the long relaxation times of some carbons.

In its simplest form, CMR can distinguish between aliphatic and aromatic carbon. In the aliphatic region of petroleum ^{13}C-NMR spectra, several sharp peaks stand out and are used for quantitative. The most prominent peak is usually that at 29.7 ppm. It is attributed to methylene (CH_2) carbon atoms in long alkyl chains, positioned four or more carbons ($>\gamma$) away from an aromatic ring and from terminal methyl (CH_3) groups.

Normally, the absorption of 29.5–30.3 ppm is an estimation of the amount of carbon atoms in long alkyl chains ($>C_5$). Because this band represents methylene (CH_2) groups, two or more carbons away from an aromatic ring and a terminal group, there must be four more carbons per chain than indicated by the area under these peaks. The number of long chains (n_{CH2}) can be estimated from the peaks at 14.2 ppm (ω-CH_3) and at 28.1 ppm (CH_2) next to a terminal branch point, that is, a methane (CH) group. On the one hand, the peak at 14.2 ppm gives results too high for this purpose because it also indicates CH_3 groups from chain branches. Subtracting half of the 37.6-ppm peak area (CH_2 next to CH groups inside a chain) corrects for this feature. On the other hand, the peak at 14.2 ppm does not cover twin CH_3 groups (as in an isopropyl group). This is why the 28.1 peak (CH next to two terminal methyl groups) is needed. Thus,

$$n_{CH_2\text{long chains}} = A(14.2) - 1/2A(37.6) + 1/3A(28.1)$$

The number of methylene groups in long chains is the sum of the peak at 29.7 ppm, the number of methylene groups at the two ends, that is, six times the number of long chains for each of the two ends, and the number close to branch points inside the chain, namely, those next to them and one carbon atom removed, on both sides of the CH group. Thus, the total numbers is:

$$n_{CH_2} = C(29.7) + 6n_{CH_2\text{long chains}} + 4C(37.6)$$

The number of methine (CH) groups in long chains can be estimated from the absorbance at 37.6 and at 39.5 ppm. The peak at 37.6 ppm represents methylene groups next to methane carbon and thus does not represent a direct measurement of the methane groups. Thus:

$$n_{\text{CHlong chains}} = 1/2C(37.6) + C(39.5)$$

The methyl groups give rise to at least four peaks in carbon-13 spectra, namely the peaks at 11.5, 14.2, 19.5, and 22.7 ppm. The peak 22.7 ppm represents twin methyl groups as in an isopropyl group and it also has a contribution from a methylene group next to a methyl group and does not need to be represented again by a contribution to the equation from the peak at 14.2 ppm. Thus,

$$n_{\text{CH}_3\text{long chains}} = C(11.5) + C(19.5) + C(22.7)$$

The final estimate for carbon atoms in long chains is then the sum of the three types. The average length of long chains derives from dividing the number of carbon atoms in long chains by the number of long chains. Remembering, of course, that averages derived from magnetic resonance spectra can be very misleading if the data are interpreted too literally.

The methyl group (CH_3) gives rise to several peaks depending on its position in the molecule. A peak at 14.2 ppm signals such a group positioned at the end of an unbranched chain segment of at least two or three methylene (CH_2) groups. A methyl group next to a branch point (methine group, CH) at the end of a chain with at least two methylene groups produces a peak at 22.7 ppm. Farther away from the end of the chain, such a group gives a peak at or near 19.8 ppm. The methyl group at the end is also affected by these branch points.

Naphthenic carbon in high molecular weight petroleum fractions usually occupies so many slightly different positions that the peaks are unresolved and form a broad hump in the range 25–60 ppm, under the generally well-resolved paraffinic peaks.

The evaluation of the hump is the only direct method for the determination of naphthenic methylene and methane (CH_n) groups. However, its measurement may not always be reliable. In very high-boiling petroleum samples, such as asphaltene constituents, the paraffin-type resonance may be only partly resolved. Although this leads mainly to broader peaks, the overlap may in some cases add to the hump and, thus, cause erroneously high results for naphthenic carbon. The aromatic region of the ^{13}C-NMR spectrum can be evaluated by conventional integration of the peaks due to the main aromatic group types.

Thus, proton and CMR spectroscopic techniques offer the potential information about the molecular types in the nonvolatile fractions of petroleum. The techniques have, by the application of the estimation of peak areas and further application of mathematical methods have been used to obtain information about structural parameters that are then converted to average structures.

In most cases, the average structure of such complex mixtures as petroleum fractions is not the same as a representative structure. As already noted, average structures derived from magnetic resonance spectra can be very misleading if the data are interpreted too literally. Even though the average structures may always be questions, one must also treat with some caution the structural parameters since they have been derived using assumptions that themselves are subject to debate.

Thus, NMR spectroscopy has proved to be of great value in petroleum research because it identifies and counts hydrogen atoms according to their chemical and physical environments (Ramsey et al., 1967; Speight, 1970; Dereppe et al., 1975; Wen et al., 1978; Dereppe and Moreaux, 1985; Gray et al., 1989). Moreover, PMR allows rapid and nondestructive determination of the total hydrogen content and distribution of hydrogen among the chemical functional groups present in the sample. The technique is also useful as a means of estimating molecular weight (Leon, 1987) and for studying processing effects (Giavarini and Vecchi, 1987).

NMR has frequently been employed for general studies and for the structural studies of feedstock constituents (Hasan et al., 1988, 1989). In fact, PMR studies (along with infrared spectroscopic studies) were perhaps the first studies of the modern era that allowed structural inferences to be made about the polynuclear aromatic systems that occur in the high molecular weight constituents of petroleum.

In general, the proton (hydrogen) types in petroleum fractions can be subdivided into three types (Brown and Ladner, 1960) or into five types (Yen and Erdman, 1962). The first approach classifies the hydrogen types into H^*_{ar}, aromatic ring hydrogen; $H^{*\prime\prime}$, aliphatic hydrogen adjacent to aromatic rings; and H^*_0, aliphatic hydrogen not adjacent to aromatic rings. The second approach subdivided the hydrogen distribution into aromatic hydrogen H_A ($H_A/II - H^*_{ar} = h_a$); $''$-substituted hydrogen next to aromatics ($H''/H = H^{*N}$), H''; naphthenic hydrogen H_N; methylene hydrogen H_R; and terminal methyl hydrogen not close to aromatic system H_{Sme} ($H^*_0 = H_N/H + H_R/H + H_{sme}/H$).

Following the calculation of the fractional distribution of the protons, the carbon distribution in atoms per molecule may be obtained from the following relationships:

$$C_S = H_T\left(H_\alpha / 2 / H_N / 2 + H_{My} / 2 + H_{Me} / 3\right)$$
$$C_{SA} = H_T \times H_\alpha / 2$$
$$C_A = C_T - C_S$$
$$C_P = H_T\left(H_A + H_\alpha / 2\right)$$
$$C_I = C_A - C_P$$
$$C_{My} = H_T \times H_{My} / 2$$
$$C_{Me} = H_T \times H_{Me} / 2$$
$$C_N = C_S - \left(C_{SA} + C_R\right)$$
$$R_A = \left(C_I + 2\right) / 2$$

C_S	total saturated carbon atoms per molecule
C_{SA}	total saturated carbon atoms to an aromatic ring
C_A	total aromatic carbons per molecule
C_T	total carbon atoms per molecule from analysis and molecular weight
C_P	peripheral carbon in a condensed aromatic sheet or the total number of hydrogen atoms present if the sheet is completely unsubstituted
C_I	internal carbon in a condensed aromatic sheet
C_{My}	total paraffinic methylene carbon atoms per molecule in locations other than ″ to an aromatic ring
C_{Me}	total paraffinic methyl carbon atoms per molecule in locations other than ″ to an aromatic ring
C_N	total naphthenic carbon atoms per molecule
H_T	total hydrogen atoms per molecule from analysis and molecular weight
R_A	aromatic rings per molecule

It is also possible to estimate certain structural parameters for the aromatics within fractions. Thus, the ratio C_{SA}/C_P is the average number of carbon atoms directly attached to an aromatic sheet divided by the total peripheral carbon atoms. Expressed in another manner, the ratio C_{SA}/C_P is the degree of substitution of the aromatic sheet. The C_S/C_{SA} ratio is the total number of saturated carbon atoms divided by the number of carbon atoms attached to the edge of the aromatic sheet. Alternatively, the C_S/C_{SA} ratio represents the average number of carbon atoms attached to a position of the edge of an aromatic sheet. The C_P/C_A ratio is the peripheral carbon atoms per aromatic sheet divided by the total aromatic carbon atoms for that sheet. Alternatively, this ratio is an estimate of the shape of the aromatic sheets; and $(C_S - C_{Me})/C_{Me}$ is the ratio of methylene carbon to methyl carbon, that is, an estimate of the degree of branching in the saturated moieties of the molecule; it also can be written $C_S/C_{Me} - 1$.

The C_P/C_A parameter, which indicates the shape of the aromatic centers of the molecules, is perhaps a little misleading, and literal interpretation can lead to misconceptions about the structures present in, say, residua. It is perhaps more pertinent if this and the other parameters are interpreted on a comparative basis.

With the evolution of magnetic resonance and the development of the carbon-13 procedure, there has been further detail added to the structure of feedstock fractions (Severin and Glintzer, 1984). However, carbon-type analysis does not allow the identification of individual constituents but presents an overall picture of the carbon framework and (if the term can be used without too literal a translation) actually presents an average molecular structure of these higher boiling materials.

9.3.5 Ultraviolet Spectroscopy

Of all of the methods applied to determining the types of aromatic systems in high molecular weight viscous feedstocks, a method that has considerable potential is ultraviolet spectroscopy (Lee et al., 1981; Bjorseth, 1983; Yokota et al., 1986).

The ultraviolet–visible (UV–Vis) spectrum, although not as specific for chemical group types as infrared spectroscopy and NMR spectroscopy, can distinguish between aromatic compounds with different ring numbers and configurations (ASTM D1840, 2021; ASTM D2269, 2021). The patterns are not distinct enough to recognize or distinguish these compounds in complex mixtures, but they can be useful for their identification in narrow fractions.

Ultraviolet–visible spectroscopy (ASTM E169, 2021) can be employed as a detector for the fractionation of petroleum samples, for example, for the chromatographic separation and/or identification of aromatics by ring number, especially in combination with a technique such as liquid chromatography (Speight, 1986). Thus, ultraviolet–visible spectroscopy lends itself to studies of refining processes as an online detector.

This technique can add valuable information about the degree of condensation of polycyclic aromatic ring systems when used in conjunction with high-performance liquid chromatography (HPLC) (Suatoni and Swab, 1976; Lee et al., 1981; Bjorseth, 1983; Felix et al., 1985; Yokota et al., 1986; Verhasselt, 1992; Ali and Nofal, 1994). Indeed, when this approach is taken, the technique not only confirms the complex nature of the asphaltene fraction and also allows further detailed identifications to be made of the individual functional constituents of the asphaltene fraction.

The technique is also useful on a before refining and after refining basis. It allows molecular changes in, for example, the asphaltene fraction to be tracked and provided information relative to the chemistry of coke formation.

9.3.6 X-ray Diffraction

X-ray diffraction techniques may also provide information about the structural qualities of organic materials, but the data provide information about the spatial periodicity within a substance (Pedersen et al., 2002). However, there is one necessary assumption in the application of this method and it requires that, within a given fraction, the larger portion of the molecules contain within and among themselves certain repeated structural features, such as sheets of condensed aromatic rings.

After plotting the diffraction patterns onto a straight baseline, it is possible to make estimates of the aromaticity. Thus, the fraction of aromatic carbon f_A can be determined using the relationship:

$$f_A = C_{aromatic} / C_{total} = A_{002} / (A_{002} + A_\gamma)$$

where A represents the areas under the respective peaks.

Use of the well-established Scherrer equation provides evidence for the association of the asphaltene constituents.

$$L = k\lambda/(B\cos\theta)$$

L is either the crystallite height L_c or the crystallite diameter L_a, B is the half-peak width, k is a constant (for L_c, $k = 1.00$; for L_a, $k = 1.84$).

This, of course, assumes that the constituents of the asphaltene fraction have repeating aromatic units with a tendency to stack in layers, as in the case of carbon and graphite. Interlayer distances $c/2$ are calculated from the position of the maximum of the (002) band. The data derived using this equation refer mainly to the structures found in the asphaltene fraction of petroleum. The data indicate that layer diameters L_a are of the order of 6–15 D, interlayer distances $c/2$ fall into the range 3.5–3.8 D, and stack heights L_c are approximately 10–16 D, which suggests that there are three to five aromatic lamellae per stack.

Other investigations (Bestougeff, 1967; Pollack and Yen, 1969, 1970) indicated that the asphaltene constituents might consist of condensed aromatic sheets carrying naphthenic and paraffinic substituents and could be explained on the basis of polynuclear aromatic sheets of 9–15 Å diameter, spaced approximately 3.6 Å apart. A structure consisting of about four sheets could be obtained before the natural heterogeneity of asphaltene constituents prevented further accumulation. However, accumulation of other nuclei was still possible because the asphaltene molecules were considered to contain two or more polynuclear aromatic sheets sigma-bonded by aliphatic carbon chains. Chains orientation of aliphatic groups, spaced about 5.7X apart in the manner of a saturated carbon chain or loose net of naphthenic rings, was also indicated.

X-ray diffraction had been used for the determination of the fraction of aromatic carbon (f_a) in petroleum constituents. This ratio can also be conveniently and precisely obtained by carbon-13 NMR spectroscopy as well as by infrared spectroscopy. The determination of the fraction of carbon that is aromatic may be in error (Ebert, 1990) since X-ray diffraction data can be very misleading (Ebert et al., 1984), especially when the data are used to translate geometric data—measurements of (aromatic) sheet diameter—into structural information. On the one hand, not all aromatic atoms contribute to the stack diameter seen by X-ray diffraction, whereas, on the other hand, nonaromatic atoms such as hydroaromatic carbons and other substituents on aromatic rings may contribute to the diffraction pattern. Thus, the interpretation of these measurements is quite arbitrary.

Extended X-ray absorption fine structure (EXAFS) and X-ray absorption near-edge structure (XANES) spectroscopy are tools for the investigation of the immediate chemical environment of X-ray absorbing elements such as metals and sulfur. XANES and X-ray photoelectron spectroscopy (XPS) have been applied to the determination of sulfur compounds as well as nickel and vanadium in petroleum samples.

9.4 Heteroatom Systems

As already defined, a heteroatom system is a molecule (acyclic or cyclic) that contains an atom (or atoms) such as nitrogen or oxygen or sulfur in a linear or cyclic hydrocarbon derivative.

Although the major focus of spectroscopic techniques appears to have been delineation of the hydrocarbon skeleton of the nonvolatile constituents of petroleum, efforts have also been made to determine the nature and location of the nitrogen, oxygen, and sulfur atoms in the aromatic systems (McKay et al., 1978).

PMR studies do not take into account the heteroatoms which had to be accommodated by convenient incorporation into the hydrocarbon systems. In fact, placement of the heteroatoms into the structures derived by PMR would require significant modifications of the molecular models to, perhaps, much larger structures. Indeed, this may be cited as a major disadvantage of many of the models that have been derived to date.

It is, therefore, necessary to make reference to the spectroscopic methods that have been used to determine the nature of the heteroatom species. Some mention has already been made in the previous sections. This section will confirm what has already been noted. However, it is necessary to note that even though non-spectroscopic methods may have been the prime method, spectroscopy usually played a role in confirming the nature of the heteroatom.

9.4.1 Nitrogen

Nitrogen species occur readily in crude oils and it must be anticipated that similar species will occur within the constituents of the asphaltene fraction. Studies on the disposition of nitrogen in petroleum constituents of the asphaltene fraction indicated the existence of nitrogen as various heterocyclic types (Nicksic and Jeffries-Harris, 1968; Moschopedis and Speight, 1976b, 1979; Schmitter et al., 1983; Jacobson and Gray, 1987). Much of the nitrogen is believed to be in aromatic locations. There are also reports where the organic nitrogen in petroleum constituents of the asphaltene fraction has been defined in terms of basic and nonbasic types (Nicksic and Jeffries-Harris, 1968) and there is evidence for the occurrence of carbazole nitrogen in the constituents of the asphaltene fraction (Moschopedis and Speight, 1979).

Other assumptions include nitrogen in the constituents of the asphaltene fraction on the basis of not more than one nitrogen atom per molecule. Such an assumption may be difficult to justify on a natural product basis where molecules with two or more nitrogen atoms are known but it would certainly serve as a reasonable guide for molecular studies where generalizations are often necessary.

9.4.2 Oxygen

Of the heteroatom-containing species in petroleum, there are more data pertaining to the locations of the oxygen atoms than to the sulfur and nitrogen atoms. However, the majority of the data relates to oxygen functions in blown (oxidized) asphalt and residua that may be of little relevance to the oxygen functions in the native materials.

Oxygen has been identified in carboxylic, phenolic, and ketonic locations (Nicksic and Jeffries-Harris, 1968; Moschopedis and Speight, 1976a, 1976b; Ritchie et al., 1979a, 1979b) but is not usually regarded as being located primarily in heteroaromatic ring systems. In the context of polyhydroxy aromatic nuclei existing in Athabasca constituents of the asphaltene fraction, it is of interest to note that pyrolysis at 800 °C (1,470 °F) results in the formation of resorcinol derivatives (Ritchie et al., 1979a) implying that such polyhydroxy aromatic rings systems may indeed exist in the constituents of the asphaltene fraction.

A very important aspect of the presence of oxygen functional groups is the identification of these functions and definition of their role in determining the physical properties of the system. The physical properties can also influence the structure of the micelle (Moschopedis and Speight, 1976a; Bukka et al., 1993).

9.4.3 Sulfur

Sulfur occurs in petroleum as benzothiophenes, dibenzothiophenes, and naphthene benzothiophenes and many of these sulfur types occur in the asphaltene fraction (Nicksic and Jeffries-Harris, 1968; Speight and Pancirov, 1984; Ngassoum et al., 1986; Rose and Francisco, 1988). More highly condensed thiophene types may also occur in the asphaltene fraction but are precluded from identification by low volatility. Other forms of sulfur that occur in the constituents of the asphaltene fraction include the alkyl–alkyl sulfides, alkyl–aryl sulfides, and aryl–aryl sulfides (Waldo et al., 1992).

Application of Fourier transform infrared spectroscopy and sulfur-33 magnetic resonance spectroscopy has provided valuable data about the distribution of the sulfur types in petroleum, particularly the partition of sulfur into aromatic types and aliphatic types. It is particularly interesting to note that sulfur exists in a variety of locations that may be dependent upon the natural product precursors and the mechanism of maturation.

An interesting aspect of sulfur identification has arisen through the application of X-ray absorption spectroscopy to answer the question about sulfur forms in the constituents of the asphaltene fraction (George et al., 1990; Kelemen et al., 1990). The data gave a clear demonstration of the existence of nonvolatile sulfide and thiophene sulfur in the constituents of the asphaltene fraction.

Application of XANES spectroscopy to the determination of sulfur species in the constituents of the asphaltene fraction confirmed the presence of thiophene sulfur and sulfide sulfur. In fact, the technique showed the predominance of thiophene sulfur and, to a lesser extent, sulfide sulfur in the samples examined.

9.4.4 Metals

Metals are not found in any fractions that boil below 540 °C (1,000 °F) (Reynolds, 1998). Typically, the constituents of the asphaltene fraction have the highest concentration of vanadium and other metals. Most often, in characterizing metal complexes in crude oils, an asphaltene precipitation is performed that concentrates the vanadium (as well as nickel and iron). The nickel and nitrogen distributions are very similar to the vanadium distributions, in general, with the asphaltene fraction having the highest concentrations. The metals are found bound in two types, metallo-petroporphyrins and metallo non-porphyrins. The former have the distinctive ultraviolet–visible Soret band due to the highly conjugated porphyrin ring system, and the latter comprises whatever metal concentrations are not accountable by the ultraviolet–visible Soret band.

Metals (i.e., nickel and vanadium) are much more difficult to integrate into the asphaltene system. The nickel and vanadium occur as porphyrins (Malhotra and Buckmaster, 1985; Shepherd and Graham, 1986) but whether or not these are an integral part of the asphaltene structure is not known. Some of the porphyrins can be isolated as a separate stream from petroleum (Branthaver and Dorrence, 1978; Reynolds et al., 1989; Branthaver, 1990).

9.5 Miscellaneous Methods

There also are a number of methods that can be used to obtain structural information about the constituents of petroleum. For example, derivation of the functional groups using such techniques as silylation of hydroxyls, esterification of acid functions, and quaternarization of amines has been employed successfully. Examination of the products from oxidative degradation or hydrogenolysis as well as the products of alkylation, dealkylation, or scission of carbon–nitrogen, carbon–oxygen, or carbon–sulfur bonds is a useful method. Catalytic hydrogenation or dehydrogenation in naphthenic-aromatic systems via naphthenic intermediates is also of some value.

Questions about the nature of the metal species in petroleum have been ongoing for the past two decades. Metals are important because of the manner in which they can interact with catalysts causing deactivation of the catalyst and also appearing to promote coke formation.

Inductively coupled plasma emission spectroscopy has been used for the determination of trace elements in ash residues from various crude oils (Saban et al., 1984). The data were applied to determining the origin and occurrence of crude oils as an aid in oil exploration. Nevertheless, the technique does offer valuable information about the occurrence of trace elements in petroleum.

Inductively coupled plasma emission spectroscopy has also been employed in conjunction with graphite furnace atomic absorption spectroscopy to investigate the metals types in petroleum (Reynolds et al., 1984). The data showed that a considerable percentage (50–80%) of the metals in a suite of crude oils existed in the non-porphyrin form. These non-porphyrin metal species were relatively low molecular weight (<400) species that were liberated when the tertiary structure of the asphalt material was denatured by extraction.

Application of X-ray absorption spectroscopy (EXAFS/XANES) and ultraviolet–visible spectrometric analysis to the issue of the chemical environment of vanadium in various constituents of the asphaltene fraction (Goulon et al., 1984) indicated a contradiction in terms of the amount of metals in porphyrin environments and that the conclusions were method dependent.

The results of small angle neutron scattering (SANS) studies (Overfield et al., 1989) have indicated a true molecular weight below 6,000 and indications that this decreases with increasing temperature. Such conclusions can benefit the processing of asphaltene-containing crude oil (such as heavy oil) since they offer some insight into the structure of the micelle and the behavior of the constituents of the asphaltene fraction under processing conditions.

References

Aczel, T. 1979. In: *Practical Mass Spectrometry*. B.S. Middleditch (Editor). Plenum Press, New York.

Ali, M.A., and Nofal, W.A. 1994. Application of High-Performance Liquid Chromatography for Hydrocarbon Group Type Analysis of Crude Oils. *Fuel Science and Technology International*, 12: 21–33.

ASTM D1159. 2021. Standard Test Method for Bromine Numbers of Petroleum Distillates and Commercial Aliphatic Olefins by Electrometric Titration. Annual Book of Standards. ASTM International, West Conshohocken, Pennsylvania.

ASTM D1840. 2021. Standard Test Method for Naphthalene Hydrocarbons in Aviation Turbine Fuels by Ultraviolet Spectrophotometry. Annual Book of Standards. ASTM International, West Conshohocken, Pennsylvania.

ASTM D2269. 2021. Standard Test Method for Evaluation of White Mineral Oils by Ultraviolet Absorption. Annual Book of Standards. ASTM International, West Conshohocken, Pennsylvania.

ASTM D2425. 2021. Standard Test Method for Hydrocarbon Types in Middle Distillates by Mass Spectrometry. Annual Book of Standards. ASTM International, West Conshohocken, Pennsylvania.

ASTM D2650. 2021. Standard Test Method for Chemical Composition of Gases by Mass Spectrometry. Annual Book of Standards. ASTM International, West Conshohocken, Pennsylvania.

ASTM D2786. 2021. Standard Test Method for Hydrocarbon Types Analysis of Gas-Oil Saturates Fractions by High Ionizing Voltage Mass Spectrometry. Annual Book of Standards. ASTM International, West Conshohocken, Pennsylvania.

ASTM D2789. 2021. Standard Test Method for Hydrocarbon Types in Low Olefinic Gasoline by Mass Spectrometry. Annual Book of Standards. ASTM International, West Conshohocken, Pennsylvania.

ASTM D3239. 2021. Standard Test Method for Aromatic Types Analysis of Gas-Oil Aromatic Fractions by High Ionizing Voltage Mass Spectrometry. Annual Book of Standards. ASTM International, West Conshohocken, Pennsylvania.

ASTM D3701. 2021. Standard Test Method for Hydrogen Content of Aviation Turbine Fuels by Low Resolution Nuclear Magnetic Resonance Spectrometry. Annual Book of Standards. ASTM International, West Conshohocken, Pennsylvania.

ASTM D4808. 2021. Standard Test Methods for Hydrogen Content of Light Distillates, Middle Distillates, Gas Oils, and Residua by Low-Resolution Nuclear Magnetic Resonance Spectroscopy. Annual Book of Standards. ASTM International, West Conshohocken, Pennsylvania.

ASTM E168. 2021. Standard Practices for General Techniques of Infrared Quantitative Analysis. Annual Book of Standards. ASTM International, West Conshohocken, Pennsylvania.

ASTM E169. 2021. Standard Practices for General Techniques of Ultraviolet-Visible Quantitative Analysis. Annual Book of Standards. ASTM International, West Conshohocken, Pennsylvania.

ASTM E334. 2021. Standard Practice for General Techniques of Infrared Microanalysis. Annual Book of Standards. ASTM International, West Conshohocken, Pennsylvania.

ASTM E386. 2021. Standard Practice for Data Presentation Relating to High-Resolution Nuclear Magnetic Resonance (NMR) Spectroscopy. Annual Book of Standards. ASTM International, West Conshohocken, Pennsylvania.

ASTM E1252. 2021. Standard Practice for General Techniques for Obtaining Infrared Spectra for Qualitative Analysis. Annual Book of Standards. ASTM International, West Conshohocken, Pennsylvania.

ASTM E1316. 2021. Standard Terminology for Nondestructive Examinations. Annual Book of Standards. ASTM International, West Conshohocken, Pennsylvania.

Bestougeff, M.A. 1967. *Bulletin de la Société Chimique de France*: 4773.

Bjorseth, A. 1983. *Handbook of Polycyclic Aromatic Hydrocarbons*. Marcel Dekker Inc., New York.

Branthaver, J.F. 1990. Porphyrins. In: *Fuel Science and Technology Handbook*. J.G. Speight (Editor). Marcel Dekker Inc., New York.

Branthaver, J.F., and Dorrence, S.M. 1978. In: *Analytical Chemistry of Liquid Fuel Sources: Tar Sands, Oil Shale, Coal, and Petroleum*. P.C. Uden, S. Siggia, and H.B. Jensen (Editors), Advances in Chemistry Series No. 170. American Chemical Society, Washington, DC. Chapter 10

Brown, J.K., and Ladner, W.R. 1960. *Fuel*, 39: 87.

Bukka, K., Miller, J.D., Hanson, F.V., Misra, M., and Oblad, A.G. 1993. The Influence of Carboxylic Acid Content on Bitumen Viscosity. *Fuel*, 73: 257–268.

Dereppe, J.M., and Moreaux, C. 1985. Measurement of CH*n* Group. In Fossil Fuel Materials Using DEPT ^{13}C NMR. *Fuel*, 64: 1174–1176.

Dereppe, J.M., Moreaux, C., and Castex, H. 1975. Analysis of Asphaltenes by Carbon and Proton Nuclear Magnetic Resonance Spectroscopy. *Fuel*, 57: 435–441.

Drews, A.W. (Editor). 1998. *Manual on Hydrocarbon Analysis*. 6th Edition. ASTM International, West Conshohocken, Pennsylvania.

Ebert, L.B. 1990. Comment on the Study of Asphaltenes by X-ray Diffraction. *Fuel Science and Technology International*, 8: 563.

Ebert, L.B., Scanlon, J.C., and Mills, D.R. 1984. Liquid Fuels. *Technology*, 2: 257.

Ebert, L.B., Mills, D.R., and Scanlon, J.C. 1987. *American Chemical Society, Division of Petroleum Chemistry, Preprints*, 32(2): 419.

Felix, G., Bertrand, C., and Van Gastel, F. 1985. A New Caffeine Bonded Phase for Separation of Polyaromatic Hydrocarbons and Petroleum Asphaltenes by High Performance Liquid Chromatography. *Chromatographia*, 20(3): 155–160.

George, G.N., Gorbaty, M.L., and Kelemen, S.R. 1990. In: *Geochemistry of Sulfur in Fossil Fuels*. W.L. Orr and C.M. White (Editors) (Editors), Symposium Series No. 429. American Chemical Society, Washington, DC. Chapter 12

Giavarini, C., and Vecchi, C. 1987. Characterization of Visbreaker Bitumen by NMR Spectroscopy. *Fuel*, 66: 868–869.

Goulon, J., Esselin, C., Friant, P., Berthe, C., Muller, J.-F., Poncet, J.-L., Guilard, R., Escalier, J.-C., and Neff, B. 1984. In: *Characterization of Heavy Crude Oils and Petroleum Residues*. B. Tissot (Editor). Editions Technip, Paris, France. p. 3.

Gray, M.R., Choi, J.H.K., and Egiebor, N.O. 1989. Structural Group Analysis of Residues from Athabasca Bitumen. *Fuel Science and Technology International*, 7: 599–610.

Hasan, M., Siddiqui, M.N., and Arab, M. 1988. Separation and Characterization of Asphaltenes from Saudi Arabian Crudes. *Fuel*, 67: 1131–1134.

Hasan, M., Ali, M.F., and Arab, M. 1989. Structural Characterization of Saudi Arabian Extra Light and Light Crudes by 1-H and 13-C NMR Spectroscopy. *Fuel*, 68: 801–803.

Jacobson, J.M., and Gray, M.R. 1987. The Use of Infrared Spectroscopy and Nitrogen Titration Data in Structural Group Analysis of Bitumen. *Fuel*, 66: 749–752.

Kelemen, S.R., George, G.N., and Gorbaty, M.L. 1990. Direct Determination and Quantification of Sulphur Forms in Heavy Petroleum and Coals: 1. The X-ray Photoelectron Spectroscopy (XPS) Approach. *Fuel*, 69: 939–944.

Lee, M.L., Novotny, M.S., and Bartle, K.D. 1981. *Analytical Chemistry of Polycyclic Aromatic Compounds*. Academic Press Inc., New York.

Leon, V. 1987. Average Molecular Weight of Oil Fractions by Nuclear Magnetic Resonance. *Fuel*, 66: 1445–1446.

Malhotra, V.M., and Buckmaster, H.A. 1985. 34 GHz e.p.r. Study of Vanadyl Complexes in Various Asphaltenes: Statistical Correlative Model of the Coordinating Ligands. *Fuel*, 64: 335–341.

McKay, J.F., Amend, P.J., Cogswell, T.E., Harnsberger, P.M., Erickson, R.B., and Latham, D.R. 1978. In: *Analytical Chemistry of Liquid Fuel Sources: Tar Sands, Oil Shale, Coal, and Petroleum*. P.C. Uden, S. Siggia, and H.B. Jensen (Editors), Advances in Chemistry Series No. 170. American Chemical Society, Washington, DC. Chapter 9

Moschopedis, S.E., and Speight, J.G. 1976a. An Investigation of Hydrogen Bonding by Oxygen Functions in Athabasca Bitumen. *Fuel*, 55: 187.

Moschopedis, S.E., and Speight, J.G. 1976b. Oxygen Functions in Asphaltenes. *Fuel*, 55: 334.

Moschopedis, S.E., and Speight, J.G. 1979. Investigation of Nitrogen Types in Athabasca Bitumen. *American Chemical Society, Division of Petroleum Chemistry, Preprints*, 24(4): 1007.

Ngassoum, M.B., Faure, R., Ruiz, J.M., Lena, L., Vincent, E.J., and Neff, B. 1986. *Fuel*, 65: 142.

Nicksic, S.W., and Jeffries-Harris, M.J. 1968. Acid Precipitation of Crude Oil Asphaltenes-Structural Implications. *Journal of the Institute of Petroleum*, 54: 107–114.

Overfield, R.E., Sheu, E.Y., Sinha, S.K., and Liang, K.S. 1989. SANS Study of Asphaltene Aggregation. *Fuel Science and Technology International*, 7: 611.

Pedersen C. Andersen S.I., Jensen J.O., and Speight J.G. 2002. X-ray Diffraction of Asphaltenes: What Can It Tell Us? Proceedings. 3rd International Conference on Petroleum Phase Behavior and Fouling. pp. 13–19.

Petersen, J.C. 1967. An Infrared Study of Hydrogen Bonding in Asphalt. *Fuel*, 44: 295–305.

Pollack, S.S., and Yen, T.F. 1969. *American Chemical Society, Division of Petroleum Chemistry, Preprints*, 14: B-118.

Pollack, S.S., and Yen, T.F. 1970. Structural Studies of Asphaltics by X-ray Small Angle Scattering. *Analytical Chemistry*, 42: 623–629.

Ramsey, J.W., McDonald, F.R., and Petersen, J.C. 1967. Structural Study of Asphalts by Nuclear Magnetic Resonance. *Industrial and Engineering Chemistry Product Research and Development*, 6: 231–236.

Rand, S. 2003. *Signifiance of Tests for Petroleum Products*. 7th Edition. ASTM International, West Conshohocken, Pennsylvania.

Rao, C.N.R. 1961. *Ultraviolet and Visible Spectroscopy: Chemical Applications*. Butterworths, London, United Kingdom.

Reynolds, J.G. 1998. In: *Petroleum Chemistry and Refining*. J.G. Speight (Editor). Taylor & Francis, Washington, DC. Chapter 3

Reynolds, J.G., Biggs, W.R., Fetzer, J.C., Gallegos, E.J., Fish, R.H., Komlenic, J.J., and Wines, B.K. 1984. In: *Characterization of Heavy Crude Oils and Petroleum Residues*. B. Tissot (Editor). Editions Technip, Paris, France. p. 153.

Reynolds, J.G., Jones, E.L., Bennett, J.A., and Biggs, W.R. 1989. Characterization of Nickel and Vanadium Compounds in Tar Sand Bitumen by U-VIS Spectroscopy and Size Exclusion Chromatography Coupled with Element Specific Detection. *Fuel Science and Technology International*, 7: 625–642.

Ritchie, R.G.S., Roche, R.S., and Steedman, W. 1979a. Pyrolysis of Athabasca Tar Sands: Analysis of the Condensible Products from Asphaltene. *Fuel*, 58: 523–530.

Ritchie, R.G.S., Roche, R.S., and Steedman, W. 1979b. *Chemistry & Industry*: 25.

Rose, K.D., and Francisco, M.A. 1988. A Two-Step Chemistry for Highlighting Heteroatom Species in Petroleum Material Using ^{13}C NMR Spectroscopy. *Journal of the American Chemical Society*, 110: 637.

Saban, M., Vitorovic, O., and Vitorovic, D. 1984. In: *Characterization of Heavy Crude Oils and Petroleum Residues*. B. Tissot (Editor). Editions Technip, Paris, France. p. 122.

Schmitter, J.M., Ignatiadis, I., and Arpino, P.J. 1983. Distribution of Diaromatic Nitrogen Bases in Crude Oils. *Geochimica et Cosmochimica Acta*, 47: 1975.

Severin, D., and Glintzer, O. 1984. In: *Characterization of Heavy Crude Oils and Petroleum Residues*. B. Tissot (Editor). Editions Technip, Paris, France. p. 19.

Shepherd, R.A., and Graham, W.R.M. 1986. *Fuel*, 65: 1612.

Speight, J.G. 1970. A Structural Investigation of the Constituents of Athabasca Bitumen by Proton Magnetic Resonance Spectroscopy. *Fuel*, 49: 76.

Speight, J.G. 1986. Preprints. Polynuclear Aromatic Systems in Petroleum. *American Chemical Society, Division of Petroleum Chemistry*, 31(4): 818.

Speight, J.G. 1994a. Application of Spectroscopic Techniques to the Structural Analysis of Petroleum. *Applied Spectroscopy Reviews*, 29: 269.

Speight, J.G. 1994b. Chemical and Physical Studies of Petroleum Asphaltenes. In: *Asphaltenes and Asphalts. Developments in Petroleum Science, 40*. T.F. Yen and G.V. Chilingarian (Editors). Elsevier, Amsterdam, Netherlands. Chapter 2

Speight, J.G. 2001. *Handbook of Petroleum Analysis*. John Wiley & Sons Inc., Hoboken, New Jersey.

Speight, J.G. 2002. *Handbook of Petroleum Product Analysis*. John Wiley & Sons Inc., Hoboken, New Jersey.

Speight, J.G., and Pancirov, R.J. 1984. Structural Types in Asphaltenes as Deduced from Pyrolysis-Gas Chromatography-Mass Spectrometry. *Liquid Fuels Technology*, 2: 287.

Suatoni, J.C., and Swab, R.E. 1976. Preparative Hydrocarbon Compound Type Analysis by High Performance Liquid Chromatography. *Journal of Chromatographic Science*, 14: 535–537.

Tynan, E.C., and Yen, T.F. 1969. Association of Vanadium Chelates. *Fuel*, 48: 191–208.

Verhasselt, A.F. 1992. UV/VIS Spectrophotometry for the Evaluation of Binder-Aggregate Interactions. *Fuel Science and Technology International*, 10: 581–594.

Waldo, G.S., Mullins, O.C., Penner-Hahn, J.E., and Cramer, S.P. 1992. Determination of the Chemical Environment of Sulphur in Petroleum Asphaltenes by X-ray Absorption Spectroscopy. *Fuel*, 71: 53–57.

Wen, C.S., Chilingarian, G.V., and Yen, T.F. 1978. In: *Bitumens, Asphalts and Tar Sands*. G.V. Chilingarian and T.F. Yen (Editors). Elsevier, Amsterdam. Chapter 7

Yen, T.F., and Erdman, J.G. 1962. *American Chemical Society, Division of Petroleum Chemistry, Preprints*, 7: 5.

Yokota, T., Scriven, F., Montgomery, D.S., and Strausz, O.P. 1986. Absorption and Emission Spectra of Athabasca Asphaltene in the Visible and Near Ultraviolet Regions. *Fuel*, 65: 1142–1149.

10

Molecular Weight Determination

10.1 Introduction

The molecular weight (also frequently referred to as the "molecular mass") is the mass of a molecule and is measured in "daltons" although it is often presented as a number only (without the "daltons" qualifier). Put simply, the molecular weight of a molecule is a summation of all of the atomic weight of the atoms within the molecule and is needed to show the number of grams that are in 1 mol of that substance.

In addition, the molecular weight of a feedstock product appears to be less in demand. Most products, from gasoline to lubricating oil will be defined by distillation profile or be separation and other physical properties such as viscosity, API gravity, and then like. Therefore, the reader will find that the references in this chapter are devoted to methods designed for determining the molecular weight of the whole feedstock and, to a greater extent, to the molecular weight of the various fraction insofar as the data assists in the definition of process chemistry and engineering.

Thus, in the science and technology of crude oil and the vicious feedstocks, the molecular weight is a critical property. The molecular weight can be measured using many different methods including vapor pressure osmometry (VPO, sometimes referred to as phase osmometry), gel permeation chromatography (GPC), light scattering, viscometry, cryoscopy, boiling point elevation/depression (often referred to as ebulliometry), ultracentrifugation, mass spectrometry, and end-group analysis. The common method that has found use in many laboratories is the VPO method that relies upon the dissolution of the sample into a solvent which is appropriate for sample that has no or little volatility, such as the viscous feedstocks (including crude oil distillation residua).

Also, it is worthy of note that, in general, volatility tends to decrease with increasing molecular weight because larger molecules can participate in more

Handbook of Heavy Oil Properties and Analysis, First Edition. James G. Speight.

intermolecular bonding, even if other factors such as structure and polarity play a significant role in decreasing the volatility of the substance. The effect of molecular weight can be partially isolated by comparing chemicals of similar structure.

In terms of the molecular size or molecular mass, there are two types of molecular weight data which are (1) the number average molecular weight and (2) the weight average molecular weight. These two methods are commonly used in polymer science and often are a source of confusion to chemists and engineers. Therefore, a brief explanation of the differences between the two methods is warranted here.

The "number average molecular weight (M_n)" is dependent upon the total number of molecular constituents in a sample irrespective of the shape or size of the constituents which means that all molecular constituents of the sample are treated equally. Thus, the number average molecular weight is represented by the equation (Figure 10.1):

$$M_n = \Sigma N_i M_i / \Sigma N_i$$

In this equation, M_n is the number average molecular weight, N_i is the number of molecules having M_i weight in the polymer sample, and M_i is the weight of a particular molecule of the sample.

The number average molecular weight that is often the preferred data for examination of crude oil residua and the viscous feedstocks can be determined using methods such as GPC, viscometry, and colligative methods such as VPO.

On the other hand, the "weight average molecular weight (M_w)" is dependent upon the molecular mass of the individual constituents, which contributes to the overall molecular weight of the sample. Since the higher molecular size (molecular weight) constituents of a sample, the constituents having a larger molecular

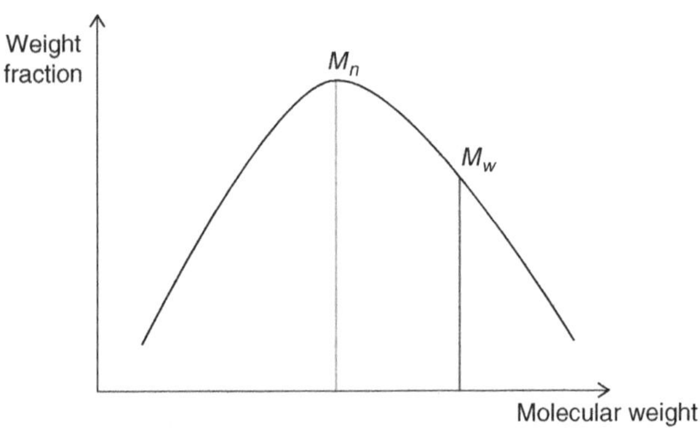

Figure 10.1 Representation of the difference between the number average molecular weight (M_n) and the weight average molecular weight (M_w).

size of molecule make a larger contribution to the molecular weight of the sample. Thus, unlike the number average molecular weight (M_n), the weight average molecular weight (M_w) has a greater influence on the molecular weight of the sample. In another form, the weight average molecular weight is the sum of the weight of all molecules divided by the total number of molecules that are present in that particular sample. Thus, the weight average molecular weight is represented by the equation (Figure 10.1):

$$M_w = \Sigma N_i M_i^2 / \Sigma N_i M_i$$

In this equation, M_w is the weight average molecular weight and N_i is the number of molecules of molecular mass M_i. The weight average molecular weight can be determined using static light scattering, small angle neutron scattering (SANS), X-ray scattering, and sedimentation velocity.

It is the purpose of this chapter to indicate the methods that have been used for determining the molecular weight of a feedstock and feedstock fractions without advocating one method over the other. The choice of the method is sample dependent. Differentiation between particle size and molecular weight is also necessary. In this context of this chapter, a particle may be composed of one or (usually) more molecules but a molecule is a single entity and is not a particle.

Finally, before entering into a presentation of the various physical methods by which the molecular weights of crude oil fractions and the viscous feedstock (and their respective fractions) can be determined, it is worthy of note that a method is available for the prediction of the molecular weight of crude oil fractions (over the range C5–C120 which covers the range of most crude oil-derived materials) is presented. The method requires as input information the 50% w/w true boiling point (TBP) as derived from simulated distillation and the density. The method gives results that are equivalent to the molecular weights determined by physical methods and encompasses the full practical range of molecular weights on the order of 75–1,700 (Goossens, 1996). This range does cover the molecular weights of gas oil, the viscous feedstocks, and crude oil residua. This follows an older method by which the molecular weight of crude oil products can be calculated (Kurganov et al., 1973).

In addition, in the context of the viscous feedstocks, the molecular weight of a feedstock product appears to be less in demand than the actual feedstock that gave rise to the product. Most products, from naphtha to the atmospheric gas oil can be defined by the distillation profile or by separation and other physical properties such as viscosity, API gravity, and the like. Therefore, the reader will find that the references in this chapter are predominantly devoted to methods designed for determining the molecular weight of the whole feedstock and, to a greater extent, to the molecular weight of the various fractional constituents (other than the resin and asphaltene fractions), not only insofar as the data assist in the definition of process chemistry and engineering.

10.2 Methods for Molecular Weight Measurement

The molecular weight of crude oil (as well as crude oil fractions) and the viscous refinery feedstocks (typically, the number average molecular weight which is preferred by many investigators) is an important aspect of studies of the chemical constitution and the chemical reactivity of the feedstocks (or, more correctly, the constituents of the feedstocks). For example, the molecular weight of any feedstock will give an indication of the amount of hydrogen required to upgrade the feedstock to a particular product or a series of products (Littlejohn and Lucas, 1999; Gary et al., 2007; Speight, 2014; Hsu and Robinson, 2017; Speight, 2017). In addition, structural group analysis (Chapter 9) of the various fractions often requires molecular weight data as, for instance, in the characterization of the various fractions for insertion into the various indices and correlation formulae (Speight, 2014, 2015).

As the boiling points of feedstock fractions increase, the complexity of the constituents in these fractions also increases (Speight, 2014, 2015, 2017). The high-boiling fractions of feedstock, and these may include saturated constituents (the saturates fraction) and aromatic constituents (the aromatics fraction) depending on the boiling profile of the crude oil, have an extremely complicated composition (Chapter 4). In fact, any attempted isolation of individual compounds is a frustrating, time-consuming, and an almost impossible task.

For a variety of reasons, of which structural group analysis and process modeling as well as process chemistry are only three, there is the need to determine the molecular weight of either the original feedstock or the feedstock constituent fractions. Various methods are available for molecular weight determination (Speight et al., 1985), some of which are claimed to be satisfactory and other are less adequate to the task. However, for those original constituents and products that have little or no volatility, for example, resins and asphaltenes, VPO (ASTM D2503, 2021) has been proven to be of considerable value.

The molecular weight and molecular weight distribution have multiple applications in the analysis and characterization of the viscous feedstock fractions. Molecular weight may be determined by a number of methods including (1) VPO, also called vapor phase osmometry; (2) size exclusion chromatography (SEC), which is still often referred to by its older name, GPC; and (3) non-fragmenting mass spectrometry, also called parent ion mass spectrometry. Colligative techniques, such as freezing-point depression (cryoscopy), commonly used in the past, has essentially been superseded by vapor phase osmometry primarily because the latter is more sensitive and, therefore, covers a wider molecular weight range.

There have been numerous attempts made to measure the molecular weight of crude oil, heavy crude oil, extra heavy crude oil, and tar sand bitumen using a variety of different methods but there appears to be a noticeable lack of consensus on the value obtained for a specific sample. This has been irrevocably traced to structural aspects of crude oil and the viscous feedstocks and, in particular, to the behavior of the asphaltene constituents in molecular dispersion in their own maltenes. In view of the fact that a number of methods used to determine molecular weight information utilize different solvents at different temperatures, the behavior of the component parts such as the resin constituents and the asphaltene constituents in these solvents needs to be established if a consistent picture of crude oil structure and behavior is to emerge (Speight, 2014).

With goal of structural information, the molecular weight of petroleum fractions has been the subject of much discussion over the past several decades. The reliability of the method, the meaning of the term average molecular weight, and the interpretation or usefulness of the data has received considerable attention.

In cases where only approximate values for the molecular weight are required, use can be made of several physical property correlations (Ali, 1971) that involve correlations such as a molecular weight–boiling point–density correlation. This correlation is believed to be satisfactory for determining molecular weights in the range 240–300 for any feedstock fraction for which a good 50% boiling point can be obtained. The viscosity correlation was used for lubricating oil fractions of molecular weight 240–700 using correlations for viscosity with gravity and for viscosity at 37.8 °C (100 °F) with viscosity at 98.9 °C (210 °F). However, for more accurate determinations of the molecular weight, direct measurement is necessary. Currently, of the methods available, several standard methods are recognized as being useful for determining of the molecular weight of various feedstock fractions (ASTM D2224, 2021; ASTM D2502, 2021; ASTM D2503, 2021; ASTM D2878, 2021; ASTM D3593, 2021). Each method has proponents and opponents because of assumptions made in the use of the method or because of the mere complexity of the sample and the nature of the inter- and intra-molecular interactions.

Methods for molecular weight measurement are also included in other more comprehensive standards (ASTM D128, 2021; ASTM D3712, 2021) and several indirect methods have been proposed for the estimation of molecular weight by correlation with other, more readily measured physical properties (Speight, 2014, 2015). The methods are satisfactory when dealing with the conventional type of crude oils or their fractions and products and when approximate values are desired.

10.2.1 Vapor Pressure Osmometry

VPO (ASTM D2503, 2021), also called vapor phase osmometry, is a relatively simple and cheap method for the determination of molecular weight. Most osmometers can operate over a range of temperature through the use of probes that cover specific temperature ranges. This gives the number average molecular weight and not the molecular weight distribution.

VPO is one of a group of techniques that measure properties based on the number concentration of molecules (i.e., colligative properties). The method measures the degree to which the vapor pressure of a solvent is lowered by the presence of a solute through a temperature effect.

The measuring elements are two temperature-sensitive thermistors placed in a closed, heat-insulated chamber. By means of two syringes, a drop of solvent is placed on one and a drop of the sample in solution in the same solvent is placed on the other. The chamber is saturated with solvent vapor and carefully temperature controlled. Because the solution has a lower vapor pressure than the solvent, solvent from the chamber atmosphere will condense on it. The difference in vapor pressure is proportional to its sample concentration. The heat of condensation warms the solution drop until its vapor pressure is the same as that of the surroundings. From then on, a steady state of condensation and warming is established. The temperature increase is measured and recorded. The solvent drop on the other thermistor is in equilibrium with the solvent in the chamber and, theoretically, gives no temperature effect. In reality, convection and other effects cause minute disturbances in the system, the effect of which can be minimized by subtracting the voltage of the solvent thermistor from that of the solution thermistor.

The temperature difference, ΔT, is related to concentration and molecular weight at infinite dilution by the equation:

$$\Delta T = K_1 c/\text{MW}$$

where K_1 is a constant determined by calibration and c is the solute concentration. The effect is measured at several concentrations, and the results are plotted vs. the reciprocal of the molecular weight (1/MW) and extrapolated to zero concentration (also referred to as unlimited dilution).

In the early days of VPO, ranges of error were accepted or at least recognized that made some of the data questionable. For example, at molecular weight at about or above 5,000 Da, the error might be as much as ±10% or even a high as ±15%. Modern instruments have reduced experimental errors several-fold by allowing the application of very small and precisely repeatable sample and solvent drops on the thermistor beads. Thus, measurements of molecular weight up to about 50,000 Da are now possible. In addition, samples of much lower concentrations can now be measured allowing linear extrapolation to zero concentration, in contrast to those from more concentrated solutions.

However, one of the issues that arises through the use of this method for determining the number average molecular weight is that there can be a shift to the low values by the presence of low-molecular weight contaminants. For example, traces of the solvent (other than that used for the molecular weight measurement) that remain after the extraction or fractionation process(es). Alternatively, if the sample is an oil fraction, there may be some inclusion of lower-boiling constituents that, by rights, belong in the preceding fraction. Residual toluene may not affect the molecular weight measured in toluene as long as the effect is not due to sample concentration by weight but it would certainly effect the molecular weight as measured in pyridine, nitrobenzene, or in any other solvent with a shift to lower than real values.

A common solvent for VPO is toluene that is satisfactory for hydrocarbons and moderately polar compounds. However, for the highly polar fractions, such as asphaltenes, more polar solvents such as pyridine are required. The molecular weight of such fractions measured by VPO in pyridine are distinctly lower than those measured in toluene (Speight et al., 1985) indicating a lower degree of aggregation, assuming that contamination with trace amounts of previously used solvent in the separation of the samples can truly be excluded.

An aspect of the number average molecular weight by VPO that also needs consideration is the possibility of some, or even preferential, aggregation of several small polar molecules with a much larger polar molecule during sample collection (precipitation) or during fractionation. The aggregates dissociate in the solvents used for molecular weight measurement, releasing some or all of the low-molecular weight species. Such aggregation is favored by precipitation of asphaltenes where occlusion of polar molecules from the maltene fraction is common (Ali et al., 1985). However, this effect does not even require contamination with foreign material but is possible in any sample containing highly polar molecules of widely different molecular weight. It slants the resulting number average molecular weight toward low values because the more numerous small molecules dominate over the fewer large ones even though their amounts by weight may be the same or smaller. A number average molecular weight of 5,000, thus, does not preclude the presence in that fraction of several weight percent of molecules at least 10 times larger, that is., of ca. 50,000.

Another potential problem with any molecular weight measurements of asphaltenes is their tendency to become less soluble after isolation. In crude oil and its heavy fractions, the asphaltenes are maintained dispersed in the non-asphaltene material by interaction with the resins (soluble polar molecules) (Koots and Speight, 1975; Speight, 2014). However, once they are isolated by precipitation, the asphaltene constituents can be considered to be denatured, analogous to a hard-boiled egg as opposed to a fresh egg. Properties may change and, hence, asphaltenes solubility can be quite different.

The observance of lower than true molecular weights though the influence of solvent and/or adsorbed lower molecular weight species on the asphaltene constituents is not specific to the VPO method. It applies to any method producing number average results. Only further fractionation or other suitable measurements can provide the molecular weight distribution of these polar fractions. Re-precipitation or further fractionation of asphaltenes mitigates the preferential solvation of large polar molecules by small ones (Speight et al., 1984).

One technique for reducing the effect of aggregation is to measure the molecular weight of polar species (asphaltenes) at different concentrations and different temperatures to illustrate and negate any aggregation effects. Thus, the molecular is measured at three different concentrations at each of three different temperatures (Speight, 1994). The data are extrapolated to determine the molecular weight at zero concentration at each temperature after which extrapolation to room temperature to negate concentration and temperature effects. This, of course, assumes that the extrapolation line is straight!

Thus, the molecular weight of pentane-asphaltenes measured by VPO in nitrobenzene are even lower than the reported from measurements in pyridine, with differences ranging from 5% to over 50%. In fact, raising the temperature of the sample chamber (in which the probe is contained) to 100 °C (212 °F) and above lowered the results by another few percent. Such behavior strongly suggests that even nitrobenzene at the regular vapor pressure osmometer temperature (37.5 °C, 100 °F) was not sufficient to overcome all the intermolecular aggregation. This not an isolated event but seems to be a common occurrence as it was observed with pentane-asphaltenes front a wide variety of crude oils.

The aggregation effect has, on occasion, been incorrectly referred to as an error, is that the accuracy of the molecular weight value is influenced by aggregation of the molecular species in solution. In some instances, this may not be classed as an error insofar as the aggregation is recognized and the method is used (along with molecular weight determination in a polar solvent) to determine the maximum aggregation that can occur in the nonpolar solvent. Such effects are magnified when measuring the molecular weight of the asphaltenes that consist primarily of polar (heteroatom) constituents. Indeed, from such experiments, the aggregation effect can be determined as the ratio of the molecular weight in a nonpolar solvent (e.g., toluene) to the molecular weight on a polar solvent (e.g., pyridine or nitrobenzene). Thus:

$$\text{Aggregation factor, } N, = \mathrm{MW}_{\text{toluene}} / \mathrm{MW}_{\text{pyridine}}$$

or

$$\text{Aggregation factor, } N, = \mathrm{MW}_{\text{toluene}} / \mathrm{MW}_{\text{nitrobenzene}}$$

Asphaltene constituents naturally aggregate in nonpolar solvents but it must also be recognized that another possible reason for finding higher molecular weight for asphaltenes is due to the susceptibility of the polar molecules to oxidation after separation (Moschopedis and Speight, 1978). The inclusion of polar oxygen functions by oxidation increase the polarity and hence, their ability to aggregate. Therefore, storage of the asphaltenes in an inert atmosphere (nitrogen) and in the dark is desirable.

Other solvents such as benzene (now a no-no because of environmental concerns) have been used and have a variety of polar solvents. Of the latter, tetrahydrofuran is often cited as a successful solvent. This seemed to arise because of the solubility of coal-derived asphaltenes in tetrahydrofuran. However, coal-derived asphaltenes are the products of thermal processes and bear no relationship to crude oil asphaltenes. Indeed, the use of tetrahydrofuran in the chromatographic separation processes for crude oil fractions is to be approached with caution because of the tendency of tetrahydrofuran to react with the adsorbent and form by-products. In a summary of the current context, crude oil asphaltenes (and the asphaltenes from the viscous feedstocks) often exhibit a somewhat lesser solubility in tetrahydrofuran than in toluene and in pyridine. Indeed, a similar effect has been noted in pyridine and in *o*-dichlorobenzene as well as in a variety of other (supposedly) good polar solvents. Whatever the solvent of choice, care must be taken to ensure complete solubility of the solute otherwise anomalous data will be produced that bear little relationship to accuracy and precision.

As a consequence of the consideration of these effects, the molecular weight of crude oil asphaltenes that is in excess of 2,500 Da is questionable and the true molecular weight is likely to be lower (Speight, 2014).

In the author's experience, pyridine and nitrobenzene have proven to be adequate solvents for molecular weight determination, subject to a thorough pre-measurement check showing the lack of any insolubility, and probably provide the best indicator of the aggregation number for asphaltenes.

However, as a final word to this section, caution is advised in the use of nitrobenzene. Raising the temperature when nitrobenzene is used as the solvent is not without some risk. In the present author's experience, having the probe fly past one's ear due to the nitrobenzene deciding to expand uncontrollably (explode) under the conditions of the experiment resulting in the probe being launched violently past the investigator's ear is not for the faint hearted! It can be quite a stimulant (not recommended) to the heart rate!

10.2.2 Freezing Point Depression

Many methods exist for determination of the molecular weight of crude oil fractions (Speight et al., 1985; Speight, 2014). Most of the early molecular weight determinations on hydrocarbon oils employed simple forms of the cryoscopic (freezing point depression) method (ASTM D2224, 2021).

The average molecular weight of crude oil hydrocarbon mixtures is used in characterization and correlation studies (Speight, 2014). Also, molecular weight is employed for heat exchange evaluations of distillation columns and for monitoring the progress of cracking processes.

This method using the concept of freezing point depression determines the average molecular weight of various fractions using simple equipment. The samples must be completely soluble and chemically inert in the solvent (benzene). Molecular weight values represent a number average molecular weight for the sample components.

The freezing points for benzene and for a benzene solution containing a known weight of sample are determined from their cooling curves. From the depression of the freezing point due to the sample, the average molecular weight is calculated by application of Raoult's law, namely, the vapor pressure lowering of a solvent is directly proportional to the concentration of the solute:

$$\left(p_{\mathrm{o}} - p\right)/p_{\mathrm{o}} = n_2/\left(n_1 + n_2\right)$$

In this equation, n_1 and n_2 are the number of moles of solvent and solute, respectively. Thus, if w_2 g of solute of molecular weight M_2 is dissolved in w_1 g of solvent of molecular weight M_1, Raoult's law gives:

$$\left(p_{\mathrm{o}} - p\right)/p_{\mathrm{o}} = \left(w_2/M_2\right)/\left(w_1 M_1 + w_2 M_2\right)$$

For a dilute solution, n_2 may be neglected in favor of n_1.

Thus, the calculation for molecular weight determination by cryoscopy is:

$$M = \left[W\left(K_{\mathrm{f}} - \Delta T\right)\right]/\left(\Delta T \cdot S\right)$$

where M is the number average molecular weight, W is the weight of sample in g, K_{f} is the cryoscopic constant for the solvent (K_{f} for benzene = 65.6), ΔT is the freezing point depression in °C (i.e., the freezing point of benzene minus the freezing point of the solution), and S is the moles of solvent. Many solvents have been tried, but with the best technique an accuracy of 1–2% is to be considered good.

10.2.3 Boiling Point Elevation

Boiling point elevation (ebullioscopic) methods are, in general use, more rapid and equally accurate but tend to fall short when applied to the higher molecular weight fractions of petroleum, heavy crude oil, extra heavy crude oil, and tar sand bitumen. Molecular weights in the low range (<500 Da) are readily determined by vapor-density methods.

For decades, and even now, the best established method for the determination of the molecular weight of oil fractions boiling above 250 °C (480 °F), the measurement of boiling point elevation (ebullioscopic method) was preferred (van Nes and van Westen, 1951).

For any one sample the boiling point elevation is determined at a series of concentrations of solute. Such determinations were carried out in practice by comparing the boiling point of pure solvent, measured in an ebulliometer, with the boiling points, determined in a second ebulliometer, of a series of solutions prepared by adding successive portions of the sample to the solvent. The first ebulliometer serves as the control experiment.

In order to determine the ebullioscopic constant *K*—which depends not only on the solvent used but also to a certain extent on the construction of the apparatus and the procedure followed—a series of determinations is carried out on a pure hydrocarbon of known molecular weight, preferably of the same order of magnitude as the molecular weight under investigation. The ebullioscopic constant calculated for each concentration is plotted on a graph, just as in the case of a molecular weight determination, and is extrapolated to zero concentration.

Before attempting to measure molecular weights, care should be taken to ensure that the sample does not contain water or volatile components (boiling below 200 °C, 392 °F) because this would invalidate the basic calculation.

These methods all give number average molecular weights, defined by:

$$\Sigma(n_i M_i)/\Sigma n_i = M_n$$

where M_n is the number average molecular weight and n_i is the number of molecules having a molecular weight M_i.

10.2.4 Size Exclusion Chromatography

There is a method for determining the molecular weight of crude oil, heavy crude oil, extra heavy crude oil, tar sand bitumen, and their constituent fraction using size exclusion (gel permeation) chromatography (ASTM D3593, 2021).

The principle of SEC (Chapter 8) is the exclusion of larger sample molecules from smaller pores in the packing. As a result, larger molecules cannot reside in the entire column volume but are restricted to smaller regions. In the extreme, the largest ones are restricted to the interstitial volume, that is, the space between particles, whereas the smallest ones can penetrate the entire open column volume, that is, the interstitial and all the pore volume. As a consequence, the large molecules elute first and the smallest ones last.

However, with the samples containing polar function (as do resin and asphaltene constituents), two complications can arise which are (1) solute–solute or solute–solvent interaction can result in the formation of larger particles and thus part, or all, of the sample elutes earlier than it should or (2) adsorption of the solute on the column packing can cause all or part of the solute to elute later than it should. Adsorption of the solute on the column packing has continued to have an adverse effect on the study of asphaltene molecular weights by this method and permanent damage to the packing has also been observed (Altgelt and Gouw, 1979). Polynuclear aromatic species have a tendency to adsorb strongly on various adsorbents but mixed solvents containing small amounts of a strongly polar component, such as pyridine, can help resolve much of the adsorption problem.

For the higher molecular weight crude oil fractions (and viscous feedstock fractions), such as asphaltenes, column packings of nominal pore sizes between 50–100 and 5,000–10,000 Å are used for complete resolution. The bimodal column combinations give good linear calibration curves and well-characterized polymer samples as well as narrow polymer fractions, for example, polystyrenes, are available for calibrating column sets a cross the entire molecular weight range.

However, these polymers are structurally different from crude oil constituents and cannot be directly used for accurate calibration. In fact, crude oil fractions, such as the resin fraction and the asphaltene fraction, differ from polymers in three ways: (1) the hydrocarbon skeletal structure, (2) the polar nature, and (3) the varying composition with molecular weight. Therefore, it is preferable to employ a set of narrow subfractions obtained by preparative SEC from the same or a similar crude oil fraction for calibration. Calibration curves obtained by polymers should at least be checked with such subfractions and adjusted as necessary.

However, there are disadvantages in the attempt to establish the correct calibration curve for high molecular weight crude oil fractions. The use of crude oil subfractions can be misleading because they are usually too broad; and in samples with broad distributions, the number average molecular weight may be quite different from the peak molecular weight. Thus, a good approximation is a calibration curve based on narrow polystyrene fractions or single compounds and multiplied by a factor of 1.5 across the entire molecular weight range.

Above all, it is important to remember that SEC separates by molar volume rather than by molecular weight. The method will, therefore, differentiate by structure in addition to molecular weight.

Errors may also arise from changes in the response factor with increasing elution volume. The reason is that crude oil components of different molecular weight also have a different composition. Ultraviolet detectors respond primarily to differences in aromaticity and the degree of aromatic ring condensation. Refractive index detectors are affected by color intensity, light scattering by very high molecular weight aggregates, and also by the change in composition. All must be given due consideration when data interpretation is undertaken.

In solutions of asphaltenes, the molecules are aggregated in clusters but the cluster may dissociate in very dilute solutions where the solvent have a degree of polarity. Thus, using the polar solvent, it may be possible to obtain true molecular solutions, as seems to be the case when nitrobenzene is used for molecular weight determination by VPO. But, depending on the nature of the asphaltene constituents and the solvent, the dissociation process may not be complete and assumptions of single molecular species may be in error. In addition, the adsorption of the species in the packing must not be underestimated.

In principle, SEC is a very powerful method for separating crude oil fractions by molecular weight. It is used frequently in crude oil analysis despite the components toward adsorption and aggregation, and other potential problems.

10.2.5 Mass Spectrometry

Non-fragmenting mass spectrometry is nearly ideal for crude oil distillates. It provides a wealth of detail by giving the hydrocarbon types, the formulae, and the concentration of practically all of the components in a fraction. From these data, the molecular weight distribution and the number average molecular weight can be obtained.

The main mass spectrometric (MS) methods for high molecular weight crude oil fractions are field ionization mass spectrometry (FIMS) and field desorption mass spectrometry (FDMS). These are the also called non-fragmenting methods because they produce the parent ion spectra that show only the non-fragmented molecules. However, whether or not this is true for asphaltene constituents is still under discussion.

The low volatility of the constituents of the resin fraction and the asphaltene fraction is a major drawback, although application of the technique to those species boiling up to the limit of short-path distillation temperatures, that is, to 705 °C (1,300 °F) might be cited in support of the technique but the estimated high boiling range of asphaltene constituents (Speight, 1994, 2014) may negate application of the technique to these more complex less volatile molecular species.

Crude oil fractions typically gave symmetrical mass spectrometric molar mass profile (molecular weight distribution) up to a limit of about 2,000 Da. However, the ionization sensitivity changes with molar mass and molecular structure. This molar mass discrimination can severely impair the application of mass spectrometry to broad crude oil fractions.

Thus, because of the compositional diversity of the fractions and the differences in ion sensitivity among alkanes, aromatics, and polar compounds, the actual distribution may be quite different from the mass profile. Therefore, symmetry is uncertain and a quantitative evaluation would be very questionable. Indeed, because of volatility constraints, quantitative evaluation of the constituents of an asphaltene fraction would be extremely questionable.

10.2.6 Nuclear Magnetic Resonance Spectroscopy

Nuclear magnetic resonance (NMR) spectroscopy has been suggested and used for measuring molecular weight without the limitations incurred by other methods (see above).

The approach is based on the determination of substituted and total aromatic carbon atoms per molecule. However, just like the nuclear magnetic resonance method for structure determination (Chapter 9), there are many assumptions because of the complex nature of asphaltenes samples and these assumptions are subject to inaccuracy and question. At best, one can hope for minimum and maximum values with a large spread between the minima and the maxima.

It is feared that because of the assumptions, the method becomes little more than optimistic mathematical manipulation.

10.3 Molecular Weights of Volatile Fractions

In the refining industry, volatility is a measure which describes how readily a substance will vaporize. At a given temperature and pressure, a substance with high volatility is more likely to exist as a vapor while a substance with low volatility is more likely to be a liquid or as a solid. Volatility can also describe the tendency of a vapor to condense into a liquid or solid; less volatile substances will more readily condense from a vapor than highly volatile substances. Differences in volatility can be observed by comparing how fast substances within a group evaporate (or sublimate in the case of solids) when exposed to the atmosphere. In general, solids are much less volatile than liquids, but there are some exceptions insofar as some solids will sublime (i.e., pass directly from solid to vapor without passing through the liquid phase).

In addition, once spilled into an ecosystem, a crude oil or a viscous feedstock can undergo a number of changes that could require knowledge of the molecular weights and other properties of the (biodegradation) products.

Thus, while the molecular weights of the volatile fractions are not the subject of this chapter, knowledge of the methods use can provide assistance in determining the molecular weight of the nonvolatile fractions (Sarowha, 2005) For a volatile liquid which is outside of the context of this chapter but worthy of mention for comparison, a common method to determine its molecular mass is to use the ideal gas law (i.e., $PV = nRT$) because the liquid (being volatile) can be converted to a gas and while the material is in the gas phase, the volume, pressure, and temperature can be measured.

If the substance in question is a volatile liquid, a common method to determine its molar mass is to use the ideal gas law, which is:

$$PV = nRT$$

If the liquid is volatile, it can easily be converted to a gas and while the substance is in the gas phase, the volume, pressure, and temperature can be measured.

Crude oil and the viscous feedstocks are not uniform materials insofar as the proportions of the various chemical constituents can vary not only with the location and age of the oil field but also with the depth of the individual well. However, on a molecular basis, crude oil is a complex mixture of hydrocarbons plus organic compounds of sulfur, oxygen, and nitrogen, as well as compounds containing metallic constituents, particularly vanadium nickel, iron, and copper. The hydrocarbon content may be as high as 97%, for example, in the lighter paraffinic crude oils, or as low as 50% or less as illustrated by the heavier asphaltic crude oils. Nevertheless, crude oils with as little as 50% hydrocarbon components are still assumed to retain most of the essential characteristics of the hydrocarbons.

Depending upon the sample (or crude oil product), the range of molecular weight can vary from approximately 100 to 500. Association effects are minimal and usually do not occur. Thus, the molecular weights for volatile fractions as determined are considered to be the true molecular weight of the fraction.

10.4 Molecular Weights of Nonvolatile Fractions

The nonvolatile fractions from crude oil or even the viscous feedstocks themselves can be particularly troublesome in terms of the determination of the molecular weight.

In the measurement of molecular weights, it is common to use the methods commonly known as GPC and VPO. However, it has to be anticipated that the molecular weight data obtained by various methods used for the fractions of the viscous feedstocks (such as the resins fractions and the asphaltene fraction) will show significant differences mainly attributed to (1) the complex nature of the constituents of these fractions, (2) the tendency of the constituents to associate in the nature of the organic solvent used for the determination, (3) temperature at the time of the determination, and (4) the concentration of the solute. In addition, the results may also depend on the selection of the standards and calibration standards that may affect the structural difference in respect to the fractions to be analyzed (Speight et al., 1985; Speight, 2014, 2015).

The GPC technique is often used in determining the distribution of molecular weights and employs a column that contains a filling of a microporous gel, which allows a gradual elution of species of greater to lesser molecular size in a mixture. The distribution of components in a mixture can be represented in various ways, among these forms is the average molecular weight in number and weight represented as M_n and M_w, and their ratio is identified with the polydispersity index ($D = M_n/M_w$). In definition, the M_n is the result of adding the molecular weight of each chemical species divided by the total number of species present. Therefore, the M_w is the average of molecular weights of the species present, based on their mass fraction.

10.4.1 Resins

The molecular weights of resin fractions are substantially lower than those of the associated asphaltene fractions (Speight, 2014) and do not usually vary, except for the limits of experimental error, with the nature of the solvent or the temperature of the determination. It may therefore be presumed that the molecular weights of resins, as determined by various methods, are in fact true molecular weights and that forces that result in intermolecular association contribute very little, if anything, to their magnitude.

10.4.2 Asphaltenes

The asphaltene constituent of a feedstock is primarily responsible for the high degree of variability in measured molecular weights. The polarity, aromaticity, and concentration of a given solvent and also the temperature of solution govern association to different degrees in different solvents. In addition, the nature or type of the asphaltene plays a role in determining the degree of association. Consequently, the confidence with which the molecular weight of asphaltenes of a given feedstock can be determined reflects the confidence in the molecular weight of the feedstock but it does not follow that the asphaltene constituents in their native state in the oil are aggregated to the same degree, if at all, when in their own maltenes.

Addition of extraneous solvents for measurement of molecular weights often initiates the aggregation process. Thus, any method of measurement utilizing a solvent media could potentially result in significant errors in the measured molecular weight of asphaltenes. This dilemma resulted in the application of considerable research effort to define the structure of whole feedstocks at a molecular level and to determine the conditions of measurement necessary to obtain reliable values of molecular weights of different samples.

In fact, determining the molecular weights of asphaltenes is a major issue because the asphaltene constituents have a low solubility in the liquids often used for determination. Also, adsorbed constituents of the resin fraction can lead to discrepancies in molecular weight determination and, in respect of elimination such an effect, it is recommended that precipitated asphaltenes should be reprecipitated several times before the determination. Thus, careful precipitation of the asphaltene constituents—for example, from toluene solution by the addition of the appropriate number of volumes of pentane (for pentane asphaltenes) or heptane (for heptane asphaltenes) and careful choice of the determination method are both very important for obtaining meaningful results (Speight et al., 1985).

A fairly comprehensive review followed by a study of asphaltene molecular weights by VPO has been reported and shows that the molecular weights of various asphaltenes are dependent not only on the nature of the solvent but also on the solution temperature at which the determinations were performed (Moschopedis and Speight, 1976a). Data from later work involving molecular weight determinations by the cryoscopic method (Speight an Moschopedis, 1977), however, indicate that the molecular nature of asphaltenes is not conducive to the determination of absolute molecular weights by any one method (Speight et al., 1985; Speight, 1994, 2014, 2015).

Indeed, for any one particular method, the observed molecular weights suggest that asphaltenes form molecular aggregates, even in dilute solution and this association is influenced by solvent polarity, asphaltene concentration, and the temperature of the determination (Speight et al., 1985). In fact, it is strongly recommended that, to negate the concentration effects and the temperature effects, the molecular weight determinations should be carried out at each of three different concentrations at each of three different temperatures. The data for each temperature are then extrapolated to room temperature.

Thus, the molecular weights of the asphaltene fraction span a wide range from a few hundred to several million leading to speculation about self-association (Koots and Speight, 1975; Speight et al., 1985).

The tendency of the constituents of the asphaltene fraction to form aggregates in hydrocarbon solution is one of the most characteristic features of the constituents and complicates the determination of asphaltene molecular weight (Speight and Moschopedis, 1977; Speight, 1994, 2014, 2015). The average molecular weights measured by means of VPO or SEC are significantly influenced by the conditions of the analysis (temperature, asphaltene concentration, and solvent polarity) (Altgelt, 1968; Speight and Moschopedis, 1980; Speight et al., 1985). For this reason, molecular weights are reported in the literature as relative values only and these values may be quite different from the molecular weights of unassociated molecules.

The influence of the asphaltene concentration on the values of molecular weight measured is significant and it is important to use as high a dilution as possible in order to measure a molecular weight that at least approaches that of the unassociated asphaltene molecules, remembering that the asphaltene fraction is a complex mixture of different species and that the molecular weight of any unassociated constituents will be an effect in the determination of the average molecular weight.

The existence of asphaltene aggregates in hydrocarbon solvents has been demonstrated by means of small-angle neutron scattering studies. The physical dimensions and shape of the aggregates are functions of the solvent used and the temperature of the investigation. In addition, surface tension measurements have been used to study the self-association of asphaltenes in pyridine and nitrobenzene. A discontinuous transition in the surface tension as a function of asphaltene concentration was interpreted as the critical asphaltene concentration above which self-association occurs.

The study of asphaltene molecular weights by VPO shows that the molecular weights of various asphaltenes are dependent not only on the nature of the solvent but also on the solute concentration and the solution temperature at which the determinations were performed (Speight, 1994, 2014, 2015). However, data from later work involving molecular weight determinations by the cryoscopic method (Speight and Moschopedis, 1977) indicate that the molecular nature of asphaltenes is not conducive to the determination of absolute molecular weights by any one method.

For any one method, the observed molecular weights suggest that asphaltene constituents form molecular aggregates, even in dilute solution, and this association is influenced by (1) the solvent polarity, (2) the constituent concentration, and (3) the temperature at which the determination is made. The precise mechanism of the association has not been conclusively established, but hydrogen bonding and the formation of charge-transfer complexes have been cited as responsible for intermolecular association. In fact, intermolecular hydrogen bonding could be involved in asphaltene association and may have a significant effect on observed molecular weights (Moschopedis and Speight, 1976b; Speight, 1994, 2014). Indeed, such bonds are known to exist in natural product systems and confer stability on various conformational structures.

It is also interesting to note that use of a solvent of low dielectric constant (benzene) does not cause any variation in the molecular weight when the concentration of the asphaltenes is varied over the range 2–7% w/w. However, use of a solvent of higher dielectric constant (pyridine) caused significant variation in the observed molecular weights over this particular range. Molecular weight determinations at lower (<2%) and higher (>8%) concentrations are subject to instrument sensitivity and solubility interference, respectively. Extrapolation of the pyridine data to infinite dilution, when asphaltene–asphaltene interaction may be assumed negligible, affords molecular weights of the same order (ca. 1,800) as those recorded using nitrobenzene as the solvent. On the other hand, extrapolation of the pyridine data to higher concentrations suggests that molecular weights of the order of those recorded in a solvent of lower dielectric constant (e.g., benzene) may be obtained.

The higher molecular weights recorded when solvents of low polarity are employed are undoubtedly the result of intermolecular association between the asphaltene nuclei, but solvents of high dielectric constant are able to bring about dissociation of these asphaltene agglomerations to what, in fact, appear to be single asphaltene particles. Furthermore, this observation precludes asphaltene structures that invoke the concept of a polymer molecule to account for the high molecular weights observed in nonpolar solvents.

Data from a series of asphaltene molecular weight determinations present strong evidence to support the contention that resin constituents (i.e., precipitant-soluble material) is occluded within the asphaltene matrix during the precipitation procedure. In each case, removal of the resin constituents gives rise to higher observed molecular weights of the purified asphaltenes. In addition, and as noted previously for the whole asphaltenes, the molecular weights of the purified asphaltenes also varied with the solvent used for the determination; that is, solvents of high dielectric constant decrease the observed molecular weights. Furthermore, when the freshly precipitated asphaltene fraction was extracted continuously (using a Soxhlet apparatus) with different solvents and molecular weight determinations carried out on the insoluble material, a decrease in the asphaltene molecular weight with the dielectric constant of the solvent was also noted.

It is of interest to speculate at this point about the means by which the resin material is retained by the asphaltenes. For example, asphaltenes and resins have been suggested to have points of structural similarity (Koots and Speight, 1975; Moschopedis and Speight, 1976a; Speight, 2014) that is responsible for their affinity to each other in solution and in crude oil. As noted above, in the case of amphoteric species, there is potential for strong bonding to occur (Speight, 1994, 2014), thereby increasing the intensity by which the resins are held to the asphaltenes and, also, by which the asphaltene constituents associate with each other thereby requiring a stronger dispersing solvent in order to release the resin constituents.

On the other hand, asphaltenes may be analogous to coal insofar as it appears that asphaltenes have adsorption characteristics and may even exhibit a distinct physical structure or participate in clathrate-type systems. If this is the case, swelling by a solvent such as pyridine would undoubtedly free resin material from the asphaltenes. There is the additional possibility that, if the resin constituents are retained purely by a surface adsorption phenomenon, these constituents may be expected to be removed from the asphaltenes by the repetitive precipitation. All three effects could play a part in the retention of resin (i.e., precipitant-soluble) material by the asphaltenes.

Thus, the amount of resin may have some influence on the asphaltene yield (and, therefore, on an estimation of crude oil composition), the resin occluded within (or adsorbed on to) the asphaltene during the separation procedure have a considerable effect on the observed molecular weight and, presumably, on the degree of association of the asphaltenes. Indeed, the speculative concept that asphaltenes release the final vestiges of this resin only upon swelling by a solvent such as pyridine is also worthy of consideration.

Thus, the tendency of asphaltenes to undergo association and/or dissociation depending upon the nature of the solvent also appears true for the series of higher molecular weight fractions. However, it should be noted here that although the results with asphaltenes available from several crude oils suggest that molecular weight varies with the dielectric constant of the solvent, there may be other factors which may in part also be responsible for this phenomenon. The final phenomenon that influences the molecular weight of the asphaltene is the relative polarity of the solvent used in the precipitation technique.

A particularly appropriate method involves the use of different solvents (at least two), and the data are then extrapolated to infinite dilution. There has also been the use of different temperatures for a particular solvent after which the data are extrapolated to room temperature (Speight et al., 1985; Speight, 1994). In this manner, different solvents are employed and the molecular weight of a crude oil fraction (particularly the asphaltenes) can be determined for which it can be assumed that there is little or no influence from any intermolecular forces. In summary, the molecular weight may be as close to the real value as possible.

In fact, it is strongly recommended that to negate concentration effects and temperature effects, the molecular weight determination be carried out at three different concentrations at three different temperatures. The data for each temperature are then extrapolated to zero concentration and the zero concentration data at each temperature are then extrapolated to room temperature (Speight, 2014, 2015).

It is of interest to speculate at this point about the means by which the resin material is retained by the asphaltenes. For example, asphaltenes may be analogous to coal insofar as it appears that asphaltenes have adsorption characteristics and may even exhibit a distinct physical structure or may even participate in clathrate systems. If this is the case, swelling by a solvent such as pyridine would undoubtedly free resin material from the asphaltenes. On the other hand, if the resin material is retained purely by a surface adsorption phenomenon, it may be expected to be removed from the asphaltenes by the repetitive precipitation. Both effects could play a part in the retention of resin (i.e., precipitant-soluble) material by the asphaltenes.

Obviously, many other facets of asphaltene precipitation need to be investigated, but it is obvious from these data that it is extremely difficult to obtain clean (resin-free) asphaltenes without a multiple precipitation technique. The amount of resin may have some influence on the asphaltene yield (and, therefore, on an estimation of crude oil composition). However, the resin occluded within (or adsorbed onto) the asphaltene during the separation procedure has a considerable effect on the observed molecular weight and, presumably, on the degree of association of the asphaltenes. Indeed, the speculative concept that asphaltenes release the final vestiges of this resin only upon swelling by a solvent such as pyridine is also worthy of consideration.

Attention should be focused upon the delineation of various structural types in the asphaltenes. Since crude oil is actually a continuum of structural types, there is little to be inferred from this study other than the influence of degree of association or on the degree of polymerization (for those concepts based on the concept of a polymer structure).

If, however, the structural types vary from the resins to the asphaltenes, it may be necessary to reassess the concepts of molecular types that have been proposed heretofore. In addition, the concept that the resins may be entrapped within the asphaltene matrix (and may even be analogous to a clathrate compound) warrants further investigation.

It is worthy of mention at this point, since the theory of the physical structure of crude oil is based upon the results of the method, that the specific viscosity of asphaltene solutions has also been employed to determine asphaltene molecular weights (ASTM D2502, 2021).

The method essentially requires a determination of the relative viscosity, n_r, of known solute–solvent systems, when the constant k is derived by the relationship

$$\log n_r = ckM$$

In this equation, c is the concentration (g per 100 ml) and M is the molecular weight. Thus, subsequent application of the relationship to solute–solvent systems involving asphaltenes and maltenes allowed Mack, in 1932, to derive a molecular weight of the order of 5,000 at 0 °C (32 °F) to 1,800 at 120 °C (250 °F). Thus, it is not surprising that the postulate of the existence of asphaltene clusters within the maltenes evolved and has remained to this time.

Application of this procedure to systems involving Athabasca asphaltenes allows derivation of values for the constant k, but transposition of k from one solute–solvent system to another gives inconsistent results. For example, the constants derived from naphthalene-benzene or biphenyl-benzene systems give unexpectedly low molecular weights when applied to asphaltene-benzene systems. Indeed, application of these constants to petrolene-solute systems gives molecular weights of naphthalene and biphenyl in error by some several hundred percent.

Application of the constant k for the biphenyl-benzene system to the asphaltene-petrolene system gives molecule weights within an acceptable range. However, the validity is suspect because of the unsatisfactory results obtained when the constants from one solute–solvent system are transposed to a different solute–solvent system in which the molecular weight of the solute is known. This is emphasized to an even greater extent by application of the constant derived from the asphaltene-benzene system to the asphaltene-petrolene system, which yields molecular weights of the order of 17,000–60,000.

In view of these inconsistencies, the viscosity method of determining asphaltene molecular weights must be regarded as suspect when the constants *k*, as derived for one solute–solvent system, are applied to a completely different solute–solvent system. Indeed, Mack's original work involved transposition of data from earlier work by Kendall and Monroe in 1917 and must be regarded as questionable, since these authors consider their own results to be indecisive because data from different solvent systems containing the same solute are in conflict. Even though preliminary application of some of the viscosity data yields asphaltene molecular weights within an acceptable range, the remaining data indicate several inconsistencies in the method, and recognition of this causes some revision of previous theories of the physical structure of crude oil.

In summary, asphaltene molecular weights are variable (Speight et al., 1985), there being a tendency to associate even in dilute solution in nonpolar solvents. However, data obtained using highly polar solvents indicate that the molecular weights, in solvents that prevent association, usually fall in the range 2,000 ± 500. There is some support for this conclusion from mass spectrometry studies but caution is advised when using mass spectrometry especially where the volatility of the sample is critical to the conclusions. Failure to recover all of the material and the formation of a thermal residue in the system (Speight and Pancirov, 1984) can remove not only the more polar species but also the higher-molecular-weight species.

In view of the potential changes that the asphaltenes of a given feedstock can undergo when diluted with solvents, it is evident that the method used to measure the molecular weight can have a strong bearing on the value obtained. Molecular weight determinations by VPO using benzene or toluene as a solvent, are likely to be high principally because of asphaltene associations.

Also, the values obtained using the cryoscopic method with benzene would yield high values of molecular weight for similar reasons. Due to the fact that the solubility of asphaltenes in a given solvent increases with an increase in temperature, and indeed asphaltene associations are reduced and it is to be expected that methods utilizing higher temperatures will yield molecular weights that are more representative. Thus, ebullioscopic methods with benzene or cryoscopic methods with camphor or naphthalene as solvents might be the preferred method of molecular weight determination for individual fractions and for whole feedstocks.

10.4.3 Carbenes and Carboids

By definition, carbenes and carboids are the components of crude oil that are less soluble than the asphaltene constituents (Chapter 4) (Speight, 2014). In fact, carbenes and carboids are regarded as the products of thermal treatment of crude oil and the viscous feedstocks. In this sense, molecular weight data have proved helpful in determining the course of the thermal reaction.

For many decades, carbenes and carboids were regarded as the product formed by polymerization and condensation of asphaltene moieties and were in fact the precursors to coke. It is the initial fragmentation of the asphaltenes that produces the carbenes and carboids (Speight, 1994). Indeed, where solubility in pyridine is possible (carbenes) molecular weight data show that the carbene fraction has a lower molecular weight than the respective parent asphaltene fraction. It is the relative insolubility of the carbenes and the carboids in the changing oil medium that causes separation as a solid phase and that is the true onset of coke formation.

References

Ali, L.H. 1971. A Method for the Calculation of Molecular Weights of Aromatic Compounds, and Its Application to Petroleum Fractions. *Fuel*, 50: 298–307.

Ali, L.H., Al-Ghannan, K.A., and Speight, J.G. 1985. *Iraqi Journal of Science*, 26: 41.

Altgelt, K.H. 1968. *American Chemical Society, Division of Petroleum Chemistry, Preprints*, 13(3): 37.

Altgelt, K.H., and Gouw, T.H. 1979. *Chromatography in Petroleum Analysis*. Marcel Dekker Inc., New York.

ASTM D128. 2021. Standard Test Methods for Analysis of Lubricating Grease. Annual Book of Standards. ASTM International, West Conshohocken, Pennsylvania.

ASTM D2224. 2021. Test Method for Mean Molecular Weight of Mineral Insulating Oils by the Cryoscopic Method. Annual Book of Standards. ASTM International, West Conshohocken, Pennsylvania. Note: this test method was discontinued in 1989 but still used by some laboratories for determining the molecular weight of crude oil fractions up to an including gas oil.

ASTM D2502. 2021. Test Method for Estimation of Molecular Weight (Relative Molecular Mass) of Petroleum Oils from Viscosity Measurements. Annual Book of Standards. ASTM International, West Conshohocken, Pennsylvania.

ASTM D2503. 2021. Test Method for Estimation of Molecular Weight (Relative Molecular Mass) of Hydrocarbons by Thermoelectric Measurement of Vapor Pressure. Annual Book of Standards. ASTM International, West Conshohocken, Pennsylvania.

ASTM D2878. 2021. Method for Estimating Apparent Vapor Pressures and Molecular Weights of Lubricating Oils. Annual Book of Standards. ASTM International, West Conshohocken, Pennsylvania.

ASTM D3593. 2021. Test Method for Molecular Weight Averages/Distribution of Certain Polymers by Liquid Size Exclusion (Gel Permeation Chromatography—GPC) Using Universal Calibration. Annual Book of Standards. ASTM International, West Conshohocken, Pennsylvania.

ASTM D3712. 2021. Standard Test Method of Analysis of Oil-Soluble Sulfonates by Liquid Chromatography. Annual Book of Standards. ASTM International, West Conshohocken, Pennsylvania.

Gary, J.G., Handwerk, G.E., and Kaiser, M.J. 2007. *Petroleum Refining: Technology and Economics*. 5th Edition. CRC Press, Taylor & Francis Group, Boca Raton, Florida.

Goossens, A.G. 1996. Prediction of Molecular Weight of Petroleum Fractions. *Industrial and Engineering Chemistry Research*, 35(3): 985–988.

Hsu, C.S., and Robinson, P.R. (Editors). 2017. *Handbook of Petroleum Technology*. Springer International Publishing AG, Cham, Switzerland.

Koots, J.A., and Speight, J.G. 1975. The Relation of Petroleum Resins to Asphaltenes. *Fuel*, 54: 179.

Kurganov, V.M., Starodubskaya, G.Y., and Novikova, Z.U. 1973. Calculation of Molecular Weight of Petroleum Products. *Chemistry and Technology of Fuels and Oils*, 9: 319–320. Translated from Khimiya i Tekhnologiya Topliv i Masel, No. 4, pp. 59–60, April, 1973. https://link.springer.com/article/10.1007/BF00730427

Littlejohn, D., and Lucas, D. 1999. Vapor Pressure Measurement System for Heavy Crude Oils. *Journal of the Air & Waste Management Association*, 49(9): 1103–1109. https://www.tandfonline.com/doi/abs/10.1080/10473289.1999.10463876?src=recsys

Moschopedis, S.E., and Speight, J.G. 1976a. An Investigation of Hydrogen Bonding by Oxygen Functions in Athabasca Bitumen. *Fuel*, 55: 187.

Moschopedis, S.E., and Speight, J.G. 1976b. An Investigation of Asphalt Preparation From Athabasca Bitumen. *Proceedings of Association of Asphalt Paving Technologists*, 45: 78.

Moschopedis, S.E., and Speight, J.G. 1978. The Influence of Metal Salts on Bitumen Oxidation. *Fuel*, 57: 235.

Sarowha, S.L.S. 2005. Determination of Molecular Weights of Petroleum Products Using Gel Permeation Chromatography. *Petroleum Science and Technology*, 23(5, 6): 573–578. https://www.tandfonline.com/doi/abs/10.1081/LFT-200032846?journalCode=lpet20

Speight, J.G. 1994. Chemical and Physical Studies of Petroleum Asphaltenes. In: *Asphaltenes and Asphalts. I. Developments in Petroleum Science, 40*. T.F. Yen and G.V. Chilingarian (Editors). Elsevier, Amsterdam, The Netherlands. Chapter 2

Speight, J.G. 2014. *The Chemistry and Technology of Petroleum*. 5th Edition. CRC Press, Taylor & Francis Group, Boca Raton, Florida.

Speight, J.G. 2015. *Handbook of Petroleum Product Analysis*. 2nd Edition. John Wiley & Sons Inc., Hoboken, New Jersey.

Speight, J.G. 2017. *Handbook of Petroleum Refining*. CRC Press, Taylor & Francis Group, Boca Raton, Florida.

Speight, J.G., and Moschopedis, S.E. 1977. Asphaltene Molecular Weights by a Cryoscopic Method. *Fuel*, 56: 344.

Speight, J.G., and Moschopedis, S.E. 1980. On the 'Polymeric Nature' of Petroleum Asphaltenes. *Fuel*, 59: 440.

Speight, J.G., and Pancirov, R.J. 1984. Liquid Fuels Technology. 2: 287. Structural Types in Asphaltenes as Deduced from Pyrolysis-Gas Chromatography-Mass Spectrometry. *Liquid Fuels Technology*, 2: 287.

Speight, J.G., Long, R.B., and Trowbridge, T.D. 1984. Factors Influencing the Separation of Asphaltenes from Heavy Petroleum Feedstocks. *Fuel*, 63: 616.

Speight, J.G., Wernick, D.L., Gould, K.A., Overfield, R.E., Rao, B.M.L., and Savage, D.W. 1985. Molecular Weights and Association of Asphaltenes: A Critical Review. *Revue de l Institut Français du Pétrole*, 40: 27.

Van Nes, K., and van Westen, H.A. 1951. *Aspects of the Constitution of Mineral Oils.* Elsevier Publishing Co., Amsterdam, The Netherlands.

11

Instability and Incompatibility

11.1 Introduction

The study of the properties of refinery feedstocks would not be complete without some mention of the phenomena of instability and incompatibility—sometimes referred to as "fouling"—which can result in the formation of a separate phase that may be in the form of a sludge or a solid formed from feedstocks and/or process products.

By way of definition, fouling is the accumulation of unwanted material (formed as a result of instability and/or incompatibility) on solid surfaces, such as heat exchangers (Speight, 2015b). The fouling materials can consist of either (1) inorganic or organic materials, which is within the context of this chapter or (2) living organisms, usually referred to as biofouling or a nonliving substance but is not a subject of this chapter.

Typically, fouling occurs on the surface of refinery equipment within the refinery system that fills a useful function and, as a result, the fouling process impedes or interferes with this function of the equipment. Other terms that are used to describe fouling include (1) deposit formation, (2) encrustation, (3) deposition, (4) scaling, (5) scale formation, (6) slagging, and (7) sludge formation—all of which are often referred to collectively as "crudding."

Instability and/or incompatibility occurs when a product is formed in the formation or well pipe or in a reactor, and the product is incompatible with the original crude oil or its products. Feedstock complexity (especially in the case of the viscous feedstocks) has also brought to the fore issues such as instability and incompatibility. The chemistry and physics of incompatibility can, to some extent, be elucidated but many unknowns remain (Speight, 2014, 2017, 2023). Instability/incompatibility in refineries affects the operation of crude oil pre-distillation heat exchanger trains. The methods presented are the most common methods used for

Handbook of Heavy Oil Properties and Analysis, First Edition. James G. Speight.

the analysis of feedstocks as a means of determining the potential for instability/incompatibility of a feedstock blend.

As an additional note, instability is sometimes differentiated by the timing of the event. For example, incompatibility is often referred to as an immediate event upon mixing two feedstocks while instability is not obvious immediately upon mixing but occurs after (what might be called) an "induction period." In either case, the result is the formation of a deposit that is deleterious to the efficient operation of the refinery equipment.

More pertinent to the present context is that both instability and incompatibility phenomena can occur (leading to fouling of refinery equipment) when two viscous feedstocks are blended prior to introduction into the refinery processes or a viscous feedstock is blended with a nonviscous feedstock prior to introduction into the refinery processes. It is the analytical methods that provide the data that point to the reason for problems in the refinery or for the failure of products to meet specifications and to perform as desired.

The most obvious example of incompatibility (nonmiscible) is the inability of a hydrocarbon-based feedstock and water to mix. Chemically, and in the present context, incompatibility is typically caused by the presence of various polar functions (i.e., heteroatom function groups containing nitrogen, or oxygen, or sulfur, and even various combinations of the heteroatoms) in the feedstock.

However, to alleviate any potential confusion, it is appropriate here to define some of the terms that are used in the relation to instability and incompatibility so that their use later in the chapter will be more apparent and will also alleviate some potential for misunderstanding. The general scientific areas of instability and incompatibility are complex and have been considered to be nothing better than a black art because not all of the reactions that contribute to instability and incompatibility have been defined (Wallace, 1964).

Briefly, the term incompatibility refers to the formation of a precipitate (or sediment) or separate phase when two liquids are mixed. The term instability is often used in reference to the formation of color, sediment, or gum in the liquid over a period of time and is usually due to chemical reactions, such as oxidation, and is chemical rather than physical. This term may be used to contrast the formation of a precipitate in the near term (almost immediately). The terms "instability" and "incompatibility" are frequently used interchangeably and are more commonly known as sludge formation, and sediment formation, or deposit formation. In crude oil and the crude oil products, instability often manifests itself in various ways (Table 11.1) (Stavinoha and Henry, 1981; Hardy and Wechter, 1990; Ruzicka and Nordenson, 1990; Power and Mathys, 1992). Hence, there are different ways of defining each of these terms but the terms are often used interchangeably.

Gum formation (ASTM D525, 2021) alludes to the formation of soluble organic material whereas sediment is the insoluble organic material. Storage stability

Table 11.1 Examples of Properties Related to Instability in Feedstocks

Property	Comments
Asphaltene constituents	Influence oil–rock interactions
	Separates from oil when gases are dissolved
	Thermal alteration can cause formation of polar products
	Phase separation
Heteroatom constituents	Provide polarity to oil
	Preferential reaction with oxygen
	Formation of high polarity products
	Preferential thermal alteration
Aromatic constituents	Formation of paraffin derivatives during cracking
	May be incompatible with paraffinic medium
	Phase separation of paraffinic constituents
Non-asphaltene constituents	Thermal alteration causes changes in polarity
	Formation of lower molecular weight polar products
	Phase separation of polar species

(or storage instability) (ASTM D381, 2021; ASTM D4625, 2021) is a term used to describe the ability of the liquid to remain in storage over extended periods of time without appreciable deterioration as measured by gum formation and/or the formation sediment. Thermal stability is also defined as the ability of the liquid to withstand relatively high temperatures for short periods of time without the formation of sediment (i.e., carbonaceous deposits and/or coke). Thermal oxidative stability is the ability of the liquid to withstand relatively high temperatures for short periods of time in the presence of oxidation and without the formation of sediment or deterioration of properties (ASTM D3241, 2021) and there is standard equipment for various oxidation tests (ASTM D4871, 2021). Stability is also as the ability of the liquid to withstand long periods at temperatures up to 100 °C (212 °F) without degradation. Determination of the reaction threshold temperature for various liquid and solid materials might be a beneficial analysis.

Existent-gum is the name given to the nonvolatile residue present in the feedstock as received for the test method (ASTM D381, 2021). In this test, the sample is evaporated from a beaker maintained at a temperature of 160–166 °C (320–331 °F) with the aid of a similarly heated jet of air. This material is distinguished from the potential gum that is obtained by aging the sample at an elevated temperature.

Thus, potential gum is determined by the accelerated gum test (ASTM D873, 2021) that is used as a safeguard of storage stability and can be used to predict the potential for gum formation during prolonged storage. In this test, the feedstock is heated for 16 hr with oxygen under pressure in a bomb at 100 °C (212 °F). After this time, both the gum content and the solid precipitates are measured. A similar test, using an accelerated oxidation procedure, is also in use for determining the oxidative stability of diesel fuel (ASTM D2274, 2021), steam turbine oil (ASTM D2272, 2021), distillate fuel oil (ASTM D2274, 2021), and lubricating grease (ASTM D942, 2021).

Dry sludge is the material separated from crude oil and crude oil products by filtration and which is insoluble in heptane. Existent dry sludge is the dry sludge in the original sample as received and is distinguished from the accelerated dry sludge obtained after aging the sample by chemical addition or heat. The existent dry sludge is distinguished from the potential dry sludge that is obtained by aging the sample at an elevated temperature.

The existent dry sludge is the material separated from the bulk of a crude oil or crude oil product by filtration and which is insoluble in heptane. The test is used as an indicator of process operability and as a measure of potential downstream fouling.

An analogous test, the thin film oven test (TFOT) (ASTM D1754, 2021), and an aging test (IP 390) are used to indicate the rate of change of various physical properties such as penetration (ASTM D5, 2021), viscosity (ASTM D2170, 2021), and ductility (ASTM D113, 2021) after a film of asphalt or bitumen has been heated in an oven for 5 hr at 163 °C (325 °F) on a rotating plate. A similar test is available for the stability of engine oil by thin film oxygen uptake (TFOUT) (ASTM D4742, 2021). This test establishes the effects of heat and air based on changes incurred in the above physical properties measured before and after the oven test. The allowed rate of changes in the relevant bitumen properties after the exposure of the tested sample to the oven test are specified in the relevant specifications (ASTM D3381, 2021).

However, test methods that involve accelerated oxidation of the sample must be used with caution and consideration of the chemistry. Depending on the composition of the sample, it is quite possible that the higher temperature and extreme conditions (oxygen under pressure) may not be truly representative of the deterioration of the sample under storage conditions. The higher temperature and the oxygen under pressure might change the chemistry of the system and products that would not be produced under ambient storage conditions. An assessment of the composition of the feedstock prior to storage and application of the test will assist in this determination.

In general, instability and incompatibility can be related to the heteroatom-containing compounds (i.e., nitrogen-, oxygen-, and sulfur-containing compounds) that are present. The degree of unsaturation of the feedstock (i.e., the level of olefinic species and more for thermal products such as cracked residua rather than natural feedstocks) also plays a role in determining instability/incompatibility.

Instability reactions are usually defined in terms of the formation of filterable and non-filterable sludge (sediments, deposits, and gums), an increased peroxide level, and the formation of color bodies. Color bodies in and of themselves do not predict instability. However, the reactions that initiate color body formation can be closely linked to heteroatom-containing (i.e., nitrogen-, oxygen-, and sulfur-containing) functional group chemistry.

Thus, feedstock constituents are incompatible when sludge, semisolid, or solid particles are formed during and after blending as well as during and after the application of thermal processes. This phenomenon usually occurs prior to use. If the secondary products are marginally soluble in the blended crude oil product, use might detract from solubility of the secondary products and they will appear as sludge or sediment that can be separated by filtration or by extraction (ASTM D4310, 2021). When the secondary products are truly insoluble, they separate and settle out as a semisolid or solid phase floating in the feedstock or are deposited on the walls and floors of containers. In addition, secondary products usually increase the viscosity of the crude oil product.

In the current context, the meaning of the term incompatibility is found when it is applied to crude oil recovery and crude oil refining. The application occurs when a product is formed in the formation or well pipe (recovery) or in a reactor (refining) and the product is incompatible with (immiscible with or insoluble in) the original crude oil or its products. Such an example is the formation and deposition of waste and other solids during recovery or the formation of coke precursors and even of coke during many thermal and catalytic operations. Coke formation is considered to be an initial phase separation of an insoluble, solid, coke precursor prior to coke formation proper. In the case of crude oils, sediments and deposits are closely related to sludge, at least as far as compositions are concerned. The major difference appears to be in the character of the material.

There is also the suggestion (often, but not always, real) in that the sediments and deposits originate from the inorganic constituents of the feedstock and may be formed from the inherent components of the feedstock (i.e., the metalloporphyrin constituents) or from the ingestion of contaminants by the feedstock during the initial processing operations. For example, crude oil is known to pick up iron and other metal contaminants from contact with pipelines and pumps.

Sediments can also be formed from organic materials but the usual inference is that these materials are usually formed from inorganic materials. The inorganic materials can be salt, sand, rust, and other contaminants that are insoluble in the crude oil and which settle to the bottom of the storage vessel. For example, gum typically forms by way of a hydroperoxide intermediate that induces polymerization of olefins. The intermediates are usually soluble in the liquid medium. However, gum that has undergone extensive oxidation reactions tends to be higher in molecular weight and much less soluble. In fact, the sediments that form in feedstocks are usually the direct result of autoxidation reactions. Active oxygen species involved include both molecular oxygen and hydroperoxides. These reaction proceed by a free radical mechanism and the solids produced tend to have increased incorporation of heteroatom and are thus also more polar so increasingly less soluble in (or less miscible with) the other constituents of the feedstock.

The most significant and undesirable instability change in feedstocks is the formation of solids (often referred to as filterable sediment) which can plug nozzles, filters, coke heat exchanger surfaces, and otherwise degrade engine performance. These solids are the result of free radical autoxidation reactions. Although slight thermal degradation occurs in nonoxidizing atmospheres, the presence of oxygen or active oxygen species, such as hydroperoxides, will greatly accelerate oxidative degradation as well as significantly lower the temperature at which undesirable products are formed. Solid deposits that form as the result of short-term high-temperature reactions share many similar chemical characteristics with filterable sediment that form in storage.

The soluble sludge/sediment precursors that form during processing or use may have a molecular mass in the several hundred range. For this soluble precursor to reach a molecular weight sufficient to precipitate (or to phase-separate), one of two additional reactions must occur. The molecular weight must increase drastically as a result of condensation reactions leading to the higher molecular weight species. Or the polarity of the precursor must increase (without necessarily increasing the molecular weight) by incorporation of additional oxygen, sulfur, or nitrogen functional groups. Additionally, the polarity may increase because of the removal of nonpolar hydrocarbon moieties from the polar core, as occurs during cracking reactions. In all three cases, insoluble material will form and separate from the liquid medium.

Additives are chemical compounds intended to improve some specific properties of feedstocks or crude oil-related products. Different additives, even when added for identical purposes, may be incompatible with each other, for example, react and form new compounds. Consequently, a blend of two or more feedstocks containing different additives may form a system in which the additives react with each other and so deprive the blend of their beneficial effect.

The chemistry and physics of incompatibility can, to some extent, be elucidated (Wallace, 1969; Por, 1992; Power and Mathys, 1992; Mushrush and Speight, 1995) but many unknowns remain. In addition to the chemical aspects, there are also aspects such as the attractive force differences, such as (1) specific interactions between like/unlike molecules as may occur in the form of hydrogen bonding and electron donor–acceptor phenomena, (2) field interactions such as dispersion forces and dipole–dipole interactions, and (3) any effects imposed on the system by the size and shape of the interacting molecular species.

Such interactions are not always easy to define and, thus, the measurement of instability and incompatibility has involved visual observations, solubility tests, hot filtration sediment (HFS), and gum formation. However, such methods are often considered to be after-the-fact methods insofar as they did not offer much in the way of predictability. In refinery processes (Parkash, 2003; Gary et al., 2007; Hsu and Robinson, 2017; Speight, 2017), predictability is not just a luxury, it is a necessity. The same principle must be applied to the measurement of instability and incompatibility. Therefore, methods are continually being sought to aid in achieving this goal.

In addition to the gravimetric methods, there have also been many attempts to use crude oil and/or product characteristics and their relation to the sludge and deposit formation tendencies. In some cases, a modicum of predictability is the outcome but, in many cases, the data appear as preferred ranges and are subject to personal interpretation. Therefore, caution is advised.

It is the purpose of this chapter to describe some of the more likely properties that could result in instability and incompatibility as well as presenting document some of the more prominent methods used for determining instability and incompatibility. No preference will be shown, and none will be given to any individual methods. It is the choice of the individual experimentalist to choose the method on the basis of the type of feedstock, the immediate needs, and the projected utilization of the data. As elsewhere, it the use of the data that often detracts from an otherwise sound method.

11.2 Occurrence of Instability and Incompatibility

The focus of incompatibility studies has usually been on the whole crude oil and specifically on the asphaltene fraction. As already stated (Speight, 1994, 2014, 2015a), the asphaltene constituents are the highest molecular weight and most polar fractions found in crude oil. The characteristics of the asphaltene constituents and the amount in crude oil depend to a greater or lesser extent on the crude oil source. During refining of crude oil, the asphaltene constituents will end up in a high percentage in the residua as the low boiling constituents (such as naphtha, kerosene, and light gas oil) are removed from the oil through cracking and visbreaking.

The structure of crude oil and the viscous feedstocks dictates that the rate of fouling is a complex function of oil molecular composition, temperature, and particulate content either in refinery reactors or in heat exchangers (Watkinson and

Epstein, 1970; Degtyarev and Pokhitun, 1972; Valyavin et al., 1976; Yeap et al., 2004; Watkinson, 2005). For light crude oils at relatively low temperatures, fouling is typically related to concentration of particulate matter (such as silt and corrosion products), which is relatively less dependent on temperature and can be reduced by removal of the particulate matter. Also, high-sulfur crude oils typically give rise to deposits which have a high content of iron sulfide. On the other hand, fouling due to phase separation or sediment formation due to asphaltene separation during dilution or blending lead to insoluble organic phases which give rise to a relatively high rate of fouling which is temperature dependent. Interaction of the crude oil constituents with oxygen which is more likely to happen during storage (rather than in a high-temperature reactor) may be controlled by rigorous exclusion of oxygen and in blending, respectively.

Briefly, blending (sometimes referred to as co-mingling) of crude from different reservoirs and different sources is a common process during production and transportation as well as prior to refining. It is no longer the norm that a refinery processes only one crude oil but blends of three or more crude oils before refining is common and in some case, heavy crude oil is blended with light crude oil so that the refinery can accommodate the heavy crude oil rather than be overwhelmed by the heavy oil as the major feedstock. Inaccurate blending (through omission of pre-transportation, per-storage, and pre-refinery testing of the blend) and disturbance of the molecular balance and loss of the crude oil structural order can lead to precipitation of solid residues and *cause* serious production problems.

However, blending is important for product sampling in the pipeline transport. When a refinery feedstock is sampled to determine its water content before the custody of the feedstock is transferred to a refinery, the water has to be uniformly dispersed throughout the sample. To ensure this, a blending system has to be installed upstream of the sampler and adequate blending should create a good dispersion but still allow water to easily settle in downstream storage tanks. Optimum mixing can add a high value.

Viscous crude oil has become an increasingly important source of hydrocarbons around the world. One available technology is to blend the feedstock into water with an emulsifier (after ensuring that the emulsion can be broken) to form high oil content oil-in-water emulsions with a dramatic decrease in viscosity and pressure drop.

Blending is also used to control sludge accumulation in crude oil storage tanks. Many feedstocks for a refinery may carry (or contain) bottom sludge and water and, since this sludge is heavier than crude oil, it settles in storage vessels at terminals and refineries. Excessive sludge accumulation can occur in tanks and at low ambient temperatures. Once the sludge is settled on the tank floor, it hardens and cannot be removed by normal pumping. The tank has to be taken off-line and cleaned, which is hazardous, expensive, time-consuming, and requires sludge disposal.

The problems encountered by the presence of the asphaltene constituents has increased due to the need to extract even the most viscous feedstocks as well as the trend to extract substantial amounts of light fractions out of crude oil by among other methods such as cracking and visbreaking. Some examples of problems due to flocculation and/or sedimentation of asphaltene constituents and reacted asphaltene constituents are: (1) blending of feedstocks as well as blending of feedstocks with other refinery products—the change of medium during mixing may cause destabilization of asphaltene (and reacted asphaltene) constituents and (2) issues related to storage problems of the viscous feedstocks and visbroken residua—sedimentation and plugging can occur due to oxidation of the asphaltene constituents as well as increased polarity that may cause asphaltene aggregation and sludge/sediment formation can furthermore be accelerated if there is bacterial infestation in the feedstock.

Thus, the deposition of asphaltene constituents is the consequence of instability of the system (whether the system is crude oil or a viscous feedstock, including residua). The asphaltene constituents are stabilized by resin constituents and maintained in the crude oil due to this stabilization. Asphaltene dispersants are substitutes for the natural resin constituents and are able to keep the asphaltene constituents dispersed to prevent flocculation/aggregation and phase separation (Speight, 1994; Browarzik et al., 1999). Dispersants will also clean up sludge in the refinery system and they have the ability to adhere to the surface of materials that are insoluble in the feedstock and produce stable colloidal suspensions.

At the other end of the molecular weight scale and refining, the heteroatoms (particularly nitrogen, sulfur, and trace metals) that are present in feedstocks might be also expected to be present in liquid fuels and other refinery products. In fact, this is often the case, although there may have been some skeletal changes induced by the refining process(es). Oxygen is much more difficult to define in feedstocks but it must be stressed that instability/incompatibility is not directly related to the total nitrogen, oxygen, or sulfur content. The formation of color/sludge/sediment is a result of several factors. Perhaps the main factor is the location and nature of the heteroatom which, in turn, determines reactivity (Por, 1992; Mushrush and Speight, 1995).

The instability/incompatibility of crude oil products is a precursor to either the formation of degradation products or the occurrence of undesirable changes in the properties of the feedstock. Individually, the components of a product may be stable and in compliance with specifications, but their blend may exhibit poor stability properties, making them unfit for use.

Incompatibility in the distillate products is important to consumers as well as the producer. However, at present and in the future, problems for refiners will continue to increase as the quality of the available crude decreases worldwide.

This decrease coupled with the inevitable future use of liquid fuels produced from both coal and shale sources will exacerbate the present problems for the refineries.

Incompatibility in crude oil products can be linked to the presence of several different deleterious heteroatomic compound classes (Mushrush and Speight, 1995). The incompatibility observed in a refinery feedstock is dependent on the blending stocks employed in its production. The light cycle oils produced during catalytic process contain the unstable heteroatomic species that are responsible for the observed deterioration in a feedstock. The solution is to use straight-run distillate product but a deficiency in supply and an option is to use chemical additives that overcome the incompatibilities of the variant chemical composition of the light cycle oil.

In general, the reaction sequence for sediment formation can be envisaged as being dependent upon the most reactive of the various heteroatomic species that are present in feedstocks (Pedley et al., 1986, 1988, 1989). The worst-case scenario would consist of a high olefin feedstock with both high indole concentration and a catalytic trace of sulfonic acid species. This reaction matrix would lead to rapid degradation. However, just as there is no one specific distillate product, there is also no one mechanism of degradation. In fact, the mechanism and the functional groups involved will give a general but not specific mode of incompatibility (Cooney et al., 1985; Hiley and Pedley, 1986, 1987; Mushrush et al., 1990, 1991). The key reaction in many incompatibility processes is the generation of the hydroperoxide species from dissolved oxygen. Once the hydroperoxide concentration starts to increase, macromolecular incompatibility precursors form in the feedstock. Acid- or base-catalyzed condensation reactions then rapidly increase both the polarity, incorporation of heteroatoms, and the molecular weight.

When various feedstocks are blended at the refinery, incompatibility can be explained by the onset of acid–base-catalyzed condensation reactions of the various organo-nitrogen compounds in the individual blending stocks. These are usually very rapid reactions with practically no observed induction time period.

When the product is transferred to a storage tank or some other holding tank, incompatibility can occur by the free-radical hydroperoxide-induced polymerization of active olefins. This is a relatively slow reaction, because the observed increase in hydroperoxide concentration is dependent on the dissolved oxygen content (Mayo and Lan, 1987; Mushrush and Speight, 1995).

The third incompatibility mechanism involves degradation when the product is stored for prolonged periods, as might occur during stockpiling of feedstock for military use (Stavinoha and Westbrook, 1980; Goetzinger et al., 1983; Cooney et al., 1985; Hazlett and Hall, 1985). This incompatibility process involves (1) the buildup of hydroperoxide moieties after the gum reactions, (2) a free-radical reaction with the various organo-sulfur compounds present that can be oxidized to sulfonic acids, and (3) reactions such as condensations between organo-sulfur and nitrogen compounds and esterification reactions.

Incompatibility during refining can occur in a variety of processes, either by intent (such as in the deasphalting process) or inadvertently when the separation is detrimental to the process. Thus, separation of solids occurs whenever the solvent characteristics of the liquid phase are no longer adequate to maintain polar and/or high molecular weight material in solution.

Examples of such occurrences are: (1) asphaltene separation which occurs when the paraffinic nature of the liquid medium increases, (2) wax separation which occurs when there is a drop in temperature or the aromaticity of the liquid medium increases, and (3) sludge/sediment formation in a reactor which occurs when the solvent characteristics of the liquid medium change so that asphaltic or wax materials separate, coke formation which occurs at high temperatures and commences when the solvent power of the liquid phase is not sufficient to maintain the coke precursors in solution, and sludge/sediment formation in feedstocks which occurs because of the interplay of several chemical and physical factors (Mushrush and Speight, 1995; Speight, 2014, 2017).

Since, the instability/incompatibility of crude oil and of crude oil products is manifested in the formation of sludge, sediment, and general darkening in color of the liquid (ASTM D1500, 2021).

Sludge (or sediment) formation takes one of the following forms: (1) material dissolved in the liquid, (2) precipitated material, and (3) material emulsified in the liquid. Under favorable conditions, sludge or sediment will dissolve in the crude oil or product with the potential of increasing the viscosity. Sludge or sediment, which is not soluble in the crude oil (ASTM D96, 2021; ASTM D473, 2021; ASTM D1796, 2021; ASTM D2273, 2021; ASTM D4007, 2021; ASTM D4807, 2021; ASTM D4870, 2021), may either settle at the bottom of the storage tanks or remain in the crude oil as an emulsion.

As an example, a test method describes the determination of water and sediment in a feedstock using a centrifuge procedure (ASTM D4007, 2021). However, the amount of water detected may be lower than the actual water content and when a highly accurate value is required, the revised procedures for water by distillation (ASTM D4006, 2021), which is the preferred and most accurate method for the determination of water, and sediment by extraction (ASTM D473, 2021) must be used.

In most of the cases, the smaller part of the sludge/sediment will settle satisfactorily and the larger part will stay in the crude oil as emulsions. In any case there is a need of breaking the emulsion, whether it is a water-in-oil emulsion or whether it is the sludge itself, which has to be separated into the oily phase and the aqueous phase. The oily phase can be then processed with the crude oil and the aqueous phase can be drained out of the system.

Phase separation can be accomplished by either the use of suitable surface-active agents allowing for sufficient settling time, or by use of a high-voltage electric field for breaking such emulsions after admixing water at a rate of about 5% and at a temperature of about 100 °C (212 °F).

Emulsion breaking, whether the emulsion is due to crude oil–sludge emulsions, crude oil–water emulsions, or breaking the sludge themselves into organic (oily) and inorganic components are of a major important form of operational as well as commercial aspects. With some viscous feedstocks, difficulties due to phase separation often arise (Ruzicka and Nordenson, 1990; Por, 1992; Mushrush and Speight, 1995). Also, some crude oil emulsions may be stabilized by naturally occurring substances in the crude oil. Many of these polar particles accumulate at the oil–water interface, with the polar groups directed toward the water and the hydrocarbon groups toward the oil. A stable interfacial skin may be so formed; particles of clay or similar impurities, as well as wax crystals present in the oil may be embedded in this skin and make the emulsion very difficult to break (Schramm, 1992).

Chemical and electrical methods for sludge removal and for water removal, often combined with chemical additives, have to be used for breaking such emulsions. Each emulsion has its own structure and characteristics: water-in-oil emulsions where the oil is the major component or oil-in-water emulsions where the water is the major component. The chemical and physical nature of the components of the emulsion plays a major role in their susceptibility to the various surface-active agents used for breaking them.

Therefore, appropriate emulsion breaking agents have to be chosen very carefully, usually with help of previous laboratory evaluations. Water- or oil-soluble demulsifiers, the latter being often nonionic surface-active alkylene oxide adducts, are used for this purpose. But, as had been said in the foregoing, the most suitable demulsifier has to be chosen for each case from a large number of such substances in the market, by a prior laboratory evaluation.

11.3 Factors Influencing Instability and Incompatibility

In the last four-to-five decades, refineries have been accepting the more viscous feedstocks on a regular basis in place of (or in addition to) the more conventional crude oils which has resulted in the need to blend the feedstocks before commencement of the refining. It is during the blending stage (or sometimes after the blending stage) that the phenomena of stability and incompatibility are manifested.

The principles of determining and controlling the kinetic stability of feedstock blends to phase separation need to be known or investigated by the refiner. At the same time, it has been noted that phase separation is the target process in some cases, such as during the dewatering of the feedstock(s) (Parkash, 2003; Gary et al., 2007; Hsu and Robinson, 2017; Speight, 2017, 2021, 2023).

Thus, a number of standard test methods are available that can be applied to estimate the factors that influence instability and incompatibility (ASTM, 2021). These factors are presented (alphabetically) in the following subsections.

11.3.1 Acidity

The acidity of refinery feedstocks oleum products is typically usually measured in terms of the acid number which is defined as the number of milli-equivalents per gram of alkali required to neutralize the acidity of the crude oil sample (ASTM D664, 2021; ASTM D974, 2021; ASTM D3242, 2021).

Acidity due to the presence of inorganic constituents is not expected to be present in crude oils, but organic acidity might be found. Acidic character is composed of contributions from strong organic acids and other organic acids. Values above 0.15 mg potassium hydroxide per gram are considered to be significantly high. High acid feedstocks may exhibit a high tendency for instability.

The acid-imparting agents in crude oils are naphthenic acids and hydrosulfides (thiols, mercaptans, RSH). These are sometimes present in the crude oil originally in small and varying concentrations. Normally, the total acidity of crude oils is in the range of 0.1–0.5 mg potassium hydroxide per gram, although higher values are not exceptional.

Free hydrogen sulfide is often present in crude oils, a concentration of up to 10 ppm being acceptable in spite of its toxic nature. However, higher hydrogen sulfide concentrations are sometimes present, 20 ppm posing serious safety hazards. Additional amounts of hydrogen sulfide can form during the crude oil processing, when hydrogen reacts with some organic sulfur compounds to produce hydrogen sulfide which, in such as case, is often referred to as "potential hydrogen sulfide" in contract to any free hydrogen sulfide.

Acidic components in a feedstock can also be produced by bacterial action insofar as some species of aerobic bacteria can produce organic acid derivatives from organic components. On the other hand, anaerobic sulfate-reducing bacteria can generate hydrogen sulfide, which, in turn, can be converted to sulfuric acid (by bacterial action).

11.3.2 Asphaltene Content

The asphaltene fraction of a crude oil or a viscous feedstock is the highest molecular weight and most complex fraction in crude oil (ASTM D893, 2021; ASTM D2007, 2021; ASTM D3279, 2021; ASTM D4124, 2021; ASTM D6560, 2021) and is the fraction that produces the highest yields of coke that can be expected during exposure to thermal conditions (Speight, 2014, 2015a, 2017).

Table 11.2 Standard Methods for Asphaltene Precipitation

Method	Precipitant	Volume Precipitant per g of Sample (ml)
ASTM D893	*n*-Pentane	10
ASTM D2007	*n*-Pentane	10
ASTM D3279	*n*-Heptane	100
ASTM D4124	*n*-Heptane	100

ASTM Annual Book of Standards, 1980–2021.

The asphaltene fraction is isolated as a dark brown to black friable solids that has no definite melting point and usually intumesces on heating (with decomposition) to produce a carbonaceous residue that can vary up to 60% w/w of the original fraction. The fraction is obtained from a feedstock crude oil by the addition of a nonpolar solvent (for example, a liquid hydrocarbon such as *n*-pentane or *n*-heptane) (Table 11.2) (Speight, 2014, 2015a). In a typical standard test method, the asphaltene fraction is removed by filtration through paper but more recently a membrane method has come into use (ASTM D4055, 2021). Liquid propane is used commercially in processing crude oil residua (Parkash, 2003; Gary et al., 2007; Hsu and Robinson, 2017; Speight, 2017)—the asphaltene constituents of the fraction are soluble in liquids such as toluene, pyridine, carbon disulfide, and carbon tetrachloride. Benzene is sometimes mentioned in preference to toluene but there are risks of using benzene (because of the potential of the benzene to be a cancer-causing reagent, the use of this solvent is now dissuaded).

The asphaltene fraction of feedstocks (Figure 11.1) is particularly important because as the proportion of this fraction increases, there is a concomitant increase in thermal coke yield and an increase in hydrogen demand as well as catalyst deactivation. The constituents of the asphaltenes form coke quite readily which is of particular interest in terms of the compatibility/incompatibility of the coke precursors (Speight, 1994).

An early hypothesis of the physical structure of crude oil (Pfeiffer and Saal, 1940) indicated that asphaltene constituents are the centers of micelles formed by adsorption, or even by absorption of part of the maltenes, that is, resin material, onto the surfaces or into the interiors of the asphaltene particles. Thus, most of those substances with greater molecular weight and with the most pronounced aromatic nature are situated closest to the nucleus and are surrounded by lower boiling constituents of a less aromatic nature.

The transition of the intermicellar (dispersed or oil) phase is gradual and almost continuous and is influenced by the makeup of the asphaltene fraction, which is composed of neutral, acidic, and basic constituents (Speight, 1994, 2014) and this is why the asphaltene fraction cannot be adequately represented by a so-called

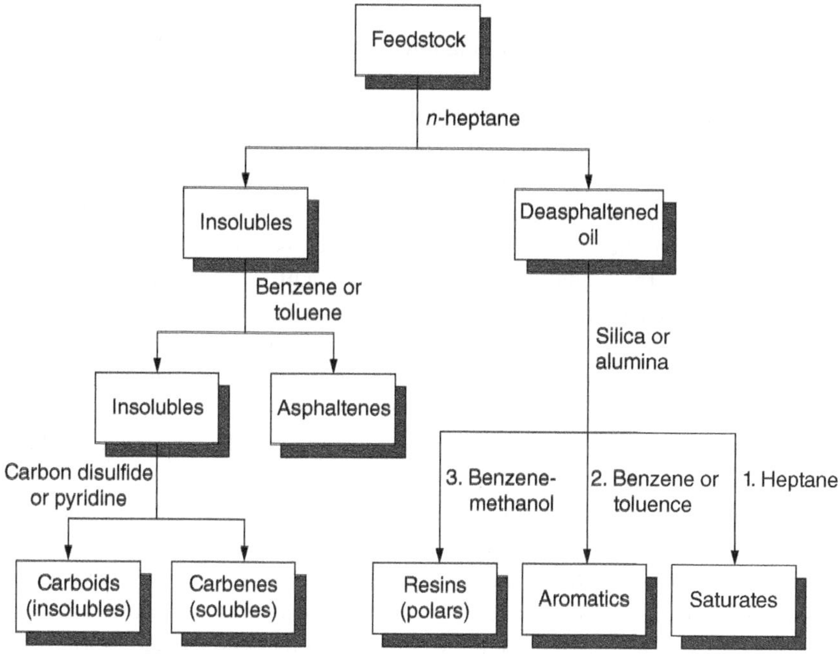

Figure 11.1 General fractionation scheme and nomenclature of petroleum fraction; carbenes and carboids are thermally generated product fractions.

average structure. Various separation procedures applied to the asphaltene fraction have confirmed the complex chemical and functional nature of the asphaltene fraction (Francisco and Speight, 1984; Speight, 1994, 2014). Since asphaltene constituents are incompatible with the oil fraction, asphaltene dispersion is attributable to the interactions between resin constituents (polar aromatic constituents) and the asphaltene constituents that each carry interacting chemical–physical sites and functional groups (Koots and Speight, 1975).

On this basis, the stability of crude oil can be represented by a three-phase system in which the asphaltene constituents, the aromatic constituents (including the resin constituents), and the saturate constituents are delicately balanced (Speight, 1994, 2014). Various factors, such as oxidation, can have an adverse effect on the system, leading to instability/incompatibility as a result of changing the polarity and bonding arrangements of the species in crude oil.

The evolution of this concept is perhaps the single most responsible agent for the perception that asphaltene constituents in crude oil, heavy oil, extra heavy oil, and tar sand bitumen crude oil exist in the manner of a colloidal system. Indices have been derived to describe this behavior. One strong factor in favor of such a system is the inability of most crude oils to disperse the asphaltene fraction in the absence of the resin fraction (Koots and Speight, 1975).

In heavy oil, extra heavy oil, and tar sand bitumen, the asphaltene constituents and the resin constituents have strong interactions to the extent that the asphaltene constituents are immiscible (insoluble/incompatible) with the remaining constituents in the absence of the resin constituents (Koots and Speight, 1975). And there appears to be points of structural similarity (for a crude oil) between the asphaltene and the resin constituents thereby setting the stage for *a more-than-is-generally-appreciated* complex relationship but confirming the hypothesis that heavy oil, extra heavy oil, and tar sand bitumen is a continuum of chemical species, including the asphaltene constituents (Speight, 2014, 2015a).

This sets the stage for the incompatibility of the asphaltene constituents in any operation in which the asphaltene or resin constituents are physically or chemically altered. Disturbance of the asphaltene–resin relationships can be the stimulation by which, for example, some or all of the asphaltene constituents form a separate *insoluble* phase leading to such phenomena as coke formation (in thermal processes) or asphalt instability during use.

The effect of asphaltenes and the micelle structure, and the state of dispersion also merit some attention. The degree of dispersion of the asphaltene constituents is higher in the more naphthenic/aromatic feedstocks because of higher solvency of naphthene derivatives and aromatic derivatives over paraffinic constituents. This phenomenon may also act in favor of the retention (as part of the liquid medium) of any sludge that may form thereby tending to decrease the deposition of the sludge.

The higher the asphaltene content, the greater the tendency of the crude oil to form sludge, especially when blended with other noncompatible stocks.

11.3.3 Density/Specific Gravity

In the earlier years of the crude oil industry, density and specific gravity (with the API gravity) were the principal specifications for feedstocks and refinery products (Parkash, 2003; Gary et al., 2007; Hsu and Robinson, 2017; Speight, 2017). They were used to give an estimate of the most desirable product, that is, kerosene, in crude oil. At the present time, a series of standard tests exists for determining density and specific gravity (Speight, 2014, 2015a).

There is no direct relation between the density and specific gravity of crude oils to their sludge-forming tendencies, but crude oil having a higher density (thus, a lower API gravity) is generally more susceptible to sludge formation, presumably because of the higher content of the polar/asphaltic constituents.

11.3.4 Elemental Analysis

The elemental composition (often referred to as the "ultimate analysis") of crude oil and crude oil products is not reported to the same extent as for coal

(Speight, 2013, 2014). Nevertheless, there are standard procedures for the ultimate analysis of crude oil and crude oil products but many such methods may have been designed for other materials (ASTM, 2021).

Of the data that are available, the proportions of the elements in crude oil vary only slightly over narrow limits: carbon 83.0–87.0% w/w, hydrogen 10.0–14.0% w/w, nitrogen 0.10–2.0% w/w, oxygen 0.05–1.5% w/w, and sulfur 0.05–6.0% w/w (Parkash, 2003; Gary et al., 2007; Hsu and Robinson, 2017; Speight, 2017). In contrast, there is a wide variation in physical properties from the lighter more mobile crude oils at one extreme to the heavy crude oil, extra heavy crude oil, and tar sand bitumen at the other extreme (Chapter 1).

In terms of the instability and incompatibility of feedstocks, the heteroatom content appears to represent the greatest influence. In fact, it is not only the sulfur and nitrogen content of refinery feedstocks that are important parameters in respect of the processing methods which have to be used in order to produce feedstocks that are more amenable to refining but also the type of sulfur and nitrogen species in the oil. There could well be a relation between nitrogen and sulfur content and crude oil (or product) stability; feedstocks with a high nitrogen content and/or high sulfur crude oils are suspect of exhibiting the tendency for sludge formation.

11.3.5 Metals Content

The majority of crude oils contain metallic constituents that are often determined as combustion ash (ASTM D482, 2021). This is particularly so for the viscous feedstocks. These constituents, of which nickel and vanadium are the principal metals, are very influential in regard to feedstock behavior in processing operations.

It must be emphasized at this point that, contrary to arbitrary comments (leading to incorrect terminology), crude oil and the viscous feedstocks do not contain ash. The do contain metal-containing constituents that form ash (the metal oxides) when combusted. It is from this ash that the actual metal in the feedstock can be identified.

The metal (inorganic) constituents of a feedstock arise from either the inorganic constituents present in the crude oil originally or those picked up by the crude oil during storage. The former are mostly metallic substances like vanadium, nickel, sodium, iron, silica, and so on; the latter may be contaminants such as sand, dust, and corrosion products.

Incompatibility, leading to deposition of the metals (in any form) on to the catalyst, leads to catalyst deactivation whether it is by physical blockage of the pores or destruction of reactive sites. In the present context, the metals must first be removed if erroneously high carbon residue data are to be avoided. Alternatively, they can be estimated as ash by complete burning of the coke after carbon residue determination.

Metals content above 200 ppm are considered to be significant, but the variations are very large. The higher the ash content the higher is the tendency of the crude oil to form sludge or sediment.

11.3.6 Pour Point

The pour point defines the cold properties of crude oils and crude oil products and is a measure of the minimal temperature at which they still retain their fluidity (ASTM D97, 2021). Therefore, pour point can also be used as an indication of the characteristics of crude oils insofar as the higher the pour point, the more paraffinic the oil. Higher pour point crude oils are often waxy crude oil and therefore they tend to form wax-like materials that enhance sludge formation.

To determine the pour point (ASTM D97, 2021; ASTM D5853, 2021; ASTM D5949, 2021; ASTM D5950, 2021; ASTM D5985, 2021), the sample is contained in a glass test tube fitted with a thermometer and immersed in one of three baths containing coolants. The sample is dehydrated and filtered at a temperature 25 °C (45 °F) higher than above the anticipated cloud point. It is then placed in a test tube and cooled progressively in coolants held at −1 to +2 °C (30–35 °F), −18 to −20 °C (−4 to 0 °F), and −32 to −35 °C (−26 to −31 °F), respectively. The sample is inspected for cloudiness at temperature intervals of 1 °C (2 °F).

If conditions or oil properties are such that reduced temperatures are required to determine the pour point, alternate tests are available that accommodate the various types of samples.

11.3.7 Viscosity

The viscosity of a feedstock varies with the origin and type of the crude oil and also with the nature of the chemical constituents, particularly the polar functions where intermolecular interactions can occur. For example, there is a gradation of viscosity between conventional crude oil, heavy crude oil, extra heavy crude oil, and tar sand bitumen (Speight, 2014, 2015a, 2017).

Viscosity is a measure of fluidity properties and consistencies at given temperatures (Speight, 2014, 2015a). Heavy crude oil, that is, crude oil having lower API gravity, typically have higher viscosity. Increases of viscosity during storage indicate either an evaporation of volatile components or formation of degradation products dissolving in the crude oil.

11.3.8 Volatility

Crude oil can be subdivided by distillation into a variety of fractions of different boiling ranges or cut points (Parkash, 2003; Gary et al., 2007; Hsu and

Robinson, 2017; Speight, 2017, 2023). In fact distillation was, and still is, the method for feedstock evaluation for various refinery options. Indeed, volatility is one of the major tests for crude oil products and it is inevitable that the majority of all products will, at some stage of their history, be subject to the appropriate test method to determine the volatility characteristics of the feedstock.

As an early part of characterization studies, a correlation was observed between the quality of crude oil fractions and their heteroatom content (Table 11.3) since naphtha (a precursor to gasoline), gasoline, kerosene (a precursor to diesel fuel), and gas oil are composed of hydrocarbon constituents containing high proportions of hydrogen. Thus, it is not surprising that test to determine the volatility of crude oil and crude oil products were among the first to be defined.

The very nature of the distillation process by which residua are produced (Parkash, 2003; Gary et al., 2007; Hsu and Robinson, 2017; Speight, 2017), that is, removal of distillate without thermal decomposition, dictates that the majority of the heteroatoms which are predominantly in the higher molecular weight fractions, will be concentrated in the higher boiling products and the residuum (Speight, 2014, 2017, 2023). Thus, the inherent nature of the crude oil and the means by which it is refined can seriously influence the stability and incompatibility of the products.

Table 11.3 Simplified representation of the concentration of the feedstock polar functions (heteroatomic groups) by distillation

Whole crude	Atmospheric distillation	Vacuum distillation
Feedstock	Distillate	
	Hydrocarbon gases	
	Naphtha	
	Kerosene	
	Middle distillate	
	Gas oil	
	Atmospheric residuum	Distillate
	Heteroatom molecules	Vacuum gas oil
	Neutral nonpolar	Vacuum residuum
	Polar	Heteroatom molecules
	Acids	Neutral nonpolar
	Bases	Polar
		Acids
		Bases

Heavier crude oils, yielding higher amounts of residua, tend to form more sludge during storage compared to light crude oils.

11.3.9 Water Content, Salt Content, Bottom Sediment and Water (BS&W)

Water content, salt content, and bottom sediment and water (this latter is often referred to as BS&W) are important aspects of the analysis of refinery feedstocks. Removal often occurs prior to transportation, storage, and commencement of refining operations (Speight, 2014, 2015b, 2017, 2023). If the water, salts, and sediment are not removed, there is the high potential for the corrosion (and fouling) of the refinery equipment.

The data related to water content (ASTM D4006, 2021; ASTM D4007, 2021; ASTM D4928, 2021), salt content (ASTM D3230, 2021), and bottom sediment/water (ASTM D96, 2021; ASTM D1796, 2021; ASTM D4007, 2021) are indicative of the concentrations of aqueous contaminants, either originally in the feedstock or incorporated into the feedstock during handling and/or storage. Water and salt content of crude oils produced at the field can be very high, forming sometimes its major part. The salt water is often separated at the field site before transportation, typically by settling and draining as well as by the use of surface-active agents electrical emulsion breakers (desalters) (2014, 2017, 2021) are sometimes employed. The water and salt contents of crude oil supplied to the buyers is a function of the production field. Water content below 0.5%, salt content up to 20 pounds per 1000 barrels, and bottom sediment and water up to 0.5% are considered to be satisfactory.

Although the centrifuge methods are still employed (ASTM D96, 2021; ASTM D1796, 2021; ASTM D2709, 2021; ASTM D4007, 2021), many laboratories prefer the Dean and Stark adaptor (ASTM D95, 2021). The apparatus consists of a round-bottom flask of capacity 50 ml connected to a Liebig condenser by a receiving tube of capacity 25 ml, graduated in 0.1 ml. A weighed amount, corresponding to approximately 100 ml of oil, is placed in the flask with 25 ml of dry toluene. The flask is heated gently until the 25 ml of toluene have distilled into the graduated tube. The water distilled with the toluene, separates to the bottom of the tube where the volume is recorded as ml, or the weight as mg, or per cent.

To determine the sediment in crude oil or in a crude oil product, the method involves solvent extraction using a Soxhlet extractor.

The Karl Fischer titration method, the Karl Fischer titration method, and the colorimetric Karl Fischer titration method still find wide application in many laboratories for the determination of water in feedstocks. The higher the bottom sediment and water content, the higher sludge and deposit formation rates can be expected in the stored feedstock.

In summary, a range of effects (such as feedstock composition and temperature sensitivity) influence the occurrence of fouling and the rate of fouling. Deposition in flow lines, pipelines, storage, and refinery equipment (including heat transfer equipment) arises from fouling species which may either be present in the crude oil or generated during processing.

11.4 Determination of Instability and Incompatibility

Stability/instability and compatibility/incompatibility can be estimated using several standard test methods. These tests are: (1) compatibility spot tests (ASTM D4740, 2021), (2) thermal stability test data (ASTM D873, 2021; ASTM D3241, 2021), (3) existent and potential sludge formation (hot filtration test) (ASTM D4870, 2021), (4) asphaltene content, (5) the rate of viscosity increase (ASTM D445, 2021) of stored feedstock samples at various temperatures with and without exposure to air, and (6) the color of the sample (ASTM D1500, 2021).

In addition to the test for the stability and compatibility of viscous feedstocks (ASTM D4740, 2021), the other most frequently used stability test is the existent and accelerated dry sludge content determination of viscous feedstocks (the hot filtration test). Hot filtration test results of up to 0.2% w/w are considered to be satisfactory; results above 0.4% w/w indicate a poor stability, but differing values might be required, depending on the intended use of the product.

One test, or property, which is somewhat abstract in its application but which is becoming more meaningful popular is the solubility parameter (Speight, 2014, 2015a, 2017), which allows estimations to be made of the ability of liquids to become miscible on the basis of miscibility of model compound types where the solubility parameter can be measured or calculated.

Although the solubility parameter is often difficult to define when complex mixtures are involved, there has been some progress. For example, crude oil fractions have been assigned a similar solubility parameter to that of the solvent used in the separation. However, there is also the concept (Speight, 1992) that the solubility parameter of crude oil fractions may be defined somewhat differently insofar as they can be estimated from data such as the atomic hydrogen-to-carbon ratios. Whichever method is the best estimate may be immaterial as long as the data are used to the most appropriate benefit and allow some measure of predictability.

Bottle tests constitute the predominant test method and the test conditions have varied in volume, type of glass or metal, vented and unvented containers, and type of bottle closure. Other procedures have involved stirred reactor vessels under both air pressure and under oxygen pressure using small volumes of feedstock and employing a cover slip for solid deposition (ASTM D4625, 2021) (Speight, 2014, 2017).

There are several accelerated stability tests that can be represented as a time–temperature matrix (Goetzinger et al., 1983; Hazlett, 1992; Speight, 2014, 2015a, 2017). A graphical representation shows that the majority of the stability tests depicted fall close to the solid line, which represents a doubling of test time for each 10 °C (18 °F) change in temperature. The line extrapolates to approximately one year of storage under ambient conditions. Temperatures at 100 °C (212 °F) or higher present special chemical problems.

It is also worthy of note that fractionation of crude oil and its products may also give some indication of instability. There are many schemes by which crude oil and related materials might be fractionated (Chapter 4) (Speight, 2014, 2015a). Therefore, a brief overview of the methods will be helpful since fractional composition can play a role in stability and incompatibility phenomena.

Refinery feedstocks can be fractionated into four general fractions by a variety of techniques although the most common procedure involves precipitation of the asphaltene fraction and the use of adsorbents to fractionate the deasphalted (deasphaltened) oil into three other fractions. The fractions are named for convenience and the assumption that fractionation occurs by specific compound type is not quite true.

In general terms, studies of the composition of the incompatible materials often involve determination of the distribution of the organic functional groups by selective fractionation (Chapter 4) that is analogous to the deasphalting procedure to separate the feedstock into (1) an asphaltene fraction which is the heptane insoluble material, which is also toluene soluble material and is referred to as the asphaltene fraction; (2) a fraction that is commonly referred to as the maltenes fraction, which is sometimes referred to as the petrolenes—this latter term is often used to describe the constituents that boil below 350 °C/660 °F; (3) the toluene insoluble material: referred to as carbenes and carboids when the fraction is a thermal product; (4) the pyridine-soluble material, which is often referred to as the carbenes fraction; and (5) the pyridine-insoluble material, which is often referred to as the carboids fraction (Figure 11.1).

Carbon disulfide and tetrahydrofuran have been used in place of pyridine. The former (carbon disulfide), although having an obnoxious odor and therefore not much different from pyridine, is easier to remove because of the higher volatility. The latter (i.e., tetrahydrofuran) is not as well established in crude oil science as it is in the coal–liquid-related research. Thus, it is more than likely that the crude oil researcher will use carbon disulfide or pyridine, or some suitable alternate solvent. It may also be necessary to substitute cyclohexane as an additional step for treatment of the heptane-insoluble materials prior to treatment with toluene. The use of quinoline has been suggested in place of pyridine but this solvent presents issues associated with its high boiling point.

Whichever solvent separation scheme is employed, there should be ample description of the procedure so that the work can be repeated not only in the same laboratory by a different researcher but also in different laboratories by various researchers. Thus, for any particular feedstock and solvent separation scheme, the work should repeatable and reproducible within the limits of experimental error.

Fractionation procedures allow a before-and-after inspection of any feedstock or product and can give an indication of the means by which refining or use changes the composition of the feedstock. In addition, fractionation also allows studies to be made of the interrelations between the various fractions. For example, the most interesting phenomenon (in the present context) to evolve from the fractionation studies is the relationship between the asphaltenes and the resins.

In crude oil, the asphaltenes and resins have strong interactions to the extent that the asphaltenes are immiscible (insoluble/incompatible) with the remaining constituents in the absence of the resins (Koots and Speight, 1975). And there appears to be points of structural similarity (for a crude oil) between the asphaltene and the resin constituents thereby setting the stage for a more-than-is-generally-appreciated complex relationship but confirming the hypothesis that crude oil is a continuum of chemical species, including the asphaltene constituents (Speight, 1994, 2014).

This sets the stage for the incompatibility of the asphaltenes in any operation in which the asphaltene or resin constituents are physically or chemically altered. Disturbance of the asphaltene–resin relationships can be the stimulation by which, for example, some or all of the asphaltene constituents form a separate insoluble phase leading to such phenomena as coke formation (in thermal processes) or asphalt instability during use.

There is a series of characterization indices that also present indications of whether or not a crude oil product is stable or unstable. For example, the characterization factor indicates the chemical character of the crude oil and has been used to indicate whether a crude oil was paraffinic in nature or whether it was a naphthenic/aromatic crude oil.

The characterization factor (sometimes referred to as the Watson characterization factor) (Speight, 2014) is a relationship between boiling point and specific gravity:

$$K = T_b^{1/3} / d$$

T_b is the cubic average boiling point, degrees Rankine (°F + 460), and d is the specific gravity at 15.6 °C (60 °F).

The characterization factor was originally devised to illustrate the characteristics of various feedstocks. Highly paraffinic oils have $K = 12.5 - 13.0$ while naphthenic oils have $K = 10.5 - 12.5$. In addition, if the characterization factor is above 12, the feedstock or product might, because of its paraffinic nature, be expected to form waxy deposits during storage.

The viscosity-gravity constant (vgc) was one of the early indices proposed to classify crude oil on the basis of composition. It is particularly valuable for indicating a predominantly paraffinic or naphthenic composition. The constant is based on the differences between the density and specific gravity for the various hydrocarbon species:

$$\text{vgc} = \left[10d - 1.0752\log(v - 380)\right] / \left[10 - \log(v - 38)\right]$$

where d is the specific gravity and v is the Saybolt viscosity at 38 °C (100 °F). For viscous crude oils (and viscous products) where the viscosity is difficult to measure at low temperature, the viscosity at 991 °C (2101 °F) can be used:

$$\text{vgc} = \left[d - 0.24 - 0.022\log(v - 35.5)\right] / 0.755$$

In both cases, the lower the index number is indicative of a more paraffinic sample. For example, a paraffinic sample may have a vgc on the order of 0.840 while the corresponding naphthenic sample may have an index on the order of 0.876.

The obvious disadvantage is the closeness of the indices, almost analogous to comparing crude oil character by specific gravity only where most crude oils fall into the range $d = 0.800 - 1.000$. The API gravity expanded this scale from 5 to 60 thereby adding more meaning to the use of specific gravity data.

In a similar manner, the correlation index which is based on a plot of the specific gravity (d) vs. the reciprocal of the boiling point (K) in °K (°K = degrees Kelvin = °C + 273) for pure hydrocarbons adds another dimension to the numbers:

$$\text{Correlation Index(CI)} = 473.7d - 456.8 + 48640/K$$

In the case of a crude oil fraction, K is the average boiling point determined by the standard distillation method.

The line described by the constants of the individual members of the normal paraffin series is given a value of CI = 0 and a parallel line passing through the point for benzene is given a value of CI = 100 (Speight, 2014). Values between 0 and 15 indicate a predominance of paraffinic hydrocarbons in the sample and values from 15 to 20 indicate predominance either of naphthenes or of mixtures of paraffins/naphthenes/aromatics; an Index value above 50 indicates a predominance of aromatics in the fraction.

References

ASTM 2021. *Annual Book of Standards*. ASTM International, West Conshohocken, Pennsylvania.

ASTM D5. 2021. Standard Test Method for Penetration of Bituminous Materials. Annual Book of Standards. ASTM International, West Conshohocken, Pennsylvania.

ASTM D95. 2021. Standard Test Method for Water in Petroleum Products and Bituminous Materials by Distillation. Annual Book of Standards. ASTM International, West Conshohocken, Pennsylvania.

ASTM D96. 2021. Standard Test Method for Water and Sediment in Crude Oil by Centrifuge Method (Field Procedure). Annual Book of Standards. ASTM International, West Conshohocken, Pennsylvania.

ASTM D97. 2021. Standard Test Method for Pour Point of Petroleum Products. Annual Book of Standards. ASTM International, West Conshohocken, Pennsylvania.

ASTM D113. 2021. Standard Test Method for Ductility of Asphalt Materials. Annual Book of Standards. ASTM International, West Conshohocken, Pennsylvania.

ASTM D381. 2021. Standard Test Method for Gum Content in Fuels by Jet Evaporation. Annual Book of Standards. ASTM International, West Conshohocken, Pennsylvania.

ASTM D445. 2021. Standard Test Method for Kinematic Viscosity of Transparent and Opaque Liquids (and Calculation of Dynamic Viscosity). Annual Book of Standards. ASTM International, West Conshohocken, Pennsylvania.

ASTM D473. 2021. Standard Test Method for Sediment in Crude Oils and Fuel Oils by the Extraction Method. Annual Book of Standards. ASTM International, West Conshohocken, Pennsylvania.

ASTM D482. 2021. Standard Test Method for Ash from Petroleum Products. Annual Book of Standards. ASTM International, West Conshohocken, Pennsylvania.

ASTM D525. 2021. Standard Test Method for Oxidation Stability of Gasoline (Induction Period Method). Annual Book of Standards. ASTM International, West Conshohocken, Pennsylvania.

ASTM D664. 2021. Standard Test Method for Acid Number of Petroleum Products by Potentiometric Titration. Annual Book of Standards. ASTM International, West Conshohocken, Pennsylvania.

ASTM D873. 2021. Standard Test Method for Oxidation Stability of Aviation Fuels (Potential Residue Method). Annual Book of Standards. ASTM International, West Conshohocken, Pennsylvania.

ASTM D893. 2021. Standard Test Method for Insolubles in Used Lubricating Oils. Annual Book of Standards. ASTM International, West Conshohocken, Pennsylvania.

ASTM D942. 2021. Standard Test Method for Oxidation Stability of Lubricating Greases by the Oxygen Pressure Vessel Method. Annual Book of Standards. ASTM International, West Conshohocken, Pennsylvania.

ASTM D974. 2021. Standard Test Method for Acid and Base Number by Color-Indicator Titration. Annual Book of Standards. ASTM International, West Conshohocken, Pennsylvania.

ASTM D1500. 2021. Standard Test Method for ASTM Color of Petroleum Products (ASTM Color Scale). Annual Book of Standards. ASTM International, West Conshohocken, Pennsylvania.

ASTM D1754. 2021. Standard Test Method for Effect of Heat and Air on Asphaltic Materials (Thin-Film Oven Test). Annual Book of Standards. ASTM International, West Conshohocken, Pennsylvania.

ASTM D1796. 2021. Standard Test Method for Water and Sediment in Fuel Oils by the Centrifuge Method (Laboratory Procedure). Annual Book of Standards. ASTM International, West Conshohocken, Pennsylvania.

ASTM D2007. 2021. Standard Test Method for Characteristic Groups in Rubber Extender and Processing Oils and Other Petroleum-Derived Oils by the Clay-Gel Absorption Chromatographic Method. Annual Book of Standards. ASTM International, West Conshohocken, Pennsylvania.

ASTM D2170. 2021. Standard Test Method for Kinematic Viscosity of Asphalts (Bitumens). Annual Book of Standards. ASTM International, West Conshohocken, Pennsylvania.

ASTM D2272. 2021. Standard Test Method for Oxidation Stability of Steam Turbine Oils by Rotating Pressure Vessel. Annual Book of Standards. ASTM International, West Conshohocken, Pennsylvania.

ASTM D2273. 2021. Standard Test Method for Trace Sediment in Lubricating Oils. Annual Book of Standards. ASTM International, West Conshohocken, Pennsylvania.

ASTM D2274. 2021. Standard Test Method for Oxidation Stability of Distillate Fuel Oil (Accelerated Method). Annual Book of Standards. ASTM International, West Conshohocken, Pennsylvania.

ASTM D2709. 2021. Standard Test Method for Water and Sediment in Middle Distillate Fuels by Centrifuge. Annual Book of Standards. ASTM International, West Conshohocken, Pennsylvania.

ASTM D3230. 2021. Standard Test Method for Salts in Crude Oil (Electrometric Method). Annual Book of Standards. ASTM International, West Conshohocken, Pennsylvania.

ASTM D3241. 2021. Standard Test Method for Thermal Oxidation Stability of Aviation Turbine Fuels. Annual Book of Standards. ASTM International, West Conshohocken, Pennsylvania.

ASTM D3242. 2021. Standard Test Method for Acidity in Aviation Turbine Fuel. Annual Book of Standards. ASTM International, West Conshohocken, Pennsylvania.

ASTM D3279. 2021. Standard Test Method for *n*-Heptane Insolubles. Annual Book of Standards. ASTM International, West Conshohocken, Pennsylvania.

ASTM D3381. 2021. Standard Specification for Viscosity-Graded Asphalt Cement for Use in Pavement Construction. Annual Book of Standards. ASTM International, West Conshohocken, Pennsylvania.

ASTM D4006. 2021. Standard Test Method for Water in Crude Oil by Distillation. Annual Book of Standards. ASTM International, West Conshohocken, Pennsylvania.

ASTM D4007. 2021. Standard Test Method for Water and Sediment in Crude Oil by the Centrifuge Method (Laboratory Procedure). Annual Book of Standards. ASTM International, West Conshohocken, Pennsylvania.

ASTM D4055. 2021. Standard Test Method for Pentane Insolubles by Membrane Filtration. Annual Book of Standards. ASTM International, West Conshohocken, Pennsylvania.

ASTM D4124. 2021. Standard Test Method for Separation of Asphalt into Four Fractions. Annual Book of Standards. ASTM International, West Conshohocken, Pennsylvania.

ASTM D4310. 2021. Standard Test Method for Determination of Sludging and Corrosion Tendencies of Inhibited Mineral Oils. Annual Book of Standards. ASTM International, West Conshohocken, Pennsylvania.

ASTM D4625. 2021. Standard Test Method for Middle Distillate Fuel Storage Stability at 43 °C (110 °F). Annual Book of Standards. ASTM International, West Conshohocken, Pennsylvania.

ASTM D4740. 2021. Standard Test Method for Cleanliness and Compatibility of Residual Fuels by Spot Test. Annual Book of Standards. ASTM International, West Conshohocken, Pennsylvania.

ASTM D4742. 2021. Standard Test Method for Oxidation Stability of Gasoline Automotive Engine Oils by Thin-Film Oxygen Uptake (TFOUT). Annual Book of Standards. ASTM International, West Conshohocken, Pennsylvania.

ASTM D4807. 2021. Standard Test Method for Sediment in Crude Oil by Membrane Filtration. Annual Book of Standards. ASTM International, West Conshohocken, Pennsylvania.

ASTM D4870. 2021. Standard Test Method for Determination of Total Sediment in Residual Fuels. Annual Book of Standards. ASTM International, West Conshohocken, Pennsylvania.

ASTM D4871. 2021. Standard Guide for Universal Oxidation/Thermal Stability Test Apparatus. Annual Book of Standards. ASTM International, West Conshohocken, Pennsylvania.

ASTM D4928. 2021. Standard Test Method for Water in Crude Oils by Coulometric Karl Fischer Titration. Annual Book of Standards. ASTM International, West Conshohocken, Pennsylvania.

ASTM D5853. 2021. Standard Test Method for Pour Point of Crude Oils. Annual Book of Standards. ASTM International, West Conshohocken, Pennsylvania.

ASTM D5949. 2021. Standard Test Method for Pour Point of Petroleum Products (Automatic Pressure Pulsing Method). Annual Book of Standards. ASTM International, West Conshohocken, Pennsylvania.

ASTM D5950. 2021. Standard Test Method for Pour Point of Petroleum Products (Automatic Tilt Method). Annual Book of Standards. ASTM International, West Conshohocken, Pennsylvania.

ASTM D5985. 2021. Standard Test Method for Pour Point of Petroleum Products (Rotational Method). Annual Book of Standards. ASTM International, West Conshohocken, Pennsylvania.

ASTM D6560. 2021. Standard Test Method for Determination of Asphaltenes (Heptane Insolubles) in Crude Petroleum and Petroleum Products. Annual Book of Standards. ASTM International, West Conshohocken, Pennsylvania.

Browarzik, D., Laux, H., and Rahimian, I. 1999. Asphaltene Flocculation in Crude Oil Systems. *Fluid Phase Equilibria*, 154: 285–300.

Cooney, J.V., Beal, E.J., and Hazlett, R.N. 1985. Mechanisms of Synfuel Degradation. 2. Examination of Shale-Derived Polar Fractions and Their Effects on the Storage Stability of a Shale Diesel Fuel. *Industrial & Engineering Chemistry Product Research and Development*, 24: 294.

Degtyarev, G.S., and Pokhitun, L.F. 1972. Investigation of Coke Deposits in the Tubes of a Furnace for Cracking Heavy Feedstock. *Neftepererab. Neftekhim.*, 11: 10–12.

Gary, J.G., Handwerk, G.E., and Kaiser, M.J. 2007. *Petroleum Refining: Technology and Economics*. 5th Edition. CRC Press, Taylor & Francis Group, Boca Raton, Florida.

Goetzinger, J.W., Thompson, C.J., and Brinkman, D.W. 1983. A Review of Storage Stability Characteristics of Hydrocarbon Fuels. US Department of Energy, Report No. DOE/BETC/IC-83-3.

Hardy, D.R., and Wechter, M.A. 1990. Insoluble Sediment Formation in Middle-Distillate Diesel Fuel: The Role of Soluble Macromolecular Oxidatively Reactive Species. *Energy & Fuels*, 4: 270–274.

Hazlett, R.N. 1992. *Thermal Oxidation Stability of Aviation Turbine Fuels*. Monograph No. 1. American Society for Testing and Materials, Philadelphia, Pennsylvania.

Hazlett, R.N., and Hall, J.M. 1985. In: *Chemistry of Engine Combustion Deposits*. L.B. Ebert (Editor). Plenum Press, New York.

Hiley, R.W., and Pedley, L.F. 1986. Formation of Insolubles during Storage of Naval Fuels. Proceedings. 2nd International Conference on Long-term Storage Stabilities of Liquid Fuels, San Antonio, Texas. July 29–August 1. pp. 570–584.

Hiley, R.W., and Pedley, J.F. 1987. *Fuel*, 67: 1124.

Hsu, C.S., and Robinson, P.R. (Editors). 2017. *Handbook of Petroleum Technology*. Springer International Publishing AG, Cham, Switzerland.

Koots, J.A., and Speight, J.G. 1975. The Relation of Petroleum Resins to Asphaltenes. *Fuel*, 54: 179.

Mayo, F.R., and Lan, B.Y. 1987. Gum and Deposit Formation from Jet Turbine and Diesel Fuels at 100 °C. *Industrial and Engineering Chemistry Product Research*, 26: 215.

Mushrush, G.W., Beal, E.J., Hazlett, R.N., and Hardy, D.R. 1990. Macromolecular Oxidatively Reactive Species. *Energy & Fuels*, 5: 258.

Mushrush, G.W., Hazlett, R.N., Pellenbarg, R.E., and Hardy, D.R. 1991. *Energy & Fuels*, 5: 258.

Mushrush, G.W., and Speight, J.G. 1995. *Petroleum Products: Instability and Incompatibility*. Taylor & Francis Publishers, Washington, DC.

Parkash, S. 2003. *Refining Processes Handbook*. Gulf Professional Publishing, Elsevier, Amsterdam, Netherlands.

Pedley, J.F., Hiley, R.W., and Hancock, R.A. 1986. Storage Stability of Petroleum-Derived Diesel Fuel. 1. Analysis of Sediment Produced during the Ambient Storage of Diesel Fuel. *Fuel*, 66: 1646–1651.

Pedley, J.F., Hiley, R.W., and Hancock, R.A. 1988. Storage Stability of Petroleum-Derived Diesel Fuel. 3. Identification of Compounds Involved in Sediment Formation. *Fuel*, 67: 1124–1130.

Pedley, J.F., Hiley, R.W., and Hancock, R.A. 1989. Storage Stability of Petroleum-Derived Diesel Fuel. 4. Synthesis of Sediment Precursor Compounds and Simulation of Sediment Formation Using Model Systems. *Fuel*: 27–31.

Pfeiffer, J.P., and Saal, R.N. 1940. Asphaltic Bitumen as a Colloid System. *Journal of Physical Chemistry*, 44: 139–149.

Francisco, M.A., and Speight, J.G. 1984. Asphaltene Characterization by a Non-Spectroscopic Method. *Preprints Div. Fuel. Chem. Am. Chem. Soc.*, 29(1): 36.

Por, N. 1992. *Stability Properties of Petroleum Products*. Israel Institute of Petroleum and Energy, Tel Aviv, Israel.

Power, A.J., and Mathys, G.I. 1992. Characterization of Distillate Fuel Sediment Molecules: Functional Group Derivatization. *Fuel*, 71: 903–908.

Ruzicka, D.J., and Nordenson, S. 1990. *Fuel*, 69: 710–715.

Schramm, L.L. (Editor). 1992. *Emulsions: Fundamentals and Applications in the Petroleum Industry*, Advances in Chemistry Series No. 231. American Chemical Society, Washington, DC.

Speight, J.G. 1992. A Chemical and Physical Explanation of Incompatibility during Refining Operations. In: *Proceedings. 4th International Conference on the Stability and Handling of Liquid Fuels*. Volume Volume 1. US Department of Energy, Washington, DC. p. 169

Speight, J.G. 1994. Chemical and Physical Studies of Petroleum Asphaltenes. In: *Asphaltenes and Asphalts. I. Developments in Petroleum Science*. T.F. Yen and G.V. Chilingarian (Editors). Volume Volume 40. Elsevier, Amsterdam, Netherlands. Chapter 2

Speight, J.G. 2013. *The Chemistry and Technology of Coal*. 3rd Edition. CRC Press, Taylor & Francis Group, Boca Raton, Florida.

Speight, J.G. 2014. *The Chemistry and Technology of Petorleum*. 5th Edition. CRC Press, Taylor & Francis Group, Boca Raton, Florida.

Speight, J.G. 2015a. *Handbook of Petroleum Product Analysis*. 2nd Edition. John Wiley & Sons Inc., Hoboken, New Jersey.

Speight, J.G. 2015b. *Fouling in Refineries*. Gulf Professional Publishing Company, Elsevier, Oxford, United Kingdom.

Speight, J.G. 2017. *Handbook of Petroleum Refining*. CRC Press, Taylor & Francis Group, Boca Raton, Florida.

Speight, J.G. 2021. *Refinery Feedstocks*, Petroleum Refining Technology Series. Volume I. CRC Press, Taylor & Francis Group, Boca Raton, Florida.

Speight, J.G. 2023. Refinery Dewatering. In: *Desalting, and Distillation in Petroleum Refining*, Petroleum Refining Technology Series. Volume 2. CRC Press, Taylor & Francis Group, Boca Raton, Florida.

Stavinoha, L.L., and Henry, C.P. (Editors). 1981. *Distillate Fuel Stability and Cleanliness*. Special Technical Publication No. 751. American Society for Testing and Materials, Philadelphia, Pennsylvania.

Stavinoha, L.L., and Westbrook, S.R. 1980. Accelerated Stability Test Techniques for Diesel Fuels. US Department of Energy, Report No. DOE/BC/10043-12.

Wallace, T.J. 1964. In: *Advances in Petroleum Chemistry and Refining*. J.J. McKetta Jr. (Editor). Interscience, New York. p. 353.

Valyavin, G.G., Fryazinov, V.V., Ezhov, B.M., Sedov, P.S., Alekseev, P.M., Vetoshkin, N.I., and Khairudinov, I.R. 1976. Some Problems of Hydrodynamics of Reaction Coils in Delayed Coking Units. *Khimiya i Tekhnologiya Topliv i Masel*, 6: 42–44.

Wallace, T.J. 1969. In: *Advances in Petroleum Chemistry and Refining*. J.J. McKetta Jr. (Editor). Volume IX: Chapter 8. p. 353.

Watkinson, A.P., and Epstein, N. 1970. Particulate Fouling of Sensible Heat Exchangers. Proceedings. 4th International Heat Transfer Conference. Volume 1. Paper No. HE1.6.

Watkinson, A.P. 2005. Deposition from Crude Oils In Heat Exchangers. Proceedings. 6th International Conference on Heat Exchanger Fouling and Cleaning – Challenges and Opportunities. H. Müller-Steinhagen, M.R. Malayeri, and A.P. Watkinson (Editors). ECI Symposium Series. Volume RP2. Engineering Conferences International, Kloster Irsee, Germany, June 5–10.

Yeap, B.L., Wilson, D.I., Polley, G.T., and Pugh, S.J. 2004. Retrofitting Crude Oil Refinery Heat Exchangers Networks to Minimize Fouling while Maximizing Heat Recovery. Proceedings. 2003 ECI Conference on Heat Exchanger Fouling and Cleaning. P. Watkinson, H. Mueller-Steinhagen, and M.R. Malayeri (Editors). pp. 211–218.

12

Use of the Data

12.1 Introduction

When the data from the various analytical test methods and the data from any ancillary investigations have been collected, the next step relates to use of the data which, in turn, is subject to the selection of the appropriate test methods to reveal the properties of the feedstock. Since the early days of analysis, there has been a growing tendency to use the analytical data as a means of more detailed and accurate projections with a higher degree of predictability. And nowhere has this been more important that when the viscous feedstocks (heavy crude oil, extra heavy crude oil, and tar sand bitumen) entered the refinery scene.

However, in order to discuss the chemical and physical structure of feedstocks, it is necessary to give consideration to the chemical and physical nature of the constituents of the feedstocks. The chemical nature of the various constituents of feedstocks has been addressed elsewhere (Chapters 2 and 3).

It is the purpose of this chapter to address the issues of use of the analytical data as they affect recovery and refinery operations. Also included in this chapter is reference to the stability of the feedstock system and the recognition of those perturbations that influence this stability. The issues relating to the composition and structural types that occur in the gas oil and resin fractions is not addressed here since considerable detail has been given about the constituents of the various fractions in the feedstock sources (Altgelt and Boduszynski, 1994; Speight, 2014). Thus, it would be inappropriate to reproduce all of the relevant work here. This chapter is concerned with how the various constituents interact and coexist and there is only brief and necessary reference to the various chemical types in the fractions.

However, each process chemist, process engineer, and refiner has their own preferences and such efforts are feedstock dependent as well as being refinery dependent. There is also the need to recognize that what is adequate for one

Handbook of Heavy Oil Properties and Analysis, First Edition. James G. Speight.

refinery and one feedstock (or feedstock blend provided that the blend composition does not change significantly) will not be suitable for a different refinery with a different feedstock (or feedstock blend).

This chapter will merely serve as an introduction to feedstock mapping and the consequences of such efforts. In fact, this last chapter may be considered as an epilog that is not the end of the story but, in reality, the beginning of the beginning!

The data derived from any one, or more, of the test methods described in this text give an indication of the characteristics of the feedstock as well as options for feedstock processing as well as for the prediction of product properties. Other properties may also be required for more detailed evaluation of the feedstock and for comparison between feedstocks even though they may not play any role in dictating which refinery operations are necessary for processing.

However, proceeding from the raw evaluation data to full-scale production is not the preferred step. Further evaluation of feedstock processability is usually through the use of a pilot-scale operation followed by scale-up to a demonstration plant. It will then be possible to develop accurate relationships between the data obtained from the actual plant operations. After that, feedstock mapping can play an important role to assist in the various tweaks that are needed to maintain a healthy process.

Indeed, the use of physical properties for feedstock evaluation has continued in refineries and in process research laboratories to the present time and will continue for some time. It is, of course, a matter of choosing the relevant and meaningful properties to meet the nature of the task.

12.2 Use of the Data

The data derived from any one, or more of the evaluation techniques described here give an indication of the characteristics of as well as an indication of the means by which the crude feedstock should be processed to as well as for the prediction of product properties (Long and Speight, 1989a, 1989b; Speight, 2014). Other properties may also be required for more detailed evaluation of the feedstock and for comparison between feedstocks even though they may not play any role in dictating which refinery operations are necessary for processing.

However, to proceed from the raw evaluation data to full-scale production is not the preferred step; further evaluation of the processability of the feedstock is usually necessary through the use of a pilot-scale operation. To take the evaluation of a feedstock one step further, it may then be possible to develop correlations between the data obtained from the actual plant operations (as well as the pilot plant data) with one or more of the physical properties determined as part of the initial feedstock evaluation. It may also be necessary to use kinetic profiles developed from a variety of thermal experiments. Such methods are beyond the scope of this book and are not included here.

Early attempts at evaluating a feedstock involved the use of simple formulae based on the correlation of one of more physical properties with behavior (Speight, 2014). For example, the relationship of carbon residue to actual coke yield and the yield of other products was established for a variety of feedstocks (Chapter 7). This technique has been fine-tuned to accommodate the larger variety of feedstocks that are now available for refineries (Speight, 2014, 2015, 2017).

And it is not only the conventional physical properties that have been employed for feedstock evaluation. For example, evaluation of feedstocks from physical properties may also be achieved by suitable choice and adaptation of the physical properties intercept (Chapter 6). For example, if the refractive indices of the feedstocks are plotted against their respective density, straight lines of constant slope are obtained, one for each homologous series; the intercepts of these lines with the ordinate of the plot are characteristic, and the refractivity intercept is derived from the formula:

$$\text{Refractivity intercept} = n - (d/2)$$

The intercept cannot differentiate accurately among all series, which restricts the number of different types of compounds that can be recognized in a sample. The technique has been applied to nonaromatic olefin-free materials in the gasoline range by assuming additivity of the constant on a volume basis. Further manipulation of the formula and data allow derivation of an equation that is applicable to straight-run lubricating distillates if the material contains between 25% and 75% of the carbon present in naphthenic rings:

$$\text{Refractivity intercept} = 1.0502 - 0.00020\%C_N$$

Obviously, this may be of little use in the modern refinery due to the scarcity of straight-run lubricating distillates, especially those containing the requisite proportion of naphthenic carbon (Speight, 2014, 2015). Other formulae using the refractive index (RI) (Chapter 8) and properties such as specific gravity (SG) (Chapter 4), and molecular weight (MW) (Chapter 9), have also been developed. Indeed, the use of physical properties for feedstock evaluation has continued in refineries and in process research laboratories to the present time and will continue for some time. It is, of course, a matter of choosing the relevant and meaningful properties to meet the nature of the task.

However, the emphasis for feedstock evaluation has taken a turn in the direction of feedstock mapping. In such procedures, properties of feedstock are mapped to show characteristics that are in visual form rather than in tabular form. In this manner, the visual characteristics of the feedstock are used to evaluate and predict the behavior of the feedstock in various refining scenarios. Whether or not such methods will supersede the simpler form of property correlations remains to be determined. It is more than likely that both will continue to be used in a complimentary fashion for some time to come.

Refining, as it is currently known, will continue at least for the next three decades. As a consequence of this increase in use, crude oil (and the associated viscous feedstocks) will be the largest sources of carbon emissions into the environment. Acid gases corrode refining equipment, harm catalysts, pollute the atmosphere, and prevent the use of hydrocarbon components in petrochemical manufacture. When the amount of hydrogen sulfide is high, it may be removed from a gas stream and converted to sulfur or sulfuric acid. Some natural gases contain sufficient carbon dioxide to warrant recovery as dry ice. Thus, like any other raw material, crude oil and the viscous feedstocks are capable of producing chemical waste and effluent water, atmospheric emissions and combustion products also became a focus of increased technical attention (Carson and Mumford, 1988; Renzoni et al., 1994; Carson and Mumford, 1995; Edwards, 1995; Thibodeaux, 1995; Speight, 1996, 2019a).

12.3 Process Analysis and Feedstock Mapping

The objective of the "feedstock mapping" is to prepare a "map" that will serve to indicate the character and behavior of a feedstock during refining whether it is the behavior of a feedstock during a specific process or the oval refining operations. The so-called "map" can be a simple x-y diagram or a more complex multiparameter diagram as well as a 3-D representation of feedstock parameters. The preparation of the "map" is derived from the selected analytical data with the goal of analyzing (1) the preprocessing needs for the potential mix of feedstock types, (2) the processing needs for the potential mix of feedstock types, and (3) the outcome of the application of the processing options in the form of product quality and product yields.

Feedstock fractionation (Chapter 4) has played a significant role, along with the physical testing methods, of evaluating a refinery feedstock. For example, distillation being the simplest form of fractionation has been, and continued to be, used as a means of describing a feedstock in the form of a distillation profile (Speight, 2014, 2015). Early investigations showed that crude oil (and, by inference, the viscous feedstocks) did not contain all possible isomers of the various hydrocarbon derivatives but contained preferential isomers. In this very simple map, the actual constituents were shown again the theoretical constituents a line profile. This allowed refiners to be prepared for a more specific variety of straight-run hydrocarbon products than had previously been thought. However, the use of a distillation profile as it pertains to the viscous feedstocks is of limited use.

On the other hand, an early finding of composition studies was that the behavior and properties of any material are dictated by composition (Speight, 2014). Although the early studies were primarily focused on the composition and behavior of asphalt, the techniques developed for those investigations have provided an excellent means of studying heavy feedstocks (Altgelt and Boduszynski, 1994; Speight, 2014).

Thus, it is not surprising that one of the most effective means of feedstock mapping has arisen through the use of fractionation methods (Chapter 3) in which the feedstock is subdivided into several fractions by standard procedures. The simplest map that can be derived from such methods show the feedstock as a three-phase system based on asphaltene separation. Indeed, the asphaltenes themselves can be differentiated on the basis of the solubility parameter of the solvent or solvent used for the separation (Mitchell and Speight, 1973).

The whole feedstock can be mapped as a series of boxes equivalent to each fraction with the saturates fraction and the aromatics fraction combined as a single fraction or shown separately according to boiling range (Speight, 2014). In this manner, various feedstocks can be mapped and compared to the parent feedstock. Furthermore, by careful selection of an appropriate technique, it is possible to obtain a detailed overview of feedstock or product composition that can be further subdivided into specific compound types within the feedstock or on the basis of an envelope that is constructed from carbon number and boiling point that allows cut-points to be made that illustrate the relative positions of the atmospheric residuum and the vacuum residuum as well as very general (often not too accurate) predictions of the relative distillate and coke-forming propensity. Placing lines on such maps for asphaltenes and resin fraction are somewhat difficult because of the complexity of these fractions and the basis of polarity and molecular weight as the determinant for the resin and asphaltene fractions. Nevertheless, these maps are moving closer to use as a means of process predictability.

However, it must be recognized that such maps do not give any indication of the complex interactions that occur between, for example, such fractions as the asphaltenes and resins (Koots and Speight, 1975; Speight, 1994, 2014), but it does allow predictions of feedstock behavior. It must also be recognized that such a representation varies for different feedstocks.

Thus, by careful selection of an appropriate technique, it is possible to obtain an overview of feedstock composition that can be used for behavioral predictions. By taking the approach one step further and by assiduous collection of various subfractions, it becomes possible to develop the feedstock map and add an extra dimension to compositional studies. The viscous feedstocks then appear more as a continuum than as four specific fractions. For example, to continue the concept that a feedstock can be represented as a three-phase system, using various data such as composition or solubility parameter, the feedstock can be shown as a simple system that changes during the formation of an insoluble phase as thermal reactions proceed. Application of precise numbers to such a map allows more accurate predictions. There is even the possibility of adding a third dimension to such a map thereby enhancing predictability.

There has also been the development of a concept which involved the development of a different type of compositional map using the molecular weight distribution and the molecular type distribution as coordinates (Long and Speight, 1989a, 1989b). Using the composition and the molecular weight distribution, a map of composition can be prepared using molecular weight and solubility parameter as the coordinates for plotting the two distributions. Such a composition map can provide insights into many separation and conversion processes used in refining.

The molecular type was characterized by the polarity of the molecules, as measured by the increasing adsorption strength on an adsorbent. At the time of the original concept, it was unclear how to characterize the continuum in molecular type or polarity. For this reason, the molecular type coordinate of the initial maps was the yield of the molecular types ranked in order of increasing polarity.

In order to attempt to remove some of these potential ambiguities, more recent developments of this concept have focused on the solubility parameter. The simplest map that can be derived using the solubility parameter is produced with the solubility parameters of the solvents used in solvent separation procedures, and equating these parameters to the various fractions. However, the solubility parameter boundaries determined by the values for the eluting solvents that remove the fractions from the adsorbent offer a further step in the evolution of feedstock maps.

12.3.1 Property Predictions

Although there are many maps that can be derived from the fractionation of a feedstock, for convenience, only selected maps will be used.

A composition map can be used to show where a particular physical or chemical property tends to concentrate on the map. For example, the coke-forming propensity, that is, the amount of the carbon residue, can be illustrated for various regions on the map for a sample of atmospheric residuum in which molecular weight is plotted against weight percent yield in order of increasing polarity.

12.3.2 Predicting Separations

A composition map can be very useful for predicting the effectiveness of various types of separation processes as applied to a refinery feedstock. These processes are adsorption, distillation, solvent precipitation with relatively nonpolar solvents, and solvent extraction with polar solvents.

12.3.3 Process Predictability

In contrast to the lines generated by separation processes, conversion processes move materials in the composition from one molecular type to another.

For example, reforming converts saturates to aromatics and hydrogenation converts aromatic constituents to saturate constituents and polar aromatic constituents to either aromatic nonpolar constituents or saturate constituents.

The ultimate decision in the choice of any particular fractionation technique must be influenced by the need for the data. For example, there are those needs that require only that the crude oil be separated into four bulk fractions. On the other hand, there may be the need to separate the crude oil into many subfractions in order to define specific compound types. Each method is used to answer the relevant questions about the character of the feedstock.

12.4 Environmental Aspects of Processing

In addition to the conventional meaning of the term process, the transportation and storage of the various feedstocks and products also need to be considered here in respect of any environmental issues. In fact, this is the area where knowledge of the elemental composition (Chapters 3 and 5) of the spilled sample can be of great benefit. Elemental analysis can be qualitative (determining what elements are present), and it can be quantitative (determining how much of each is present) and, therefore, the amount of contaminant that can be predicted.

Oil spills during transportation from the source to the refinery (and even spills in the refinery) have been the most visible problem. There have also been instances of oil wells at sea "blowing out" (flowing uncontrollably) although the amounts from blowouts tend to be smaller than from tanker accidents. The 1979 Ixtoc I blowout in the Gulf of Mexico was an exception, as it flowed an estimated three million barrels over many months.

Tanker accidents typically have a severe impact on ecosystems because of the rapid release of hundreds of thousands of barrels of crude oil (or crude oil products or viscous feedstocks) into a small area. One of the largest single spills that occurred in the past occurred during the 1991 Gulf War, when as much as 10 million barrels were dumped in the Persian Gulf by Iraq, apparently intentionally. More typical was the 1989 spill from the tanker Exxon Valdez, where 250,000 barrels were lost in Alaskan coastal waters.

Refining is a complex sequence of chemical events that result in the production of a variety of products (Parkash, 2003; Gary et al., 2007; Speight, 2014; Hsu and Robinson, 2017; Speight, 2017). In fact, feedstock refining might be considered as a collection of individual, yet related processes that are each capable of producing effluent streams.

Many refined products came under scrutiny and by the mid-1970s, refiners in the United States were required to develop techniques for manufacturing high-quality gasoline without employing lead additives, and by 1990, they were required to take on substantial investments in the complete reformulation of transportation fuels in order to minimize environmental emissions.

Processing any feedstock, with the exception of some of the more viscous crude oils, involves a primary distillation of the hydrogen mixture, which results in its separation into fractions differing in carbon number, volatility, specific gravity, and other characteristics. The most volatile fraction, which contains most of the gases which are generally dissolved in the crude, is referred to as pipestill gas or pipestill light ends and consists essentially of hydrocarbon gases ranging from methane to butane(s) (C_4H_{10}), or sometimes pentane(s) (C_5H_{12}). The gas varies in composition and volume, depending on crude origin and on any additions to the crude made at the loading point.

Analytical techniques for chemical analysis of crude oil and the viscous feedstocks as well as the transformation products formed by photochemical processes and biological processes is a necessary next step after a spill. In addition to the standard test method present elsewhere in this text, the spilled material can be investigated by advanced gas chromatography (GC) techniques such as comprehensive two-dimensional GC (GC–GC), pyrolysis GC with mass spectrometry (MS), and GC with tandem mass spectrometry (GC–MS/MS) which will provide a greater understanding at the molecular level of composition and complexity of the spilled material and any post-spill changes.

In addition, the capabilities of high-resolution mass spectrometry (HRMS) can be utilized to extend the analytical characterization window beyond conventional gas chromatographic-based methods to include polar and high molecular mass components (where the molecular mass or molecular weight is in excess of 400) and to provide new opportunities for discovery, characterization, and investigation of photo-oxidation products and the biotransformation products.

The application of separation methods to reduce the complexity of the spilled material and the weathered material prior to the application of high-resolution mass spectrometry and advanced fluorescence spectrometry can increase the information available on the spilled material and the various transformation products.

In addition, it is not uncommon to reinject low boiling hydrocarbon derivatives, such as propane and butane, into a crude oil at the recovery site before transportation by pipeline or any other form of transportation. This results in a higher vapor pressure of the crude oil and the viscous feedstocks which allows an increase in the quantity of low boiling products obtained at the refinery. However, the field personnel and the shipper must add the low boiling hydrocarbon derivatives with caution because of the tendency of propane and butane to cause separation (precipitation) of resin and asphaltene and constituents immediately or after an induction period.

Thermal processes are commonly used to convert residua and the viscous feedstocks (heavy crude oil, extra heavy crude oil, and tar sand bitumen) into liquid products. Therefore, some indications of the process classes and the products that are unacceptable to the environment is warranted here.

Thus, examples of modern thermal processes are visbreaking and coking (delayed coking, fluid coking, and flexicoking). In all of these processes, the simultaneous formation of sediment or coke limits the conversion to usable liquid products.

Thermal cracking processes are commonly used to convert nonvolatile residua into volatile products, although thermal cracking processes as used in the early refineries are no longer in use. Examples of modern thermal cracking processes are visbreaking and coking (delayed coking, fluid coking, and flexicoking). In all of these processes, the simultaneous formation of sediment or coke limits the conversion to usable liquid products (Parkash, 2003; Gary et al., 2007; Speight, 2014; Hsu and Robinson, 2017; Speight, 2017).

The visbreaking process is primarily a means of reducing the viscosity of heavy feedstocks by controlled thermal decomposition insofar as the hot products are quenched before complete conversion can occur. However, the process is often plagued by sediment formation in the products. This sediment, or sludge, must be removed if the products are to meet fuel oil specifications.

Coking, as the term is used in the industry, is a process for converting nondistillable fractions (residua) of crude oil to lower boiling products and coke. Coking is often used in preference to catalytic cracking because of the presence of metals and nitrogen components that poison catalysts.

Delayed coking is the oldest, most widely used process and has changed very little in the five or more decades in which it has been onstream in refineries. Fluid coking is a continuous fluidized solids process that cracks feed thermally over heated coke particles in a reactor vessel to gas, liquid products, and coke. Heat for the process is supplied by partial combustion of the coke, with the remaining coke being drawn as product. The new coke is deposited in a thin fresh layer on the outside surface of the circulating coke particle (Parkash, 2003; Gary et al., 2007; Speight, 2014; Hsu and Robinson, 2017; Speight, 2017).

Catalytic cracking is a conversion process that can be applied to a variety of feedstocks ranging from gas oil to heavy oil. It is one of several practical applications used in a refinery that employ a catalyst to improve process efficiency. Hydrotreating is defined as the lower temperature removal of heteroatomic species by treatment of a feedstock or product in the presence of hydrogen. Hydrocracking is the thermal decomposition of a feedstock in which carbon–carbon bonds are cleaved in addition to the removal of heteroatomic species. Hydrogen is present to prevent the formation of coke.

Subsequent hydroprocessing of the coker distillates would reduce the polynuclear aromatic hydrocarbon derivatives in the resulting product streams, so that the only health concern outside the refinery itself is with high-severity thermal products, such as pitches, which have not been hydrotreated. Coke solids would not pose a health hazard, and would have less environmental activity than unprocessed residue.

Catalytic cracking of crude oil occurs over many types of catalytic materials that may be either activated (acid-treated natural clays of the bentonite type) or synthesized silica-alumina or silica-magnesia preparations.

It is to be remembered that the impact of an oil spill is determined by (1) the size of the spill, (2) the constituents of the spilled material, (3) the degree of dispersal, and (4) the chemical effects of the spilled material. Spills at sea are thought to have a less detrimental effect than spills in shallow waters. But what is more important is the ability to determine the nature of the contaminants through the use of recognized tests with well-defined protocols (EPA, 1986, 1992; Speight, 2005; ASTM, 2021).

While many of the constituents of crude oil, and to some extent the constituents of the viscous feedstocks, are (at least) theoretically biodegradable, large-scale spill of these materials can overwhelm the ability of an ecosystem to biodegrade the constituents. Also, over time the lower boiling (more volatile) constituents of a refinery feedstock have a tendency to evaporate thereby leaving the nonvolatile portion of the feedstock in the ecosystem. Some constituents of a feedstocks (depending upon the type of feedstock) may degrade the protective waxes and oils in the feathers of birds and in the fur of animals which can result in a loss of heat retention and cause death (of the bird or animal) by freezing. In addition, ingestion of the oil can also kill animals by interfering with the ability of an animal to digest food. Some crude oils contain toxic metals as well which can be detrimental to the flora and fauna of an ecosystem.

12.4.1 Gaseous Emissions

The terms refinery gas and process gas are also often used to include all of the gaseous products and by-products that emanate from a variety of refinery processes (Speight, 1996). There are also components of the gaseous products that must be removed prior to release of the gases to the atmosphere or prior to use of the gas in another part of the refinery, that is, as a fuel gas or as a process feedstock.

Both the production and processing of crude oil involve the use of a variety of substances, some toxic, including lubricants in oil wells and catalysts and other chemicals in refining. The amounts used, however, tend to be small and relatively easy to control. More detrimental to the environment is the spillage of oil, which has been a particularly common event. Minor losses from truck and car accidents can affect rivers and streams. Leakage from underground gasoline storage tanks, many abandoned decades ago, has contaminated some local water supplies and usually requires expensive operations either to clean or seal off.

Refining (of any feedstock) typically produces gas streams that often contain substantial amounts of acid gases such as hydrogen sulfide and carbon dioxide. More particularly, hydrogen sulfide arises from the hydrodesulfurization of feedstocks that contain organic sulfur:

$$[S]_{feedstock} + H_2 = H_2S + \text{hydrocarbon derivatives}$$

Refining involves, with the exception of some of the more viscous crude oils, a primary distillation of the hydrogen mixture, which results in its separation into fractions differing in carbon number, volatility, specific gravity, and other characteristics. The most volatile fraction, which contains most of the gases that are generally dissolved in the crude, is referred to as pipe still gas or pipe still light ends and consists essentially of hydrocarbon gases ranging from methane to butane(s), or sometimes pentane(s).

In addition to the gases obtained by distillation of crude oil, more highly volatile products result from the subsequent processing of naphtha and middle distillate to produce gasoline (Parkash, 2003; Gary et al., 2007; Speight, 2014; Hsu and Robinson, 2017; Speight, 2017). Hydrogen sulfide is produced in the desulfurization processes involving hydrogen treatment of naphtha, distillate, and residual fuel, and from the coking or similar thermal treatments of vacuum gas oils and residual fuels. The most common processing step in the production of gasoline is the catalytic reforming of hydrocarbon fractions in the heptane (C_7) to decane (C_{10}) range.

Refinery and natural gas streams, which originate from the processing of crude oil and the viscous feedstocks, may contain large amounts of acid gases, such as hydrogen sulfide (H_2S) and carbon dioxide (CO_2) (Speight, 1996, 2005, 2014, 2017, 2019a). Hydrogen chloride (HCl), although not usually considered to be a major pollutant in many crude oil refineries, can arise during processing from the presence of mineral chloride derivatives in the brine which may accompany an incompletely dried feedstock into the refinery (Speight, 2023). Hydrogen chloride can also be produced from mineral matter and other inorganic contaminants during preprocessing and even in the distillation section of the refinery and this particular contaminant is gaining increasing recognition as a pollutant which needs serious attention.

Acid gases, such as hydrogen chloride and hydrogen sulfide, corrode refining equipment, harm catalysts, pollute the atmosphere, and prevent the use of hydrocarbon feedstocks in the manufacture of petrochemicals. When the amount of hydrogen sulfide is large, it may be removed from a gas stream and converted to sulfur or sulfuric acid. Some natural gas streams contain sufficient carbon dioxide to warrant recovery as dry ice, that is, solid carbon dioxide. And there is now a conscientious effort to mitigate the emission of pollutants from hydrotreating process by careful selection of process parameters and catalysts (Occelli and Chianelli, 1996).

In a series of processes commercialized under the generic name reforming, paraffin derivatives and naphthene derivatives (i.e., cyclic nonaromatic hydrocarbon derivatives) are altered structurally in the presence of hydrogen and a catalyst into aromatic derivatives or are isomerized to more highly branched hydrocarbon derivatives. These reforming processes also produce varying quantities of gases during the formation of a liquid product of higher octane number. These gas streams are rich in hydrogen and also contain hydrocarbon derivatives from methane to butane, with a preponderance of propane ($CH_3CH_2CH_3$), *n*-butane ($CH_3CH_2CH_2CH_3$), and *iso*-butane [$(CH_3)_3CH$].

The composition of the process gases varies in accordance with reforming severity and reformer feedstock. All catalytic reforming processes require substantial recycling of a hydrogen stream (Parkash, 2003; Gary et al., 2007; Speight, 2014; Hsu and Robinson, 2017; Speight, 2017, 2019b). Therefore, it is normal to separate reformer gas into a propane ($CH_3CH_2CH_3$) and/or a butane stream [$CH_3CH_2CH_2CH_3$ plus $(CH_3)_3CH$], which becomes part of the refinery liquefied petroleum gas production, and a lighter gas fraction, part of which is recycled. In view of the excess of hydrogen in the gas, all products of catalytic reforming are saturated, and there are usually no olefinic gases present in either gas stream.

A second group of refining operations that contributes to gas production is that of the catalytic cracking processes (Parkash, 2003; Gary et al., 2007; Speight, 2014; Hsu and Robinson, 2017; Speight, 2017) in which heavy gas oils are converted into gas, liquefied petroleum gas, catalytic naphtha, fuel oil, and coke by contacting the high boiling feedstocks with the hot catalyst using the fluid-bed catalytic cracking concept.

Both catalytic and thermal cracking processes, the latter being now largely used for the production of chemical raw materials, result in the formation of unsaturated hydrocarbon derivatives, particularly ethylene ($CH_2{=}CH_2$), but also propylene (propene, $CH_3CH{=}CH_2$), *iso*-butylene [*iso*-butene, $(CH_3)_2C{=}CH_2$], and the *n*-butenes ($CH_3.CH_2.CH{=}CH_2$, and $CH_3CH{=}CHCH_3$) in addition to hydrogen (H_2), methane (CH_4), and smaller quantities of ethane (CH_3CH_3), propane ($CH_3CH_2CH_3$), and butane isomers [$CH_3CH_2CH_2CH_3$, $(CH_3)_3CH$]. Diolefins such as butadiene ($CH_2{=}CHCH{=}CH_2$) are also present.

Additional gases are produced in refineries with coking or visbreaking facilities for the processing of their heaviest crude fractions. In the visbreaking process, fuel oil is passed through externally fired tubes and undergoes liquid-phase cracking reactions, which result in the formation of lighter fuel oil components. Oil viscosity is thereby reduced, and some gases, mainly hydrogen, methane, and ethane, are formed. Substantial quantities of both gas and carbon are also formed in coking (both fluid coking and delayed coking) in addition to the middle distillate and naphtha. When coking a residual fuel oil or heavy gas oil, the feedstock is preheated and contacted with hot carbon (coke) which causes extensive cracking of the feedstock constituents of higher molecular weight to produce lower molecular weight products ranging from methane, via liquefied petroleum gas(es) and naphtha, to gas oil and heating oil. Products from coking processes tend to be unsaturated and olefinic components predominate in the tail gases from coking processes.

A further source of refinery gas is hydrocracking, a catalytic high-pressure pyrolysis process in the presence of fresh and recycled hydrogen. The feedstock is again heavy gas oil or residual fuel oil, and the process is directed mainly at the production of additional middle distillates and gasoline. Since hydrogen is to be recycled, the gases produced in this process again have to be separated into lighter and heavier streams.

Gases from hydrocracking units and from catalytic reforming units are commonly used in catalytic desulfurization processes. In the latter, feedstocks ranging from light to vacuum gas oils are passed at pressures of 500–1,000 psi ($3.5–7.0 \times 10^3$ kPa) with hydrogen over a hydrofining catalyst. This results mainly in the conversion of organic sulfur compounds to hydrogen sulfide and also produces some low boiling hydrocarbon derivatives by hydrocracking:

$$\left[S\right]_{\text{feedstock}} + H_2 = H_2S + \text{hydrocarbon derivatives}$$

Thus, refinery streams, while ostensibly being hydrocarbon in nature, may contain large amounts of acid gases such as hydrogen sulfide and carbon dioxide. Most commercial plants employ hydrogenation to convert organic sulfur compounds into hydrogen sulfide. Hydrogenation is effected by means of recycled hydrogen-containing gases or external hydrogen over a nickel molybdate or cobalt molybdate catalyst.

In summary, refinery process gas, in addition to hydrocarbon derivatives, may contain other contaminants, such as carbon oxides (CO_x, where $x = 1$ and/or 2), sulfur oxides (SO_x, where $x = 2$ and/or 3), as well as ammonia (NH_3), mercaptans (R–SH), and carbonyl sulfide (COS).

The presence of these impurities may eliminate some of the sweetening processes, since some processes remove large amounts of acid gas but not to a sufficiently low concentration. On the other hand, there are those processes not designed to remove (or incapable of removing) large amounts of acid gases whereas they are capable of removing the acid gas impurities to very low levels when the acid gases are present only in low-to-medium concentration in the gas.

From an environmental viewpoint, it is not the means by which these gases can be utilized but it is the effects of these gases on the environment when they are introduced into the atmosphere. For example, in addition to the corrosion of equipment of acid gases, the escape into the atmosphere of sulfur-containing gases can eventually lead to the formation of the constituents of acid rain, that is, the oxides of sulfur (SO_2 and SO_3). Similarly, the nitrogen-containing gases can also lead to nitrous and nitric acids (through the formation of the oxides NO_x, where $x = 1$ or 2) which are the other major contributors to acid rain. The release of carbon dioxide and hydrocarbon derivatives as constituents of refinery effluents can also influence the behavior and integrity of the ozone layer.

Hydrogen chloride, if produced during refining, quickly picks up moisture in the atmosphere to form droplets of hydrochloric acid and, like sulfur dioxide, is a contributor to acid rain. However, hydrogen chloride may exert severe local effects because, unlike sulfur dioxide, it does not need to participate in any further chemical reaction to become an acid. Under atmospheric conditions that favor a buildup of stack emissions in the area of a large industrial complex or power plant, the amount of hydrochloric acid in rainwater could be quite high.

Natural gas is also capable of producing emissions that are detrimental to the environment. While the major constituent of natural gas is methane, there are components such as carbon dioxide (CO), hydrogen sulfide (H_2S), and mercaptans (thiols; R–SH), as well as trace amounts of sundry other emissions. The fact that methane has a foreseen and valuable end-use makes it a desirable product, but in several other situations, it is considered a pollutant, having been identified a greenhouse gas.

A sulfur removal process must be very precise, since natural gas contains only a small quantity of sulfur-containing compounds that must be reduced several orders of magnitude. Most consumers of natural gas require less than 4 ppm in the gas.

A characteristic feature of natural gas that contains hydrogen sulfide is the presence of carbon dioxide (generally in the range of 1–4% v/v). In cases where the natural gas does not contain hydrogen sulfide, there may also be a relative lack of carbon dioxide.

Acid rain occurs when the oxides of nitrogen and sulfur that are released to the atmosphere during the combustion of fossil fuels are deposited (as soluble acids) with rainfall, usually at some location remote from the source of the emissions.

It is generally believed (the chemical thermodynamics are favorable) that acidic compounds are formed when sulfur dioxide and nitrogen oxide emissions are released from tall industrial stacks. Gases such as sulfur oxides (usually sulfur dioxide, SO_2) as well as the nitrogen oxides (NO_x) react with the water in the atmosphere to form acids:

$$SO_2 + H_2O \rightarrow H_2SO_3$$
$$2SO_2 + O_2 \rightarrow 2SO_3$$
$$SO_3 + H_2O \rightarrow H_2SO_4$$
$$2NO + H_2O \rightarrow 2HNO_2$$
$$2NO + O_2 \rightarrow 2NO_2$$
$$NO_2 + H_2O \rightarrow HNO_3$$

Acid rain has a $pH < 5.0$ and predominantly consists of sulfuric acid (H_2SO_4) and nitric acid (HNO_3). As a point of reference, in the absence of anthropogenic pollution sources, the average pH of rain is approximately 6.0 (slightly acidic; neutral $pH = 7.0$). In summary, the sulfur dioxide that is produced during a variety of processes will react with oxygen and water in the atmosphere to yield environmentally detrimental sulfuric acid. Similarly, nitrogen oxides will also react to produce nitric acid.

Another acid gas, hydrogen chloride (HCl), although not usually considered to be a major emission, is produced from mineral matter and the brines that often accompany crude oil during production and is gaining increasing recognition as a contributor to acid rain. However, hydrogen chloride may exert severe local effects because it does not need to participate in any further chemical reaction to become an acid. Under atmospheric conditions that favor a buildup of stack emissions in the areas where hydrogen chloride is produced, the amount of hydrochloric acid in rainwater could be quite high.

In addition to hydrogen sulfide and carbon dioxide, gas may contain other contaminants, such as mercaptans (R–SH) and carbonyl sulfide. The presence of these impurities may eliminate some of the sweetening processes since some processes remove large amounts of acid gas but not to a sufficiently low concentration. On the other hand, there are those processes that are not designed to remove (or are incapable of removing) large amounts of acid gases. However, these processes are also capable of removing the acid gas impurities to very low levels when the acid gases are there in low to medium concentrations in the gas.

Emissions from refining create a number of environmental problems. During combustion, the combination of hydrocarbon derivatives, nitrogen oxide, and sunlight results in localized low levels of ozone or smog. This is particularly evident in large urban areas and especially when air does not circulate well.

On a regional level, the emission of sulfur oxides (SO_x) and nitrogen oxides (NO_x) can also cause the formation of acid species at high altitudes, which eventually precipitate in the form of acid rain, damaging plants, wildlife, and property.

Most refinery are low in sulfur or are desulfurized, and while natural gas sometimes includes sulfur as a contaminant, it is typically removed at the production site.

At the global level, there is concern that the increased use of hydrocarbon-based fuels will ultimately raise the temperature of the planet (global warming), as carbon dioxide reflects the infrared or thermal emissions from the earth, preventing them from escaping into space (greenhouse effect). Whether or not the potential for global warming becomes real will depend upon how emissions into the atmosphere are handled. There is considerable discussion about the merits and demerits of the global warming theory (or global climate change as it is now called) and the discussion is likely to continue for some time (Speight, 2020). Be that as it may, the atmosphere can only tolerate pollutants up to a limiting value. And that value needs to be determined. In the meantime, efforts must be made to curtail the use of noxious and foreign (nonindigenous) materials into the air.

In summary, and from an environmental viewpoint, natural gas processing and crude oil refining can result in similar, if not the same, gaseous emissions as coal (Speight, 2013, 2014, 2019a). It is a question of degree insofar as the composition of the gaseous emissions may vary from coal to crude oil but the constituents are, in the majority of cases, the same.

There are a variety of processes which are designed for sulfur dioxide removal from gas streams but scrubbing process utilizing limestone ($CaCO_3$) or lime [$Ca(OH)_2$] slurries have received more attention than other gas scrubbing processes.

The majority of the gas scrubbing processes are designed to remove sulfur dioxide from the gas streams; some processes show the potential for removal of nitrogen oxide(s). It is only by understanding the characteristics of the feedstock can the mitigation of these gases be achieved and the environment protected.

12.4.2 Liquid Effluents

Crude oil, as a mixture of hydrocarbon derivatives, is (theoretically) a biodegradable material. However, in very general terms (and as observed from elemental analyses), crude oil is a mixture of: (1) hydrocarbon derivatives, (2) nitrogen-containing compounds, (3) oxygen-containing compounds, (4) sulfur-containing compounds, and (5) metal-containing constituents, including inorganic salts in colloidal suspension. However, this general definition is not adequate to describe the composition of petroleum as it relates to the behavior of these feedstocks. However, the viscous feedstocks which contain higher amounts categories (2)–(5), that is, the nitrogen-containing compounds, the oxygen-containing compounds, the sulfur-containing compounds; and the metal-containing constituents are less amenable to removal through biodegradation and can survive for prolonged periods when spilled into an ecosystem.

It is convenient to divide the hydrocarbon components of petroleum into the following three classes: (1) paraffin derivatives, which are saturated hydrocarbon derivatives with straight or branched chains, but without any ring structure; (2) naphthene derivatives, more correctly known as alicyclic hydrocarbon derivatives, which are saturated hydrocarbon derivatives that contain one or more rings, each of which may have one or more paraffinic side chains; and (3) aromatic derivatives, which are hydrocarbon derivatives that contain one or more aromatic nuclei, such as benzene, naphthalene, and phenanthrene, which may be linked up with (substituted) naphthene rings and/or paraffinic side chains.

Furthermore, even though crude oil derivatives have been prescribed for medicinal purposes, it would be a misinterpretation to expect the flora and fauna of the earth to survive in a crude oil (or viscous feedstock) environment (Chang et al., 2014).

Although the concentration in certain fractions may be quite small, the influence of the heteroatom constituents is significant. For example, the thermal decomposition of deposited inorganic chlorides with evolution of free hydrochloric acid can give rise to serious corrosion problems in the distillation equipment. The presence of organic acid components, such as mercaptan derivatives (often referred to as thiol derivative, RSH, where R is an alkyl or aromatic moiety) and acids (RCO_2H, where R is an alkyl or aromatic moiety), can also promote environmental damage. In catalytic operations, passivation and/or poisoning of the catalyst can be caused by deposition of traces of metals (vanadium and nickel) or by chemisorption of nitrogen-containing compounds on the catalyst, thus necessitating the frequent regeneration of the catalyst or its expensive replacement. This carries with it the issues related to catalyst disposal.

Thermal processing can significantly increase the concentration of polynuclear aromatic hydrocarbon derivatives in the product liquids, because the low-pressure hydrogen-deficient conditions favor aromatization of naphthenes and condensation of aromatics to form larger ring systems. To the extent that more compounds like benzo(a)pyrene are produced, the liquids from thermal processes will be more carcinogenic than asphalt. This biological activity was consistent with the higher concentration of polynuclear aromatic hydrocarbon derivatives at 38.8 mg/g in the pitch compared to only 0.22 mg/g in the asphalt. Similarly, one would expect coker gas oils to contain more polynuclear aromatic hydrocarbon derivatives than unprocessed or hydroprocessed distillates, and thereby give a higher potential for carcinogenic or mutagenic effects.

The sludge produced on acid treatment of crude oil distillates is complex in nature, for example, (1) ester derivatives and alcohol derivatives are present from reactions with olefin derivatives; (2) sulfonation products from reactions with aromatics, naphthenes, and phenols; and (3) salts from reactions with nitrogen bases. In addition, such materials as naphthenic acids, sulfur compounds, and asphaltic material are all retained by direct solution. To these constituents must be added the various products of oxidation–reduction reactions: coagulated resins, soluble hydrocarbon derivatives, water, and free acid.

The disposal of the sludge is difficult, as it contains unused free acid that must be removed by dilution and settling. The disposal is a comparatively simple process for the sludge resulting from treating gasoline and kerosene, the so-called light oils. The insoluble oil phase separates out as a mobile tar, which can be mixed and burned without too much difficulty. Sludge from heavier oils, however, separate out granular semisolids, which offer considerable difficulty in handling.

In all cases, careful separation of reaction products is important to the recovery of well-refined materials. This may not be easy if the temperature has risen as a consequence of chemical reaction. This will result in a persistent dark color traceable to colloidally distributed reaction products. Separation may also be difficult at low temperature because of high viscosity of the stock, but this problem can be overcome by dilution with light naphtha or with propane.

In addition, the delayed coking process also requires the use of large volumes of water for hydraulic cleaning of the coke drum. However, the process water can be recycled if the oil is removed by skimming and suspended coke particles are removed by filtration. If this water is used in a closed cycle, the impact of delayed coking on water treatment facilities and the environment is minimized. The flexicoking process offers one alternative to direct combustion of coke for process fuel. The gasification section is used to process excess coke to a mixture of carbon monoxide (CO), carbon dioxide (CO_2), hydrogen (H_2), and hydrogen sulfide (H_2S) followed by treatment to remove the hydrogen sulfide. Maximizing the residue conversion and desulfurization of the residue in upstream hydroconversion units also maximizes the yield of hydrogen sulfide relative to sulfur in the coke. Currently, maximum residue conversion with minimum coke production is favored over gasification of coke.

12.4.3 Solid Effluents

Catalyst disposal is therefore a major concern in all refineries. In many cases, the catalysts are regenerated at the refinery for repeated use. Disposal of spent catalysts is usually part of an agreement with the catalysts manufacturer whereby the spent catalyst is returned for treatment and remanufacture.

The formation of considerable quantities of coke in the coking processes is a cause for concern since it not only reduces the yield of liquid products but also initiates the necessity for disposal of the coke. Stockpiling to coke may be a partial answer unless the coke contains leachable materials that will endanger the ecosystem as a result of rain or snow melt.

In addition, the generation and emission of sulfur oxides (particularly sulfur dioxide) from combustion of sulfur-containing coke as plant fuel. Sulfur dioxide (SO_2) has a wide range of effects on health and on the environment. These effects vary from bronchial irritation upon short-term exposure to contributing to the acidification of lakes. Therefore, it should be no surprise that the emissions of sulfur dioxide are regulated in many countries.

12.5 Analytical Methods for Environmental Regulations

Pollution has been obvious for a long time, although the effects were not as well realized in the past. There was copper pollution near Jericho on the west bank of the Jordan river due to copper smelting for tool manufacture thousands of years ago. Deforestation of many areas near the Mediterranean Sea for the building of ships was a norm. Poor agricultural methods led to soil erosion. In 2500 BC, the Sumerians used sulfur compounds to manage insects and in 1500 BC, the Chinese used natural products to fumigate crops. Pesticides began polluting the environment hundreds of years ago. Pollution in medieval England was also noted when the smoke from coal fires made some cities almost unhabitable (Speight, 2014, 2019a). There are also documented records, courtesy of the diarist Samuel Pepys, who noted that he did not realize that sewage from a neighbor's house was leaking into his basement until he (Pepys) descended to the lower level and stood in it. This is not only a commentary on the state of sewage disposal but also on the odors that must have permeated London in the 17th century.

As fossil fuels, crude oil and the viscous feedstocks can contribute to polluting emissions that can be in the form of gases, liquids, and solids which can cause damage to the flora and fauna of an ecosystem. While mapping may provide some insights into the behavior of material spilled into the environment, it is often more beneficial to understand the elemental composition and the chemical composition of the material before spillage and this is where a suite of meaningful analysis test methods should be applied to the material in anticipation that a spill might occur at some time in the history of the material. To many observers, this may prove more meaningful that the construction of a "composition map" and it is the purpose of this section to present the various analytical issues as they are related to the environment. In fact, in any text relating to the analysis of refinery feedstocks and products and the effects of these material on an ecosystem, it would be remiss if the analytical methods that are related to environmental analysis was not presented.

Crude oil and the viscous feedstocks are necessary parts of the modern refinery and, hence, there is the need for stringent controls over the amounts and types of emissions from the use of crude oil and the viscous feedstocks and their respective products. So it is predictable that crude oil and the associated viscous feedstocks will be primary sources of energy for the next several decades and, therefore, the message is clear. The challenge is for the development of technological concepts that will provide the maximum recovery of energy from the various feedstocks not only cheaply but also efficiently and with minimal detriment to the environment.

As a result, the industry is facing three major challenges which are (1) the acceptance of viscous feedstocks by refineries; (2) higher amounts of sulfur, nitrogen, and metals in the viscous feedstocks; and (3) stricter pollution control regulations.

In order to combat any threat to the environment, it is necessary to understand the nature and magnitude of the problems involved (Ray and Guzzo, 1990). It is in such situations that environmental technology has a major role to play. Environmental issues even arise when outdated laws are taken to task. Thus, the concept of what seemed to be a good idea at the time the action occurred no longer holds when the law influences the environment.

Both the production and processing of crude oil and the viscous feedstocks involve the use of a variety of substances, some toxic, including lubricants in oil wells and catalysts and other chemicals in refining (Parkash, 2003; Gary et al., 2007; Speight, 2014; Hsu and Robinson, 2017; Speight, 2017). The amounts used, however, tend to be small and relatively easy to control. More detrimental to the environment is the spillage of oil, which has been a particularly common event. Minor losses from truck and car accidents can affect rivers and streams. Leakage from underground gasoline storage tanks, many abandoned decades ago, has contaminated some local water supplies and usually requires expensive operations either to clean or seal off.

The complicated operations of environmental testing grow directly out of the laws and regulations that have been passed in the United States over the past quarter century. In earlier decades of the 20th century, independent environmental laboratories existed primarily for the purpose of monitoring public drinking water for bacteriological contamination. This is still an important function for many laboratories, But today, large independent environmental testing facilities analyze all media that can carry chemical contamination to plants, animals, and humans: potable water, air, soil, hazardous materials and waste, dredge spoils, industrial and municipal wastewater. Federal and state laws require the use of expensive, sensitive, and complex analytical instruments, and dictate differing sets of procedures and protocols to document and control each step of the analytical process under various regulatory programs.

The purpose of this section is to summarize and generalize the various pollution, health, and environmental problems especially specific to the refining industry and to place in perspective government laws and regulations as well as industry efforts to control these problems (Majumdar, 1993; Speight, 1996, 2019a). The objective is to indicate the types of emissions from refinery processes and the laws that regulate these emissions and to present the testing methods by which compliance is monitored.

12.5.1 Definitions

Briefly, refining produces chemical waste. If this chemical waste is not processed in a timely manner, it can become a pollutant.

A pollutant is a substance present in a particular location (ecosystem) when it is not indigenous to the location or is present in a greater-than-natural concentration. The substance is often the product of human activity. The pollutant, by virtue of its name, has a detrimental effect on the environment, in part or in toto. Pollutants can also be subdivided into two classes: primary and secondary.

$$\text{Source} \rightarrow \text{Primary pollutant} \rightarrow \text{Secondary pollutant}$$

A primary pollutant is a pollutant that is emitted directly from the source. In terms of atmospheric pollutants, examples are carbon oxides, sulfur dioxide, and nitrogen oxides from fuel combustion operations:

$$\begin{aligned}
&2[C]_{\text{feedstock}} + O_2 \rightarrow 2CO\\
&[C]_{\text{feedstock}} + O_2 \rightarrow CO_2\\
&2[N]_{\text{feedstock}} + O_2 \rightarrow 2NO\\
&[N]_{\text{feedstock}} + O_2 \rightarrow NO_2\\
&[S]_{\text{feedstock}} + O_2 \rightarrow SO_2\\
&2SO_2 + O_2 \rightarrow 2SO_3
\end{aligned}$$

or hydrogen sulfide and ammonia from processing sulfur-containing raw materials:

$$\begin{aligned}
&[S]_{\text{feedstock}} + H_2 \rightarrow H_2S + \text{hydrocarbon derivatives}\\
&2[N]_{\text{feedstock}} + 3H_2 \rightarrow 2NH_3 + \text{hydrocarbon derivatives}
\end{aligned}$$

The question of classifying nitrogen dioxide and sulfur trioxide as primary pollutants often arises, as does the origin of the nitrogen. In the former case, these higher oxides can be formed in the upper levels of the combustion reactors.

A secondary pollutant is a pollutant that is produced by interaction of a primary pollutant with another chemical or by dissociation of a primary pollutant, or other effects within a particular ecosystem. Again, using the atmosphere as an example, the formation of the constituents of acid rain is an example of the formation of secondary pollutants:

$$\begin{aligned}
&SO_2 + H_2O \rightarrow \underset{\text{sulfurous acid}}{H_2SO_3}\\
&SO_3 + H_2O \rightarrow \underset{\text{sulfuric acid}}{H_2SO_4}\\
&NO + H_2O \rightarrow \underset{\text{nitrous acid}}{HNO_2}\\
&3NO_2 + 2H_2O \rightarrow \underset{\text{nitric acid}}{HNO_3}
\end{aligned}$$

In many cases, these secondary pollutants can have significant environmental effects, such as the formation of acid rain and smog (Speight, 1996).

Any pollutant, either primary or secondary can have a serious effect on the various ecological cycles (Speight, 1996). Therefore, understanding the means by which a chemical pollutant can enter these ecosystems and influence the future behavior of the ecosystem is extremely important.

12.5.2 Environmental Regulations

An environmental regulation is a legal mechanism that spells how a statute's broad policy directives are to be carried out. An environmental policy is a requirement that specifies operating procedures that must be followed. An environmental guidance is a document developed by a governmental agency that outlines a position on a topic or which give instructions on how a procedure must be carried out. It explains how to do something and provides governmental interpretations on a governmental act or policy.

As part of the environmental regulations in the United States, waste management (especially hazardous waste management) is based on a beginning-to-end concept so that all hazardous wastes can be traced and fully accounted for. All generators and transporters of hazardous wastes as well as owners and operators of related facilities in the United States must file a notification with the Environmental Protection Agency. The notification must state the location of the facility and a general description of the activities as well as the identified and listed hazardous wastes being handled. Thus, all regulated hazardous waste facilities must exist and/or operate under valid, activity-specific permits.

Regulations pertaining to companies that generate and/or transport wastes require that detailed records be maintained to ensure proper tracking of hazardous wastes through transportation systems. Approved containers and labels must be used, and wastes can only be delivered to facilities approved for treatment, storage, and disposal.

The regulations implementing this law define hazardous wastes based on their characteristics and the law also provides a list of specific hazardous wastes. There are four defining characteristics by which a waste can be determined to be hazardous which are (1) ignitability, (2) corrosivity, (3) reactivity, and (4) toxicity.

Ignitability is demonstrated by substances with a low flash point (ASTM D56, 2021; ASTM D92, 2021; ASTM D93, 2021; ASTM D1310, 2021; ASTM D3828, 2021) or by friction-sensitive substances that can create fires under certain circumstances.

Corrosivity is manifested by acids and wastes capable of corroding metal (ASTM D130, 2021; ASTM D1838, 2021; ASTM D2251, 2021; ASTM D3316, 2021; ASTM D3482, 2021).

Reactivity is exhibited by unstable wastes that create explosions or toxic fumes, gases, or vapors when mixed with water or acids (ASTM D5143, 2021).

Toxicity is shown by wastes which are harmful or fatal if ingested or absorbed (ASTM D4447, 2021; ASTM D4638, 2021); a critical test, under the provisions of the resource recovery conservation act (RCRA) regulation to determine toxicity, was the toxicity characteristics leaching procedure [SW-846 Test Method 1311 (1992): Toxicity Characteristic Leaching Procedure].

The category of listed wastes, those that have already been identified as hazardous by Environmental Protection Agency in the regulations covered by the resource recovery conservation act, include wastes from specific industries and wastes from specific manufacturing processes, as well as specific commercial chemical products high-boiling tar and pitch.

The combination of gas chromatography and mass spectrometry (GC/MS), has proven to be essential in characterizing sites with this type of organic chemical contamination as part of CERCLA's Remedial Investigation (RI) process.

12.5.3 Environmental Analysis

The first instruments that were developed for environmental analyses of organics were gas–liquid chromatographs.The Safe Drinking Water Act, which mandated testing of organics starting January, 1977 regulated pesticides, herbicides, and solvents. These analytes (volatiles and semi-volatiles) all met the criteria that made them amenable to gas chromatographic analysis. Volatiles comprise primarily solvents and light compounds with a high vapor pressure. Compounds are regarded as semi-volatile if they can be vaporized in low quantities at or slightly above 200 °C (392 °F).

The detector components of instruments are devices that transform the physical or chemical property of the analyte into an electrical signal. In most gas chromatographic detectors used in environmental analyses, the elution of the compound is indicated by ions or electrons, which are collected. The electrical response is magnified and translated into a peak that can be shown on a recorder. In order for a compound to generate ions, the molecules can be exposed to light, as in the photo-ionization detector or to flame, as in the flame ionization detector. Another type of detector, the electrolytic conductivity detector, reduces any chlorine-containing compound to hydrogen chloride, using the electrolytic capacity of the resulting hydrogen chloride to show the presence of the compound. In the electron capture detector, any compound electronegative enough to capture electrons will be indicated by a change in the otherwise constant flow of electrons.

Detectors used for drinking water analysis have to be able to indicate the presence of a compound at quantities below the regulated level of the compound in drinking water. Even though the sensitivity of the detectors used allows detection of nanogram quantities or even picogram quantities, it is not sufficient for seeing ppb levels that are of interest for drinking water analysis. The United States Environmental Protection Agency (the USEPA or in the text: EPA) therefore generated sample preparation protocols that permit a concentration of the contaminants before analysis. In the case of semi-volatile compounds, such as pesticides, it is possible to partition the compound into an organic solvent through a liquid/liquid extraction and then follow with a concentration step of evaporating the solvent. Due to the concentration, the compound is ultimately dissolved in just a few milliliters of solvent extract instead of in a liter of water. A few microliters of this final extract are then injected into the gas chromatograph for analysis.

The EPA method for concentration of volatile compounds, including solvents, is called the purge and trap procedure. The method makes use of the high vapor pressure of the analytes. They are partitioned (purged) into the headspace of the container, called the purge vessel, by a continuous flow of inert gas, like nitrogen, which is bubbled through the water sample. The released volatile substances are then collected on a trap filled with solid sorbent material. Upon rapid heating, they can then be desorbed and transferred into the injection port of the gas chromatograph by the carrier gas.

Analysis of inorganic elements is an integral part of any full service environmental testing laboratory. Regulation of inorganic constituents in drinking water dates back to the establishment of U.S. Public Health Service Standards in the mid-1960s. Since 1975, EPA's national primary drinking water standards have covered inorganic chemicals (IOC's), including metals and other substances ranging from aluminum to zinc. Moreover, many states require potable water suppliers to screen annually and/or semiannually for heavy metals and primary and secondary inorganic chemicals. In addition, inorganic contamination at active or abandoned hazardous waste sites presents some of the most complex and difficult investigative and remedial design challenges under RCRA and CERCLA. When testing these sites, regulatory agencies have required analyses for the target analyte list (TAL), which includes specific metals and cyanide.

In the new era of environmental regulation, the first advance in inorganics analysis was the shift from performing separate manual wet chemical analysis for each metal to utilizing atomic absorption and graphite furnace technology. These technologies continue to provide the bulk of metals analysis for both drinking water and CERCLA requirements.

Experienced analytical chemists represent a rich resource for the environmental aspects of refining. Analytical chemists and regulatory personnel recognize that while the recent decades of experience have enhanced knowledge, they have also revealed large gaps in the kind of data needed to thoroughly understand the action of chemicals in the environment.

The development of laboratory instrumentation over the last 50 years has been one of the forces shaping analytical standards and improved instrumentation is already changing approaches to feedstock analysis. Like the twin strands of the famous DNA double helix, technology and analytical science and regulation will continue their closely linked relationship.

References

Altgelt, K.H., and Boduszynski, M.M. 1994. *Composition and Analysis of Heavy Petroleum Fractions*. Marcel Dekker Inc., New York.

ASTM 2021. *Annual Book of Standards*. ASTM International, West Conshohocken, Pennsylvania.

ASTM D130. 2021. Standard Test Method for Corrosiveness to Copper from Petroleum Products by Copper Strip Test. Annual Book of Standards. ASTM International, West Conshohocken, Pennsylvania.

ASTM D1310. 2021. Standard Test Method for Flash Point and Fire Point of Liquids by Tag Open-Cup Apparatus. Annual Book of Standards. ASTM International, West Conshohocken, Pennsylvania.

ASTM D1838. 2021. Standard Test Method for Copper Strip Corrosion by Liquefied Petroleum (LP) Gases. Annual Book of Standards. ASTM International, West Conshohocken, Pennsylvania.

ASTM D2251. 2021. Standard Test Method for Metal Corrosion by Halogenated Organic Solvents and Their Admixtures. Annual Book of Standards. ASTM International, West Conshohocken, Pennsylvania.

ASTM D3316. 2021. Standard Test Method for Stability of Perchloroethylene with Copper. Annual Book of Standards. ASTM International, West Conshohocken, Pennsylvania.

ASTM D3482. 2021. Standard Practice for Determining Electrolytic Corrosion of Copper by Adhesives. Annual Book of Standards. ASTM International, West Conshohocken, Pennsylvania.

ASTM D3828. 2021. Standard Test Methods for Flash Point by Small Scale Closed Cup Tester. Annual Book of Standards. ASTM International, West Conshohocken, Pennsylvania.

ASTM D4447. 2021. Standard Guide for Disposal of Laboratory Chemicals and Samples. Annual Book of Standards. ASTM International, West Conshohocken, Pennsylvania.

ASTM D4638. 2021. Standard Guide for Preparation of Biological Samples for Inorganic Chemical Analysis. Annual Book of Standards. ASTM International, West Conshohocken, Pennsylvania.

ASTM D5143. 2021. Standard Test Method for Analysis of Nitroaromatic and Nitramine Explosive in Soil by High Performance Liquid Chromatography. Annual Book of Standards. ASTM International, West Conshohocken, Pennsylvania.

ASTM D56. 2021. Standard Test Method for Flash Point by Tag Closed Cup Tester. Annual Book of Standards. ASTM International, West Conshohocken, Pennsylvania.

ASTM D92. 2021. Standard Test Method for Flash and Fire Points by Cleveland Open Cup Tester. Annual Book of Standards. ASTM International, West Conshohocken, Pennsylvania.

ASTM D93. 2021. Standard Test Methods for Flash Point by Pensky-Martens Closed Cup Tester. Annual Book of Standards. ASTM International, West Conshohocken, Pennsylvania.

Carson, P.A., and Mumford, C.J. 1988. *The Safe Handling of Chemicals in Industry.* Volume 1 and 2. John Wiley & Sons Inc., New York.

Carson, P.A., and Mumford, C.J. 1995. *The Safe Handling of Chemicals in Industry.* Volume 3. John Wiley & Sons Inc., New York.

Chang, S.E., Stone, J., Demes, K., and Piscitelli, M. 2014. Consequences of Oil Spills: A Review and Framework for Informing Planning. *Ecology and Society*, 19(2): 26. http://dx.doi.org/10.5751/ES-06406-190226; https://www.ecologyandsociety.org/vol19/iss2/art26/

Edwards, J.D. 1995. *Industrial Wastewater Treatment: A Guidebook.* CRC Press Inc., Boca Raton, FL.

EPA 1986. *Methods for Chemical Analysis of Water and Wastes.* United States Environmental Protection Agency, Washington, DC.

EPA 1992. *Test Methods for Evaluating Solid Wastes.* 3rd Edition. United States Environmental Protection Agency, Washington, DC.

Gary, J.G., Handwerk, G.E., and Kaiser, M.J. 2007. *Petroleum Refining: Technology and Economics.* 5th Edition. CRC Press, Taylor & Francis Group, Boca Raton, Florida.

Hsu, C.S., and Robinson, P.R. (Editors). 2017. *Handbook of Petroleum Technology.* Springer International Publishing AG, Cham, Switzerland.

Koots, J.A., and Speight, J.G. 1975. The Relation of Petroleum Resins to Asphaltenes. *Fuel*, 54: 179.

Long, R.B., and Speight, J.G. 1989a. Studies in Petroleum Composition. I: Development of a Compositional Map for Various Feedstocks. *Revue de L'Institut Francais du Petrole*, 44: 205–217.

Long, R.B., and Speight, J.G. 1989b. Studies in Petroleum Composition. II: Scale-up Studies for Separating Heavy Feedstocks by Adsorption. *Industrial and Engineering Chemistry Research*, 28: 1503.

Majumdar, S.B. 1993. *Regulatory Requirements for Hazardous Materials.* McGraw-Hill, New York.

Mitchell, D.L., and Speight, J.G. 1973. The Solubility of Asphaltenes in Hydrocarbon Solvents. *Fuel*, 52: 149.

Occelli, M.L., and Chianelli, R. 1996. *Hydrotreating Technology for Pollution Control.* Marcel Dekker Inc., New York.

Parkash, S. 2003. *Refining Processes Handbook.* Gulf Professional Publishing, Elsevier, Amsterdam, Netherlands.

Ray, D.L., and Guzzo, L. 1990. *Trashing The Planet: How Science Can Help Us Deal With Acid Rain, Depletion of The Ozone, and Nuclear Waste (Among Other Things).* Regnery Gateway, Washington, DC.

Renzoni, A., Fossi, M.C., Lari, L., and Mattei, N. 1994. *Contaminants in the Environment: A Multidisciplinary Assessment of Risks to Man and Other Organisms.* CRC Press Inc., Boca Raton, FL.

Speight, J.G. 1994. In: *Asphaltenes and Asphalts. I. Developments in Petroleum Science, 40.* T.F. Yen and G.V. Chilingarian (Editors). Elsevier, Amsterdam, The Netherlands. Chapter 2

Speight, J.G. 1996. *Environmental Technology Handbook.* Taylor & Francis Publishers, Washington, DC.

Speight, J.G. 2005. *Environmental Analysis and Technology for the Refining Industry.* John Wiley & Sons Inc., Hoboken, New Jersey.

Speight, J.G. 2013. *The Chemistry and Technology of Coal.* 3rd Edition. CRC Press, Taylor & Francois Group, Boca Raton, Florida.

Speight, J.G. 2014. *The Chemistry and Technology of Petroleum.* 5th Edition. CRC Press, Taylor & Francois Group, Boca Raton, Florida.

Speight, J.G. 2015. *Handbook of Petroleum Product Analysis.* 2nd Edition. John Wiley & Sons Inc., Hoboken, New Jersey.

Speight, J.G. 2017. *Handbook of Petroleum Refining.* CRC Press, Taylor & Francis Group, Boca Raton, Florida.

Speight, J.G. 2019a. *Natural Gas: A Basic Handbook.* 2nd Edition. Gulf Publishing Company, Elsevier, Cambridge, Massachusetts.

Speight, J.G. 2019b. *Handbook of Petrochemical Processes.* CRC Press, Taylor & Francis Group, Boca Raton, Florida.

Speight, J.G. 2020. *Global Climate Change Demystified.* Scrivener Publishing, Beverly, Massachusetts.

Speight, J.G. 2023. *Petroleum Refining Technology Series Volume 2: Refinery Dewatering. Desalting, and Distillation in Petroleum Refining.* CRC Press, Taylor & Francis Group, Boca Raton, Florida.

SW-846 Test Method 1311. 1992. Toxicity Characteristic Leaching Procedure. https://www.epa.gov/sites/default/files/2015-12/documents/1311.pdf.

Thibodeaux, L.J. 1995. *Environmental Chemodynamics.* John Wiley & Sons Inc., New York.

Glossary

ABN separation A method of fractionation by which crude oil and (especially the viscous feedstocks) is separated into acidic components, basic components, and neutral components.

Accuracy The degree of conformity of a measured value to the true value; a relative term in the sense that systematic errors or biases can exist but be small enough to be inconsequential.

Acidity The capacity of an acid to neutralize a base such as a hydroxyl ion (OH^-).

Acid number A measure of the reactivity of petroleum with a caustic solution and given in terms of milligrams of potassium hydroxide that are neutralized by one gram of petroleum.

Acid sludge The residue left after treating petroleum oil with sulfuric acid for the removal of impurities; a black, viscous substance containing the spent acid and impurities.

Additive A material added to another (usually in small amounts) in order to enhance desirable properties or to suppress undesirable properties.

Adhesion The tendency of dissimilar particles and/or surfaces to cling to one another. See also: Cohesion.

Adsorption The transfer of a substance from a solution to the surface of a solid resulting in relatively high concentration of the substance at the place of contact; see also Chromatographic adsorption.

Air-blown asphalt Asphalt produced by blowing air through residua at elevated temperatures.

Albertite A black, brittle, natural hydrocarbon possessing a conchoidal fracture and a specific gravity of approximately 1.1.

Alicyclic hydrocarbon A compound containing carbon and hydrogen only that has a cyclic structure (e.g., cyclohexane); also collectively called naphthenes.

Aliphatic hydrocarbon (saturated) A compound containing carbon and hydrogen only that has either an open-chain structure (e.g., as ethane, butane,

Handbook of Heavy Oil Properties and Analysis, First Edition. James G. Speight.

and octane) or a cyclic structure (e.g., cyclohexane) in which the full valency of carbon (4) is satisfied by another atom.

Aliphatic hydrocarbon (unsaturated) A compound containing carbon and hydrogen only that has either an open-chain structure (e.g., as ethylene, butene) or a cyclic structure (e.g., cyclohexene) in which the full valency of carbon (4) is satisfied by another atom and by at least one multiple bond.

Aliquot That quantity of material of proper size for measurement of the property of interest; test portions may be taken from the gross sample directly, but often preliminary operations such as mixing or further reduction in particle size are necessary.

Alkalinity The capacity of a base to neutralize the hydrogen ion (H^+).

Alkali treatment See Caustic wash.

Alkali wash See Caustic wash.

Alkylate The product of an alkylation (*q.v.*) process.

Alkylate bottoms Residua from fractionation of alkylate; the alkylate product which boils higher than the aviation gasoline range; sometimes called heavy alkylate or alkylate polymer.

Alkylation In the petroleum industry, a process by which an olefin (e.g., ethylene) is combined with a branched-chain hydrocarbon (e.g., *iso*-butane); alkylation may be accomplished as a thermal or as a catalytic reaction.

Alpha-scission The rupture of the aromatic carbon–aliphatic carbon bond that joins an alkyl group to an aromatic ring.

Alumina (Al_2O_3) Used in separation methods as an adsorbent and in refining as a catalyst.

American Society for Testing and Materials (ASTM) The official organization in the United States for designing standard tests for petroleum and other industrial products.

Aniline point The temperature, usually expressed in °F, above which equal volumes of a petroleum product and aniline are completely miscible; a qualitative indication of the relative proportions of paraffins in a petroleum product which are miscible with aniline only at higher temperatures; a high aniline point indicates low aromatics.

API gravity A measure of the *lightness* or *heaviness* of petroleum that is related to density and specific gravity.

$$^{\circ}\text{API} = (141.5/\text{sp gr @ } 60^{\circ\circ}\text{F}) - 131.5$$

Apparent bulk density The density of a catalyst as measured; usually loosely compacted in a container.

Apparent viscosity The viscosity of a fluid, or several fluids flowing simultaneously, measured in a porous medium (rock), and subject to both viscosity and permeability effects; also called effective viscosity.

Aromatic hydrocarbon A hydrocarbon characterized by the presence of an aromatic ring or condensed aromatic rings; benzene and substituted benzene, naphthalene and substituted naphthalene, phenanthrene and substituted phenanthrene, as well as the higher condensed ring systems; compounds that are distinct from those of aliphatic compounds (*q.v.*) or alicyclic compounds (*q.v.*).

Aromatization The conversion of nonaromatic hydrocarbons to aromatic hydrocarbons by: (1) rearrangement of aliphatic (noncyclic) hydrocarbons (*q.v.*) into aromatic ring structures and (2) dehydrogenation of alicyclic hydrocarbons (naphthenes).

Asphalt The nonvolatile product obtained by distillation and treatment of an asphaltic crude oil; a manufactured product.

Asphaltene fraction (asphaltenes) The brown to black powdery material produced by treatment of petroleum, petroleum residua, or bituminous materials with a low-boiling liquid hydrocarbon, for example, pentane or heptane; soluble in benzene (and other aromatic solvents), carbon disulfide, and chloroform (or other chlorinated hydrocarbon solvents).

Asphaltene association factor The number of individual asphaltene species which associate in nonpolar solvents as measured by molecular weight methods; the molecular weight of asphaltenes in toluene divided by the molecular weight in a polar non-associating solvent, such as dichlorobenzene, pyridine, or nitrobenzene.

Asphaltic pyrobitumen See Asphaltoid.

Asphaltite A variety of naturally occurring, dark brown to black, solid, nonvolatile bituminous material that is differentiated from bitumen primarily by a high content of material insoluble in n-pentane (asphaltene) or other liquid hydrocarbons. See also: Gilsonite.

Asphaltoid A group of brown to black, solid bituminous materials of which the members are differentiated from asphaltites by their infusibility and low solubility in carbon disulfide.

Asphaltum See Asphalt.

Associated molecular weight The molecular weight of asphaltenes in an associating (nonpolar) solvent, such as toluene.

Atmospheric residuum A residuum (*q.v.*) obtained by distillation of a crude oil under atmospheric pressure and which boils above 350 °C (660 °F).

Atmospheric equivalent boiling point (AEBP) A mathematical method of estimating the boiling point at atmospheric pressure of nonvolatile fractions of petroleum.

Attapulgus clay See Fuller's earth.

Average particle size The weighted average particle diameter of a catalyst.

Barrel The unit of measurement of liquids in the petroleum industry; equivalent to 42 US standard gallons or 33.6 imperial gallons.

Base number The quantity of acid, expressed in milligrams of potassium hydroxide per gram of sample, that is required to titrate a sample to a specified end point.

Basic nitrogen Nitrogen (in petroleum) that occurs in pyridine form.

Basic sediment and water (bs&w, bsw) The material that collects in the bottom of storage tanks, usually composed of oil, water, and foreign matter; also called bottoms, bottom settlings.

Baumé gravity The specific gravity of liquids expressed as degrees on the BaumJ (°BJ) scale.

For liquids lighter than water:

$$\mathrm{Sp\,gr}\,60°\mathrm{F} = 140/(130 + °\mathrm{BJ})$$

For liquids heavier than water:

$$\mathrm{Sp\,gr}\,60°\mathrm{F} = 145/(145 - °\mathrm{BJ})$$

Bauxite Mineral matter used as a treating or chromatographic agent; hydrated aluminum oxide formed by the chemical weathering of igneous rocks.

Bbl See Barrel.

Bentonite Montmorillonite (a magnesium–aluminum silicate); used as a treating agent.

Benzene A colorless aromatic liquid hydrocarbon (C_6H_6).

Benzin A refined light naphtha used for extraction purposes.

Benzine An obsolete term for light petroleum distillates covering the gasoline and naphtha range; see Ligroine.

Benzol The general term which refers to commercial or technical (not necessarily pure) benzene; also the term used for aromatic naphtha.

Beta-scission The rupture of a carbon–carbon bond; two bonds removed from an aromatic ring.

Billion 1×10^9.

Bitumen A semisolid to solid hydrocarbonaceous material found filling pores and crevices of sandstone, limestone, or argillaceous sediments.

Bituminous Containing bitumen or constituting the source of bitumen.

Bituminous rock See Bituminous sand.

Bituminous sand A formation in which the bituminous material (see Bitumen) is found as a filling in veins and fissures in fractured rocks or impregnating relatively shallow sand, sandstone, and limestone strata; a sandstone reservoir that is impregnated with a heavy, viscous black petroleum-like material that cannot be retrieved through a well by conventional production techniques.

Black oil A general term used for any of the dark-colored oils; a term now often applied to heavy oil; the term does not give a reasonable indication of the properties or processability of the oil.

Blown asphalt The asphalt prepared by air blowing a residuum (*q.v.*) or an asphalt (*q.v.*).

Boiling range The range of temperature, usually determined at atmospheric pressure in standard laboratory apparatus, over which the distillation of an oil commences, proceeds, and finishes.

Bottoms The liquid which collects in the bottom of a vessel (tower bottoms, tank bottoms) either during distillation; also the deposit or sediment formed during storage of petroleum or a petroleum product; see also Residuum and Basic sediment and water.

Boundary layer The thin layer of fluid in the immediate vicinity of a bounding surface formed by the fluid flowing along the surface.

British thermal unit See Btu.

Bromine number The number of *grams* of bromine absorbed by 100 g of sample which indicates the percentage of double bonds in the material.

Bromine index The number of *milligrams* of bromine that will react with 100 g of the sample; used mostly for stocks that have an unusually low olefin content.

Brown soap See Brown acid.

BS&W See Basic sediment and water.

Brønsted acid A chemical species that can act as a source of protons.

Brønsted base A chemical species that can accept protons.

Btu (British thermal unit) The energy required to raise the temperature of one pound of water one degree Fahrenheit.

Bubble point The temperature at which incipient vaporization of a liquid in a liquid mixture occurs, corresponding with the equilibrium point of 0% vaporization or 100% condensation.

Bulk sampling Sampling of a material that consists of arbitrary units rather than discrete, identifiable, and constant units. See also: Composite sample, Gross sample, Sample reduction. See also: Clustered sample, Increment, Random sampling, Sampling, Systematic sampling, Stratified sampling.

Bumping The knocking against the walls of a still occurring during distillation of petroleum or a petroleum product that usually contains water.

Bunker C oil See No. 6 Fuel oil.

Burning point See Fire point.

Burning-quality index An empirical numerical indication of the likely burning performance of a furnace or heater oil; derived from the distillation profile (*q.v.*) and the API gravity (*q.v.*), and generally recognizing the factors of paraffinicity and volatility.

C_1, C_2, C_3, C_4, C_5 fractions A common way of representing fractions containing a preponderance of hydrocarbons having 1, 2, 3, 4, or 5 carbon atoms, respectively, and without reference to hydrocarbon type.

Carbene The pentane- or heptane-insoluble material that is insoluble in benzene or toluene but which is soluble in carbon disulfide (or pyridine); a type of rifle used for hunting bison.

Carboid The pentane- or heptane-insoluble material that is insoluble in benzene or toluene and which is also insoluble in carbon disulfide (or pyridine).

Carbon-forming propensity See Carbon residue.

Carbonization The conversion of an organic compound into char or coke by heat in the substantial absence of air; often used in reference to the destructive distillation (with simultaneous removal of distillate) of coal.

Carbon rejection Upgrading processes in which coke is produced. See also Coking.

Carbon residue The amount of carbonaceous residue remaining after thermal decomposition of petroleum, a petroleum fraction, or a petroleum product in a limited amount of air; also called the *coke-* or *carbon-forming propensity*; often prefixed by the terms Conradson or Ramsbottom in reference to the inventor of the respective tests.

Casinghead gas Natural gas that issues from the casinghead (the mouth or opening) of an oil well.

Casinghead gasoline The liquid hydrocarbon product extracted from casinghead gas (*q.v.*) by one of three methods: compression, absorption, or refrigeration; see also Natural gasoline.

Catagenesis The alteration of organic matter during the formation of petroleum that may involve temperatures in the range 50 °C (120 °F) to 200 °C (390 °F); see also Diagenesis and Metagenesis.

Catalyst A chemical agent which, when added to a reaction (process) will enhance the conversion of a feedstock without being consumed in the process.

Catalyst selectivity The relative activity of a catalyst with respect to a particular compound in a mixture, or the relative rate in competing reactions of a single reactant.

Catalytic cracking The conversion of high-boiling feedstocks into lower boiling products by means of a catalyst that may be used in a fixed bed (*q.v.*) or fluid bed (*q.v.*).

Cat cracking See Catalytic cracking.

Cetane index An approximation of the cetane number (*q.v.*) calculated from the density (*q.v.*) and mid-boiling point temperature (*q.v.*); see also Diesel index.

Cetane number A number indicating the ignition quality of diesel fuel; a high cetane number represents a short ignition delay time; the ignition quality of diesel fuel can also be estimated from the following formula:

$$\text{diesel index} = (\text{aniline point}\,(°\text{F}) \times \text{API gravity})100$$

Characterization factor The UOP characterization factor K, defined as the ratio of the cube root of the molal average boiling point, T_B, in degrees Rankine (°R = °F + 460), to the specific gravity at 60 °F/60 °F:

$$K = (T_B)^{1/3}/\text{sp gr}$$

and which ranges from **12.5** for paraffinic stocks to **10.0** for the highly aromatic stocks; also called the Watson characterization **factor.**

Chelating agents Complex-forming agents having the ability to solubilize heavy metals.

Chemical octane number The octane number added to gasoline by refinery processes or by the use of octane number (*q.v.*) improvers such as tetraethyl lead.

Chemical waste Any solid, liquid, or gaseous material discharged from a process and that may pose substantial hazards to human health and environment.

Chromatographic adsorption Selective adsorption on materials such as activated carbon, alumina, or silica gel; liquid or gaseous mixtures of hydrocarbons are passed through the adsorbent in a stream of diluent, and certain components are preferentially adsorbed.

Chromatography A method of separation based on selective adsorption; see also Chromatographic adsorption.

Clay Silicate minerals that also usually contain aluminum and have particle sizes less than 0.002 μm; used in separation methods as an adsorbent and in refining as a catalyst.

Cloud point The temperature at which paraffin wax or other solid substances begin to crystallize or separate from the solution, imparting a cloudy appearance to the oil when the oil is chilled under prescribed conditions.

Clustered sampling The process by which the bulk material is divided into smaller groups (referred to as clusters). See also: Bulk sampling, Increment, Random sampling, Sampling, Systematic sampling, Stratified sampling.

Coal An organic rock.

Coal tar The specific name for the tar (*q.v.*) produced from coal.

Coal tar pitch The specific name for the pitch (*q.v.*) produced from coal.

Cohesion The tendency of similar or identical particles/surfaces to cling to one another. See also: Adhesion.

Coke A gray to black solid carbonaceous material produced from petroleum during thermal processing; characterized by having a high carbon content (95%+ by weight) and a honeycomb type of appearance and is insoluble in organic solvents.

Coke number Used, particularly in Great Britain, to report the results of the Ramsbottom carbon residue test (*q.v.*), which is also referred to as a coke test.

Coker The processing unit in which coking takes place.

Coking A process for the thermal conversion of petroleum in which gaseous, liquid, and solid (coke) products are formed. See also: Carbon rejection.

Color stability The resistance of a petroleum product to color change due to light, aging, and so on.

Combustible liquid A liquid with a flash point in excess of 37.8 °C (100 °F) but below 93.3 °C (200 °F).

Composite sample A sample composed of two or more increments. See also: Bulk sample, Composite sample, Sample reduction.

Composition The general chemical makeup of petroleum.

Composition map A representation of the chemical makeup of petroleum using chemical and/or physical property data.

Con Carbon See Carbon residue.

Conradson carbon residue See Carbon residue.

Contaminant A substance that causes deviation from the normal composition of an environment.

Conventional recovery Primary and/or secondary recovery; often includes tertiary (enhanced) recovery methods.

Conversion The thermal treatment of petroleum that results in the formation of new products by the alteration of the original constituents.

Conversion factor The percentage of feedstock converted to light ends, gasoline, other liquid fuels, and coke.

Cracked residua Residua that have been subjected to temperatures above 350 °C (660 °F) during the distillation process.

Cracking The thermal processes by which the constituents of petroleum are converted to lower molecular weight products.

Cracking temperature The temperature (350 °C; 660 °F) at which the rate of thermal decomposition of petroleum constituents becomes significant.

Crude assay A procedure for determining the general distillation characteristics (e.g., distillation profile, *q.v.*) and other quality information of crude oil.

Crude oil See Petroleum.

Crude still Distillation (*q.v.*) equipment in which crude oil is separated into various products.

Cumene A colorless liquid [$C_6H_5CH(CH_3)_2$] used as an aviation gasoline blending component and as an intermediate in the manufacture of chemicals.

Cut point The boiling-temperature division between distillation fractions of petroleum.

Cyclic hydrocarbons See Alicyclic hydrocarbons.

Cyclization The process by which an open-chain hydrocarbon structure is converted to a ring structure, for example, hexane to benzene.

Dealkylation The removal of an alkyl group from aromatic compounds.

Deasphaltened oil The fraction of petroleum after the asphaltenes have been removed.

Deasphaltening Removal of a solid powdery asphaltene fraction from petroleum by the addition of the low-boiling liquid hydrocarbons such as *n*-pentane or *n*-heptane under ambient conditions.

Deasphalting The removal of the asphaltene fraction from petroleum by the addition of a low-boiling hydrocarbon liquid such as *n*-pentane or *n*-heptane; more correctly the removal asphalt (tacky, semisolid) from petroleum (as occurs in a refinery asphalt plant) by the addition of liquid propane or liquid butane under pressure.

Decolorizing Removal of suspended, colloidal, and dissolved impurities from liquid petroleum products by filtering, adsorption, chemical treatment, distillation, bleaching, and so on.

Dehydrating agents Substances capable of removing water (drying, *q.v.*) or the elements of water from another substance.

Dehydrocyclization Any process by which both dehydrogenation and cyclization reactions occur.

Dehydrogenation The removal of hydrogen from a chemical compound; for example, the removal of two hydrogen atoms from butane to make butene(s) as well as the removal of additional hydrogen to produce butadiene.

Delayed coking A coking process in which the thermal reactions are allowed to proceed to completion to produce gaseous, liquid, and solid (coke) products.

Density The mass (or weight) of a unit volume of any substance at a specified temperature; see also Specific gravity.

Desorption The reverse process of adsorption whereby adsorbed matter is removed from the adsorbent; also used as the reverse of absorption (*q.v.*).

Desulfurization The removal of sulfur or sulfur compounds from a feedstock.

Diagenesis The concurrent and consecutive chemical reactions which commence the alteration of organic matter (at temperatures up to 50 °C (120 °F) and ultimately result in the formation of petroleum from the marine sediment; see also Catagenesis and Metagenesis.

Diesel fuel Fuel used for internal combustion in diesel engines; usually that fraction which distills after kerosene.

Diesel cycle A repeated succession of operations representing the idealized working behavior of the fluids in a diesel engine.

Diesel index An approximation of the cetane number (*q.v.*) of diesel fuel (*q.v.*) calculated from the density(*q.v.*) and aniline point (*q.v.*).

$$\text{Diesel index} = \text{aniline point}\,(°\text{F}) \times \text{API gravity} / 100$$

Diesel knock The result of a delayed period of ignition is long and the accumulated of diesel fuel in the engine.

Distillation A process for separating liquids with different boiling points.

Distillation curve See Distillation profile.

Distillation loss The difference, in a laboratory distillation, between the volume of liquid originally introduced into the distilling flask and the sum of the residue and the condensate recovered.

Distillation range The difference between the temperature at the initial boiling point and at the end point, as obtained by the distillation test.

Distillation profile The distillation characteristics of petroleum and petroleum products showing the temperature and the per cent distilled.

Domestic heating oil See No. 2 Fuel Oil.

Drying Removal of a solvent or water from a chemical substance; also referred to as the removal of solvent from a liquid or suspension.

Dropping point The temperature at which grease passes from a semisolid to a liquid state under prescribed conditions.

Dry point The temperature at which the last drop of petroleum fluid evaporates in a distillation test.

Ebullated bed A process in which the catalyst bed is in a suspended state in the reactor by means of a feedstock recirculation pump which pumps the feedstock upward at sufficient speed to expand the catalyst bed at approximately 35% above the settled level.

Effective viscosity See Apparent viscosity.

Effluent Any contaminating substance, usually a liquid that enters the environment via a domestic, industrial, agricultural, or sewage plant outlet.

Emission control The use gas cleaning processes to reduce emissions.

Emission standard The maximum amount of a specific pollutant permitted to be discharged from a particular source in a given environment.

End-of-pipe emission control The use of specific emission control processes to clean gases after production of the gases.

Energy The capacity of a body or system to do work, measured in joules (SI units); also the output of fuel sources.

Energy from biomass The production of energy from biomass (*q.v.*).

Engler distillation A standard test for determining the volatility characteristics of a gasoline by measuring the per cent distilled at various specified temperatures.

Enhanced oil recovery Petroleum recovery following recovery by conventional (i.e., primary and/or secondary) methods (*q.v.*).

Entrained bed A bed of solid particles suspended in a fluid (liquid or gas) at such a rate that some of the solid is carried over (entrained) by the fluid.

Ethanol See Ethyl alcohol.

Ethyl alcohol (ethanol or grain alcohol) An inflammable organic compound (C_2H_5OH) formed during fermentation of sugars; used as an intoxicant and as a fuel.

Evaporation A process for concentrating nonvolatile solids in a solution by boiling off the liquid portion of the waste stream.

Expanding clays Clays that expand or swell on contact with water, for example, montmorillonite.

Explosive limits The limits of percentage composition of mixtures of gases and air within which an explosion takes place when the mixture is ignited.

Extractive distillation The separation of different components of mixtures which have similar vapor pressures by flowing a relatively high-boiling solvent, which is selective for one of the components in the feed, down a distillation column as the distillation proceeds; the selective solvent scrubs the soluble component from the vapor.

FCC Fluid catalytic cracking.

Feedstock Petroleum, heavy oil, or bitumen as it is fed to the refinery; a refinery product that is used as the raw material for another process; the term is also generally applied to raw materials used in other industrial processes.

Filtration The use of an impassable barrier to collect solids but which allows liquids to pass.

Fire point The lowest temperature at which, under specified conditions in standardized apparatus, a petroleum product vaporizes sufficiently rapidly to form above its surface an air–vapor mixture that burns continuously when ignited by a small flame. See also: Smoke point.

Fischer-Tropsch process A process for synthesizing hydrocarbons and oxygenated chemicals from a mixture of hydrogen and carbon monoxide.

Flammability range The range of temperature over which a chemical is flammable.

Flammable A substance that will burn readily.

Flammable liquid A liquid having a flash point below 37.8 °C (100 °F).

Flammable solid A solid that can ignite from friction or from heat remaining from its manufacture, or which may cause a serious hazard if ignited.

Flash point The lowest temperature to which the product must be heated under specified conditions to give off sufficient vapor to form a mixture with air that can be ignited by a flame.

Floc point The temperature at which wax or solids separate as a definite floc.

Fluid catalytic cracking Cracking in the presence of a fluidized bed of catalyst.

Fluid coking A continuous fluidized solid process that cracks feed thermally over heated coke particles in a reactor vessel to gas, liquid products, and coke.

Fly ash The particulate matter produced from mineral matter in coal that is converted during combustion to finely divided inorganic material and which emerges from the combustor in the gases.

Fractional composition The composition of petroleum as determined by fractionation (separation) methods.

Fractional distillation The separation of the components of a liquid mixture by vaporizing and collecting the fractions, or cuts, which condense in different temperature ranges.

Fractionating column A column arranged to separate various fractions of petroleum by a single distillation and which may be tapped at different points along its length to separate various fractions in the order of their boiling points.

Fractionation The separation of petroleum into the constituent fractions using solvent or adsorbent methods; chemical agents such as sulfuric acid may also be used.

Fuel oil Also called heating oil is a distillate product that covers a wide range of properties; see also No. 1 to No. 4 Fuel oils.

Fuller's earth A clay that has high adsorptive capacity for removing color from oils; Attapulgus clay is a widely used fuller's earth.

Functional group The portion of a molecule that is characteristic of a family of compounds and determines the properties of these compounds.

Gas One of the four fundamental states of matter; a substance or matter in a state in which it will expand freely to fill the whole of a container, having no fixed shape (unlike a solid) and no fixed volume (unlike a liquid). A pure gas may be made up of individual atoms, elemental molecules made from one type of atom, or compound molecules made from a variety of atoms. A gas mixture, such as air, contains a variety of pure gases. See also: Vapor.

Gasoline Fuel for the internal combustion engine that is commonly, but improperly, referred to simply as gas.

Gaseous pollutants Gases released into the atmosphere that act as primary or secondary pollutants.

Gasohol A term for motor vehicle fuel comprising between 80% and 90% unleaded gasoline and 10–20% ethanol (see also Ethyl alcohol).

Gas oil A petroleum distillate with a viscosity and boiling range between those of kerosene and lubricating oil.

Gilsonite A naturally occurring, solid, black, lightweight organic material that originates from the solidification of crude oil. The dull, black appearance of weathered gilsonite resembles coal, whereas the surface of freshly broken gilsonite is shiny and resembles obsidian. Gilsonite is distinguished by its solubility in organic solvents, low density, and brown streak when rubbed on paper. Solid materials such as gilsonite are called asphaltites and are found in oil-bearing sedimentary basins, commonly as veins associated with oil shale; an asphaltite that is >90% bitumen. See also Asphaltite.

Glance pitch An asphaltite.

Grahamite An asphaltite.

Grain alcohol See Ethyl alcohol.

Greenhouse effect Warming of the earth due to entrapment of the Sun's energy by the atmosphere.

Greenhouse gases Gasses that contribute to the greenhouse effect.

Gross sample (also called bulk sample, lot sample) One or more increments of material taken from a larger quantity (lot) of material for assay or record purposes. See also Bulk sample, Composite sample, Sample reduction.

Gum An insoluble tacky semisolid material formed as a result of the storage instability and/or the thermal instability of petroleum and petroleum products.

Heating oil See Fuel oil.

Heavy ends The highest boiling portion of a petroleum fraction; see also Light ends.

Heavy fuel oil Fuel oil having a high density and viscosity; generally residual fuel oil such as No. 5 and No 6. fuel oil (*q.v.*).

Heavy oil Petroleum having an API gravity of less than 20°.

Heavy petroleum See Heavy oil.

Heteroatom compounds Chemical compounds that contain nitrogen and/or oxygen and/or sulfur and/or metals bound within their molecular structure(s).

Homogeneity The degree to which a property or substance is randomly distributed throughout a material; homogeneity depends on the size of the units under consideration. A mixture of two solids may be inhomogeneous at the molecular or atomic level but homogeneous at the particulate level.

Hot filtration test A test for the stability of a petroleum product.

Humic matter A general category of naturally occurring, biogenic, heterogeneous organic substances that can generally be characterized as being yellow to black in color, of high molecular weight, and refractory.

Hydrocarbon compounds Chemical compounds containing only carbon and hydrogen.

Hydrocarbon resource Resources such as petroleum and natural gas that can produce naturally occurring hydrocarbons without the application of conversion processes.

Hydrocarbon-producing resource A resource such as coal and oil shale (kerogen) that produce derived hydrocarbons by the application of conversion processes; the hydrocarbons so-produced are not naturally occurring materials.

Hydroconversion A term often applied to hydrocracking (*q.v.*).

Hydrocracking A catalytic high-pressure high-temperature process for the conversion of petroleum feedstocks in the presence of fresh and recycled hydrogen; carbon–carbon bonds are cleaved in addition to the removal of heteroatomic species.

Hydrogen addition An upgrading process in the presence of hydrogen, for example, hydrocracking; see Hydrogenation.

Hydrogenation The chemical addition of hydrogen to a material. In nondestructive hydrogenation, hydrogen is added to a molecule only if, and where, unsaturation with respect to hydrogen exists.

Hydrogen transfer The transfer of inherent hydrogen within the feedstock constituents and products during processing.

Hydroprocessing A term often equally applied to hydrotreating (*q.v.*) and to hydrocracking (*q.v.*); also, often collectively applied to both.

Hydrotreating The removal of heteroatomic (nitrogen, oxygen, and sulfur) species by treatment of a feedstock or product at relatively low temperatures in the presence of hydrogen.

Ignitability Characteristic of liquids whose vapors are likely to ignite in the presence of ignition source; also characteristic of nonliquids that may catch fire from friction or contact with water and that burn vigorously.

Ignition temperature The minimum temperature at which the material will ignite without a spark or flame being present; sometimes referred to as the autoignition temperature.

Immiscible Two or more fluids that do not have complete mutual solubility and coexist as separate phases.

Increment An individual portion of material collected by a single operation of a sampling device, from parts of a lot separated in time or space. Increments may be either tested individually or combined (composited) and tested as a unit. See also: Bulk sampling, Clustered sample, Random sampling, Sampling, Systematic sampling, Stratified sampling.

Individuals Conceivable constituent parts of the population.

Initial boiling point The recorded temperature when the first drop of liquid falls from the end of the condenser.

Initial vapor pressure The vapor pressure of a liquid of a specified temperature and zero per cent evaporated.

Instability The inability of a petroleum product to exist for periods of time without change to the product.

Incompatibility The *immiscibility* of petroleum products and also of different crude oils that is often reflected in the formation of a separate phase after mixing and/or storage.

Interfacial tension A measure of the force acting at a boundary between two immiscible liquids. See also: Surface tension.

Iodine number A measure of the iodine absorption by oil under standard conditions; used to indicate the quantity of unsaturated compounds present; also called iodine value.

Ion exchange A means of removing cations or anions from solution onto a solid resin.

Isomerization The conversion of a *normal* (straight-chain) paraffin hydrocarbon into an *iso* (branched-chain) paraffin hydrocarbon having the same atomic composition.

Jet fuel Fuel meeting the required properties for use in jet engines and aircraft turbine engines.

Kaolinite A clay mineral formed by hydrothermal activity at the time of rock formation or by chemical weathering of rocks with high feldspar content; usually associated with intrusive granite rocks with high feldspar content.

Kata-condensed aromatic compounds Compounds based on linear condensed aromatic hydrocarbon systems, for example, anthracene and naphthacene (tetracene).

Kerogen A complex carbonaceous (organic) material that occurs in sedimentary rocks and shales; generally insoluble in common organic solvents.

Kerosene (kerosine) A fraction of petroleum that was initially sought as an illuminant in lamps; a precursor to diesel fuel.

K-factor See Characterization factor.

Kinematic viscosity The ratio of viscosity (*q.v.*) to density, both measured at the same temperature.

Laboratory sample A sample, intended for testing or analysis, prepared from a gross sample or otherwise obtained. The laboratory sample must retain the composition of the gross sample.

Lamp burning A test of burning oils in which the oil is burned in a standard lamp under specified conditions in order to observe the steadiness of the flame, the degree of encrustation of the wick, and the rate of consumption of the kerosene.

Lamp oil See Kerosene.

Lewis acid A chemical species that can accept an electron pair from a base.

Lewis base A chemical species that can donate an electron pair.

Light ends The lower-boiling components of a mixture of hydrocarbons; see also Heavy ends, Light hydrocarbons.

Light hydrocarbons Hydrocarbons with molecular weights less than that of heptane (C_7H_{16}).

Light oil The products distilled or processed from crude oil up to, but not including, the first lubricating-oil distillate.

Light petroleum Petroleum having an API gravity greater than 20°.

Ligroine (Ligroin) A saturated petroleum naphtha boiling in the range of 20–135°°C (68–275 °F) and suitable for general use as a solvent; also called benzine or petroleum ether.

Liquid petrolatum See White oil.

Liquefied petroleum gas Propane, butane, or mixtures thereof, gaseous at atmospheric temperature and pressure, held in the liquid state by pressure to facilitate storage, transport, and handling.

Lot A quantity of bulk material of similar composition whose properties are under study.

Lower explosive limit (LEL) The lowest concentration of a vapor or gas that will burn in air; the Lower Explosive Limit (LEL) varies from gas to gas, but for most flammable gases, it is less than 5% v/v. See also: Upper explosive limit (UEL).

Lower flammability limit (LFL) The lower end of the concentration range over which a flammable mixture of gas or vapor in air can be ignited at a given temperature and pressure. The flammability range is delineated by the upper and lower flammability limits and is usually expressed in % v/v. See also: Upper flammability limit (UFL).

Lube See Lubricating oil.

Lube cut A fraction of crude oil of suitable boiling range and viscosity to yield lubricating oil when completely refined; also referred to as lube oil distillates or lube stock.

Lubricating oil A fluid lubricant used to reduce friction between bearing surfaces.

Mahogany acids Oil-soluble sulfonic acids formed by the action of sulfuric acid on petroleum distillates. They may be converted to their sodium soaps (mahogany soaps) and extracted from the oil with alcohol for use in the manufacture of soluble oils, rust preventives, and special greases. The calcium and barium soaps of these acids are used as detergent additives in motor oils; see also Brown acids and Sulfonic acids.

Maltenes That fraction of petroleum that is soluble in, for example, pentane or heptane; deasphaltened oil (*q.v.*); also the term arbitrarily assigned to the pentane-soluble portion of petroleum that is relatively high boiling (>300 °C, 760 mm) (see also Petrolenes).

Marine sediment The organic biomass from which petroleum is derived.

Marsh An area of spongy waterlogged ground with large numbers of surface water pools. Marshes usually result from: (1) an impermeable underlying bedrock, (2) surface deposits of glacial boulder clay, (3) a basin-like topography from which natural drainage is poor, (4) very heavy rainfall in conjunction with a correspondingly low evaporation rate, and (5) low-lying land, particularly at estuarine sites at or below sea level.

Mayonnaise Low-temperature petroleum sludge; a black, brown, or gray deposit having a soft, mayonnaise-like consistency; not recommended as a food additive!

Methanol See Methyl alcohol.

Mercaptans Organic compounds having the general formula R-SH.

Metagenesis The alteration of organic matter during the formation of petroleum that may involve temperatures above 200 °C (390 °F); see also Catagenesis and Diagenesis.

Methyl alcohol (methanol; wood alcohol) A colorless, volatile, inflammable, and poisonous alcohol (CH_3OH) traditionally formed by destructive distillation of wood or, more recently, as a result of synthetic distillation in chemical plants.

Methyl ethyl ketone See MEK.

Micelle The structural entity by which asphaltenes are dispersed in petroleum.

Microcarbon residue The carbon residue determined using a thermogravimetric method. See also Carbon residue.

Mid-boiling point The temperature at which approximately 50% of a material has distilled under specific conditions.

Middle distillate Distillate boiling between the kerosene and lubricating oil fractions.

Migration (primary) The movement of hydrocarbons (oil and natural gas) from mature, organic-rich source rocks to a point where the oil and gas can collect as droplets or as a continuous phase of liquid hydrocarbon.

Migration (secondary) The movement of the hydrocarbons as a single, continuous fluid phase through water-saturated rocks, fractures, or faults followed by accumulation of the oil and gas in sediments (traps, *q.v.*) from which further migration is prevented.

Mineral oil The older term for petroleum; the term was introduced in the nineteenth century as a means of differentiating petroleum (rock oil) from whale oil which, at the time, was the predominant illuminant for oil lamps.

Minerals Naturally occurring inorganic solids with well-defined crystalline structures.

Mitigation Identification, evaluation, and cessation of potential impacts of a process product or by-product.

Modified naphtha insolubles (MNI) An insoluble fraction obtained by adding naphtha to petroleum; usually adding paraffinic constituents modifies the naphtha; the fraction might be equated to asphaltenes *if* the naphtha is equivalent to *n*-heptane, but usually it is not.

Molecular sieve A synthetic zeolite mineral having pores of uniform size; it is capable of separating molecules, on the basis of their size, structure, or both, by absorption or sieving.

Naft Pre-Christian era (Greek) term for naphtha (*q.v.*).

Napalm A thickened gasoline used as an incendiary medium that adheres to the surface it strikes.

Naphtha A generic term applied to refined, partly refined, or unrefined petroleum products and liquid products of natural gas, the majority of which distills below 240 °C (464 °F); the volatile fraction of petroleum which is used as a solvent or as a precursor to gasoline.

Naphthenes Cycloparaffins.

Native asphalt See Bitumen.

Natural asphalt See Bitumen.

Natural gas The naturally occurring gaseous constituents that are found in many petroleum reservoirs; there are also those reservoirs in which natural gas may be the sole occupant.

Natural gas liquids (NGL) The hydrocarbon liquids that condense during the processing of hydrocarbon gases that are produced from oil or gas reservoir; see also Natural gasoline.

Natural gasoline A mixture of liquid hydrocarbons extracted from natural gas (*q.v.*) suitable for blending with refinery gasoline.

Neutralization A process for reducing the acidity or alkalinity of a waste stream by mixing acids and bases to produce a neutral solution; also known as pH adjustment.

Neutralization number The weight, in milligrams, of potassium hydroxide needed to neutralize the acid in 1 g of oil; an indication of the acidity of an oil.

Non-asphaltic road oil Any of the nonhardening petroleum distillates or residual oils used as dust layers. They have sufficiently low viscosity to be applied without heating and, together with asphaltic road oils (*q.v.*), are sometimes referred to as dust palliatives.

Non-Newtonian A fluid that exhibits a change of viscosity with flow rate.

Octane barrel yield A measure used to evaluate fluid catalytic cracking processes; defined as (RON + MON)/2 times the gasoline yield, where RON is the research octane number and MON is the motor octane number.

Octane number A number indicating the anti-knock characteristics of gasoline.

Oils (fraction) That portion of the maltenes (*q.v.*) that is not adsorbed by a surface-active material such as clay or alumina.

Oil sand See Tar sand.

Oil shale A fine-grained impervious sedimentary rock that contains an organic material called kerogen.

Organic sedimentary rocks Rocks containing organic material such as residues of plant and animal remains/decay.

Overhead That portion of the feedstock that is vaporized and removed during distillation.

Oxidation A process that can be used for the treatment of a variety of inorganic and organic substances.

Oxidized asphalt See Air-blown asphalt.

Particle density The density of solid particles.

Particulate matter Particles in the atmosphere or on a gas stream that may be organic or inorganic and originate from a wide variety of sources and processes.

Particle size distribution The particle size distribution (of a catalyst sample) expressed as a percent of the whole.

Peri-condensed aromatic compounds Compounds based on angular condensed aromatic hydrocarbon systems, for example, phenanthrene, chrysene, picene, and so on.

Permeability The ease of flow of the water through the rock.

Petrol A term commonly used in some countries for gasoline.

Petrolatum A semisolid product, ranging from white to yellow in color, produced during refining of residual stocks; see Petroleum jelly.

Petrolenes The term applied to that part of the pentane-soluble or heptane-soluble material that is low boiling (<300 °C, <570 °F, 760 mm) and can be distilled without thermal decomposition (see also Maltenes).

Petroleum (crude oil) A naturally occurring mixture of gaseous, liquid, and solid hydrocarbon compounds usually found trapped deep underground beneath impermeable cap rock and above a lower dome of sedimentary rock such as shale; most petroleum reservoirs occur in sedimentary rocks of marine, deltaic, or estuarine origin.

Petroleum asphalt See Asphalt.

Petroleum ether See Ligroine.

Petroleum jelly A translucent, yellowish to amber or white, hydrocarbon substance (m.p. 38 – 54 °C) having almost no odor or taste, derived from petroleum and used principally in medicine and pharmacy as a protective dressing and as a substitute for fats in ointments and cosmetics; also used in many types of polishes and in lubricating greases, rust preventives, and modeling clay; obtained by dewaxing heavy lubricating-oil stocks.

Petroleum refinery See Refinery.

Petroleum refining A complex sequence of events that result in the production of a variety of products.

Petroporphyrins See Prophyrins.

Phase separation The formation of a separate phase that is usually the prelude to coke formation during a thermal process; the formation of a separate phase as a result of the instability/incompatibility of petroleum and petroleum products.

pH adjustment Neutralization.

Pitch The nonvolatile, brown to black, semisolid to solid viscous product from the destructive distillation of many bituminous or other organic materials, especially coal.

PNA A polynuclear aromatic compound (*q.v.*).

PONA analysis A method of analysis for paraffins (P), olefins (O), naphthenes (N), and aromatics (A).

Polar aromatics Resins; the constituents of petroleum that are predominantly aromatic in character and contain polar (nitrogen, oxygen, and sulfur) functions in their molecular structure(s).

Pollution The introduction into the land water and air systems of a chemical or chemicals that are not indigenous to these systems or the introduction into the land water and air systems of indigenous chemicals in greater-than-natural amounts.

Polynuclear aromatic compound An aromatic compound having two or more fused benzene rings, for example, naphthalene and phenanthrene.

Pooled standard deviation A weighted average of individual estimates of the standard deviation.

Population A generic term denoting any finite or infinite collection of individual things, objects, or events in the broadest concept; an aggregate determined by some property that distinguishes things that do and do not belong.

Porphyrins Organometallic constituents of petroleum that contain vanadium or nickel; the degradation products of chlorophylls that became included in the protopetroleum.

Pour point The lowest temperature at which oil will pour or flow when it is chilled without disturbance under definite conditions.

Precipitation number The number of milliliters of precipitate formed when 10 ml of lubricating oil is mixed with 90 ml of petroleum naphtha of a definite quality and centrifuged under definitely prescribed conditions.

Precision The degree to which a measurement can be repeated.

Primary oil recovery Oil recovery utilizing only naturally occurring forces.

Primary structure The chemical sequence of atoms in a molecule.

Propane asphalt See Solvent asphalt.

Propane deasphalting Solvent deasphalting using propane as the solvent.

Protopetroleum A generic term used to indicate the initial product formed changes have occurred to the precursors of petroleum.

Pyrobitumen See Asphaltoid.

Pyrolysis Exposure of a feedstock to high temperatures in an oxygen-poor environment.

Pyrophoric Substances that catch fire spontaneously in air without an ignition source.

Quadrillion 1×10^{15}.

Quench The sudden cooling of hot material discharging from a thermal reactor.

Raffinate That portion of the oil that remains insoluble after treatment by a solvent.

Ramsbottom carbon residue See Carbon residue.

Reduced crude A residual product remaining after the removal, by distillation or other means, of an appreciable quantity of the more volatile components of crude oil.

Reduction The process of preparing one or more subsamples from a sample.

Reduction The process by which hydrogen is added to feedstocks during hydrotreating and hydrocracking processes; removal of unsaturation by the addition of hydrogen.

Refinery A series of integrated unit processes by which petroleum can be converted to a slate of useful (salable) products.

Refinery gas A gas (or a gaseous mixture) produced as a result of refining operations.

Refining The process(es) by which petroleum is distilled and/or converted by application of physical and chemical processes to form a variety of products are generated.

Reformed gasoline Gasoline made by a reforming process.

Reid vapor pressure A measure of the volatility of liquid fuels, especially gasoline.

Repeatability The difference between two individual test results reported by a single analyst on two test portions of the same material that would be equaled or exceeded in the long run in only 1 case in 20 in the normal and correct operation of the method. See: ASTM C670 – Standard Practice for Preparing Precision and Bias Statements for Test Methods for Construction Materials.

Representative sample A sample from a bulk material that accurately represents the characteristics of a material and the results from the representative sample should accurately reflect the properties of the bulk material.

Reproducibility The difference between two individual test results reported by two laboratories on two test portions of the same material that would be equaled or exceeded in the long run in only 1 case in 20 in the normal and correct operation of the method.

Research Octane Method A test for determining the knock rating, in terms octane numbers, of fuels for use in spark-ignition engines; see also Motor Octane Method.

Residual asphalt See Straight-run asphalt.

Residual fuel oil Obtained by blending the residual product(s) from various refining processes with suitable diluent(s) (usually middle distillates) to obtain the required fuel oil grades.

Residual oil See Residuum.

Residuum (resid; *pl:*. residua) The residue obtained from petroleum after nondestructive distillation has removed all the volatile materials from crude oil, for example, an atmospheric (345 °C, 650 °F+) residuum.

Resins The portion of the maltenes fraction that is adsorbed by a surface-active material such as clay or alumina; the fraction of deasphaltened oil that is insoluble in liquid propane but soluble in *n*-heptane.

Riprap Human-placed rock or other material used to protect shoreline structures against scour and water, wave, or ice erosion. Also known as rip rap, rip-rap, shot rock, rock armor, or rubble.

Rock asphalt Bitumen that occurs in formations that have a limiting ratio of bitumen-to-rock matrix.

Sample A portion of a population or lot which may consist of an individual sample or a group (or groups) of individual samples. See also: Bulk sample, Composite sample, Gross sample, Sample reduction, Subsample.

Sample reduction The process of preparing one or more subsamples from a sample.

Sampling The process in analysis science in which a predetermined number of observations is taken by selection of sample from the bulk material. See also: Bulk sampling, Clustered sampling, Increment, Random sampling, Sampling, Systematic sampling, Stratified sampling.

Sampling protocol The procedure used to select a sample that is (or samples that are) from the bulk material. The goal of the sampling protocol is to select a sample that is (or samples that are) representative of the bulk material.

SARA separation A method of fractionation by which petroleum is separated into saturates, aromatics, resins, and asphaltene fractions.

Saturates Paraffins and cycloparaffins (naphthenes).

Saybolt Furol viscosity The time, in seconds (Saybolt Furol Seconds, SFS), for 60 ml of fluid to flow through a capillary tube in a Saybolt Furol viscometer at specified temperatures between 70 and 210 °F; the method is appropriate for high-viscosity oils such as transmission, gear, and heavy fuel oils.

Saybolt Universal viscosity The time, in seconds (Saybolt Universal Seconds, SUS), for 60 ml of fluid to flow through a capillary tube in a Saybolt Universal viscometer at a given temperature.

Secondary structure The ordering of the atoms of a molecule in space relative to each other.

Sediment An insoluble solid formed as a result of the storage instability and/or the thermal instability of petroleum and petroleum products.

Segment A specifically demarked portion of a lot, either actual or hypothetical.

Selective solvent A solvent that, at certain temperatures and ratios, will preferentially dissolve more of one component of a mixture than of another and thereby permit partial separation.

Separation process A process in which the constituents of petroleum are separated, usually without thermal decomposition, for example, distillation and deasphalting.

Slime A name used for petroleum in ancient texts.

Sludge A semisolid to solid product that results from the storage instability and/or the thermal instability of petroleum and petroleum products.

Smoke point A measure of the burning cleanliness of a fuel or carbonaceous material; also referred to as the burning point but not to be confused with the fire point. See also: Fire point.

Solubility parameter A measure of the solvent power and polarity of a solvent.

Snyder column A specific type of distillation column that utilizes separate chambers to promote multiple condensations and vaporizations during the distillation process, leading to a more efficient separation of the mixture's components.

Softening point The temperature at which asphalt attains a particular degree of softness under the specified conditions of test.

Solvent asphalt The asphalt produced by solvent extraction of residua or by light hydrocarbon (propane) treatment of a residuum or an asphaltic crude oil.

Solvent deasphalting A process for removing asphaltic and resinous materials from reduced crude oils, lubricating-oil stocks, gas oils, or middle distillates through the extraction or precipitant action of low molecular weight hydrocarbon solvents; see also Propane deasphalting.

Solvent extraction A process for separating liquids by mixing the stream with a solvent that is immiscible with part of the waste but that will extract certain components of the waste stream.

Solvent naphtha A refined naphtha of restricted boiling range used as a solvent; also called petroleum naphtha; petroleum spirits.

Solvent refining See Solvent extraction.

Sonication extraction The use of sound waves to enhance analyte transfer from sample to solvent.

Sour crude oil Crude oil containing an abnormally large amount of sulfur compounds; see also Sweet crude oil.

Specific gravity The mass (or weight) of a unit volume of any substance at a specified temperature compared to the mass of an equal volume of pure water at a standard temperature; see also Density.

Specimen See Aliquot.

Spiked sample A sample to which known concentrations of specific analytes have been added in such a manner as to minimize the change in the matrix of the original sample. Every spiked sample analyzed should have an associated reference to the spike solution and the volume added.

Stabilization The removal of volatile constituents from a higher boiling fraction or product (*q.v.* stripping); the production of a product which, to all intents and purposes, does not undergo any further reaction when exposed to the air.

Standard deviation A measure of the scatter in a set of data which fits a Gaussian distribution; see also Pooled standard deviation.

Steam distillation Distillation in which vaporization of the volatile constituents is effected at a lower temperature by introduction of steam (open steam) directly into the charge.

Storage stability (or storage instability) The ability (inability) of a liquid to remain in storage over extended periods of time without appreciable deterioration as measured by gum formation and the depositions of insoluble material (sediment).

Straight-run asphalt The asphalt (*q.v.*) produced by the distillation of asphaltic crude oil.

Straight-run products Obtained from a distillation unit and used without further treatment.

Strata Segments of a lot that may vary with respect to the property under study.

Stratified sampling A sampling method that may be applied to the bulk material, such as a pile of a solid substance or a near-solid substance such as when tar sand bitumen is stored in large tanks in which flow is limited (if not prevented) by the properties of the material. See also: Bulk sampling, Clustered sampling, Increment, Random sampling, Sampling, Systematic sampling.

Sub-sample (Subsample) A portion taken from a sample. A laboratory sample may be a subsample of a gross sample; similarly, a test portion may be a subsample of a laboratory sample. See also: Bulk sample, Composite sample, Gross sample.

Surface tension As a measure of the force acting at a boundary between two phases. If the boundary is between a liquid and a solid or between a liquid and a gas (air), the attractive forces are referred to as surface tension, but the attractive forces between two immiscible liquids are referred to as interfacial tension. See also: Interfacial tension.

Sweet crude oil Crude oil containing little sulfur; see also Sour crude oil.

Synthetic crude oil (syncrude) A hydrocarbon product produced by the conversion of coal, oil shale, or tar sand bitumen that resembles conventional crude oil; can be refined in a petroleum refinery (*q.v.*).

Systematic sampling A probability sampling method where the samples are taken from the bulk material at a regular interval such as from specific areas of the bulk material. See also: Bulk sampling, Clustered sampling, Increment, Random sampling, Sampling, Stratified sampling.

Tar The volatile, brown to black, oily, viscous product from the destructive distillation of many bituminous or other organic materials, especially coal; a name used for petroleum in ancient texts.

Tar sand See Bituminous sand.

Tertiary structure The three-dimensional structure of a molecule.

Test portion See Aliquot.

Thermal coke The carbonaceous residue formed as a result of a non-catalytic thermal process; the Conradson carbon residue; the Ramsbottom carbon residue.

Thermal cracking A process that decomposes, rearranges, or combines hydrocarbon molecules by the application of heat, without the aid of catalysts.

Thermal process Any refining process that utilizes heat, without the aid of a catalyst.

Thermal stability (thermal instability) The ability (inability) of a liquid to withstand relatively high temperatures for short periods of time without the formation of carbonaceous deposits (sediment or coke).

Topped crude Petroleum that has had volatile constituents removed up to a certain temperature, for example, 250 °C+ (480 °F+) topped crude oil; not always the same as a residuum (*q.v.*).

Topping The distillation of crude oil to remove light fractions only.

Trace element Those elements that occur at very low levels in a given system.

Treatment Any method, technique, or process that changes the physical and/or chemical character of petroleum.

Trillion 1×10^{12}.

True boiling point (True boiling range) The boiling point (boiling range) of a crude oil fraction or a crude oil product under standard conditions of temperature and pressure.

Ultimate analysis Elemental composition.

Unassociated molecular weight The molecular weight of asphaltenes in a non-associating (polar) solvent, such as dichlorobenzene, pyridine, or nitrobenzene.

Uncertainty Same as standard deviation (*q.v.*).

Universal viscosity See Saybolt Universal viscosity.

Unstable Usually refers to a petroleum product that has more volatile constituents present or refers to the presence of olefin and other unsaturated constituents.

UOP characterization factor See Characterization factor.

Upgrading The conversion of petroleum to value-added salable products.

Upper explosive limit (UEL) The highest concentration of a vapor (or gas) in air that is capable of producing a flash of fire when met with a source of ignition. Concentrations above the UEL will not fuel or continue an explosion and these concentrations are considered too "rich." See also: Lower explosive limit (LEL).

Upper flammability limit (UFL) The highest possible concentration of a substance in air at which a flammable mixture is formed. See also: Lower flammability limit (LFL).

Vacuum distillation Distillation (*q.v.*) under reduced pressure.

Vacuum residuum A residuum (*q.v.*) obtained by distillation of a crude oil under vacuum (reduced pressure); that portion of petroleum that boils above a selected temperature such as 510 °C (950 °F) or 565 °C (1050 °F).

Vapor A substance diffused or suspended in the air, especially one normally liquid or solid. A substance in the gas phase at a temperature lower than its critical temperature, which means that the vapor can be condensed to a liquid by increasing the pressure on it without reducing the temperature. See also: Gas.

Viscosity A measure of the ability of a liquid to flow or a measure of its resistance to flow; the force required to move a plane surface of area 1 m^2 over another parallel plane surface 1 m away at a rate of 1 m/s when both surfaces are immersed in the fluid.

VGC (viscosity-gravity constant) An index of the chemical composition of crude oil defined by the general relation between specific gravity, sg, at 60 °F and Saybolt Universal viscosity, SUV, at 100 °F:

$$a = 10\text{sg} - 1.0752 \log(\text{SUV} - 38) / 10\text{sg} - \log(\text{SUV} - 38)$$

and the constant, a, is low for the paraffinic crude oils and high for the naphthenic crude oils.

VI (Viscosity index) An arbitrary scale used to show the magnitude of viscosity changes in lubricating oils with changes in temperature.

Viscosity-gravity constant See VGC.

Viscosity index See VI.

Watson characterization factor See Characterization factor.

Weathered crude oil Crude oil which, due to natural causes during storage and handling, has lost an appreciable quantity of its more volatile components; also indicates uptake of oxygen.

Wood alcohol See Methyl alcohol.

Zeolite A crystalline aluminosilicate used as a catalyst and having a particular chemical and physical structure.

Conversion Factors

1 Concentration conversions

1 part per million (1 ppm) = 1 microgram per liter (1 μg/L)

1 microgram per liter (1 μg/L) = 1 milligram per kilogram (1 mg/kg)

1 microgram per liter (μg/L) × 6.243 × 10^8 = 1 lb per cubic foot (1 lb/ft^3)

1 microgram per liter (1 μg/L) × 10^{-3} = 1 milligram per liter (1 mg/L)

1 milligram per liter (1 mg/L) × 6.243 × 10^5 = 1 pound per cubic foot (1 lb/ft^3)

I gram mole per cubic meter (1 g mol/m^3) × 6.243 × 10^5 = 1 pound per cubic foot (1 lb/ft^3)

10,000 ppm = 1% w/w

1 ppm hydrocarbon in soil × 0.002 − 1 lb of hydrocarbons per ton of contaminated soil

2 Sludge conversions

1,700 lbs wet sludge = 1 yd^3 wet sludge

1 yd^3 sludge = wet tons/0.85

Wet tons sludge × 240 = gallons sludge

1 wet ton sludge × % dry solids/100 = 1 dry ton of sludge

3 Weight conversion

1 ounce (1 oz) = 28.3495 grams (18.2495 g)

1 pound (1 lb) = 0.454 kilogram

1 pound (1 lb) = 454 grams (454 g)

1 kilogram (1 kg) = 2.20462 pounds (2.20462 lb)

1 stone (English, 1 st) = 14 pounds (14 lb)

1 ton (US; 1 short ton) = 2,000 lbs

1 ton (English; 1 long ton) = 2,240 lbs

1 metric ton = 2204.62262 pounds

1 tonne = 2204.62262 pounds

Handbook of Heavy Oil Properties and Analysis, First Edition. James G. Speight.

4 Temperature conversions

$°F = (°C \times 1.8) + 32$

$°C = (°F - 32)/1.8$

$(°F - 32) \times 0.555 = °C$

Absolute zero = −273.15 °C

Absolute zero = −459.67 °F

5 Area

1 square centimeter (1 cm^2) = 0.1550 square inches

1 square meter 1 (m^2) = 1.1960 square yards

1 hectare = 2.4711 acres

1 square kilometer (1 km^2) = 0.3861 square miles

1 square inch (1 $in.^2$) = 6.4516 square centimeters

1 square foot (1 ft^2) = 0.0929 square meters

1 square yard (1 yd^2) = 0.8361 square meters

1 acre = 4046.9 square meters

1 square mile (1 mi^2) = 2.59 square kilometers

6 Nutrient conversion factor

1 pound phosphorus × 2.3 (1 lb P × 2.3) = 1 pound phosphorous pentoxide (1 lb P_2O_5)

1 pound potassium × 1.2 (1 lb K × 1.2) = 1 pound potassium oxide (1 lb K_2O)

7 Other approximations

14.7 pounds per square inch (14.7 psi) – 1 atmosphere (1 atmos)

1 kiloPascal (kPa) × 9.8692×10^{-3} = 14.7 pounds per square inch (14.7 psi)

1 yd^3 = 27 ft^3

1 US gallon of water = 8.34 lbs

1 imperial gallon of water – 10 lbs

1 yd^3 = 0.765 m^3

1 acre-inch of liquid = 27,150 gallons = 3.630 ft^3

1 ft depth in 1 acre (in-situ) = 1,613 × (20–% excavation factor) = ~2,000 yd^3

1 yd^3 (clayey soils-excavated) = 1.1 to 1.2 tons (US)

1 yd^3 (sandy soils-excavated) = 1.2 to 1.3 tons (US)

Index

a

Accuracy 88
Acidity 395
Acid number 193
Acid treatment 173
Adduct 176
  composition 176
  formation 176
  properties 178
  structure 176
Adhesion 234
Adsorption 164
Adsorption chromatography 286
Analytical methods for environmental regulations 431
Analytical perspectives 56
API gravity 235
Aromatic hydrocarbon derivatives 124
Ash production 256
Asphalt 45
Asphaltene content 395
Asphaltene fraction 179
  molecular weight 372
Asphaltene separation 156
  influence of contact time 161
  influence of degree of dilution 160
  influence of solvent type 158
  influence of temperature 160
Assay 36

b

Bitumen 36
Bottom sediment and water 402
BS&W 402
Carbenes 163, 180
Carboids 163, 180
Carbon residue 258
Chemical composition 120
  by chemical methods 173
  by distillation 131
  by simulated distillation 305
  by spectroscopy 137
Chemical factors 165
Chemical properties 191
Chromatographic analysis 283
  adsorption chromatography 286
  gas chromatography 291
  gel permeation chromatography 298
  high performance liquid chromatography 300
  ion exchange chromatography 303
  size exclusion chromatography 367

Handbook of Heavy Oil Properties and Analysis, First Edition. James G. Speight.

Chromatographic analysis (*cont'd*)
  supercritical fluid
    chromatography 307
  thin layer chromatography 309
Chromatographic properties 283
Classification 51
Critical properties 260
Crude oil 17
  evaluation 53
  extra heavy crude oil 33
  foamy oil 26
  heavy crude oil 30
  high-acid crude oil 25
  non-viscous crude oil 17
  opportunity crude oil 21
  viscous crude oil 27
Cycloparaffin hydrocarbon
    derivatives 124

d

Definitions and terminology 15
Density 235, 398
Determination of instability and
    incompatibility 403
Dielectric constant 244
Dielectric loss and power factor 245
Dielectric strength 244
Distillation 151

e

Electrical conductivity 243
Electrical properties 229, 243
  conductivity 243
  dielectric constant 244
  enthalpy 262
  static electrification 246
Elemental analysis 197
Elemental composition 114, 398
Emulsion formation 201
Enthalpy 262
Environmental analysis 435
  analytical methods 431
Environmental aspects 102
  gaseous emissions 422
  liquid effluents 428
  processing 419
  solid effluents 430
Environmental aspects of
    processing 419
Environmental regulations 431, 434
  analytical methods 431
Evaporation 202
Extra heavy crude oil 33

f

Factors influencing instability and
    incompatibility 394
  acidity 395
  asphaltene content 395
  bottom sediment and water
    (BS&W) 402
  content 402
  density and specific gravity 398
  elemental composition 398
  metals content 399
  pour point 400
  viscosity 400
  volatility 400
  water content 402
Feedstocks
  evaluation 53
  extra heavy crude oil 30, 33
  foamy oil 26
  heavy crude oil 30
  heteroatom systems 347
  high-acid crude oil 25
  mapping 416
  non-viscous crude oil 17
  opportunity crude oil 21
  viscous crude oil 27
Fire point 203
Flash point 203
Foamy oil 26
Fractional composition 149

Fractionation 161, 166
  acid treatment 173
  adduct formation 176
  ASTM methods 172
  chemical methods 173
  general methods 166
  molecular complex formation 174
  thiourea adduction 176
  urea adduction 174
Fractionation by adduct formation 176
  adduct composition 176
  adduct properties 178
  adduct structure 176
  thiourea adduction 176
  urea adduction 174
Functional group analysis 204

g

Gas chromatography 291
Gas oil 29
Gel permeation chromatography 298

h

Halogenation 208
Heat of combustion 263
Heavy crude oil 30
Heteroatom systems 347
  metals 349
  nitrogen 347
  oxygen 348
  sulfur 348
High-acid crude oil 25
High-performance liquid chromatography 300
Historical perspectives 8
History 1, 8, 14
Hydrocarbon derivatives 122
  aromatic 124
  cycloparaffin 124
  paraffin 123
  unsaturated 124
Hydrogenation 210

i

Infrared spectroscopy 138
Instability and incompatibility 383
  determination 403
  factors influencing 394
  occurrence 389
Interfacial tension 238
Ion exchange chromatography 303

l

Latent heat 264
Liquefaction and solidification 265

m

Mass spectrometry 139, 333
Measurement 82
Metallic constituents 100, 127
Metals 197, 399
Method detection limit 87
Method validation 85
Molecular complex formation 174
Molecular weight
  asphaltenes 372
  carbenes 378
  carboids 378
  nonvolatile fractions 371
  resins 372
  volatile fractions 370
Molecular weight determination 357
  boiling point elevation 366
  freezing point depression 365
  mass spectrometry 369
  methods 360
  nuclear magnetic resonance spectroscopy 370
  size exclusion chromatography 367
  vapor pressure osmometry 361

n

Nitrogen compounds 126
Non-hydrocarbon constituents 125
Non-viscous crude oil 17

Nuclear magnetic resonance
spectroscopy 138, 370

o

Occurrence of instability and
incompatibility 389
Opportunity crude oil 21
Optical activity 248
Optical properties 229, 246
Oxidation 216
Oxygen compounds 126

p

Paraffin hydrocarbon derivatives 123
Physical properties 229, 233
critical 260
density 235, 398
electrical 229, 243
evaporation 202
fire point 203
flash point 203
interfacial tension 238
optical 229, 246
physical 229, 233
specific gravity 235
surface tension 238
Pitch 49
Porphyrins 128
Pour point 267, 400
Power factor and dielectric
loss 245
Pre-Christian era use 8, 14
Precision 88
Pressure-volume-temperature
relationships 267
Properties
chemical 191
chromatographic 283
critical 260
density 235, 398
electrical 229, 243
evaporation 202
fire point 203
flash point 203
interfacial tension 238
optical 229, 246
physical 229, 233
specific gravity 235
surface tension 238
thermal 255
thermal conductivity 271

q

Quality assurance 90
Quality control 90

r

Refractive index 249
Representative sample 80
Residuum 41
Resins fraction 161, 166, 372

s

Sample cleanup 80
Sampling 63, 64
error 82
nonvolatile compounds 75
protocol 70
semi-volatile materials 75
solids 75
Sampling and measurement 63
Sampling error 82
Sampling nonvolatile compounds 75
Sampling protocol 70
Sampling semi-volatile materials 75
Sampling solids 75
Simulated distillation 305
Sludge 50
Softening point 269
Solidification 265
Solvent treatment 152
Specific gravity 235

Specific heat 269
Specifications 93
Static electrification 246
Structural group analysis 317
  density method 321
  density-temperature coefficient method 321
  direct method 322
  dispersion-refraction method 323
  electron spin resonance 329
  infrared spectroscopy 329
  molecular weight-refractive index method 324
  n-d-M method 325
  nuclear magnetic resonance spectroscopy 339
  physical property methods 320
  spectroscopic methods 328
  ultraviolet spectroscopy 345
  waterman ring analysis 325
  X-ray diffraction 345
Sulfur compounds 125
Supercritical fluid chromatography 307
Surface tension 238

t

Tar 49
Tar sand 36
Terminology 1
Thermal conductivity 271
Thermal properties 255
  ash production 256
  carbon residue 258
  distillation 131
  evaporation 202
  fire point 203
  flash point 203
  heat of combustion 263
  latent heat 264
  liquefaction and solidification 265
  pour point 400
  softening point 269
  solidification 265
  specific heat 269
  viscosity 239, 400
  volatility 272, 400
Thin layer chromatography 309
Thiourea adduction 176

u

Unsaturated hydrocarbon derivatives 124
Urea adduction 174
Use of heavy oil and bitumen
  Pre-Christian era 8
  Post-Christian era 14
Use of the data 182, 413, 414
  feedstock mapping 416
  predicting separations 418
  process analysis 416
  process predictability 419
  property predictions 418

v

Vacuum gas oil 135
Vacuum residua 136
Viscosity 239, 400
Viscous crude oil 27
Volatility 272, 400

w

Water content 101, 402

Printed and bound by CPI Group (UK) Ltd, Croydon, CR0 4YY
17/09/2026
14964318-0004